MW01277460

ANTIBODIES IN DIAGNOSIS AND THERAPY

Studies in Medicinal Chemistry
A series of books presenting various aspects of the research on medicinal chemistry and providing comprehensive accounts of recent and important developments in the field.

Edited by Atta-ur-Rahman, H.E.J. Research Institute of Chemistry, University of Karachi, Pakistan

---

Volume 1
Progress in Medicinal Chemistry
edited by M. Iqbal Choudhary

Volume 2
Biological Inhibitors
edited by M. Iqbal Choudhary

Volume 3
Antibodies in Diagnosis and Therapy – Technologies, Mechanisms and Clinical Data
edited by Siegfried Matzku and Rolf A. Stahel

Volumes in preparation

The Guanidine Entity in Medicinal Chemistry and in Molecular Recognition
I. Agranat and S. Topiol

Immunosuppressive and Immunomodulating Agents
M. Iqbal Choudhary

This book is part of a series. The publisher will accept continuation orders which may be cancelled at any time and which provide for automatic billing and shipping of each title in the series upon publication. Please write for details.

ANTIBODIES IN DIAGNOSIS AND THERAPY

TECHNOLOGIES, MECHANISMS AND CLINICAL DATA

Edited by

Siegfried Matzku

Department of Gene Expression and Biotechnology
Merck KGaA
Darmstadt, Germany

and

Rolf A. Stahel

Laboratory of Oncology
University Hospital Zürich
Switzerland

harwood academic publishers
Australia • Canada • China • France • Germany • India • Japan
Luxembourg • Malaysia • The Netherlands • Russia • Singapore
Switzerland

Amsteldijk 166
1st Floor
1079 LH Amsterdam
The Netherlands

British Library Cataloguing in Publication Data

Antibodies in diagnosis and therapy :
technologies, mechanisms and clinical data. – (studies in medicinal chemistry; v. 3)
1. Monoclonal antibodies – Therapeutic use 2. Monoclonal antibodies – Diagnostic use
I. Matzku, Siegfried II. Stahel, Rolf A.
616'.0798

ISBN: 90-5702-310-5
ISSN: 1024-8056

CONTENTS

PREFACE TO THE SERIES

The first two volumes in the series **Studies in Medicinal Chemistry** were published under the editorship of M. Iqbal Choudhary. The present volume edited by Siegfried Matzku and Rolf A. Stahel covers the important areas of the science of monoclonal antibodies and other related important developments including selective drug delivery using targeted enzymes for prodrug activation, immunoscintigraphy, radioimmunotherapy and clinical studies in oncology, and acute and chronic inflammation. The articles are written by leading experts in the fields and present recent developments in these important areas.

It is hoped that this volume will be received with the same enthusiasm by the readers as were the previous two volumes of this series.

Atta-ur-Rahman
Series Editor

PREFACE

Although still being in its adolescence, the science of monoclonal antibodies has already had a turbulent history. This begins with their ancestors, the polyclonal antisera and immunoglobulin fractions derived thereof. These produced a veritable hype of novel diagnostic and therapeutic approaches during the fifties and sixties. Surprisingly enough, the experience accumulated during this period has essentially been lost, some aspects of immunolocalization with labelled proteins being the major exception.

Hence, there was a brisk new start quite rapidly after the seminal paper of Köhler and Milstein in 1975, which described the generation of hybridomas as well as monoclonal antibodies. Only some four years later, first clinical studies with this novel class of molecules were reported on (see Chapters 5, 6). As is often the case with innovative approaches, the first round of testing occurred in the area of oncology. This did not only reflect the lack of curative options for many tumors, but also the rather unique situation that monoclonal antibodies provided revolutionary new tools for the identification of tumor antigens, while at the same time they could be explored as new vehicles for diagnostic and therapeutic targeting approaches. However, the second hype with regard to diagnostic and therapeutic immunoglobulins was also bound to fade out, as is so often the case when expectations are raised too high with initial enthusiasm. Yet, there are important reasons why (monoclonal) antibodies continue to have a place in the development of therapeutic strategies in an impressive number of clinical conditions. The discussion of these reasons is the dominating theme of this book. In addition, the authors of the individual chapters are giving us their view of how the exploration of monoclonal antibodies in the respective fields evolved and what the next generation of antibody-based drug constructs will look like.

ANTIBODIES AS UNIVERSAL BUILDING BLOCKS

Due to the domain architecture of immunoglobulins, constant domains being responsible for antibody function and variable domains being responsible for antigen binding, there is ample opportunity for shuffling, fusing and combining such domains (Chapters 1, 2). Furthermore, the rigid architecture of the domains allows them to graft exposed elements, e.g., binding loops, from one domain to another. Thereby it is possible to alter the species-related costume of the protein and, hence, its immunogenicity. Using the binding domains in an evolutionary setting, e.g. antibody gene libraries displayed on the surface of phages or bacteria, it is possible to create and fine-tune antibodies and fragments virtually at will. Such technology may even be used to create antibodies detecting self-structures which are normally excluded from the armamentarium of nature.

ANTIBODIES AS UNIVERSAL LIGANDS

Due to the seemingly unrestricted capacity of antibodies to react with and bind to almost any structure and due to novel technologies for creating such antibodies, it was possible to obtain antibodies detecting and reacting with almost any receptor. In a number of instances, antibodies could be selected which are either agonistic, or antagonistic, or simply binding. Hence, the generation of antibodies or fragments provides access to an important pool of surrogate ligands. By nature, antibodies are also receptors, albeit of specialized cells, namely of B cells and – indirectly – of cells endowed with a receptor for the C-terminal domain of antibodies. The latter aspect is utilized for therapeutic intervention because it is linked with a potent mechanism of cytotoxicity (see Chapter 5). The logic consequence of the Janus-like temper of antibodies is that they are recognizing themselves in a network fashion, which goes by the term of idiotypic network. The physiological and therapeutic implications of this element of complexity again remain to be defined and exploited.

ANTIBODIES AS UNIVERSAL CARRIERS

Being given the multi-domain nature of antibodies which may be used as building blocks, a number of coupling procedures exist which allow for the linkage of a plethora of extrinsic pharmacophores to the scaffold of antibodies or antibody fragments. The binding function of the N terminus being conserved, such conjugation or coupling results in the generation of targeted diagnostics or therapeutics. The entities engaged in the linkage are as diverse as radionuclides, cytotoxic drugs, toxins, enzymes, cytokines, and antibodies themselves (Chapters 2, 3, 5). However, unlike the highly cited 'magic bullets', monoclonal antibodies do not know where they are going nor does somebody manage to guide their way. In fact, they hit their target, i.e. the antigen, incidentally. Once this happens, they stick to it for some time and deploy their action so that it seems appropriate to think of targeted retention rather than targeted transport. The most recent version of targeted action is displayed by intra-cellularly generated antibody (fragments) which may interfere with cellular processes *in situ*.

ANTIBODIES AS VERSATILE PHARMACOKINETIC ENTITIES

Regarding intact antibodies of different classes and derived from different species, half life *in vivo* – in the blood stream or else – varies between hours and weeks. This time span is expanded towards the shorter end by truncation of the antibody, the smallest binding fragment with intact, univalent binding function, i.e. the Fv fragment, ranging around several minutes of half life. Hence, antibodies are having a quasi-inbuilt principle of half-life variation. Together with molecular biology methods of varying the binding strength (avidity), an unprecedented span of properties can be offered (Chapter 1). However, there are two clear limitations with respect to bioavailability and biodistribution: due to their macromolecular, proteinaceous nature, antibodies and their constructs must be delivered parenterally. Furthermore, the penetration into solid tissue is kinetically limited in most instances. There are marked exceptions to the rule, e.g., antibodies which are aggregated or damaged otherwise are most efficiently transported to the mononuclear-phagocytic system, this being part of the physiological process of scavenging altered components of the blood.

ANTIBODIES AS THE MULTI-FACETED DRUG

Different antibody classes are endowed with widely differing intrinsic functions. Most notable are cytotoxic functions residing in the Fc domains such as complement activation and antibody dependent cellular cytotoxicity. Antibodies may elicit secondary antibodies in a network fashion (see Chapter 5). Antibodies may 'modulate' cell surface molecules in that these are crosslinked and internalized. Antibody-derived epitopes, either directly or via the idiotype pathway, may induce T cell reactions.

Siegfried Matzku
Rolf A. Stahel
Volume Editors

CONTRIBUTORS

Nadine Beauger
Division of Hematology-Immunology
Maisonneuve-Rosemont Hospital Research Center
5415 L'Assumption Blvd
Montreal, Quebec
Canada H1T 2M4

Angelika Bischof Delaloye
Service de Médecine Nucléaire
Centre Hospitalier Universitaire Vaudois
Rue du Bugnon
1011 Lausanne
Switzerland

Martin Gyger
Division of Hematology-Immunology
Maisonneuve-Rosemont Hospital Research Center
5415 L'Assumption Blvd
Montreal, Quebec
Canada H1T 2M4

Sergey M. Kipriyanov
Recombinant Antibody Research Group (0445)
Diagnostics and Experimental Therapy Program
German Cancer Research Center (DKFZ)
Im Neuenheimerfeld 280
69120 Heidelberg
Germany

Melvyn Little
Recombinant Antibody Research Group (0445)
Diagnostics and Experimental Therapy Program
German Cancer Research Center (DKFZ)
Im Neuenheimerfeld 280
69120 Heidelberg
Germany

Denis C. Roy
Division of Hematology-Immunology
Maisonneuve-Rosemont Hospital Research Center
5415 L'Assumption Blvd
Montreal, Quebec
Canada HIT 2M4

Wolff Schmiegel
Medizinische Klinik
Ruhruniversität Bochum
Knappschaftskrankenhaus
In der Schornau 23-25
44892 Bochum
Germany

Jan Schmielau
Medizinische Klinik
Ruhruniversität Bochum
Knappschaftskrankenhaus
In der Schornau 23-25
44892 Bochum
Germany

Giuseppe Sconocchia
National Institutes of Health
National Cancer Institute
Experimental Immunology Branch
Bethesda MD 20892-1360
USA

David M. Segal
National Institutes of Health
National Cancer Institute
Experimental Immunology Branch
Bethesda MD 20892-1360
USA

Peter D. Senter
Bristol Myers Squibb Pharmaceutical Research Institute
3005 First Avenue
Seattle WA 98121
USA

Nathan O. Siemers
Bristol Myers Squibb Pharmaceutical Research Institute
3005 First Avenue
Seattle WA 98121
USA

Barbara A. Vance
National Institutes of Health
National Cancer Institute
Experimental Immunology Branch
Bethesda MD 20892-1360
USA

Winfried Wels
Institute for Experimental Cancer Research
Tumor Biology Center
Breisacher Strasse 117
79106 Freiburg
Germany

Janice C. Wherry
Clinical Project Director
Clinical Immunology and Infectious Disease
Schering-Plough Research Institute
2015 Galloping Hill Road
Kenilworth NJ 07033
USA

Uwe Zangemeister-Wittke
Laboratory of Oncology
Department of Internal Medicine
University Hospital Zürich
Haeldeliweg 4
8044 Zürich
Switzerland

1. RECOMBINANT ANTIBODIES: CONSTRUCTION AND PRODUCTION

MELVYN LITTLE and SERGEY M. KIPRIYANOV

German Cancer Research Center (DKFZ), Im Neuenheimerfeld 280, 69120 Heidelberg, Germany

INTRODUCTION

Antibodies are capable of highly specific interactions with a wide variety of ligands including tumor-associated markers, viral coat proteins and lymphocyte cell surface glycoproteins. They are therefore potentially very useful agents for the diagnosis and treatment of human diseases (see review by Riethmüller *et al.*, 1993). However, the use of rodent antibodies for therapy poses a number of problems. One of these is the immunogenicity of monoclonal antibodies. Repeated doses of rodent antibodies elicit an anti-immunoglobulin response, referred to as HAMA (human anti-murine antibody; Jaffers *et al.*, 1986; Khazaeli *et al.*, 1994). The use of human monoclonal antibodies would alleviate this problem but only a restricted range of specificities are available. Furthermore, human hybridomas are often unstable and/or poor producers, thus making the production of large amounts of the human immunoglobulin very difficult (James, 1994).

In the past few years, some of the limitations of monoclonal antibodies as therapeutic agents have been addressed by genetic engineering. Such an approach is particularly suitable because of the domain structure of the antibody molecule, where functional domains carrying antigen-binding activities (Fabs or Fvs) or effector functions (Fc) can be exchanged between antibodies (Figure 1). On the basis of sequence variation, the residues in the variable domains are assigned either to the hypervariable complementarity-determining regions (CDRs) or to framework (FR) regions (Wu and Kabat, 1970). It is possible to replace much of the rodent-derived sequence of an antibody with sequences derived from human immunoglobulins without loss of function. This new generation of "chimeric" and "humanized" antibodies represents an alternative to human hybridoma derived antibodies and should be less immunogenic than their rodent counterparts. Furthermore, genetically truncated versions of the antibody may be produced ranging in size from the smallest antigen-binding unit or Fv through Fab′ to $F(ab')_2$ fragments. To stabilize the association of the recombinant V_H and V_L domains, they have been joined in single-chain Fv (scFv) constructs with a short peptide linker (Huston *et al.*, 1988; Bird *et al.*, 1988). More recently it has become possible to produce totally human recombinant antibodies derived from antibody libraries or single immune B cells.

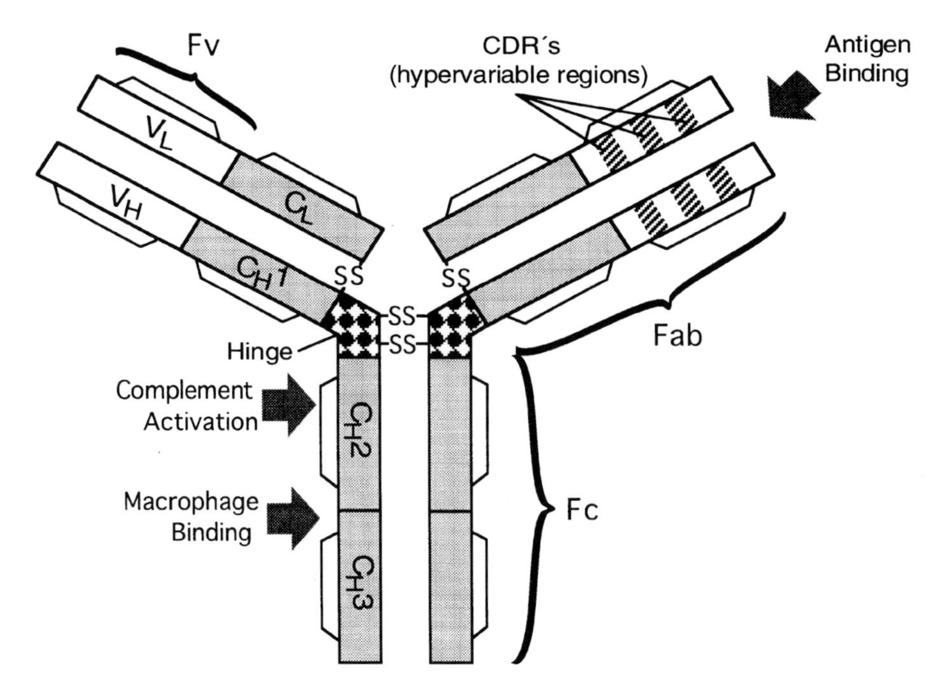

Figure 1 Schematic representation of the domain structure of an IgG molecule.

CLONING THE ANTIBODY VARIABLE REGIONS

Significant advances have been made in the *in vitro* immunization of human B cells (Borrebaeck *et al.*, 1988) and in the development of transgenic mice containing human immunoglobulin loci (Brüggemann *et al.*, 1989, 1991; Sarvetnick *et al.*, 1993; Davies *et al.*, 1993; Fishwild *et al.*, 1996). Recombinant DNA technology can also be employed to generate human monoclonal antibodies from human lymphocyte mRNA. The genetic information for antibody variable regions is generally retrieved from a hybridoma-derived cDNA library by hybridization with oligonucleotides that bind to the constant domain genes (Adair *et al.*, 1994; Pulito *et al.*, 1996) or from total cDNA preparations using the polymerase chain reaction (PCR) with antibody-specific primers (Orlandi *et al.*, 1989; Dübel *et al.*, 1994). The PCR fragments can be directly sequenced and/or ligated into the appropriate expression vector. However, problems are often encountered when attempting to retrieve the correct V region genes from the hybridoma cDNA. For example, a cell line that was monoclonal in terms of secreted antibody proved to be oligoclonal in terms of mRNA species (Jiang *et al.*, 1994). Furthermore, hybridomas in which the immortalizing fusion partner is derived from MOPC-21 may express a V_L kappa transcript that is aberrantly rearranged at the VJ recombination site, and which therefore encodes a non-functional light chain (Cabilly and Riggs, 1985; Carroll *et al.*, 1988). Cellular levels of this transcript may exceed that generated from the productive V_L gene or the primer may be more homologous to its cDNA, so that a large proportion of the PCR product may not encode a functional light chain. Some errors can also be introduced by PCR itself.

Single bacterial colonies expressing antigen-specific antibody fragments can be identified by colony screening using antigen coated membranes (Dreher *et al.*, 1991). The appropriate V_H/V_L combination may also be selectively enriched from an scFv phage library through a series of immunoaffinity steps referred to as "library panning" (McCafferty *et al.*, 1990; Marks *et al.*, 1991b). However, for these two approaches it is necessary to have sufficient amounts of soluble antigen for screening and for characterization of the selected clones by ELISA. In most cases, for antibodies directed against cell surface molecules, e.g. CD (cluster of differentiation) antigens, bio-panning and filter-screening procedures are not suitable.

In an attempt to circumvent the problems outlined above, a screening procedure for a hybridoma-derived scFv against the human transferrin receptor (TFR) was proposed (Nicholls *et al.*, 1993b). The method is based on the specific cytotoxicity of scFv fused to a binding-deficient form of *Pseudomonas* exotoxin A (PE40). Small amounts of fusion protein were expressed in a rabbit reticulocyte lysate system and the specific toxicity was checked on a eukaryotic cell line expressing the TFR (Nicholls *et al.*, 1993a,b). However, this protein synthesis inhibition assay only provides a qualitative assessment of antigen binding for a limited subset of scFv (i.e. scFv directed against a cell surface antigen that becomes internalized).

Another system for the rapid evaluation of functional antibody fragments directed against cell surface antigens employed cytotoxic bispecific antibodies (Sanna *et al.*, 1995a). DNA coding for Fab fragments was fused to the gene for *Staphylococcus aureus* protein A (SpA) and expressed in *E. coli*. These Fab–SpA fusions were then mixed with a T-cell activating antibody specific for CD3 to form a bispecific cytotoxic complex. On the basis of an *in vitro* cell cytotoxicity assay, one could assess the antigen binding activity of the antibody fragment. This system is, of course, not suitable for antibody fragments specific for T-cell surface molecules.

Recently, we developed a rapid procedure for testing the function of hybridoma-derived antibody fragments. The method is based on a flow cytometric analysis using a fluorescence activated cell scanner (FACScan) of single-chain Fv molecules produced in small scale *E. coli* cultures binding to eukaryotic cells (Kipriyanov *et al.*, 1996b). A simple extraction of the soluble periplasmic content from 5 ml cultures yields sufficient amounts of antibody fragment for an evaluation of its cell-specific binding. Constructs that have incorporated the wrong V_L kappa transcript are readily identified and the scFv can be tested for functional binding without investing considerable time and effort in large scale expression. The proposed procedure was successfully used for cloning a scFv specific for human CD19 on the surface of B-lymphocytes.

A simple procedure for the generation of human antibody fragments directly from single B cells or B-cell clones was recently described (Lagerkvist *et al.*, 1995). The procedure is based on the antigen-specific selection of single human B cells using antigen-coated magnetic beads and a cellular amplification step followed by PCR amplification of V-region genes. Using this approach, it is possible to avoid the cumbersome hybridoma technology and to obtain human antibody fragments with the original V_H/V_L pairing.

GENETICALLY ENGINEERED MONOCLONAL ANTIBODIES

Chimeric Antibodies with Human Constant Regions

The first generation of recombinant monoclonal antibodies consisted of the rodent-derived variable regions fused to human constant regions. It is thought that the most immunogenic regions of antibodies are the conserved constant domains (Khazaeli *et al.*, 1994). Because the antigen binding site of the antibody is localized within the V regions, the chimeric molecules retain their binding affinity for the antigen and acquire the function of the substituted C regions. For example, mouse::human IgG1 chimeric antibodies have been shown to mediate the lysis of tumor cells in the presence of human complement (Liu *et al.*, 1987).

Clinical experience with rodent monoclonal and mouse::human chimeric antibodies indicates that for some antibodies a strong anti-idiotypic reaction is elicited. For example, 60% of the renal allograft recipients treated with anti-human CD3 murine mAb OKT3 developed anti-idiotypic antibodies (Jaffers *et al.*, 1986). Therefore, to reduce the murine content even further, procedures have been developed for humanizing the Fv regions.

Humanization by CDR Grafting

CDRs build loops close to the antibody's N-terminus, where they form a continuous surface. Crystallographic analysis of several antibody/antigen complexes and other studies have shown that antigen binding mainly involves this surface (although some framework residues have also been found to be involved in the interaction with antigen). Thus, the antigen-binding specificity of an antibody is mainly defined by the topography of its CDR surface and by the chemical characteristics of this surface; these in turn are determined by the conformation of the individual CDRs, by the relative disposition of the CDRs, and by the nature and disposition of the side chains of the amino acids comprising the CDRs (Padlan, 1994).

A large decrease in the immunogenicity of an antibody was achieved by grafting only the CDRs of xenogenic antibodies onto human framework and constant regions (Jones *et al.*, 1986; Verhoeyen *et al.*, 1988). However, CDR grafting *per se* may not result in the complete retention of antigen-binding properties. Indeed, it is frequently found that some framework residues from the original antibody need to be preserved in the humanized molecule if significant antigen-binding affinity is to be recovered (Queen *et al.*, 1989; Co *et al.*, 1991; Ohtomo *et al.*, 1995).

Two major approaches have been used to produce effective CDR-grafted antibodies (Co and Queen, 1991). In one approach (Queen *et al.*, 1989), human V regions showing the greatest sequence homology to murine V regions are chosen from a database in order to provide the human framework. The selection of human FRs can be made either from human consensus sequences or from individual human antibodies (Kolbinger *et al.*, 1993). In some rare examples, simply transferring CDRs onto the most identical human V-region frameworks is sufficient for retaining the binding affinity of the original murine mAb (Roguska *et al.*, 1996). However, in most cases, the successful design of high-affinity CDR-grafted antibodies requires that key murine residues be substituted into the human acceptor framework to preserve the CDR conformations. Computer modeling of the antibody is used to identify such structurally important residues which are then included in order to achieve a higher binding affinity. This approach was successfully used for humanizing an anti-Tac mAb specific for the human IL-2 receptor (Queen *et al.*, 1989), antibodies against the gB and gD glycoproteins of herpes virus (Co *et al.*, 1991), a mAb 425 specific to an external domain of the human epidermal growth factor (EGF) receptor (Kettleborough *et al.*, 1991), a mAb 4D5 against the product of protooncogene *HER2* (Carter *et al.*, 1992b), a mouse mAb against human IgE (Kolbinger *et al.*, 1993), mAb CC49 against tumor-associated glycoprotein TAG-72 (Kashmiri *et al.*, 1995) and a number of antibodies against human cell surface antigens such as CD3 (Routledge *et al.*, 1991; Shalaby *et al.*, 1992; Rodrigues *et al.*, 1992; Adair *et al.*, 1994), CD4 (Pulito *et al.*, 1996), CD18 (Hsiao *et al.*, 1994), CD22 (Leung *et al.*, 1995) and CD64 (FcγRI; Graziano *et al.*, 1995). However, the process of identifying the rodent framework residues to be retained is generally unique for each reshaped antibody and can therefore be difficult to predict (Kolbinger *et al.*, 1993).

In the alternative approach, murine CDRs are grafted onto a given human V region. If the humanized antibody has reduced or no binding affinity, new constructs are made that incorporate one to three additional mouse residues near the CDRs, and so on iteratively until binding is restored. While minimizing the amount of murine sequence required for binding, the full biological activity of the original antibody may not be recovered. This approach was used for the humanization of rat anti-CAMPATH-1 mAb recognizing the human cell-surface glycoprotein CDw52 (Riechmann *et al.*, 1988) and a murine mAb against the human respiratory syncytial virus (RSV; Tempest *et al.*, 1991).

Humanization by Resurfacing (Veneering)

A statistical analysis of unique human and murine immunoglobulin heavy and light chain variable regions revealed that the precise patterns of exposed residues are different in human and murine antibodies, and most individual surface positions have a strong preference for a small number of different residues (Padlan, 1991; Pedersen *et al.*, 1994). Therefore, it may be possible to reduce the immunogenicity of a non-human Fv, while preserving its antigen-binding properties, by simply replacing exposed residues in its framework regions that differ from those usually found in human antibodies (Padlan, 1991). This would humanize the surface of the xenogenic antibody while retaining the interior and contacting residues which influence its antigen-binding characteristics and interdomain contacts. Since protein antigenicity can be correlated with surface accessibility, replacement of the surface residues may be sufficient to render the mouse variable region "invisible" to the human immune system. This procedure of humanization is referred to as veneering since only the outer surface of the antibody is altered, the supporting residues remain undisturbed (Mark and Padlan, 1994).

Variable domain resurfacing maintains the core murine residues of the Fv sequences, and it is likely to minimize CDR-framework incompatibilities. This procedure was recently successfully used for the humanization of murine mAb N901 against the CD56 surface molecule of natural killer (NK) cells and mAb anti-B4 against CD19 (Roguska *et al.*, 1994, 1996). A direct comparison of engineered versions of N901 humanized either by CDR grafting or by resurfacing showed no difference in binding affinity for the native antigen (Pedersen *et al.*, 1994; Roguska *et al.*, 1996). For anti-B4 antibody, the best CDR-grafted version required three murine residues at surface positions to maintain binding, while the best resurfaced version needed only one surface murine residue (Roguska *et al.*, 1996). Thus, even though the resurfaced version of anti-B4 has 36 murine residues in the Fv core, it is conceivable that it would be less immunogenic than a CDR-grafted version with nine murine residues in the Fv core because it has a pattern of surface residues that is more identical to a human surface pattern.

Humanization by Framework Exchange

This approach used by Benhar *et al.* (1994) for humanization of carcinoma-specific mAb B3 combines, with some differences, the principles of CDR grafting and resurfacing. The variable domains of the heavy and light chain were aligned with their best human homologues to identify framework residues that differ. Six V_H and five V_L residues that differ were not changed because they were buried in the interdomain interface, or found to result in decreased affinity when mutated. This basic design resulted in 20-fold loss of activity. Changing V_L residues at the interdomain interfacial position 100 and at the buried position 104 to the human sequence increased the activity 8-fold. Changing a V_H residue at position 82b from the human sequence back to that of the mouse restored the activity 2- to 3-fold. As a result, humanized B3(Fv) fused to the Pseudomonas exotoxin PE38 retained antigen-binding activity but lost immunogenic epitopes recognized by sera from monkeys that had been immunized with B3(Fv)-PE38 (Benhar *et al.*, 1994).

An analogous approach was used by Studnicka *et al.* (1994) for humanizing mAb H65 specific for the human CD5 antigen. The authors classified each amino acid position in the variable region according to the benefit of achieving a more human-like antibody versus the risk of decreasing or abolishing specific binding affinity. Substitutions of human residues at low-risk positions (exposed to solvent but not contributing to antigen binding or antibody structure) are likely to decrease immunogenicity with little or no effect on binding affinity. Changes at high-risk positions (directly involved in antigen binding, CDR stabilization or internal packing) are avoided to preserve the biological activity of the antibody. Moderate-risk changes were made with caution. The new "human-engineered" H65 antibody containing 20 low-risk human consensus substitutions (expressed as either IgG or Fab) retained the full binding avidity of parental murine and chimeric H65 antibodies (Studnicka *et al.*, 1994). Surprisingly, the further addition of 14 moderate-risk substitutions enhanced the avidity by 3- to 7-fold. This method should be generally applicable to the design of other human-engineered antibodies with therapeutic potential.

Choice of Constant Region

The construction of chimeric and humanized antibodies offers the opportunity of tailoring the constant region to the requirements of the antibody. Each immunoglobulin subclass differs in its ability to interact with Fc receptors and complement and thus to trigger cytolytic events. Human IgG1, for example, would be the constant region of choice for mediating antigen-dependent cytotoxicity (ADCC) and probably also complement-dependent cytotoxicity (CDC); mouse::human IgG1 chimeric antibodies have been shown to mediate the lysis of tumor cells in the presence of human mononuclear cells (Liu *et al.*, 1987). On the other hand, if the antibody is required simply to activate or block a receptor then human IgG2 or IgG4 would probably be more appropriate. Humanized versions of the immunosuppressive anti-human CD3 mAb OKT3 were prepared as IgG4 antibodies (Woodle *et al.*, 1992; Adair *et al.*, 1994).

To avoid the unwanted side effects of a particular isotype it is possible to remove or modify effector functions by site-directed mutagenesis. A single amino acid substitution of leucine by glutamic acid at position 235 in the Fc receptor binding segment of a humanized anti-CD3 mAb OKT3 IgG4 led to the retention of its immunosuppressive properties, but markedly reduced unwanted T cell activation *in vitro* (Alegre *et al.*, 1992). Similarly, another humanized version of OKT3 with two amino acid substitutions in the C_H2 portion (from Phe–Leu to an Ala–Ala at positions 234–235) retained its immunosuppressive properties, but did not activate T cells (Alegre *et al.*, 1994). In the latter case, a significant prolongation of a human allograft survival was achieved using a SCID mice model. Thus, the use of mutated Fc variants of humanized antibodies should result in fewer side effects for transplant recipients while maintaining clinical efficacy.

EXPRESSION SYSTEMS FOR RECOMBINANT ANTIBODY PRODUCTION

Several expression systems are available for the production of antibody and antibody fragments including bacteria, yeast, plants, insect cells and mammalian cells. Each has advantages, potential applications, and limitations. Bacteria cannot assemble whole glycosylated antibodies but they are very suitable hosts for the production of antibody fragments. Complete antibodies have been expressed successfully in yeast, but they contain the high-mannose, multiple-branched oligosaccharides and were shown to be defective in effector functions such as complement-mediated lysis. Antibodies produced in insect cells via baculovirus vectors also contain carbohydrate structures very different from those produced by mammalian cells. Intact, effector function-competent antibodies have been successfully expressed in myeloma cells and also in nonlymphoid mammalian cells, which possess the mechanisms required for correct immunoglobulin assembly, posttranslational modification, and secretion.

Mammalian Cell Expression

Secreted antibody production

Immunoglobulin genes can be expressed by transfection into nonsecreting myeloma or hybridoma cell lines (transfectomas) using vectors for expression of the heavy and light chains that are able to integrate into cell chromosomes and carry a selection marker (Liu *et al.*, 1987; Sun *et al.*, 1987). Usually, the rate of antibody secretion is considerably lower than those of the parent hybridomas. Selection of transfected cells using either the antibiotic G418 (*neo* marker) or mycophenolic acid (*gpt* marker) or hygromycin (*hyg* marker) give rise to clones producing recombinant chimeric antibodies with yields from 17–700 μg/l (Liu *et al.*, 1987) to 5–20 μg/ml (Sun *et al.*, 1987; Tempest *et al.*, 1991; Hsiao *et al.*, 1994). A glutamine synthetase (GS) selectable marker gene was introduced into myeloma cells by selecting for the ability of transfectants to grow in a glutamine-free medium containing asparagine after gradual depletion of the glutamine (Bebbington *et al.*, 1992). One of the selected cell lines secreted 10–15 μg/10^6 cells/day of a chimeric B72.3 IgG4 antibody during exponential growth and accumulated 560 mg/l antibody in a fed-batch air-lift fermentation system.

Cloned immunoglobulin genes can also be expressed in non-lymphoid cells so that highly efficient expression systems can be used, e.g. amplification of vectors containing the gene for dihydrofolate reductase (*dhfr*) in Chinese hamster ovary (CHO) cells deficient for DHFR production (CHO dhfr$^-$). With this system, recombinant Ig were produced at the level of about 100 μg/10^6 cells/day (Page and Sydenham, 1991; Fouser *et al.*, 1992). The successful co-amplification of transfected Ig genes linked to *dhfr* genes in mouse myeloma cell lines has also been reported. Antibody production increased to 25 μg/10^6 cells/day (Dorai and Moore, 1987) and even 70–100 μg/10^6 cells/day (Shitara *et al.*, 1994).

As a rule, it takes about 2–3 months to establish transfectant clones stably producing chimeric antibodies (Shitara *et al.*, 1994). For the fast analysis of recombinant antibody variants obtained during the process of mAb humanization, a transient expression system in COS cells was used (Alegre *et al.*, 1992; Kolbinger *et al.*, 1993; Adair *et al.*, 1994; Roguska *et al.*, 1994; Pulito *et al.*, 1996). Antibody expression was achieved by co-transfection of both heavy and light chain expression vectors containing the SV40 *ori* of replication into COS cells. The culture supernatants containing secreted antibodies were harvested 3–6 days after transfection.

To avoid problems associated with the expression of two independent vectors in a single cell, Shitara *et al.* (1994) constructed an improved expression vector in which both of the chimeric heavy and light chain gene transcription units and a *dhfr* gene transcription unit were inserted. Alternatively, Shu *et al.* (1993) designed a molecule with covalently linked V_H, V_L and Fc domains encoded in a single gene. The single-chain protein was produced by murine myeloma cells and secreted into the tissue culture fluid as a dimer with a yield of 4 μg/ml (Shu *et al.*, 1993). This single-gene construct approach provides a means of generating an immunoglobulin-like molecule that retains the specificity, binding properties, and cytolytic activity of the chimeric mAb.

Other antibody constructions expressed in mammalian cells include IgM-like polymers of IgG that possess both the Fcγ receptor binding properties of IgG and the more potent complement activity of IgM (Smith and Morrison, 1994). A bispecific single-chain molecule consisting of the first two N-terminal CD4 domains fused to a single chain Fv against the human CD3 complex has been produced in myeloma cells (Traunecker *et al.*, 1991). A similar humanized bispecific immunoadhesin-antibody molecule expressed in human embryonic kidney cells was able to target cytotoxic T cells for killing HIV-infected cells (Chamow *et al.*, 1994; Ridgway *et al.*, 1996). Antibody fragments can also be efficiently produced in mammalian cells, e.g. $F(ab')_2$ (De Sutter *et al.*, 1992), Fc (Young *et al.*, 1995), Fv (King *et al.*, 1993), scFv (Jost *et al.*, 1994) and scFv fusion proteins with carboxyl-terminal effector domains, including IL-2, the B domain of staphylococcal protein A, the S-peptide of ribonuclease S and a hexa-histidine sequence for metal chelate chromatography (Dorai *et al.*, 1994). In a mammalian expression system, it was found that glycosylated scFvs were secreted faster than their nonglycosylated counterparts (Jost *et al.*, 1994). It is also possible to produce single-chain $(Fv)_2$ molecules consisting of the two variable domains of an antibody joined through an oligopeptide linker either transiently in COS cells (Jost *et al.*, 1996) or in transfected CHO cells. In the latter case, a yield of 12–15 mg/l was obtained (Mack *et al.*, 1995).

Intracellular expression of antibody fragments

Eukaryotic cells can produce antibodies that function intracellularly. Such intracellular antibodies represent a powerful and promising approach to modulate the function of selected intracellular gene products in higher organisms (phenotypic knock-out) or to protect the cell from infectious agents (intracellular immunization). The intracellular antibodies targeted to the endoplasmic reticulum (ER) can be used to capture specific proteins as they enter the ER, preventing their transport to the cell surface. For example, the intracellular expression of the anti-Tac scFv was found to completely abrogate cell surface expression of the α subunit of the high-affinity interleukin-2 receptor in stimulated Jurkat T cells (Richardson *et al.*, 1995). Another scFv specific to the p21 protein product of the *ras* proto-oncogene blocked the ensuing meiotic maturation of *Xenopus laevis* oocytes when expressed in the cytosol (Biocca *et al.*, 1994).

Cytoplasmic expression in BSR cells (a clone of baby hamster kidney cells) of scFv against the envelope protein of the tick-borne flavivirus was shown to significantly reduce the infectivity of the louping ill virus (Jiang *et al.*, 1995).

Analogously, the expression of scFv that recognizes the CD4 binding region of the HIV-1 envelope protein in ER of COS cells inhibited *env* protein maturation and led to the reduction of envelope protein-mediated syncytia formation (Marasco *et al.*, 1993). Recombinant antibody fragments can be targeted to different subcellular compartments: to the cytoplasm by removal of the hydrophobic amino acid core sequence of the signal peptide (Biocca *et al.*, 1995), to the endoplasmic reticulum by adding a C-terminal ER retention signal (Richardson *et al.*, 1995), to the nucleus by adding the nuclear localization signal to the N-terminus of the antibody fragment (Biocca *et al.*, 1995) and even to mitochondria by fusion with the N-terminal presequence of cytochrome *c* oxidase (Biocca *et al.*, 1995). The expression levels of the mutated antibody domains vary, according to the type of targeting signal, from high expression levels for those fragments localized in ER to much lower levels for the cytosolic and nuclear scFvs, with the mitochondrial scFv displaying intermediate levels (Biocca *et al.*, 1995).

Surface expression of antibody fragments

To direct and fix a single-chain antibody to the cell surface, nucleotide sequences coding for the signal peptide and transmembrane domain of the interleukin-6 receptor were fused to the 5′ and 3′ ends, respectively, of DNA coding for a single-chain antibody against the hapten 4-ethoxy-methylene-2-phenyl-2-oxazoline-5-one (phOx). After expressing the DNA in eucaryotic cells using a retroviral vector, the antibody was presented at the cell surface as shown by its ability to bind fluorescently labeled proteins conjugated to the phOx hapten (Rode *et al.*, 1996). This system thus provides a universal cell surface coupling system, since phOx can be easily conjugated to primary amino and sulfhydryl groups. For example, monoclonal antibodies, costimulatory molecules, ligands and even mixtures of different molecules could be bound in this way to cell surfaces.

Insect Cell Expression using Baculovirus Vectors

DNA can be efficiently expressed in cultures of insect cells infected with baculovirus using expression vectors containing the promoter of the gene for polyhedrin. Invertebrate cells are capable of signal peptide cleavage, N- and O-linked glycosylation, proper cellular compartmentalization and extracellular secretion. The expression of recombinant mAb in insect cells offers several advantages with respect to post-translational modifications, stability, yields and applicability. An anti-human CD29 human::mouse chimeric antibody was recently produced in Sf9 cells with a yield of 10–15 mg/10^9 cells, higher than the level achieved by the parental mouse hybridoma (Poul *et al.*, 1995). The chimeric heavy and light chains were correctly processed and assembled into a normal immunoglobulin that was secreted into the culture medium. The chimeric antibody was found to be glycosylated and reproduced *in vitro* the functional properties of the parental mAb, including binding affinity and inhibition of lymphocyte proliferation.

These findings suggest that the baculovirus expression system is very suitable for the production of large amounts of intact antibodies for diagnostic or even therapeutic purposes.

Plant Expression

The most commonly used method for transforming plant cells employs the Ti plasmid of *Agrobacterium tumefaciens* as an agent for introducing recombinant DNA to the plant cell nucleus. Not all plants, however, are amenable to the manipulations required for the stable introduction of foreign DNA. Tobacco is the most commonly used plant since it is easily transformed and regenerated and was the first plant shown to accumulate assembled mammalian antibodies (Hiatt *et al.*, 1989).

In the first report of antibody expression in plants, the heavy and light chain cDNAs from the 6D4 hybridoma antibody were introduced individually into different tobacco plants and then sexually crossed to obtain progeny expressing both chains. A surprisingly high level of accumulation of functional antibody was observed. In the case of 6D4, greater than 1% of the total extractable protein was found to be functional antibody (Hiatt *et al.*, 1989). The subsequent expression of other antibodies in tobacco using the same strategy resulted in similar levels of accumulation (During *et al.*, 1990).

The expression of antibodies in plants potentially offers many agricultural benefits such as resistance to certain plant pathogens and a relatively inexpensive production system. For example, Tavladoraki *et al.* (1993) showed that the constitutive expression in transgenic plants of a scFv antibody directed against the plant icosahedral tombusvirus artichoke mottled crinkle virus causes a reduction of infection incidence and a delay in symptom development. The functional expression of scFv antibody fragments in developing seeds and accumulation in ripe seeds of transgenic tobacco has also been demonstrated (Fiedler and Conrad, 1995). Seeds stored for long periods showed no loss of scFv protein or its antigen-binding activity.

Antibody Expression in Eukaryotic Microorganisms

The yeast *Saccharomyces cerevisiae* can synthesize, process and secrete higher eukaryotic proteins. Wood *et al.* (1985) expressed immunoglobulin λ and μ cDNA specific for a small hapten in *S. cerevisiae* under the control of the yeast phosphoglycerate kinase (PGK) promoter. Heavy and light chains were produced that bound antigen, although the assembly and secretion was inefficient with about 75% of the total immunoglobulin remaining within the cell. Horwitz *et al.* (1988) used a similar strategy for constructing yeast strains that secrete a functional mouse::human chimeric antibody and its Fab fragment into the culture medium. To express the chimeric whole antibody, cDNA copies of the light and heavy chain genes of an anti-tumor antigen antibody were inserted into vectors containing the yeast PGK promoter, the invertase signal sequence and the PGK polyadenylation signal. The simultaneous expression of these genes resulted in the secretion of approximately 100 μg/ml of light chain and 50–80 μg/ml of heavy chain, with 50–70% of the heavy chains associated with light chains. This chimeric antibody bound to target cancer cells and exhibited antibody-dependent cellular cytotoxicity but not complement-dependent cytotoxicity. The failure to activate complement may result from the altered carbohydrate structure present on the yeast-derived antibody.

The methylotrophic yeast *Pichia pastoris* has been shown to be suitable for the high-level expression of various heterologous proteins, either intracellular or secreted into the culture supernatant (see review by Cregg *et al.*, 1993). Recently, Ridder *et al.* (1995) demonstrated the applicability of the *Pichia* expression system for the secretion of antibody fragments. A rabbit scFv against the human leukemia inhibitory factor (hLIF) was expressed in *Pichia* cells using the expression vector pPIC9 that provides the α-mating factor signal sequence for secretion and the *HIS4* gene for selection of the recombinant yeast clones. The yield was about 100 mg per liter of shake flask culture.

Another eukaryotic microorganism, the filamentous fungus *Trichoderma reesei*, was also able to produce and secrete genetically engineered forms of antibodies (Nyyssönen *et al.*, 1993). The genes coding for the light chain of the murine anti-2-phenyloxazolone IgG1, its Fab fragment and *T. reesei* cellulase (CBHI) fused to the Fab were introduced into a vector containing a cbh1 expression cassette consisting of the *cbh1* promoter, the 17 amino acid *cbh1* signal sequence and the transcription terminator. *T. reesei* transformants selected for resistance to phleomycin were shown to produce the antibody light chain, Fab and the CBHI-Fab fusion protein with yields of 0.2 mg/l, 1 mg/l and 40 mg/l, respectively. When the Fab and CBHI-Fab producers were cultivated in cellulase inducing medium in a 15 l fermentor, the amount of immunoreactive antibodies was 1 mg/l and 150 mg/l, respectively. Functional fragments were separated from CBHI by fungal protease *in vivo* or by α-chymotrypsin *in vitro* (Nyyssönen *et al.*, 1993).

Bacterial Expression

Although whole immunoglobulins have been expressed in bacterial cells (*E. coli*), the yield of active antibody was poor (Boss *et al.*, 1984) and they were not glycosylated. In contrast, bacterial systems are widely used for the production of antibody fragments such as Fab, Fv and scFv where only the antigen binding regions are required. Furthermore, these regions are usually not glycosylated.

Apart from the production of a variety of antibody fragments in Gram-negative *Escherichia coli* (see next part of review), the secretion of anti-digoxin antibody fragments has been described in Gram-positive *Bacillus subtilis* (Wu *et al.*, 1993). Secretion of V_H alone resulted in low levels of production, possibly because of exposed hydrophobic regions on the surface of V_H causing aggregation. No improvement in V_H production with co-expression of V_L could be observed. In *B. subtilis*, both polypeptides are secreted directly into the culture medium. Their assembly into Fv is therefore inhibited by dilution. This problem was solved using a peptide linker between the V_H and V_L domains: active secreted scFv protein could be produced in a shake flask culture with a yield of 5 mg/l.

ANTIBODIES FROM *ESCHERICHIA COLI*

A rapid growth in the field of antibody engineering occurred after it was shown that functional antibody fragments could be secreted into the periplasmic space and even into the medium of *E. coli* by fusing a bacterial signal peptide to the antibody's N-terminus (Skerra and Plückthun, 1988; Better *et al.*, 1988). These findings opened the way for transferring the principles of the immune system for producing specific antibodies to a given antigen into a bacterial system (Fuchs *et al.*, 1992). It was now possible to establish antibody libraries in *E. coli* that could be directly screened for binding to antigen. This was accomplished at first by transforming *E. coli* with plasmids containing PCR-amplified immunoglobulin families from the lymphocytes of immunized mice. Immunogen-reactive recombinant antibodies were then selected by an ELISA of the bacterial supernatant from isolated bacterial colonies (Ward *et al.*, 1989). This procedure was subsequently improved upon by inserting the antibody operon into bacteriophage λ. Antibody libraries were then able to be efficiently transfected into *E. coli* and plaque lifts of lysed bacterial colonies on nitrocellulose could be screened for reactivity to a radioactive labeled immunogen (Huse *et al.*, 1989; Caton and Koprowski, 1990; Mullinax *et al.*, 1990).

However, in order to screen extremely large antibody libraries containing at least 10^8 individual members, it was necessary to develop a selection system as efficient as that of the immune system, in which the antibody receptor is bound to the surface of a B-lymphocyte. After binding its antigen, the B-lymphocyte is stimulated to proliferate and mature into an IgG-producing plasma cell. A similar selection system could be imitated in microorganisms by expressing antibodies on their surface. Millions of microorganisms could then be simultaneously screened for binding to an immobilized antigen followed by the propagation and amplification of the selected microorganism. Although display systems have been developed for eukaryotic systems, e.g. retroviral (Russel *et al.*, 1993) and baculoviral (Boublik *et al.*, 1995), the most successful surface expression system in microorganisms has been obtained using filamentous bacteriophages of the M13 family.

Phage Display

The structure of the virion of filamentous phage is shown in Figure 2 (Webster and Lopez, 1985; Smith, 1988). Its single-stranded DNA is enclosed in a tubular array of approximately 2700 molecules of the major coat protein pVIII. Four minor proteins are found at the tips of the phage, including the protein pIII, which initiates phage infection of *E. coli* by binding to their F-pili.

Coat protein pIII is a relatively flexible and accessible molecule composed of two functional domains: an N-terminal domain that binds to the F-pilus of male

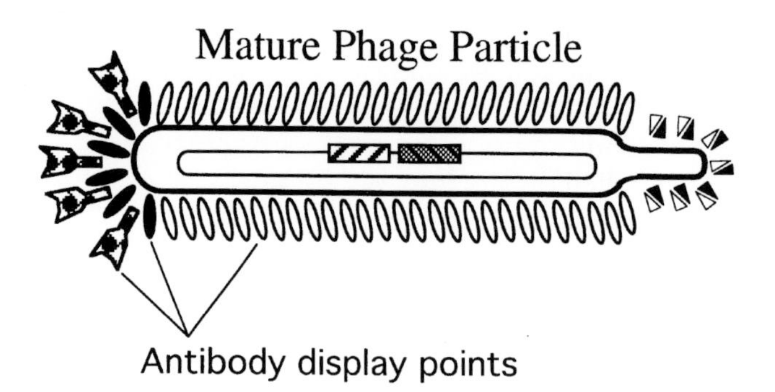

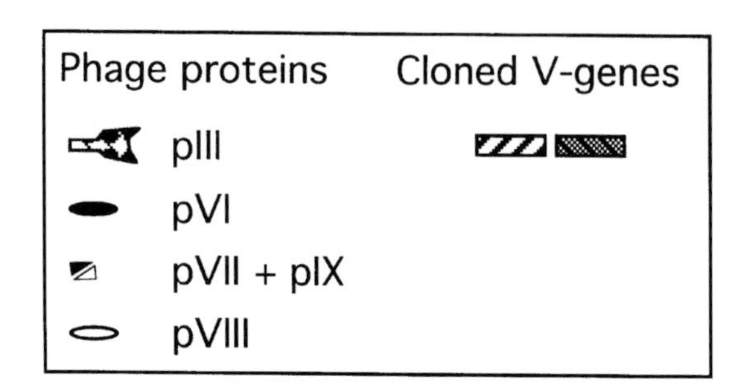

Figure 2 Schematic representation of a filamentous phage showing the capsid proteins and possible antibody display points.

bacteria and a C-terminal domain buried within the virion that is important for morphogenesis. Peptides can be inserted between the two domains of pIII (Smith, 1985) or near the N-terminus (Parmley and Smith, 1988) without destroying its functions in morphogenesis and infection. Parmley and Smith (1988) were also able to show that peptides inserted at the N-terminus of pIII could bind phages to immobilized antibodies. These findings led to the generation of phage peptide libraries that can be easily screened for binding to ligands and antibodies (Scott and Smith, 1990; Devlin *et al.*, 1990; Cwirla *et al.*, 1990).

In an analogous approach, antibodies have been attached to the surface of filamentous phages by inserting antibody DNA in the 5' end of gene III in the phage genome (McCafferty *et al.*, 1990). A major disadvantage, however, is that the expression of the antibody cannot be regulated. Furthermore, since large inserts were shown by Parmley and Smith (1988) to have adverse effects on infectivity, there is a high risk that antibody libraries would tend to accumulate the more rapidly propagating deletion mutants.

Phagemid pIII display systems

To overcome these problems, DNA coding for an antibody fragment was incorporated into a phagemid, which is basically a plasmid with a phage intergenic region that contains the packaging signal (Figure 3A; Breitling *et al.*, 1991; Hoogenboom *et al.*, 1991; Barbas *et al.*, 1991). Here, the expression of the fusion protein DNA is regulated by a bacterial promoter under the control of a *lac* operator, so that relatively large amounts of antibody can be obtained for analysis after induction with IPTG.

To display the antibody on the phage surface, the phagemid must be packaged with the proteins supplied by a helper phage such as M13K07, which packages its own DNA less efficiently than the phagemid DNA. Usually 10 helper phages are added per bacterium. The superinfection of bacteria with helper phage might possibly be inhibited by the binding of the pIII fusion protein to the pilin subunit of F-pili, which is why some workers prefer to delete the pIII N-terminal domain and fuse the antibody to the remaining C-terminal domains (Ørum *et al.*, 1993). Another minor problem might be caused by the leakiness of the *lac* operator, which could increase the selection pressure for the production of deletion mutants. To tighten repression, Breitling *et al.* (1991) employed a double *lac* operator and used *E. coli* transformed with a plasmid expressing *lac* repressor. In an alternative system, Ørum *et al.* (1993) included the gene coding for *lac* repressor in the phagemid.

The selection of reactive antibodies is mediated by successive rounds of elution of the bound phage, using either acid or alkali (Marks *et al.*, 1991b). Alternatively, proteolytic cleavage sites for trypsin (Breitling *et al.*, 1991; Ørum *et al.*, 1993) or for Genenase I (Ward *et al.*, 1996) have been introduced between the antibody domain and pIII. The competitive elution of bound phage-displayed antibodies with a monoclonal antibody has also been described (Meulemans *et al.*, 1994). The characterization of the selected antibody was facilitated by including an amber stop codon between the DNA coding for the antibody and pIII. Only the antibody domain will then be produced in non-suppressor strains of *E. coli* (Hoogenboom *et al.*, 1991). To follow the production of recombinant antibody, marker peptides recognized by monoclonal antibodies have been included either at the C-terminus (Ward *et al.*, 1989) or within the linker peptide between the heavy and light chain variable domains of a single chain antibody (Breitling *et al.*, 1991). Antibodies to the N-terminus have also proved to be very useful for following processing during secretion (Breitling *et al.*, 1991; Dübel *et al.*, 1992). In a more recent report, several of the above sequences for the improved expression, display and identification of recombinant antibodies were incorporated into a single expression phagemid (Hayashi *et al.*, 1995). This vector also provides the antibodies with a hexahistidine sequence for purification by metal chelate chromatography and a C-terminal cysteine to facilitate their conjugation with other molecules.

pIII display system for very large libraries

The display of antibodies on filamentous phages has proved to be an excellent system for selecting specific clones. The library size, however, depends on the efficiency of *E. coli* transformation. To overcome this restriction in *in vitro* packaging, *E. coli* have been transformed with plasmids encoding a heavy chain library and then infected with phagemid particles encoding a light chain library (Waterhouse *et al.*, 1993). To ensure that the heavy and light chain genes are packaged within the same particle, the *lox*-Cre site-specific recombination system of bacteriophage P1 was used to bring them together on the same phagemid. Another proposed system for the construction of multicombinatorial antibody libraries by association of the heavy- and light-chain genes from two different vectors within an infected bacterium is based on the use of the λ phage site-specific integration system (Geoffroy *et al.*, 1994). These methods could theoretically generate a phage display library containing as many different antibodies as the number of *E. coli* cells in culture. Such large libraries would be particularly useful for identifying unknown antigens of clinical interest. It might be possible to identify tumor associated antigens by the differential screening of normal and neoplastic cells. Antibodies to blood group antigens, for example, have been selected by a similar means from a phage display library (Marks *et al.*, 1993).

pIII display for antibody-mediated infection of E. coli

An elegant and perhaps more efficient method of selecting antibodies would be to use an evolutionary-based screening system combining bacterial display with phage display (Little *et al.*, 1993a). Antigens displayed on the surface of bacteria could provide a port of entry to a phage carrying a complementary antibody or *vice versa*. The presence of a selection marker would then lead to the amplification of only those clones that bound antigen. Those phage antibodies with the highest affinity should eventually dominate the culture ("survival of the fittest"). This principle has recently been demonstrated by fusing a single-chain antibody to the C-terminal domain of pIII and a corresponding antigen to the N-terminal domain of pIII. After uniting the two domains by antibody-antigen binding, the phage particle was then able to infect *E. coli* (Duenas and Borrebaeck, 1994; Duenas and Borrebaeck, 1995). This SAP approach (Selection and Amplification of Phages) makes it possible to discriminate between the kinetic parameters of phage-displayed antibody fragments using different selection strategies (Duenas *et al.*, 1996).

Theoretically, three to five pIII-antibody fusion proteins can be attached to each phagemid particle. In practice, however, the conditions usually employed for screening ensure that maximally one functional antibody per particle is displayed, thus allowing a distinction to be made between high and low affinity binders (Bass *et al.*, 1990; Ørum *et al.*, 1993). The avidity of a phagemid particle with several low-affinity antibodies can be as high as one with a single high affinity antibody. This problem is particularly acute when the antibody is displayed along the phage particle attached to the major coat protein pVIII.

Phagemid pVIII display systems

In view of the paracrystalline array of the major coat protein pVIII and the lack of pVIII mutants (Kuhn and Wickner, 1985), it is surprising that several hundred peptides fused to the N-terminus of pVIII can be incorporated into the phage coat without preventing phage assembly (Felici *et al.*, 1991; Ilyichev *et al.*, 1992). Larger polypeptides, however, do not appear to be readily incorporated. Kang *et al.* (1991a) reported that from one to twenty Fd fragments (heavy chain variable domain plus the first constant domain) fused to pVIII were incorporated along the length of the phagemid particle. Free Fab fragments can be generated, as described above for pIII fusions, by incorporating an amber mutation between the antibody and pVIII domains (Huse *et al.*, 1992). Although the pVIII systems proved to be useful for isolating antibodies with low affinities, they are less efficient than the pIII system for screening antibody libraries (Gram *et al.*, 1992). Direct comparison of pIII and pVIII systems demonstrated a much higher display efficiency of scFv fused to pIII on the phage surface as judged by ELISA. Furthermore, more scFv::pIII than scFv::pVIII phage could be recovered in biopanning experiments (Kretzschmar and Geiser, 1995).

Phagemid pVI display system

The product of filamentous phage gene VI, like pIII, is present in about five copies per phage (Simons *et al.*, 1981). Structural data indicate that pVI connects the phage body to pIII and that its C-terminus is exposed to the environment (Makowski, 1992). In a recent report, it was shown that a cDNA library inserted at the 3′ end of gene VI could be expressed and displayed as pVI fusion proteins on the surface of phage particles (Jespers *et al.*, 1995). Using this system, protease inhibitors were isolated that could bind to trypsin or factor Xa. Whereas this pVI system does not appear to be as efficient as the pIII system for displaying fusion proteins, it offers an attractive alternative for the display of proteins that may need to be fused by their N-termini rather than by their C-termini as with the pIII system. The pVI system has not yet been tested for screening antibody libraries.

Phage production

The transformation of *E. coli* with phagemids is usually accomplished by electroporation. Approximately 10^8 *E. coli* can be transformed (under favorable conditions!) in a cuvette containing 400 μl of bacterial suspension. To increase the efficiency of transformation, the phagemid DNA has been incorporated into a λ vector using the filamentous phage replication machinery to circularize its DNA (Hogrefe *et al.*, 1993). Cloning efficiencies were reported to be about four times higher than those achieved with electroporation.

Phage presentation of antibodies usually involves either single-chain antibodies (scFvs, see below) consisting of only the variable domains of antibody heavy and light chains joined by a flexible peptide linker or Fabs. In both cases, the heavy or light chain is fused to pIII or pVIII and the two chains dimerize to form an antigen-binding antibody fragment in the periplasmic space. The C-terminal domain of phage pIII or pVIII remains attached to the cytoplasmic membrane until it is incorporated into the secreted phage particle. To increase the copy number of each antibody, GroE chaperonins have been coexpressed during phage packaging (Söderlind *et al.*, 1993). This led to an almost two hundred-fold increase in the phage titer from $\sim 4 \times 10^{11}$ cfu/ml to 7×10^{13} cfu/ml.

Bacterial Cell Surface Display

Bacterial surfaces would appear to be more suitable than the pVIII system for presenting large numbers of antibodies and they may provide an alternative means of screening antibody libraries (Little *et al.*, 1993a). For example, antigens labeled with fluorescent tags could be used to select single cells in one step from antibody libraries using a fluorescence activated cell sorter (FACS). A selection could also be made with magnetic beads or with a highly sensitive colony screening assay. The latter procedure could prove to be particularly useful in the search for catalytic antibodies.

A major problem for the export of antibodies or any other proteins to the cell surface of Gram-negative bacteria such as *E. coli* is the double membrane. After secretion through the cytoplasmic membrane and cleavage of their signal peptide, they are then confronted by the largely impermeable outer membrane, which is characterized by an outer leaf of lipopolysaccharides. A rigid peptidoglycan network in the periplasmic space between the inner and outer membranes forms the cell wall. It is apparently anchored to the outer membrane by lipoproteins such as the major lipoprotein, Lpp, modified at its N-terminus by a lipid moiety that is integrated into the outer membrane. A large number of trimeric protein complexes formed by a channel-forming class of proteins known as the porins are also integrated into the outer membrane, where they permit the diffusion of molecules up to a molecular weight of about 500 Da. The porins and several other outer membrane proteins, including the abundant outer membrane protein A (OmpA), appear to contain several membrane-spanning β-sheets joined by surface loops.

Lipoproteins and OmpA have been employed to target recombinant antibodies to the cell surface of *E. coli*. In the first report of bacterial cell surface antibody presentation (Fuchs *et al.*, 1991), a single-chain antibody was fused to the N-terminus of the minor lipoprotein PAL (*p*eptidoglycan *a*ssociated *l*ipoprotein). In the presence of Tris-EDTA to perturb the outer membrane, the antibody was visualized by binding fluorescently labeled antibodies to a marker peptide in the linker between the V_H and V_L domains and by the binding of fluorescently labeled antigen. The bacterial cells expressing antibody fragment fused to PAL protein can be specifically selected by FACS using fluorescently labeled antigens (Fuchs *et al.*, 1996).

A second method for displaying single antibodies on the surface of *E. coli* employs a large N-terminal fragment of OmpA. The C-terminus of this fragment, normally found in a loop on the cell surface, was fused to the N-terminus of a single-chain antibody and the N-terminus of OmpA was fused to the signal sequence and the first nine amino acids of Lpp (Francisco *et al.*, 1993). Using a fluorescently labeled antigen, bacteria displaying the OmpA-antibody fusion protein were able to be selected with a FACS machine.

A third method exploits the secretion machinery of the IgA protease from *Neisseria*. Integration of the β-domain of the IgA protease precursor into the outer membrane is accompanied by the transport of the protease domain across the membrane, followed by autoproteolytic release into the medium. After transferring the *iga* gene to *E. coli*, this system was used to present the β-subunit of cholera toxin on the cell surface by fusing it to the β-domain of the IgA protease (Klauser *et al.*, 1990) and more recently this system has been used to display recombinant antibodies (personal communication, Thomas Meyer).

An interesting bacterial display system has recently been described that uses the protein flagellin as a carrier. It was found that the protein thioredoxin could be inserted into a large, solvent-exposed non-essential domain of flagellin and be exported to the cell surface, where it was assembled into partially functional flagella (Lu *et al.*, 1995). Random libraries of peptides were then inserted into a loop at the active site of thioredoxin and successfully screened for binding to antibodies. It would also be interesting to investigate whether the flagellin system could be used for targeting antibodies to the cell surface.

These bacterial surface display systems appear to be very promising for selecting binding proteins and ligands. For example, specific antibodies displayed on phage particles were efficiently selected by attachment to antigens inserted into the LamB protein on the surface of *E. coli* (Bradbury *et al.*, 1993). However, reliable routine procedures for screening antibody libraries displayed on bacteria still remain to be established.

Antibody Libraries

As a rule combinatorial antibody libraries were derived either from mouse or human immunoglobulin genes, although libraries generated from rabbit (Ridder *et al.*, 1995) and chicken (Davies *et al.*, 1995) have been described.

Antibody libraries from immunized donors

Antibody libraries were first generated from the lymphocyte cDNA of immunized mice or from "immunized" human donors using PCR and sets of primers homologous to the heavy (γ) and light chain (κ and λ) of Fab or Fv fragments, respectively (Orlandi *et al.*, 1989; Sastry *et al.*, 1989; Mullinax *et al.*, 1990; Persson *et al.*, 1991; Marks *et al.*, 1991a; Ørum *et al.*, 1993). An extensive study of primers suitable for amplifying the various human antibody subgroups has been made by Marks *et al.* (1991a). These primers were based on the available nucleotide sequence data at that time. To include the even more extensive protein sequence data, Welschof *et al.* (1995) designed a primer set starting from the amino acid sequence of immunoglobulins.

After PCR, the heavy and light chain partial libraries are randomly combined in an appropriate expression vector. These libraries are particularly rich in combinations that bind to the immunogen, since IgGs are secreted in relatively large amounts by clones of activated mature plasma B cells. If, for example, one in a 1000 of the mRNAs code for a chain of a binding antibody, then one original binding chain pair would be present per million clones. The proportion of mRNAs coding for antibodies specific for a particular antigen in immunized individuals, however, is usually higher and, in addition, one or both of the heavy and light chains can form functional binders with other partners. For example, one in 1200 clones (Mullinax *et al.*, 1990) and one in 5000 clones (Persson *et al.*, 1991) from donors immunized with tetanus toxin produced antibodies against the immunogen.

Using phage display selection systems, clinically useful antibodies against human carcinoembryonic antigen (CEA; Chester *et al.*, 1994) and anaphylatoxin C5a (Ames *et al.*, 1994) were isolated from immunized mice. Recombinant human antibodies have also been obtained against a variety of pathogens such as HIV (Burton *et al.*, 1991; Barbas *et al.*, 1992a, 1993a; Pilkington *et al.*, 1996), respiratory syncytial virus (Barbas *et al.*, 1992b), hepatitis B (Zebedee *et al.*, 1992), herpes simplex virus (Williamson *et al.*, 1993; Burioni *et al.*, 1994; Sanna *et al.*, 1995b), tetanus toxoid (Lang *et al.*, 1995), cytomegalovirus and varicella zoster virus (Williamson *et al.*, 1993). An autoantibody recognizing human placental DNA was isolated from a Fab phage display library prepared from a systemic lupus erythematosus (SLE) donor (Barbas *et al.*, 1995).

To obtain several antibodies to different epitopes of the gp120 protein of HIV-1, the most immunogenic epitopes were masked with antibodies corresponding to the dominant clones (Ditzel *et al.*, 1995). It was then possible to select antibodies from a phage display library to the more weakly immunogenic epitopes, thus providing a broader range of specific antibodies for possible prophylactics and therapy.

Naive (non-immune) libraries

Ideally, antibodies to any given antigen could be selected from universal antibody libraries (Little *et al.*, 1993b, 1995). For example, it is estimated that about 10^6–10^8 antibodies with different binding properties are present in the immune system at any one time. This number is sufficient to bind most antigens including synthetic compounds such as dinitrophenol that the immune system has never encountered. Considerable efforts have therefore been made in the last few years to generate libraries with large numbers of diverse antibodies.

One approach has been to amplify and randomly combine chains of the naive IgM repertoire (Marks *et al.*, 1991b). IgMs are expressed mainly on the surface of unactivated B-lymphocytes. After attachment of antigen and interaction with T-helper cells, the B cell is stimulated to proliferate and eventually matures into an IgG-secreting plasma cell. Using primers based on each of the human heavy and light chain gene families, a human non-immune antibody library was generated containing at least 10^7 members (Marks *et al.*, 1991a). After four rounds of panning, antibodies were selected that bind turkey egg-white lysozyme and 2-phenyloxazol-5-one (phOx) with affinities of K_a $10^7 M^{-1}$ and K_a $2 \times 10^6 M^{-1}$, respectively. Furthermore, it was also possible to obtain antibodies against some human self-antigens, such as tumor necrosis factor (TNF), soluble CD4 and CEA antigens (Griffiths *et al.*, 1993) and against the extracellular domain (ECD) of the tumor antigen c-*erb*B-2 (Schier *et al.*, 1995) from this library.

Higher affinity antibodies (K_a $10^9 M^{-1}$) can be obtained either by mimicking the *in vivo* process of somatic hypermutation by multiple cycles of random mutation in a bacterial mutator strain and selection (Low *et al.*, 1996) or by random or directed mutagenesis *in vitro* of hypervariable regions (CDR walking mutagenesis) followed by increasingly stringent rounds of selection with the antigen (Hawkins *et al.*, 1992; Riechmann and Weill, 1993; Balint and Larrick, 1993; Barbas *et al.*, 1994; Deng *et al.*, 1995; Yang *et al.*, 1995). It is also possible to pair a given heavy or light chain with a library of complementary chains to select improved binders (chain shuffling; Kang *et al.*, 1991; Marks *et al.*, 1992; Schier *et al.*, 1996). In one report the library for chain shuffling was generated by CDR mutagenesis (Thompson *et al.*, 1996). Chain shuffling can also be used to humanize animal antibodies by systematically pairing the animal heavy and light chains with the complementary domains from a human antibody library (Jespers *et al.*, 1994). Apart from the affinity enhancement, both CDR mutagenesis and chain shuffling can also result in a drift in epitope recognition (Ohlin *et al.*, 1996) and even in a change of antibody fine specificity (Kussie *et al.*, 1994; Ward, 1995).

For the direct selection of human antibodies with sub-nanomolar affinities, an extremely large library (1.4×10^{10} scFv repertoire) has been created from the human lymphocyte repertoire of naturally-rearranged V-genes (Vaughan *et al.*, 1996).

Semi-synthetic and synthetic libraries

A second approach creating universal antibody libraries has been to introduce random sequences into some or all of the hypervariable regions. Several positions within these regions, however, are fairly conserved and some others are not very variable (Chothia *et al.*, 1987, 1989, 1992; Kabat *et al.*, 1991). A recent analysis of the structural repertoire of immunoglobulins demonstrated that only ten combinations out of the 300 possible combinations of known permissible canonical structures account for 87% of 381 sequences analyzed (Vargas-Madrazo *et al.*, 1995). These positions should therefore be taken into account when designing randomized sequences.

The first synthetic libraries randomized only the third CDR of the heavy chain, since this region has been shown to be the most variable and the only CDR showing significant length differences. It also appears to make the largest contribution to the surface area available for contact with the antigen (Chothia *et al.*, 1987). Using this approach, it was found that an antibody that originally had a specificity for tetanus toxin could be transformed into an antibody with a relatively high affinity for fluorescein (Barbas *et al.*, 1992c). Binding motifs can also be incorporated into the randomized CDR3 regions. For example, integrin-binding antibodies were able to be isolated from such libraries containing the Arg–Gly–Asp (RGD) integrin recognition motif (Barbas *et al.*, 1993b). Moreover, it was then possible to isolate integrin-binding antibodies without the RGD motif by fixing the flanking regions and randomizing the RGD sequence (Smith *et al.*, 1994). Further antibody diversity has been achieved by randomizing the CDR3 of both the heavy and light chains using a mixture of different lengths (Barbas *et al.*, 1993c). Using these libraries, it was also possible to isolate antibodies that coordinate metals (Barbas *et al.*, 1993d) and catalyze the hydrolysis of thioesters (Janda *et al.*, 1994). Numerous antibodies to a variety of antigens were also obtained from libraries of antibodies that combined random heavy chain CDR3 sequences of varying lengths with the heavy chain CDR1 and CDR2 regions from around fifty germline sequences (Hoogenboom and Winter, 1992; Akamatsu *et al.*, 1993; Nissim *et al.*, 1994; de Kruif *et al.*, 1995a, 1995b). The high-affinity human antibodies (K_d up to 4 nM) to a number of antigens were isolated directly from a large semi-synthetic repertoire (6.5×10^{10}) of a Fab phage library created using the Cre-*lox in vitro* recombination system (Griffiths *et al.*, 1994).

To produce synthetic libraries containing random sequences in four of the hypervariable regions, the framework regions of oligonucleotides containing the random sequences were hybridized onto an antibody DNA template to facilitate their ligation (Garrard and Henner, 1993). However, to increase the efficiency of gene synthesis and to avoid any preferential incorporation of specific random sequences, we developed an alternative method using overlapping sets of oligonucleotides that are extended under PCR conditions to full length genes (Hayashi *et al.*, 1994). In this way it was possible to mutagenize all of the six hypervariable regions simultaneously in a few steps. Sequence analyses showed that while the method is basically successful, deletions were observed in the longer random sequences which were probably caused by problems in the automatic synthesis of long degenerate oligonucleotides.

Production of Functional Antigen-Binding Fragments in *E. coli*

Refolding from cytoplasmic inclusion bodies

Unlike glycosylated whole antibodies, Fab, Fv or scFv can be easily produced in bacterial cells as functional antigen-binding molecules. There are two basic strategies to obtain recombinant antibody fragments from *Escherichia coli*. The first is to produce antibody proteins as cytoplasmic inclusion bodies followed by refolding *in vitro* (Huston *et al.*, 1988, 1991; Buchner and Rudolph, 1991). In this case the protein is expressed without a signal sequence under a strong promoter. The inclusion bodies contain the recombinant protein in a non-native and non-active conformation. To obtain functional antibody, the recombinant polypeptide chains have to be dissolved and folded into the right shape. The renaturation of antibody fragments comprises several steps. Starting from the completely unfolded and reduced state, the polypeptide chains have to fold, the disulfide bridges within the domains have to be formed, the folded and oxidized molecules must associate correctly and the heterodimers (in case of Fab) must be linked covalently via a disulfide bridge. As in the case of many other multidomain or disulfide bridged proteins, side reactions leading to wrongly structured molecules may occur (Buchner and Rudolph, 1991). The percentage of misfolded molecules is often dependent on the renaturation conditions chosen. Most of the current refolding protocols consist of dilution refolding, redox refolding and disulfide-restricted refolding (Huston *et al.*, 1991).

Dilution refolding (renaturation prior to disulfide bond formation) relies on the observation that fully reduced and denatured antibody fragments can refold and their specific binding activity can be restored when the denaturant and reducing agent are removed by dialysis (Huston *et al.*, 1991). Dilution refolding involves a two-stage refolding process, with renaturation of the reduced protein followed by air oxidation. The yield of an active anti-digoxin scFv refolded by this method was 12.6% (Huston *et al.*, 1988).

Redox refolding (renaturation and disulfide bond formation simultaneously) utilizes a glutathione redox coupling system (reduced glutathione/oxidized glutathione or dithioerythritol/oxidized glutathione) to catalyze disulfide interchange as the protein refolds into its native state (Buchner and Rudolph, 1991; Buchner *et al.*, 1992). By systematic variation of the experimental parameters, such as the redox conditions, solubilizing additives in non-denaturing concentrations, concentration of denatured protein upon initiation of renaturation and duration and temperature of reconstitution, Buchner and Rudolph (1991) developed a strategy for renaturating up to 40% of the recombinant protein to functional Fab fragments.

Disulfide-restricted refolding (disulfide bond formation prior to renaturation) is a method which takes advantage of the increased efficiency of antibody refolding when disulfides bonds are kept intact. The presence of correct disulfide bonds should enhance the recovery of the native structure, while those chains with incorrect disulfide pairs will produce incorrectly folded species on removal of the denaturant. Recently, Kurucz *et al.* (1995a) demonstrated that scFvs solubilized in a particular detergent solution (2% sodium *N*-lauroylsarcosine) formed correct intrachain disulfide bonds upon oxidation in air. After formation of disulfide bridges, the recombinant protein was refolded by slow removal of denaturing agent. This procedure resulted in monomeric active scFv with a yield of 50% (Kurucz *et al.*, 1995a) and in monomeric bispecific $(scFv)_2$ with a yield of 18% (Kurucz *et al.*, 1995b).

The reconstitution of functional Ig fragments and fusion proteins containing antibody fragments from cytoplasmic inclusion bodies has been described by several research groups (for review see Skerra, 1993). This procedure is recommended particularly for the production of antibody-derived fusion proteins such as immunotoxins that might be toxic to bacterial cells or unstable due to intracellular degradation when expressed in a soluble or secreted form (Buchner *et al.*, 1992; Brinkmann *et al.*, 1992; Newton *et al.*, 1994, 1996). In the case of heterodimeric antibody fragments (i.e. Fv, Fab and disulfide-stabilized Fv), the two chains are expressed, purified and solubilized separately, and then combined in a renaturation buffer from which the correctly refolded protein is purified (Buchner and Rudolph, 1991; Brinkmann *et al.*, 1993b; Choe *et al.*, 1994; Reiter *et al.*, 1995; Webber *et al.*, 1995). The amounts of active material obtained from

1 liter of bacterial fermentation culture by refolding *in vitro* were 1.6 mg for anti-Tac scFv (Webber *et al.*, 1995), 7 mg for anti-Tac disulfide-stabilized (ds) Fv (Webber *et al.*, 1995), 10–37 mg for scFv-PE38KDEL immunotoxin (Reiter *et al.*, 1994c), 30–70 mg for dsFv-PE38KDEL immunotoxin (Reiter *et al.*, 1994c) and 110 mg for anti-digoxin scFv fused to the C-terminus of fragment B of staphylococcal protein A (Tai *et al.*, 1990).

Expression of functional antibody fragments in E. coli *by secretion*

A second approach for obtaining functional antibody fragments is to imitate the situation in the eukaryotic cell for secreting a correctly folded antibody. In *E. coli*, the secretion machinery directs proteins carrying a specific signal sequence to the periplasm (Pugsley, 1993). The vector pHOG21 (Kipriyanov *et al.*, 1996b) for the secretion of antibody fragments in *E. coli* is shown in Figure 3B. Periplasmic expression has permitted the functional testing of a wide variety of antibody fragments with different antigen-binding specificities (for review see Skerra, 1993). The antibody fragments are usually correctly processed in the periplasm, they contain intramolecular disulfide bonds and are soluble (Glockshuber *et al.*, 1990). However, the high-level expression of a recombinant protein with a bacterial signal peptide in *E. coli* often results in the accumulation of insoluble antibody fragments after transport to the periplasm (Whitlow and Filpula, 1991; Kipriyanov *et al.*, 1994), presumably via the aggregation of a folding intermediate.

It is now recognized that aggregation *in vivo* is not a function of the solubility and stability of the native state of the protein, but of those of its folding intermediates in their particular environment (Hockney, 1994; Plückthun, 1994). However, the overexpression of some enzymes of the *E. coli* folding machinery such as GroES/L chaperonins, disulfide-isomerase DSbA and proline-*cis–trans*-isomerase (PPIase) did not increase the yield of soluble antibody fragments (Knappik *et al.*, 1993; Duenas *et al.*, 1994). The degree of successful *in vivo* folding appears to depend very much on the primary sequence of the variable domains (Plückthun, 1994; Knappik and Plückthun, 1995).

Unlike cytoplasmic inclusion bodies, a significant part of the periplasmic protein aggregates has correctly formed disulfide bridges. This conclusion was derived from the observation that the redox reshuffling of disulfide bridges in scFv isolated from periplasmic inclusion bodies did not increase the yield of soluble and functional product after refolding (Kipriyanov *et al.*, 1994). The driving force behind periplasmic aggregation might be hydrophobic interactions between the molecules of folding intermediates (Dill, 1990). In this case, high protein concentrations of the secreted antibody fragment in the periplasmic space would favor the formation of insoluble aggregates over correct folding (Bowden and Georgiou, 1990; Kiefhaber *et al.*, 1991). This explanation is supported by the fact that increasing the promoter strength leads to a higher intracellular accumulation of recombinant antibodies, but mainly in an insoluble form (Kipriyanov *et al.*, 1997).

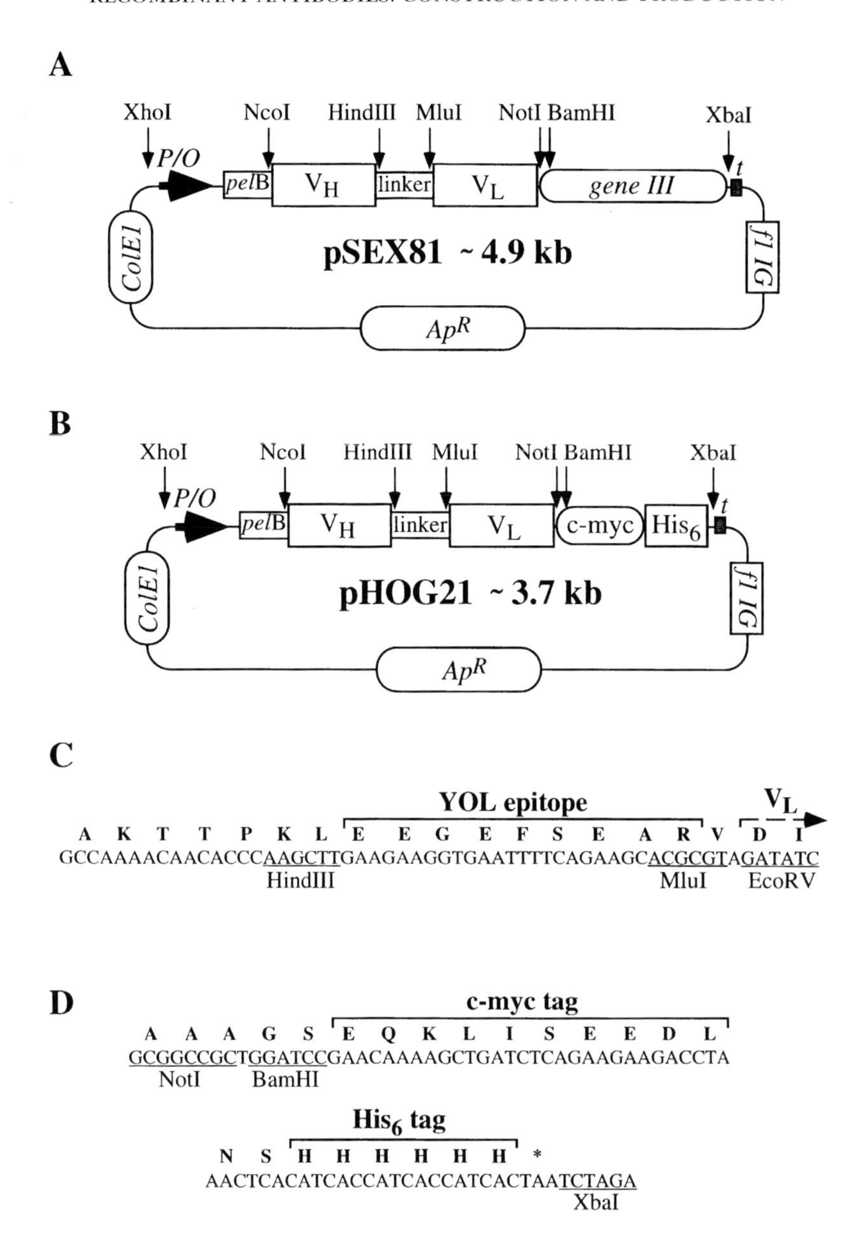

Figure 3 Schematic representation of the phage display vector pSEX81 (A) and scFv expression vector pHOG21 (B). Ap^R, ampicillin resistance-encoding gene; *c-myc*, a sequence encoding an epitope recognized by the monoclonal antibody 9E10; *ColE1*, origin of DNA replication; *f1IG*, intergenic region of phage f1; *gene III*, gene coding for pIII of M13 phage; His_6, a sequence encoding six C-terminal histidine residues; linker, a sequence encoding 17 amino acids connecting the V_H and V_L domains; *pel*B, signal peptide sequence of bacterial pectate lyase; P/O, wt *lac* promoter/operator. The nucleotide and deduced amino acid sequences of the linker region indicating the YOL epitope (C) as well as the sequence coding for the *c-myc* epitope and six histidines in the carboxy terminal part of the scFv (D) are shown underneath.

The strategy for refolding antibody fragments from the periplasm while retaining the disulfide bonds usually consists of solubilizing and denaturating the periplasmic inclusion bodies with 6 M guanidine hydrochloride or 8 M urea, followed by 100–200 fold dilution into a renaturation buffer (Pantoliano *et al.*, 1991; Whitlow and Filpula, 1991) or renaturation by slow removal of the denaturing agent by dialysis at low protein concentrations (Malby *et al.*, 1993; George *et al.*, 1994; Kipriyanov *et al.*, 1994, 1995a). After refolding, the active monomeric protein can be isolated either by affinity chromatography (George *et al.*, 1994) or by gel-filtration (Kipriyanov *et al.*, 1994, 1995a). This procedure usually yields 3–5 mg of active scFv from 1 liter of a shake-flask culture (George *et al.*, 1994; Kipriyanov *et al.*, 1994) and up to 70 mg/l in a fermenter (Milenic *et al.*, 1991).

Lowering the bacterial growth temperature has been shown to decrease periplasmic aggregation and increase the yield of soluble antibody protein (Skerra and Plückthun, 1991; Plückthun, 1994). Some antibody fragments appear to perturb the integrity of the outer membrane and are released into the medium (Takkinen *et al.*, 1991; Sizman *et al.*, 1993). This membrane leakiness, of which the molecular cause is still unknown, also seems to depend on the antibody primary sequence. For example, in identical *E. coli* host/vector systems, 1–5 mg/l of anti-glycophorin A 1C3 scFv was found in the culture medium (Lilley *et al.*, 1994) compared to 0.1–0.5 mg/l of anti-DNP U7.6 scFv and no anti-transferrin receptor OKT9 scFv (George *et al.*, 1994). Therefore, protein engineering was recently proposed as a strategy for improving folding and the yield of functional periplasmic proteins (Knappik and Plückthun, 1995). However, this approach is laborious and can only be used for a few of the most important antibodies.

The aggregation of recombinant proteins in the *E. coli* periplasm can also be reduced by growing the cells in the presence of certain non-metabolized sugars, e.g. raffinose or sucrose (Bowden and Georgiou, 1990; Sawyer *et al.*, 1994). Recently, we showed that the addition of 0.4 M sucrose to the growth medium gives a 15–25 fold increase in the yield of soluble scFv for bacterial shake-tube cultures and an 80–150 fold increase for shake-flask cultures (Kipriyanov *et al.*, 1997). We also found that the scFv could be made to accumulate in the periplasm or be secreted into the medium by simply changing the incubation conditions and the concentration of the inducer (Kipriyanov *et al.*, 1997).

The yield of soluble antibody fragments in bacteria can sometimes be improved by modifying the secondary mRNA structure (Stemmer *et al.*, 1993a; Duenas *et al.*, 1995) and by choosing more suitable *E. coli* strains (Sizmann *et al.*, 1993). Interestingly, a significant increase in the amount of a Fab antibody secreted from *E. coli* was obtained by exchanging the C_{κ} domain with a C_{λ} domain (MacKenzie *et al.*, 1994). Depending on the antibody, relatively high amounts of antibody fragments can be secreted in *E. coli* with yields of Fv from up to 40 mg/l in shake flasks and 450 mg/l in fermenters (King *et al.*, 1993), yields of scFv from up to 70 mg/l in shake flasks (Sizmann *et al.*, 1993) and 200 mg/l in a fermentor (Pack *et al.*, 1993) and yields of Fab′ from up to 1–2 g/l under conditions of high cell density fermentation (Carter *et al.*, 1992a).

Antibody Fragments

The Fv fragment consisting only of the V_H and V_L domains is the smallest Ig fragment available that carries the whole antigen-binding site (Figure 4). However, Fvs appear to have a lower interaction energy of their two chains than Fab fragments which are also held together by the constant domains C_H1 and C_L (Glockshuber *et al.*, 1990). To stabilize the association of the V_H and V_L domains, several methods have been devised.

Single-chain Fv fragments

The V_H and V_L domains have been linked in single-chain Fv constructs with a short peptide that bridges the approximately 3.5 nm between the carboxy terminus of one domain and the amino terminus of the other (Huston *et al.*, 1988; Bird *et al.*, 1988; Figure 4). Either the orientation V_H-linker-V_L or V_L-linker-V_H can be used (summarized in Huston *et al.*, 1993). ScFv derived from the antibody McPC603 with both orientations showed no significant difference in their expression level and in the free energy of folding (Knappik *et al.*, 1993).

The linker design has been based on various considerations summarized by Huston *et al.* (1993). For example, the $(Gly_4Ser)_3$ (Huston *et al.*, 1988) was designed *de novo* to prevent interference with domain folding and intercalation between V regions by avoiding side chains larger than the hydroxymethyl groups of serine. An NMR comparison of the unlinked Fv fragment of the antibody McPC603 with the corresponding scFv containing a V_H-$(Gly_4Ser)_3$-V_L linker showed no perturbation of the folding of the variable domains by the linker (Freund *et al.*, 1993, 1994).

The long linker (28 amino acids) of Takkinen *et al.* (1991) was specifically derived from the interdomain segment of a fungal cellulase, known from its

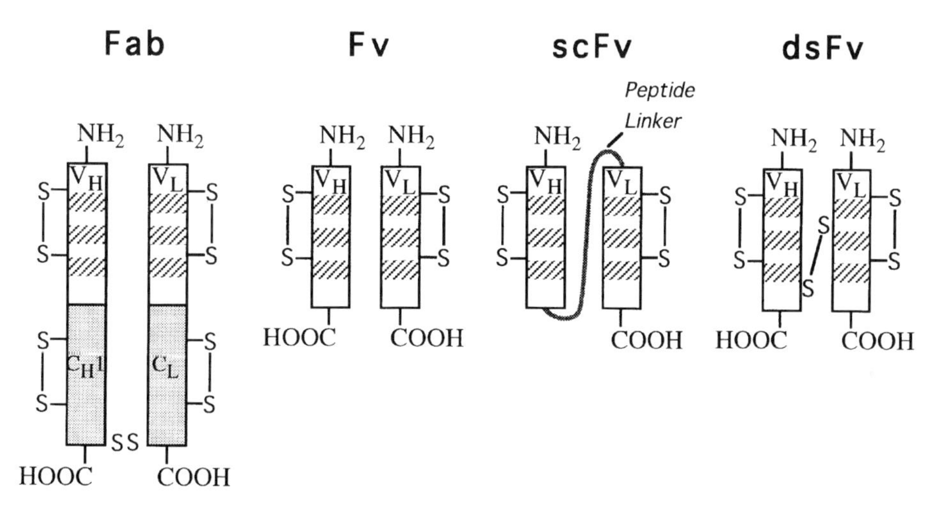

Figure 4 Monovalent antibody fragments expressed in *E. coli*.

crystal structure to be highly flexible. An anti-2-phenyloxazolone (phOx) scFv containing this long hydrophilic linker bound its antigen with an affinity comparable to the parent antibody and it was almost as stable to pH and temperature (Takkinen *et al.*, 1991). However, this relatively long linker proved to be quite sensitive to proteolytic degradation (Alfthan *et al.*, 1995).

In another approach, libraries of three-dimensional peptide structures were searched for linkers of the correct molecular dimensions (Bird *et al.*, 1988). Amino acids interfering with the Fv structure were discarded and replaced with glycine, serine, threonine and charged residues. Using a similar method, a number of flexible helical and (non-helical) linkers with a minimum length of 12 residues were constructed (Whitlow and Filpula, 1991). Sequence differences between linkers of similar length seemed to affect the stability and antigenicity of the linker itself rather than the affinity of the scFv (Whitlow and Filpula, 1991; Whitlow *et al.*, 1993). In a few cases, the linker instability appeared to be quite beneficial. For example, Essig *et al.* (1993) demonstrated that the introduction of a proteolytic clip into the linker region yields an Fv fragment that did not aggregate at the high concentrations necessary for crystallization. Another linker was designed to make the scFv more labile while retaining its activity (Solar and Gershoni, 1995). The rationale was that by creating tension in the molecule, an scFv might be generated that would be extremely sensitive to its environment, releasing its antigen upon the slightest perturbation. This could be useful for analytical systems or for recognition components in biosensors (Solar and Gershoni, 1995). A predominantly negatively charged linker containing an epitope recognized by the rat mAb YOL1/34, that is suitable for immunodetection appeared to function just as satisfactorily as the $(Gly_4Ser)_3$ linker (Breitling *et al.*, 1991 and unpublished results).

A random approach for linker construction instead of molecular modeling was recently described by Stemmer *et al.* (1993b). Inverse PCR mutagenesis was used to construct a library of 3×10^5 Fv mutants in which the C-terminus of the light chain was connected to the N-terminus of the heavy chain by a 15-amino acid peptide linker of variable composition. After plating, active mutant colonies were identified by screening colony filter lifts with a radiolabeled antigen. About 0.2% of the mutants were positive, and a selected scFv clone was shown to have an affinity similar to that of the parent antibody.

A number of workers have noticed a tendency of the V_H and V_L domains of some scFv fragments not only to associate intramolecularly but also intermolecularly (Griffiths *et al.*, 1993; Whitlow *et al.*, 1994; Desplancq *et al.*, 1994; Alfthan *et al.*, 1995). This appears to depend mainly on the length of the linker: the multimerization decreases with increasing linker length (Desplancq *et al.*, 1994). Holliger *et al.* (1993) used this phenomenon to force the formation of dimers, which they named “diabodies”, by shortening the linker (Figure 5B).

The small scFvs are particularly interesting for clinical applications (for review see Huston *et al.*, 1993). They are only half the size of Fabs and thus have lower retention times in non-target tissues, more rapid blood clearance and better tumor penetration (Milenic *et al.*, 1991; Yokota *et al.*, 1992; Adams *et al.*, 1993). ScFvs therefore represent potentially very useful molecules for the targeted

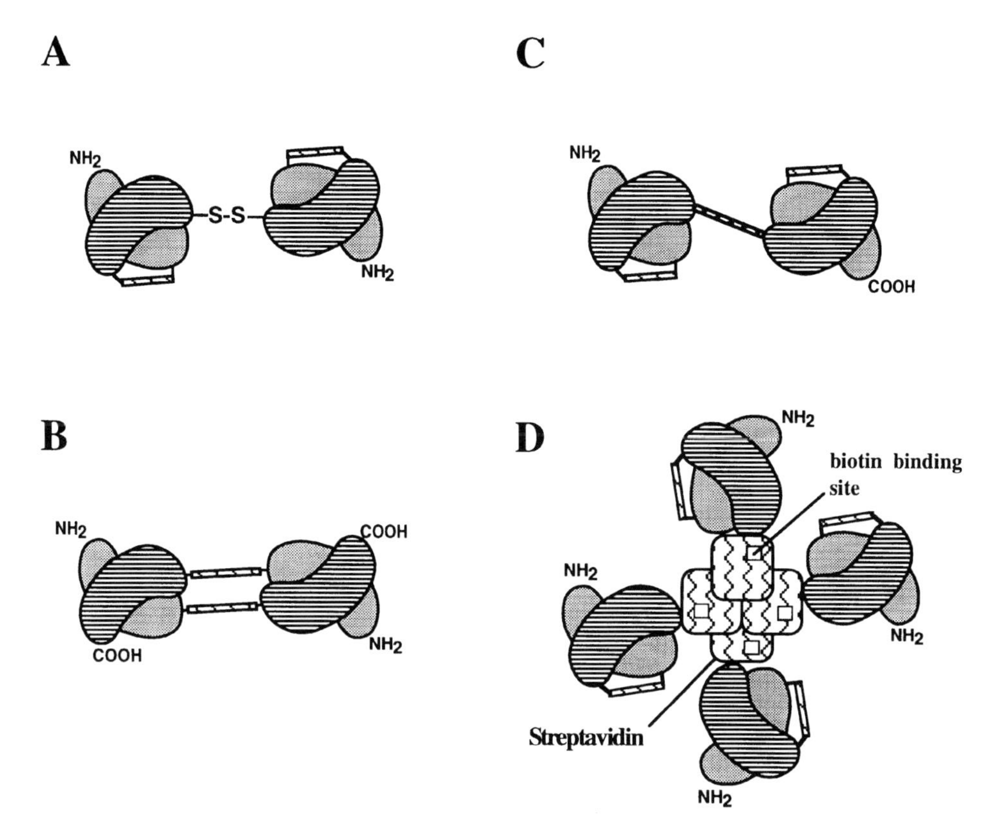

Figure 5 Bivalent and tetravalent antibody fragments expressed in *E. coli*. (A) Covalently bound $(scFv')_2$ dimers spontaneously formed in bacterial cells (Kipriyanov *et al.*, 1994) or by chemical coupling *in vitro*. (B) Noncovalent scFv dimers (diabody). (C) Bivalent single-chain $(scFv)_2$. (D) ScFv-streptavidin tetramers.

delivery of drugs, toxins, or radionuclides to a tumor site (Colcher *et al.*, 1990; Brinkmann *et al.*, 1993a; Goshorn *et al.*, 1993).

Disulfide-stabilized Fv fragments

Another strategy for linking V_H and V_L domains has been to design an intermolecular disulfide bond (Glockshuber *et al.*, 1990; Figure 4). The disulfide-stabilized (ds) Fv fragment appeared to be much more resistant to irreversible denaturation caused by storage at 37°C than the unlinked Fv. It was more stable than the scFv fragment and a chemically cross-linked Fv (Glockshuber *et al.*, 1990).

In this particular antibody, the disulfide bond was designed to connect a cysteine in the CDR3 region of the heavy chain with a cysteine in the CDR2 region of the light chain without interfering with antigen binding. However, detailed models of CDR without crystallographic data are difficult of make due to the large amount of structural variability (Chothia *et al.*, 1989). Molecular modeling was therefore performed by Jung *et al.* (1994) to identify possible interchain disulfide bond sites in the framework regions of the Fv fragment of the anti-cancer murine mAb B3. A model of the B3 Fv structure was obtained from the known structure of the most homologous Fv of mAb McPC603. The two most promising sites for disulfide bridges appeared to be V_H44-V_L100 connecting FR2 of the heavy chain with FR4 of the light chain and V_H105-V_L43 that links FR4 of the heavy chain with FR2 of the light chain (numbering scheme of Kabat *et al.*, 1991).

This prediction was evaluated experimentally by constructing disulfide-linked dsFv-immunotoxins that were expressed in *E. coli* (Brinkmann *et al.*, 1993b; Reiter *et al.*, 1994a, 1995). The antigen-binding activity, specificity and anti-tumor effect of V_H44-V_L100 B3(dsFv)-immunotoxin was indistinguishable from its single-chain Fv counterpart (Brinkmann *et al.*, 1993b; Reiter *et al.*, 1994b). Disulfide-linked Fv-immunotoxin was more stable at 37°C in human serum and more resistant to thermal and chemical denaturation than the scFv-immunotoxin. Disulfide bridges at other possible sites resulted in preparations of dsFvs with much lower activities (Reiter *et al.*, 1994a, 1995).

To determine whether the stabilization of Fvs by disulfides at positions V_H44-V_L100 is generally applicable, Reiter *et al.* (1994c) constructed and analyzed the dsFv-containing immunotoxins derived from two other mAbs specific for tumor antigens. Their results indicated that dsFvs have at least the same binding properties as scFvs (Webber *et al.*, 1995), and in one case the dsFv was significantly better (Reiter *et al.*, 1994c). However, it would be premature to assume that position V_H44-V_L100 is a universal site for introducing disulfide bridges on the basis of only three examples. A computer search of the approximately 100 available antibody crystal structures found only two structures where a disulfide bond would be permitted in position V_H44-V_L100 (Dr. A. Martin, personal communication). Nevertheless, this position appears to be reasonably promising and needs to be tested with a sufficiently large number of different antibodies to evaluate its general applicability for linking the chains of an Fv fragment.

A disulfide-linked Fv has recently been engineered for the humanized anti-p185^{HER2} antibody humAb4D5-8 (Rodrigues *et al.*, 1995). On the basis of the known crystal structure of antibody humAb4D5-8, three Fv variants were designed with an interchain disulfide bond buried at the V_L/V_H interface and expressed in *E. coli*. One variant (V_H101-V_L46) had an affinity for the antigen similar to the wild-type Fv and was used to construct a fusion protein with β-lactamase. The fusion protein retained both antigen-binding activity in murine serum and could efficiently activate a cephalosporin-based doxorubicin prodrug.

To test whether the more stable disulfide-linked Fv fragments can be used to advantage in a phage-display system, Brinkmann *et al.* (1995) constructed a dsFv of the anti-Tac antibody and a dsFv library for phage display. The dsFv phage could specifically bind antigen although their titer in supernatants was less than that of scFv phage. Nevertheless, since the dsFv phages were more stable than those expressing scFv, they could prove to be a useful alternative and addition to scFv-phage display.

Single antibody domains

To obtain even smaller antigen binding fragments than those described above, antigen binding V_H domains were isolated from the lymphocytes of immunized mice (Ward *et al.*, 1989). However, one problem of V_H domains is their "sticky patch" for interactions with V_L domains. Since naturally occurring camel antibodies lack light chains (Hamers-Casterman *et al.*, 1993; Muyldermans *et al.*, 1994), the solubility of human V_H domains has been improved by mimicking camelid heavy chain sequences (Davies and Riechmann, 1994). To make even smaller antigen-binding fragments, Pessi *et al.* (1993) removed the CDR3 and flanking regions of the V_H domains. Synthetic repertoires of both the camelized and mini V_H fragments generated through randomization of the hypervariable loops have been displayed on phage and successfully selected for specific target recognition (Martin *et al.*, 1994; Davies and Riechmann, 1995).

The advantage of single antibody domains compared to multidomain antibody fragments is not only their size but also their potentially higher stability (Davies and Riechmann, 1996).

Bi- and Multivalent Fvs

One disadvantage of antibody fragments such as Fab, scFv and dsFv is the monovalency of the product, which precludes an increased avidity due to polyvalent binding. Several therapeutically important antigens have repetitive epitopes resulting in a higher avidity for antibodies and antibody fragments with two or more antigen binding sites (Johnson *et al.*, 1986). Examples of various constructs that have been made to increase the valency of scFv are shown in Figure 5.

Multivalent scFv can be generated when the variable domains of one scFv molecule pair with the complementary domains of another (Whitlow *et al.*, 1993). The amounts of dimers and higher multimers formed are linker dependent and can be increased under certain conditions (Desplancq *et al.*, 1994; Whitlow *et al.*, 1994). The stabilities of such multivalent Fv, however, differ significantly depending on the specific variable domains from which they are formed (Whitlow *et al.*, 1994). Bispecific diabodies can be made from two single chain fusion products consisting of the V_H domain from one antibody connected by a short linker to the V_L domain of another antibody (Zhu *et al.*, 1996; Holliger *et al.*, 1993, 1996). Two antigen-binding domains have been shown by crystallographic analysis to be on opposite sides of the complex (Perisic *et al.*, 1994).

In another approach, recombinant antibody fragments containing C-terminal cysteines were dimerized by chemical coupling (Carter *et al.*, 1992a; Cumber *et al.*, 1992; Adams *et al.*, 1993) or by gentle air oxidation of reduced scFv′ (McCartney *et al.*, 1995). Dimeric and trimeric Fab′ and scFv molecules were constructed by chemical coupling with linkers containing two or three maleimide groups (King *et al.*, 1994). Cross-linkers have also been designed to contain a 12-*N*-4 macrocycle capable of chelating the radioactive label ^{90}Y.

We have recently shown that disulfide stabilized dimers can be formed in the periplasm of *E. coli* by the site-specific dimerization of scFv containing a

C-terminal cysteine (Kipriyanov *et al.*, 1994). Spontaneous assembly of bivalent antibody fragments in *E. coli* was also demonstrated for scFv fused to the kappa light chain constant domain (McGregor *et al.*, 1994) and for Fab′ containing a modified hinge region (Rodrigues *et al.*, 1993).

The genetic engineering of two scFvs linked with a third polypeptide linker (Figure 5C), as initially suggested by Huston *et al.* (1991) has now been carried out in several laboratories for the production of bispecific single chain antibody segments (Gruber *et al.*, 1994; Mallender and Voss, 1994; Neri *et al.*, 1995a; De Jonge *et al.*, 1995; Kurucz *et al.*, 1995) as well as a bispecific single-chain antibody-toxin (Schmidt *et al.*, 1996). Linear bivalent $F(ab')_2$ fragments, comprising tandem repeats of a heavy chain fragment V_H-C_H1-V_H-C_H1 cosecreted with two light chains have also been produced in *E. coli* (Zapata *et al.*, 1995).

Another interesting method for creating bivalent antibody Fv fragments is to fuse them with peptides that dimerize. For example, the C-termini of scFvs have been fused to dimerization domains such as a four-helix bundle and a leucine zipper (Pack and Plückthun, 1992; Pack *et al.*, 1993). Amphipathic helices can also be used to form tetravalent antibody fragments (Pack *et al.*, 1995). $F(ab)_2$ conjugates with alkaline phosphatase (PhoA) are bivalent due to the association of the PhoA subunits (Ducancel *et al.*, 1993) . An artificial antibody system was created using an Fv fragment fused with one or two Fc binding domains of protein A from *Staphylococcus aureus* (Ito and Kurosawa, 1993). When these fusion proteins were mixed with IgG molecules, complexes with up to three antigen binding sites were formed. In a recent report, a bifunctional product was formed between a single chain antibody fused to calmodulin and a maltose binding protein fused to a calmodulin-binding peptide ligand (Neri *et al.*, 1995b).

To develop artificial antibodies with multiple valency, we have fused single-chain antibody Fv fragments to streptavidin (Dübel *et al.*, 1995). The purified tetrameric scFv-streptavidin complexes (Figure 5D) demonstrated both antigen- and biotin-binding activity, were stable over a wide range of pH and did not dissociate at high temperatures (up to 70°C). The apparent affinity constant was found to be 35 times higher for pure scFv-streptavidin tetramers compared to monomeric single-chain antibodies (Kipriyanov *et al.*, 1995b, 1996a). We could also show that most of biotin binding sites were accessible and not blocked by biotinylated *E. coli* proteins or free biotin from the medium. These sites should therefore facilitate the construction of bispecific multivalent antibodies by the addition of biotinylated ligands.

Some of the constructs described above for making multivalent Fvs can, of course, also be used for producing multispecific Fvs. The streptavidin-scFv fusion protein, for example (Dübel *et al.*, 1995; Kipriyanov *et al.*, 1996a), can be made bi- or multispecific by mixing appropriate ratios of denatured scFvs together and then renaturing. Alternatively, biotinylated antibodies can be easily conjugated to an scFv-streptavidin tetramer. This procedure facilitates the production of a more homogenous bispecific product containing equal amounts of the two antibodies. Further conjugations can be made through C-terminal cysteines which can also be biotinylated for coupling to other streptavidin complexes (Kipriyanov *et al.*, 1994). Biotinylation of the scFv can also be performed *in vivo* by fusing the biotin carboxyl carrier protein (BCCP) to its C-terminus (Weiss *et al.*, 1994). Such bifunctional constructs could provide a facile means of coupling stimulatory ligands to mammalian cell surfaces. Gotter *et al.* (1995), for example, described a simple method for anchoring ligands and proteins to the surface of cells that have been incubated for a short time with Newcastle disease virus using an scFv-streptavidin fusion protein. The scFv was directed against the haemagglutinin-neuraminidase of the virus envelope after fusion with the cell membrane. If this single chain antibody is fused to streptavidin, biotinylated molecules can then be attached to the cell surface (unpublished results).

Multivalent complexes can also be formed by attaching recombinant antibodies to liposomes. This can be accomplished either by chemical coupling through a C-terminal cysteine (Park *et al.*, 1995) or by incorporation of biosynthetically lipid-tagged scFv antibody fragments into liposomes (Laukkanen *et al.*, 1994). The immunoliposomes with anchored recombinant antibody fragments demonstrated high-affinity antigen binding and could be used for delivering anti-tumor drugs (Park *et al.*, 1995) or for increasing the sensitivity of fluoroimmunoassays (Laukkanen *et al.*, 1995).

CONCLUSION AND PERSPECTIVES

Recombinant antibody technology is opening up new perspectives for the development of novel therapeutic and diagnostic agents. For example, human antigen binding fragments have been derived from lymphocyte gene repertoires and from libraries of synthetic antibodies. To screen such libraries containing many millions of different clones, a selection system is required with an efficiency comparable to that of the immune system. This has been achieved by displaying antibodies on the surface of phage particles containing the antibody's gene, analogous to the expression of the IgM antigen receptor on the surface of inactivated B-lymphocytes. Millions of different clones from complex libraries can be panned simultaneously over an immobilized substrate followed by amplification of the adherent phages. This process might be made even more efficient in future by exploiting the antibody-mediated infection of *E. coli*, an evolutionary principle similar to that governing clonal selection in the immune system. The size of the antibody libraries displayed on phage particles has, until recently, been restricted by the efficiency of *in vitro* packaging in *E. coli*. Using the Cre-*lox* site-specific recombination system described above (Waterhouse *et al.*, 1993; Griffiths *et al.*, 1994), a phage display library could theoretically be generated containing as many different antibodies as the number of *E. coli* in culture.

Bacterial surface display systems may eventually facilitate the selection of single cells carrying a fluorescently labeled antigen in a single step by a FACS. However, it will probably be necessary to develop more sophisticated scanning machines or other techniques in order to exploit the full potential of libraries on the bacterial surface.

Cloned antibody genes have been expressed and produced in a number of eukaryotic and prokaryotic systems, where they are readily amenable to genetic manipulation and fusions with proteins and peptides. In addition, a variety of techniques has been established for creating dimeric and multimeric reagents with more than one function. Several recombinant antibody constructs are now undergoing clinical trials.

In spite of the rapid advances of the last few years, several problems such as the routine production of experimental amounts of stable recombinant antibodies from selected clones and the best means of generating high quality universal antibody libraries need to be resolved. The novel multivalent and multifunctional recombinant antibodies need to be tested in a clinical setting. Initial optimistic estimates that this technology would make hybridoma technology redundant almost overnight have now been modified. It will take somewhat longer. New perspectives, however, have already been opened for employing recombinant antibody constructs in a variety of therapeutic and diagnostic applications.

REFERENCES

Adair, J.R., Athwal, D.S., Bodmer, M.W., Bright, S.M., Collins, A.M., Pulito, V.L., Rao, P.E., Reedman, R., Rothermel, A.L., Xu, D., Zivin, R.A. and Jolliffe, L.K. (1994) Humanization of the murine anti-human CD3 monoclonal antibody OKT3. *Hum. Antibod. Hybridomas*, **5**, 41–47.

Adams, G.P., McCartney, J.E., Tai, M.-S., Oppermann, H., Huston, J.S., Stafford, W.F., Bookman, M.A., Fand, I., Houston, L.L. and Weiner, L.M. (1993) Highly specific *in vivo* tumor targeting by monovalent and divalent forms of 741F8 anti-c-*erb*-B-2 single-chain Fv. *Cancer Res.*, **53**, 4026–4034.

Akamatsu, Y., Cole, M.S., Tso, J.Y. and Tsurushita, N. (1993) Construction of a human Ig combinatorial library from genomic V segments and synthetic CDR3 fragments. *J. Immunol.*, **151**, 4651–4659.

Alegre, M.-L., Collins, A., Pulito, V.L., Brosius, R.A., Olson, W.C., Zivin, R.A., Knowles, R., Thistlethwaite, J.R., Jolliffe, L.K. and Bluestone, J.A. (1992) Effect of a single amino acid mutation on the activating and immunosuppressive properties of a "humanized" OKT3 monoclonal antibody. *J. Immunol.*, **148**, 3461–3468.

Alegre, M.-L., Peterson, L.J., Xu, D., Sattar, H.A., Jeyarajah, D.R., Kowalkowski, K., Thistlethwaite, J.R., Zivin, R.A., Jolliffe, L.K. and Bluestone, J.A. (1994) A non-activating "humanized" anti-CD3 monoclonal antibody retains immunosuppressive properties *in vivo*. *Transplantation*, **57**, 1537–1543.

Alfthan, K., Takkinen, K., Sizmann, D., Söderlund, H. and Teeri, T.T. (1995) Properties of a single-chain antibody containing different linker peptides. *Protein Engng.*, **8**, 725–731.

Amas, R.S., Tornetta, M.A., Jones, C.S. and Tsui, P. (1994) Isolation of neutralizing anti-C5a monoclonal antibodies from a filamentous phage monovalent Fab display library. *J. Immunol.*, **152**, 4572–4581.

Balint, R.F. and Larrick, J.W. (1993) Antibody engineering by parsimonious mutagenesis. *Gene*, **137**, 109–118.

Barbas, C.F., Kang, A.S., Lerner, R.A. and Benkovic, S.J. (1991) Assembly of combinatorial antibody libraries on phage surfaces: the gene III site. *Proc. Natl. Acad. Sci. USA*, **88**, 7978–7982.

Barbas, C.F., Björling, E., Chiodi, F., Dunlop, N., Cababa, D., Jones, T.M., Zebedee, S.L., Persson, M.A.A., Nara, P.L., Norrby, E. and Burton, D.R. (1992a) Recombinant human Fab fragments neutralize human type I immunodeficiency virus *in vitro*. *Proc. Natl. Acad. Sci. USA*, **89**, 9339–9343.

Barbas, C.F., Crowe, J.E., Cababa, D., Jones, T.M., Zebedee, S.L., Murphy, B.R., Chanock, R.M. and Burton, D.R. (1992b) Human monoclonal Fab fragments derived from a combinatorial library bind to respiratory syncytial virus F glycoprotein and neutralize infectivity. *Proc. Natl. Acad. Sci. USA*, **89**, 10164–10168.

Barbas, C.F., Bain, J.D., Hoekstra, D.M. and Lerner, R.A. (1992c) Semisynthetic combinatorial antibody libraries: A chemical solution to the diversity problem. *Proc. Natl. Acad. Sci. USA*, **89**, 4457–4461.

Barbas, C.F., Collet, T.A., Amberg, W., Roben, P., Binley, J.M., Hoekstra, D., Cababa, D., Jones, T.M., Williamson, R.A., Pilkington, G.R., Haigwood, N.L., Cabezas, E., Satterthwait, A.C., Sanz, I. and Burton, D.R. (1993a) Molecular profile of an antibody response to HIV-1 as probed by combinatorial libraries. *J. Mol. Biol.*, **230**, 812–823.

Barbas, C.F., Languino, L.R. and Smith, J.W. (1993b) High-affinity self-reactive human antibodies by design and selection: Targeting the integrin ligand binding site. *Proc. Natl. Acad. Sci. USA*, **90**, 10003–10007.

Barbas, C. F., Amberg, W., Simoncsits, A., Jones, T. M. and Lerner, R.A. (1993c) Selection of human anti-hapten antibodies from semisynthetic libraries. *Gene*, **137**, 57–62.

Barbas, C.F., Rosenblum, J.S. and Lerner, R.A. (1993d) Direct selection of antibodies that coordinate metals from semisynthetic combinatorial libraries. *Proc. Natl. Acad. Sci. USA*, **90**, 6385–6389.

Barbas, C.F., Hu, D., Dunlop, N., Sawyer, L., Cababa, D., Hendry, R.M., Nara, P.L. and Burton, D.R. (1994) *In vitro* evolution of a neutralizing human antibody to human immunodeficiency virus type 1 to enhance affinity and broaden strain cross-reactivity. *Proc. Natl. Acad. Sci. USA*, **91**, 3809–3813.

Barbas, S.M., Ditzel, H.J., Salonen, E.M., Yang, W.-P., Silverman, G.J. and Burton, D.R. (1995) Human autoantibody recognition of DNA. *Proc. Natl. Acad. Sci. USA*, **92**, 2529–2533.

Bass, S., Greene, R. and Wells, J.A. (1990) Hormone phage: An enrichment method for variant proteins with altered binding properties. *Proteins*, **8**, 309–314.

Bebbington, C.R., Renner, G., Thompson, S., King, D., Abrams, D. and Yarranton, G.T. (1992) High-level expression of a recombinant antibody from myeloma cells using a glutamine synthetase gene as an amplifiable selectable marker. *Bio/Technology*, **10**, 169–175.

Benhar, I., Padlan, E.A., Jung, S.-H., Lee, B. and Pastan, I. (1994) Rapid humanization of the Fv of monoclonal antibody B3 by using framework exchange of the recombinant immunotoxin B3(Fv)-PE38. *Proc. Natl. Acad. Sci. USA*, **91**, 12051–12055.

Better, M., Cheng, C.P., Robinson, R.R. and Horowitz, A.H. (1988) *E. coli* secretion of an active chimeric antibody fragment. *Science*, **240**, 1041–1043.

Biocca, S., Pierandrei-Amaldi, P., Campioni, N. and Cattaneo, A. (1994) Intracellular immunization with cytosolic recombinant antibodies. *Bio/Technology*, **12**, 396–399.

Biocca, S., Ruberti, F., Tafani, M., Pierandrei-Amaldi, P. and Cattaneo, A. (1995) Redox state of single chain Fv fragments targeted to the endoplasmic reticulum, cytosol and mitochondria. *Bio/Technology*, **13**, 1110–1115.

Bird, R.E., Hardman, K.D., Jacobson, J.W., Johnson, S., Kaufman, B.M., Lee, S.-M., Lee, T., Pope, S.H., Riordan, G.S. and Whitlow, M. (1988) Single-chain antigen-binding proteins. *Science*, **242**, 423–426.

Borrebaeck, C.A.K., Danielsson, L. and Moller, S.A. (1988) Human monoclonal antibodies produced by primary *in vitro* immunization of peripheral blood lymphocytes. *Proc. Natl. Acad. Sci. USA*, **85**, 3995–3999.

Boss, M.A., Kenten, J.H., Wood, C.R. and Emtage, J.S. (1984) Assembly of functional antibodies from immunoglobulin heavy and light chains synthesized in *E. coli*. *Nucl. Acids Res.*, **12**, 3791–3806.

Boublik, Y., Di Bonito, P. and Jones, I.M. (1995) Eukaryotic virus display: engineering the major surface glycoprotein of the *Autographa californica* nuclear polyhedrosis virus (AcNPV) for the presentation of foreign proteins on the virus surface. *Bio/Technology*, **13**, 1079–1084.

Bowden, G.A. and Georgiou, G. (1990) Folding and aggregation of β-lactamase in the periplasmic space of *Escherichia coli*. *J. Biol. Chem.*, **265**, 16760–16766.

Bradbury, A., Persic, L., Werge, T. and Cattaneo, A. (1993) Use of living columns to select specific phage antibodies. *Bio/Technology*, **11**, 1565–1569.

Breitling, F., Dübel, S., Seehaus, T., Klewinghaus, I. and Little, M. (1991) A surface expression vector for antibody screening. *Gene*, **104**, 147–153.

Brinkmann, U., Buchner, J. and Pastan, I. (1992) Independent domain folding of *Pseudomonas* exotoxin and single-chain immunotoxins: influence of interdomain connections. *Proc. Natl. Acad. Sci. USA*, **89**, 3075–3079.

Brinkmann, U., Gallo, M., Brinkmann, E., Kunwar, S. and Pastan, I. (1993a) A recombinant immunotoxin that is active on prostate cancer cells and that is composed of the Fv region of monoclonal antibody PR1 and a truncated form of *Pseudomonas* exotoxin. *Proc. Natl. Acad. Sci. USA*, **90**, 547–551.

Brinkmann, U., Reiter, Y., Jung, S.-H., Lee, B. and Pastan, I. (1993b) A recombinant immunotoxin containing a disulfide-stabilized Fv fragment. *Proc. Natl. Acad. Sci. USA*, **90**, 7538–7542.

Brinkmann, U., Chowdhury, P.S., Roscoe, D.M. and Pastan, I. (1995) Phage display of disulfide-stabilized Fv fragments. *J. Immunol. Meth.*, **182**, 41–50.

Brüggemann, M., Caskey, H.M., Teale, C., Waldmann, H., Williams, G.T., Surani, M.A. and Neuberger, M.S. (1989) A repertoire of monoclonal antibodies with human heavy chains from transgenic mice. *Proc. Natl. Acad. Sci. USA*, **86**, 6709–6713.

Brüggemann, M., Spicer, C., Buluwela, L., Rosewell, I., Barton, S., Surani, M.A. and Rabbitts, T.H. (1991) Human antibody production in transgenic mice: expression from 100 kb of the human IgH locus. *Eur. J. Immunol.*, **21**, 1323–1326.

Buchner, J. and Rudolph, R. (1991) Renaturation, purification and characterization of recombinant Fab-fragments produced in *Escherichia coli*. *Bio/Technology*, **9**, 157–162.

Buchner, J., Pastan, I. and Brinkmann, U. (1992) A method for increasing the yield of properly folded recombinant fusion proteins: single-chain immunotoxins from renaturation of bacterial inclusion bodies. *Anal. Biochem.*, **205**, 263–270.

Burioni, R., Williamson, R.A., Sanna, P.P., Bloom, F.E. and Burton, D.R. (1994) Recombinant human Fab to glycoprotein D neutralizes infectivity and prevents cell-to-cell transmission of herpes simplex viruses 1 and 2 *in vitro*. *Proc. Natl. Acad. Sci. USA*, **91**, 355–359.

Burton, D.R., Barbas, C.F., Persson, M.A.A., Koenig, S., Chanock, R.M. and Lerner, R.A. (1991) A large array of human monoclonal antibodies to type 1 human immunodeficiency virus from combinatorial libraries of asymptomatic seropositive individuals. *Proc. Natl. Acad. Sci. USA*, **88**, 10134–10137.

Cabilly, S. and Riggs, A.D. (1985) Immunoglobulin transcripts and molecular history of a hybridoma that produces antibody to carcinoembryonic antigen. *Gene*, **40**, 157–161.

Carroll, W.L., Mendel, E. and Levy, S. (1988) Hybridoma fusion cell lines contain an aberrant kappa transcript. *Mol. Immunol.*, **25**, 991–995.

Carter, P., Kelley, R.F., Rodrigues, M.L., Snedecor, B., Covarrubias, M., Velligan, M.D., Wong, W.L.T., Rowland, A.M., Kotts, C.E., Carver, M.E., Yang, M., Bourell, J.H., Shepard, H.M. and Henner, D. (1992a) High level *Escherichia coli* expression and production of a bivalent humanized antibody fragment. *Bio/Technology*, **10**, 163–167.

Carter, P., Presta, L., Gorman, C.M., Ridgway, J.B.B., Henner, D., Wong, W.L.T., Rowland, A.M., Kotts, C., Carver, M.E. and Shepard, H.M. (1992b) Humanization of an anti-p185^{HER2} antibody for human cancer therapy. *Proc. Natl. Acad. Sci. USA*, **89**, 4285–4289.

Caton, A.J. and Koprowski, H. (1990) Influenza virus hemagglutinin-specific antibodies isolated from a combinatorial expression library are closely related to the immune response of the donor. *Proc. Natl. Acad. Sci. USA*, **87**, 6450–6454.

Chamow, S.M., Zhang, D.Z., Tan, X.Y., Mhatre, S.M., Marsters, S.A., Peers, D.H., Byrn, R.A., Ashkenazi, A. and Junghans, R.P. (1994) A humanized, bispecific immunoadhesin-antibody that retargets $CD3^+$ effectors to kill HIV-1-infected cells. *J. Immunol.*, **153**, 4268–4280.

Chester, K.A., Begent, R.H.J., Robson, L., Keep, P., Pedley, R.B., Boden, J.A., Boxer, G., Green, A., Winter, G., Cochet, O. and Hawkins, R.E. (1994) Phage libraries for generation of clinically useful antibodies. *Lancet*, **343**, 455–456.

Choe, M.H., Webber, K.O. and Pastan, I. (1994) B3(Fab)-PE38^{M}: a recombinant immunotoxin in which a mutant form of *Pseudomonas* exotoxin is fused to the Fab fragment of monoclonal antibody B3. *Cancer Res.*, **54**, 3460–3467.

Chothia, C. and Lesk, A.M. (1987) Canonical structures for the hypervariable regions of immunoglobulins. *J. Mol. Biol.*, **196**, 901–917.

Chothia, C., Lesk, A.M., Tramontano, A., Levitt, M., Smith-Gill, S.J., Air, G., Sheriff, S., Padlan, E.A., Davies, D., Tulip, W.R., Colman, P.M., Spinelli, S., Alzari, P.M. and Poljak, R.J. (1989) Conformations of immunoglobulin hypervariable regions. *Nature*, **342**, 877–883.

Chothia, C., Lesk, A.M., Gherardi, E., Tomlinson, I.M., Walter, G., Marks, J.D., Llewelyn, M.B. and Winter, G. (1992) Structural repertoire of the human V_H segments. *J. Mol. Biol.*, **227**, 799–817.

Co, M.S., Deschamps, M., Whitley, R.J. and Queen, C. (1991) Humanized antibodies for antiviral therapy. *Proc. Natl. Acad. Sci. USA*, **88**, 2869–2873.

Co, M.S. and Queen, C. (1991) Humanized antibodies for therapy. *Nature*, **351**, 501–502.

Colcher, D., Bird, R., Roselli, M., Hardman, K.D., Johnson, S., Pope, S., Dodd, S.W., Pantoliano, M.W., Milenic, D. and Schlom, J. (1990) *In vivo* tumor targeting of a recombinant single-chain antigen-binding protein. *J. Natl. Cancer Inst.*, **82**, 1191–1197.

Cregg, J.M., Vedvick, T.S. and Raschke, W.C. (1993) Recent advances in the expression of foreign genes in *Pichia pastoris*. *Bio/Technology*, **11**, 905–910.

Cumber, A.J., Ward, E.S., Winter, G., Parnell, G. and Wawrzynczak, E.J. (1992) Comparative stabilities *in vitro* and *in vivo* of a recombinant mouse antibody FvCys fragment and a bisFvCys conjugate. *J. Immunol.*, **149**, 120–126.

Cwirla, S.E., Peters, E.A., Barrett, R.W. and Dower, W.J. (1990) Peptides on phage: A vast library of peptides for identifying ligands. *Proc. Natl. Acad. Sci. USA*, **87**, 6378–6382.

Davies, J. and Riechmann, L. (1994) "Camelising" human antibody fragments: NMR studies on VH domains. *FEBS Lett.*, **339**, 285–290.

Davies, J. and Riechmann, L. (1995) Antibody V_H domains as small recognition units. *Bio/Technology*, **13**, 475–479.

Davies, J. and Riechmann, L. (1996) Single antibody domains as small recognition units: design and *in vitro* antigen selection of camelized, human V_H domains with improved protein stability. *Protein Engng*, **9**, 531–537.

Davies, E.L., Smith, J.S., Birkett, C.R., Manser, J.M., Anderson-Dear, D.V. and Young, J.R. (1995) Selection of specific phage-display antibodies using libraries derived from chicken immunoglobulin genes. *J. Immunol. Methods*, **186**, 125–135.

Davies, N.P., Rosewell, I.R., Richardson, J.C., Cook, G.P., Neuberger, M.S., Brownstein, B.H., Norris, M.L. and Brüggemann, M. (1993) Creation of mice expressing human antibody light chains by introduction of a yeast artificial chromosome containing the core region of the human immunoglobulin κ locus. *Bio/Technology*, **11**, 911–914.

De Jonge, J., Brissinck, J., Heirman, C., Demanet, C., Leo, O., Moser, M. and Thielemans, K. (1995) Production and characterization of bispecific single-chain antibody fragments. *Mol. Immunol.*, **32**, 1405–1412.

de Kruif, J., Boel, E. and Logtenberg, T. (1995a) Selection and application of human single chain Fv antibody fragments from a semi-synthetic phage antibody display library with designed CDR3 regions. *J. Mol. Biol.*, **248**, 97–105.

de Kruif, J., Terstappen, L., Boel, E. and Logtenberg, T. (1995b) Rapid selection of cell subpopulation-specific human monoclonal antibodies from a synthetic phage antibody library. *Proc. Natl. Acad. Sci. USA*, **92**, 3938–3942.

Deng, S.-J., MacKenzie, C.R., Hirama, T., Brousseau, R., Lowary, T.L., Young, N.M., Bundle, D.R. and Narang, S.A. (1995) Basis for selection of improved carbohydrate-binding single-chain antibodies from synthetic gene libraries. *Proc. Natl. Acad. Sci. USA*, **92**, 4992–4996.

Desplancq, D., King, D.J., Lawson, A.D.G. and Mountain, A. (1994) Multimerization behaviour of single chain Fv variants for the tumour-binding antibody B72.3. *Protein Engng*, **7**, 1027–1033.

Devlin, J.J., Panganiban, L.C. and Devlin, P.E. (1990) Random peptide libraries: a source of specific protein binding molecules. *Science*, **249**, 404–406.

De Sutter, K., Feys, V., Van de Voorde, A. and Fiers, W. (1992) Production of functionally active murine and murine::human chimeric $F(ab')_2$ fragments in COS-1 cells. *Gene*, **113**, 223–230.

Dill, K.A. (1990) Dominant forces in protein folding. *Biochemistry*, **29**, 7133–7155.

Ditzel, H.J., Binley, J.M., Moore, J.P., Sodroski, J., Sullivan, N., Sawyer, L.S.W., Hendry, R.M., Yang, W.P., Barbas, C.F. and Burton, D.R. (1995) Neutralizing recombinant human antibodies to a conformational V2- and CD4-binding site-sensitive epitope of HIV-1 gp 120 isolated by using an epitope-masking procedure. *J. Immunol.*, **154**, 893–906.

Dorai, H. and Moore, G.P. (1987) The effect of dihydrofolate reductase-mediated gene amplification on the expression of transfected immunoglobulin genes. *J. Immunol.*, **139**, 4232–4241.

Dorai, H., McCartney, J.E., Hudziak, R.M., Tai, M.-S., Laminet, A.A., Houston, L.L., Huston, J.S. and Oppermann, H. (1994) Mammalian cell expression of single-chain Fv (sFv) antibody proteins and their C-terminal fusions with interleukin-2 and other effector domains. *Bio/Technology*, **12**, 890–897.

Dreher, M.L., Gherardi, E, Skerra, A. and Milstein, C. (1991) Colony assays for antibody fragments expressed in bacteria. *J. Immunol. Methods*, **139**, 197–205.

Dübel, S., Breitling F., Klewinghaus, I. and Little, M. (1992) Regulated secretion and purification of recombinant antibodies in *E. coli*. *Cell Biophys.*, **21**, 69–79.

Dübel, S., Breitling, F., Fuchs, P., Zewe, M., Gotter, S., Welschof, M., Moldenhauer, G. and Little, M. (1994) Isolation of IgG antibody Fv-DNA from various mouse and rat hybridoma cell lines using the polymerase chain reaction with a simple set of primers. *J. Immunol. Methods*, **175**, 89–95.

Dübel, S., Breitling, F., Kontermann, R., Schmidt, T., Skerra, A. and Little, M. (1995) Bifunctional and multimeric complexes of streptavidin fused to single chain antibodies (scFv). *J. Immunol. Meth.*, **178**, 201–209.

Duenas, M. and Borrebaeck, C. (1994) Clonal selection and amplification of phage displayed antibodies by linking antigen recognition and phage replication. *Bio/Technology*, **12**, 999–1002.

Duenas, M., Vàzquez, J., Ayala, M., Söderlind, E., Ohlin, M., Pérez, L., Borrebaeck, C.A.K. and Gavilondo, J.V. (1994) Intra- and extracellular expression of an scFv antibody fragment in *E. coli*: Effect of bacterial strains and pathway engineering using GroES/L chaperonins. *BioTechniques*, **16**, 476–483.

Duenas, M. and Borrebaeck, C.A.K. (1995) Novel helper phage design: Intergenic region affects the assembly of bacteriophages and the size of antibody libraries. *FEMS Microbiol. Lett.*, **125**, 317–321.

Duenas, M., Ayala, M., Vàzquez, J., Ohlin, M., Söderlind, E., Borrebaeck, C.A.K. and Gavilondo, J.V. (1995) A point mutation in a murine immunoglobulin V-region strongly influences the antibody yield in *Escherichia coli*. *Gene*, **158**, 61–66.

Duenas, M., Malmborg, A.-C., Casalvilla, R., Ohlin, M. and Borrebaeck, C.A.K. (1996) Selection of phage displayed antibodies based on kinetic constants. *Mol. Immunol.*, **33**, 279–285.

During, K., Hippe, S., Kreuzaler, E. and Schell, J. (1990) Synthesis and self-assembly of a functional antibody in transgenic *Nicotiana tabacum*. *Plant Mol. Biol.*, **15**, 281–293.

Essig, N.Z., Wood, J.F., Howard, A.J., Raag, R. and Whitlow M. (1993) Crystallization of single-chain Fv proteins. *J. Mol. Biol.*, **234**, 897–901.

Felici, F., Castagnoli, L., Musacchio, A., Jappelli, R. and Cesareni, G. (1991) Selection of antibody ligands from a large library of oligopeptides expressed on a multivalent exposition vector. *J. Mol. Biol.*, **222**, 301–310.

Fiedler, U. and Conrad, U. (1995) High-level production and long-term storage of engineered antibodies in transgenic tabacco seeds. *Bio/Technology*, **13**, 1090–1093.

Fishwild, D.M., O'Donnell, S.L., Bengoechea, T., Hudson, D.V., Harding, F., Bernhard, S.L., Jones, D., Kay, R.M., Higgins, K.M., Schramm, S.R. and Lonberg, N. (1996) High-avidity human IgGκ monoclonal antibodies from a novel strain of minilocus transgenic mice. *Nature Bio/Technology*, **14**, 845–851.

Fouser, L.A., Swanberg, S.L., Lin, B.-Y., Benedict, M., Kelleher, K., Cumming, D.A. and Riedel, G.E. (1992) High level expression of a chimeric anti-ganglioside GD2 antibody: genomic kappa sequences improve expression in COS and CHO cells. *Bio/Technology*, **10**, 1121–1127.

Francisco, J.A., Campbell, R., Iverson, B.L. and Georgiou, G. (1993) Production and fluorescence-activated cell sorting of *E. coli* expressing a functional antibody fragment on the external surface. *Proc. Natl. Acad. Sci. USA*, **90**, 10444–10448.

Freund, C., Ross, A., Guth, B., Plückthun, A. and Holak, T. A. (1993) Characterization of the linker peptide of the single-chain Fv fragment of an antibody by NMR spectroscopy. *FEBS Lett.*, **320**, 97–100.

Freund, C., Ross, A., Plückthun, A. and Holak, T.A. (1994) Structural and dynamic properties of the Fv fragment and the single-chain Fv fragment of an antibody in solution investigated by heteronuclear three-dimensional NMR spectroscopy. *Biochemistry*, **33**, 3296–3303.

Fuchs, P., Breitling, F., Dübel, S., Seehaus, T. and Little, M. (1991) Targeting recombinant antibodies to the surface of *E. coli*: Fusion to a peptidoglycan associated lipoprotein. *Bio/Technology*, **9**, 1369–1372.

Fuchs, P., Dübel, S., Breitling, F., Braunagel, M., Klewinghaus, I. and Little, M. (1992) Recombinant human monoclonal antibodies: Basic principles of the immune system transferred to *E. coli*. *Cell Biophysics*, **21**, 81–92.

Fuchs, P., Weichel, W., Dübel, S., Breitling, F. and Little, M. (1996) Separation of E. coli expressing functional cell-wall bound antibody fragments by FACS. *Immunotechnology*, **2**, 97–102.

Garrard, L.J. and Henner, D.J. (1993) Selection of an anti-IGF-1 Fab from a Fab phage library created by mutagenesis of multiple CDR loops. *Gene*, **128**, 103–109.

George, A.J.T., Titus, J.A., Jost, C.R., Kurucz, I., Perez, P., Andrew, S.M., Nicholls, P.J., Huston, J.S. and Segal, D.M. (1994) Redirection of T-cell mediated cytotoxicity by a recombinant single-chain Fv molecule. *J. Immunol.*, **152**, 1802–1811.

Glockshuber, R., Malia, M., Pfitzinger, I. and Plückthun, A. (1990) A comparison of strategies to stabilize immunoglobulin Fv-fragments. *Biochemistry*, **29**, 1362–1367.

Gotter, S., Haas, C., Kipriyanov, S., Dübel, S., Breitling, F., Khazaie, K., Schirrmacher, V. and Little, M. (1995) A single-chain antibody for coupling ligands to tumor cells infected with Newcastle disease virus. *Tumor Targeting*, **1**, 107–114.

Goshorn, S.C., Svensson, H.P., Kerr, D.E., Somerville, J.E., Senter, P.D. and Fell, H.P. (1993) Genetic construction, expression, and characterization of a single-chain anti-carcinoma antibody fused to β-lactamase. *Cancer Res.*, **53**, 2123–2127.

Graziano, R.F., Tempest, P.R., White, P., Keler, T., Deo, Y., Ghebremariam, H., Coleman, K., Pfefferkorn, L.C., Fanger, M.W. and Guyre, P.M. (1995) Construction and characterization of a humanized anti-γ-Ig receptor type I (FcγRI) monoclonal antibody. *J. Immunol.*, **155**, 4996–5002.

Griffiths, A.D., Malmqvist, M., Marks, J.D., Bye, J.M., Embleton, M.J., McCafferty, J., Baier, M., Holliger, K.P., Gorick, B.D., Hughes-Jones, N.C., Hoogenboom, H.R. and Winter, G. (1993) Human anti-self antibodies with high specificity from phage display libraries. *EMBO J.*, **12**, 725–734.

Griffiths, A.D., Williams, S.C., Hartley, O., Tomlinson, I.M., Waterhouse, P., Crosby, W.L., Kontermann, R.E., Jones, P.T., Low, N.M., Allison, T.J., Prospero, T.D., Hoogenboom, H.R., Nissim, A., Cox, J.P.L., Harrison, J.L., Zaccolo, M., Gherardi, E. and Winter, G. (1994) Isolation of high affinity human antibodies directly from large synthetic repertoires. *EMBO J.*, **13**, 3245–3260.

Gram, H., Marconi, L.-A., Barbas, C.A., Collet, T.A., Lerner, R.A. and Kang, A.S. (1992) *In vitro* selection and affinity maturation of antibodies from a naive combinatorial immunoglobulin library. *Proc. Natl. Acad. Sci. USA*, **89**, 3576–3580.

Gruber, M., Schodin, B.A., Wilson, E.R. and Kranz, D.M. (1994) Efficient tumor cell lysis mediated by a bispecific single chain antibody expressed in *Escherichia coli*. *J. Immunol.*, **152**, 5368–5374.

Hamers-Casterman, C., Atarhouch, T., Muyldermans, S., Robinson, G., Hamers, C., Songa, E.B., Bendahman, N. and Hamers, R. (1993) Naturally occurring antibodies devoid of light chains. *Nature*, **363**, 446–448.

Hawkins, R.E., Russell, S.J. and Winter G. (1992) Selection of phage antibodies by binding affinity. Mimicking affinity maturation. *J. Mol. Biol.*, **226**, 889–896.

Hayashi, N., Welschof, M., Zewe, M., Braunagel, M., Dübel, S., Breitling, F. and Little, M. (1994) Simultaneous mutagenesis of antibody CDR regions by overlap extension and PCR. *BioTechniques*, **17**, 310–314.

Hayashi, N., Kipriyanov, S., Fuchs, P., Welschof, M., Dörsam, H. and Little, M. (1995) A single expression system for the display, purification and conjugation of single-chain antibodies. *Gene*, **160**, 129–130.

Hiatt, A.C., Cafferkey, R. and Bowdish, K. (1989) Production of antibodies in transgenic plants. *Nature*, **342**, 76–78.

Hockney, R.C. (1994) Recent developments in heterologous protein production in *Escherichia coli*. *Trends Biotechnol.*, **12**, 456–463.

Hogrefe, H.H., Mullinax, R.L., Lovejoy, A.E., Hay, B.N. and Sorge, J.A. (1993) A bacteriophage lambda vector for the cloning and expression of immunoglobulin Fab fragments on the surface of filamentous phage. *Gene*, **128**, 119–126.

Holliger, P., Prospero., T. and Winter, G. (1993) "Diabodies": small bivalent and bispecific antibody fragments. *Proc. Natl. Acad. Sci. USA*, **90**, 6444–6448.

Holliger, P., Brissinck, J., Williams, R.L., Thielemans, K. and Winter, G. (1996) Specific killing of lymphoma cells by cytotoxic T-cells mediated by a bispecific diabody. *Protein Engng*, **9**, 299–305.

Hoogenboom, H.R., Griffiths, A.D., Johnson, K.S., Chiswell, D.J., Hudson, P. and Winter, G. (1991) Multi-subunit proteins on the surface of filamentous phage: methodologies for displaying antibody (Fab) heavy and light chains. *Nucl. Acids Res.*, **19**, 4133–4137.

Hoogenboom, H.R. and Winter, G. (1992) By-passing immunisation. Human antibodies from synthetic repertoires of germline V_H gene segments rearranged *in vitro*. *J. Mol. Biol.*, **227**, 381–388.

Horwitz, A.H., Chang, C.P., Better, M., Hellstrom, K.E. and Robinson, R.R. (1988) Secretion of functional antibody and Fab fragment from yeast cells. *Proc. Natl. Acad. Sci. USA*, **85**, 8678–8682.

Hsiao, K.-c., Bajorath, J. and Harris, L.J. (1994) Humanization of 60.3, an anti-CD18 antibody; importance of the L2 loop. *Protein Engng*, **7**, 815–822.

Huse, W. D., Sastry, L., Iverson, S. A., Kang, A. S., Alting-Mees, M., Burton, D. R., Benkovic, S. J. and Lerner, R. A. (1989) Generation of a large combinatorial library of the immunoglobin repertoire in phage lambda. *Science* , **246**, 1275–1281.

Huse, W. D., Stinchcombe, T. J., Glaser, S. M., Starr, L., MacLean, M., Hellström, K. E., Hellström, I. and Yelton, D. E. (1992) Application of a filamentous phage pVIII fusion protein system suitable for efficient production, screening, and mutagenesis of F(ab) antibody fragments. *J. Immunol.*, **149**, 3914–3920.

Huston, J.S., Levinson, D., Mudgett-Hunter, M., Tai, M.-S., Novotny, J., Margolies, M.N., Ridge, R.J., Bruccolery, R.E., Haber, E., Crea, R. and Oppermann, H. (1988) Protein engineering of antibody binding sites: recovery of specific activity in an anti-digoxin single-chain Fv analogue produced in *E. coli.*. *Proc. Natl. Acad. Sci. USA*, **85**, 5879–5883.

Huston, J.S., Mudgett-Hunter, M., Tai, M.-S., McCartney, J., Warren, F., Haber, E. and Oppermann, H. (1991) Protein engineering of single-chain Fv analogs and fusion proteins. *Methods Enzymol.*, **203**, 46–88.

Huston, J.S., McCartney, J., Tai, M.-S., Mottola-Hartshorn, C., Jin, D., Warren, F., Keck, P. and Oppermann, H. (1993) Medical applications of single-chain antibodies. *Int. Rev. Immunol.*, **10**, 195–217.

Ilyichev, A.A., Minenkova, O.O., Kishchenko, G.P., Tat'kov, S.I., Karpishev, N.N., Eroshkin, A.M., Ofitzerov, V.I., Akimenko, Z.A., Petrenko, V.A. and Sandakhchiev, L.S. (1992) Inserting foreign peptides into the major coat protein of bacteriophage M13. *FEBS Lett.*, **301**, 322–324.

Ito, W. and Kurosawa, Y. (1993) Development of an artificial antibody system with multiple valency using an Fv fragment fused to a fragment of protein A. *J. Biol. Chem.*, **268**, 20668–20675.

Jaffers, G.J., Fuller, T.C., Cosimi, A.B., Russell, P.S., Winn, H.J. and Colvin, R.B. (1986) Monoclonal antibody therapy. *Transplantation*, **41**, 572–578.

James, K. (1994) Human monoclonal antibody technology. In M. Rosenberg, and G.P. Moore, (eds.), *Handbook of Experimental Pharmacology*, Vol. 113, *The Pharmacology of Monoclonal Antibodies*, Springer-Verlag, Berlin, Heidelberg, pp. 3–22.

Janda, K.D., Lo, C.L., Li, T., Barbas, C.F., Wirsching, P. and Lerner, R.A. (1994) Direct selection for a catalytic mechanism from combinatorial antibody libraries. *Proc. Natl. Acad. Sci. USA*, **91**, 2532–2536.

Jespers, L.S., Roberts, A., Mahler, S.M., Winter, G. and Hoogenboom, H.R. (1994) Guiding the selection of human antibodies from phage display repertoires to a single epitope of an antigen. *Bio/Technology*, **12**, 899–903.

Jespers, L.S., Messens, J.H., De Keyser, A., Eeckhout, D., Van Den Brande, I., Gansemans, Y.G., Lauwereys, M.J., Vlasuk, G.P. and Stanssens, E. (1995) Surface Expression and ligand-based selection of cDNAs fused to filamentous phage gene VI. *Bio/Technology*, **13**, 378–382.

Jiang, W., Bonnert, T.P., Venugopal, K. and Gould, E.A. (1994) A single chain antibody fragment expressed in bacteria neutralizes tick-born flaviviruses. *Virology*, **200**, 21–28.

Jiang, W., Venugopal, K. and Gould, E.A. (1995) Intracellular interference of tick-borne flavivirus infection by using a single-chain antibody fragment delivered by recombinant Sindbis virus. *J. Virol.*, **69**, 1044–1049.

Johnson, V.G., Schlom, J., Paterrson, A.J., Bennett, J., Magnany, J.L. and Colcher, D. (1986) Analysis of a human tumor-associated glycoprotein (TAG-72) identified by a monoclonal antibody B72.3. *Cancer Res.*, **46**, 850–857.

Jones, P.T., Dear, P.H., Foote, J., Neuberger, M.S. and Winter, G. (1986) Replacing the complementarity-determing regions in a human antibody with those from a mouse. *Nature*, **321**, 522–525.

Jost, C.R., Kurucz, I., Jacobus, C.M., Titus, J.A., George, A.J.T. and Segal, D.M. (1994) Mammalian expression and secretion of functional single-chain Fv molecules. *J. Biol. Chem.*, **269**, 26267–26273.

Jost, C.R., Titus, J.A., Kurucz, I. and Segal, D.M. (1996) A single-chain bispecific Fv2 molecule produced in mammalian cells redirects lysis by activated CTL. *Mol. Immunol.*, **33**, 211–219.

Jung, S.-H., Pastan, I. and Lee, B. (1994) Design of intrachain disulfide bonds in the framework region of the Fv fragment of the monoclonal antibody B3. *Proteins: Struct. Funct. Genet.*, **19**, 35–47.

Kabat, E.A., Wu, T.T., Perry, H.M., Gottesmann, K.S. and Foeller, C. (1991) In *Sequences of proteins of immunological interest*, 5th ed. U.S. Department of Health and Human Services, NIH publication No. 91-3242. National Institutes of Health, Bethesda, MD.

Kang, A.S., Barbas, C.F., Janda, K.D., Benkovic, S.J. and Lerner, R.A. (1991a) Linkage of recognition and replication functions by assembling combinatorial antibody Fab libraries along phage surfaces. *Proc. Natl. Acad. Sci. USA*, **88**, 4363–4366.

Kang, A.S., Jones, T.M. and Burton, D.R. (1991b) Antibody redesign by chain shuffling from random combinatorial immunoglobulin libraries. *Proc. Natl. Acad. Sci. USA*, **88**, 11120–11123.

Kashmiri, S.V., Shu, L., Padlan, E.A., Milenic, D.E., Schlom, J. and Hand, P.H. (1995) Generation, characterization, and *in vivo* studies of humanized anti-carcinoma antibody CC49. *Hybridoma*, **14**, 461–473.

Kettleborough, C.A., Saldanha, J., Heath, V.J., Morrison, C.J. and Bendig, M.M. (1991) Humanization of a mouse monoclonal antibody by CDR-grafting: the importance of framework residues on loop conformation. *Protein Engng*, **4**, 773–783.

Khazaeli, M.B., Conry, R.M. and LoBuglio, A.F. (1994) Human immune response to monoclonal antibodies. *J. Immunotherapy*, **15**, 42–52.

Kiefhaber, T., Rudolf, R., Kohler, H.-H. and Buchner, J. (1991) Protein aggregation *in vitro* and *in vivo*: a quantitative model of the kinetic competition between folding and aggregation. *Bio/Technology*, **9**, 825–829.

King, D.J., Byron, O.D., Mountain, A., Weir, N., Harvey, A., Lawson, A.D.G., Proudfoot, K.A., Baldock, D., Harding, S.E., Yarranton, G.T. and Owens, R.J. (1993) Expression, purification and characterization of B72.3 Fv fragments. *Biochem. J.*, **290**, 723–729.

King, D.J., Turner, A., Farnsworth, A.P.H., Adair, J.R., Owens, R.J., Pedley, R.B., Baldock, D., Proudfoot, K.A., Lawson, A.D.G., Beeley, N.R.A., Millar, K., Millican, T.A., Boyce, B.A., Antoniw, P., Mountain, A., Begent, R.H.J., Shochat, D. and Yarranton, G.T. (1994) Improved tumor targeting with chemically cross-linked recombinant antibody fragments. *Cancer Res.*, **54**, 6176–6185.

Kipriyanov, S.M., Dübel, S., Breitling, F., Kontermann, R.E. and Little, M. (1994) Recombinant single-chain Fv fragments carrying C-terminal cysteine residues: production of bivalent and biotinylated miniantibodies. *Mol. Immunol.*, **31**, 1047–1058.

Kipriyanov, S.M., Dübel, S., Breitling, F., Kontermann, R.E., Heymann, S. and Little, M. (1995a) Bacterial expression and refolding of single-chain Fv fragments with C-terminal cysteines. *Cell Biophys.*, **26**, 187–204.

Kipriyanov, S.M., Breitling, F., Little, M. and Dübel, S. (1995b) Single-chain antibody streptavidin fusions: tetrameric bifunctional scFv-complexes with biotin binding activity and enhanced affinity to antigen. *Hum. Antibod. Hybridomas*, **6**, 93–101.

Kipriyanov, S.M., Little, M., Kropshofer, H., Breitling, F., Gotter, S. and Dübel, S. (1996a) Affinity enhancement of a recombinant antibody: formation of complexes with multiple valency by a single-chain Fv fragment-core streptavidin fusion. *Protein Engng*, **9**, 203–211.

Kipriyanov, S.M., Kupriyanova, O.A., Little, M. and Moldenhauer, G. (1996b) Rapid detection of recombinant antibody fragments directed against cell-surface antigens by flow cytometry. *J. Immunol. Meth.*, **196**, 51–62.

Kipriyanov, S.M., Moldenhauer, G. and Little, M. (1997) High level production of soluble single chain antibodies in small-scale *E. coli* cultures. *J. Immunol. Meth.*, **200**, 69–77.

Klauser, T., Pohlner, J. and Meyer, T.F. (1990) Extracellular transport of cholera toxin B subunit using *Neisseria* IgA protease *β*-domain: conformation-dependent outer membrane translocation. *EMBO J.*, **9**, 1991–1999.

Knappik, A., Krebber, C. and Plückthun, A. (1993) The effect of folding catalysts on the *in vivo* folding process of different antibody fragments expressed in *Escherichia coli*. *Bio/Technology*, **11**, 77–83.

Knappik, A. and Plückthun, A. (1995) Engineered turns of a recombinant antibody improve its *in vivo* folding. *Protein Engng*, **8**, 81–89.

Kolbinger, F., Saldanha, J., Hardman, N. and Bendig, M.M. (1993) Humanization of a mouse anti-human IgE antibody: a potential therapeutic for IgE-mediated allergies. *Protein Engng*, **6**, 971–980.

Kretzschmar, T. and Geiser, M. (1995) Evaluation of antibodies fused to minor coat protein III and major coat protein VIII of bacteriophage M13. *Gene*, **155**, 61–65.

Kuhn, A. and Wickner, W. (1985) Isolation of mutants in M13 coat protein that affect its synthesis, processing and assembly into phage. *J. Biol. Chem.*, **260**, 15907–15913.

Kurucz, I., Titus, J.A., Jost, C.R. and Segal, D.M. (1995a) Correct disulfide pairing and efficient refolding of detergent solubilized single-chain Fv proteins from bacterial inclusion bodies. *Mol. Immunol.*, **32**, 1443–1452.

Kurucz, I., Titus, J.A., Jost, C.R., Jacobus, C.M. and Segal, D.M. (1995b) Retargeting of CTL by an efficiently refolded bispecific single-chain Fv dimer produced in bacteria. *J. Immunol.*, **154**, 4576–4582.

Kussie, P.H., Parhami-Seren, B., Wysocki, L.J. and Margolies, M.N. (1994) A single engineered amino acid substitution changes antibody fine specificity. *J. Immunol.*, **152**, 146–152.

Lagerkvist, A.C.S., Furebring, C. and Borrebaeck, C.A.K. (1995) Single, antigen-specific B cells used to generate Fab fragments using CD40-mediated amplification or direct PCR cloning. *BioTechniques*, **18**, 862–869.

Lang, A.B., Vogel, M., Viret, J.-F. and Stadler, B.M. (1995) Polyclonal preparations of anti-tetanus toxoid antibodies derived from a combinatorial library confer protection. *Bio/Technology*, **13**, 683–685.

Laukkanen, M.-L., Alfthan, K. and Keinänen, K. (1994) Functional immunoliposomes harboring a biosynthetically lipid-tagged single-chain antibody. *Biochemistry*, **33**, 11664–11670.

Laukkanen, M.-L., Orellana, A. and Keinänen, K. (1995) Use of genetically engineered lipid-tagged antibody to generate functional europium chelate-loaded liposomes. Application in fluoroimmunoassay. *J. Immunol. Meth.*, **185**, 95–102.

Leung, S.-O., Goldenberg, D.M., Dion, A., Pellegrini, M.C., Shevitz, J., Shih, L.B. and Hansen, H.J. (1995) Construction and characterization of a humanized, internalizing, B-cell (CD22)-specific, leukemia/lymphoma antibody, LL2. *Mol. Immunol.*, **32**, 1413–1427.

Lilley, G.G., Dolezal, O., Hillyard, C.J., Bernard, C. and Hudson, P. (1994) Recombinant single-chain antibody peptide conjugates expressed in *Escherichia coli* for the rapid diagnosis of HIV. *J. Immunol. Meth.*, **171**, 211–226.

Little, M., Fuchs, P., Breitling, F. and Dübel, S. (1993a) Bacterial surface presentation of proteins and peptides: an alternative to phage technology? *Trends in Biotechnology*, **11**, 3–5.

Little, M., Breitling, F., Dübel, S., Fuchs, P., Braunagel, M., Seehaus, T. and Klewinghaus, I. (1993b) Universal antibody libraries on phage and bacteria. In C. Terhorst, F. Malavasi, and A. Albertini (eds.), *Generation of antibodies by cell and gene immortalization. Year Immunol.* Karger, Basel, vol. 7, pp. 50–55.

Little, M., Breitling, F., Dübel, S. and Braunagel, M. (1995) Human antibody libraries in *Escherichia coli*. *J. Biotechnology*, **41**, 187–195.

Liu, A.Y., Robinson, R.R., Hellström, K.E., Murray, E.D., Chang, C.P. and Hellström, I. (1987) Chimeric mouse-human IgG1 antibody that can mediate lysis of cancer cells. *Proc. Natl. Acad. Sci. USA*, **84**, 3439–3443.

Low, N.M., Holliger, P. and Winter, G. (1996) Mimicking somatic hypermutation: affinity maturation of antibodies displayed on bacteriophage using a bacterial mutator strain. *J. Mol. Biol.*, **260**, 359–368.

Lu, Z., Murray, K.S., Van Cleave, V., LaVallie, E.R., Stahl, M. and McCoy, J.M. (1995) Expression of thioredoxin random peptide libraries on the *Escherichia coli* cell surface as functional fusions to flagellin: a system designed for exploring protein-protein interactions. *Bio/Technology*, **13**, 366–372.

Mack, M., Riethmüller, G. and Kufer, P. (1995) A small bispecific antibody construct expressed as a functional single-chain molecule with high tumor cytotoxicity. *Proc. Natl. Acad. Sci. USA*, **92**, 7021–7025.

MacKenzie, C.R., Sharma, V., Brummell, D., Bilous, D., Dubuc, G., Sadowska, J., Young, N.M., Bundle, D.R. and Narang, S.A. (1994) Effect of C_λ -C_κ domain switching on Fab activity and yield in *E. coli*: synthesis and expression of genes encoding two anti-carbohydrate Fabs. *Bio/Technology*, **12**, 390–395.

Makowski, L. (1992) Terminating a macromolecular helix: structural model for the minor proteins of bacteriophage M13. *J. Mol. Biol.*, **228**, 885–892.

Malby, R.L., Caldwell, J.B., Gruen, L.C., Harley, V.R., Ivancic, N., Kortt, A.A., Lilley, G.G., Power, B.E., Webster, R.G., Colman, P.M. and Hudson, P.J. (1993) Recombinant antineuraminidase single chain antibody: expression, characterization, and crystallization in complex with antigen. *Proteins: Struct. Funct. Genet.*, **16**, 57–63.

Mallender, W.D. and Voss, E.W. (1994) Construction, expression, and activity of a bivalent bispecific single-chain antibody. *J. Biol. Chem.*, **269**, 199–206.

Mark, G.E. and Padlan, E.A. (1994) Humanization of monoclonal antibodies. In M. Rosenberg, and G.P. Moore, (eds.), *Handbook of Experimental Pharmacology*, Vol. 113, *The Pharmacology of Monoclonal Antibodies*, Springer-Verlag, Berlin, Heidelberg, pp. 105–134.

Marks, J.D., Tristem, M., Karpas, A. and Winter, G. (1991a) Oligonucleotide primers for polymerase chain reaction amplification of human immunoglobulin variable genes and design of family-specific oligonucleotide probes. *Eur. J. Immunol.*, **21**, 985–991.

Marks, J. D., Hoogenboom, H. R., Bonnert, T. P., McCafferty, J., Griffiths, A. D. and Winter, G. (1991b) By-passing immunization: Human antibodies from V-gene libraries displayed on phage. *J. Mol. Biol.*, **222**, 581–597.

Marks, J.D., Griffiths, A.D. , Andrew, D., Malmquist, M., Clackson, T., Bye, J. and Winter, G. (1992) By-passing immunisation; Building high affinity human antibodies by chain shuffling. *Bio/Technology*, **10**, 779–783.

Marks, J.D., Ouwehand, W.H., Bye, J.M., Finnern, R., Gorick, B.D., Voak, D., Thorpe, S.J., Hughes-Jones, N.C. and Winter, G. (1993) Human antibody fragments specific for human blood group antigens from a phage display library. *Bio/Technology*, **11**, 1145–1149.

Marasco, W.A., Haseltine, W.A. and Chen, S.Y. (1993) Design, intracellular expression, and activity of a human anti-human immunodeficiency virus type 1 gp120 single-chain antibody. *Proc. Natl. Acad. Sci. USA*, **90**, 7889–7893.

Martin, F., Toniatti, C., Salvati, A.L., Venturini, S., Ciliberto, G., Cortese, R. and Sollazzo, M. (1994) The affinity-selection of a minibody polypeptide inhibitor of human interleukin-6. *EMBO J.*, **13**, 5303–5309.

McCafferty, J., Griffiths, A.D., Winter, G. and Chiswell, D.J. (1990) Phage antibodies: filamentous phage displaying antibody variable domains. *Nature*, **348**, 552–554.

McCartney, J.E., Tai, M.-S., Hudziak, R.M., Adams, G.P., Weiner, L.M., Jin, D., Stafford, W.F., Liu, S., Bookman, M.A., Laminet, A.A., Fand, I., Houston, L.L., Oppermann, H. and Huston, J.S. (1995) Engineering disulfide-linked single-chain Fv dimers [(sFv')$_2$] with improved solution and targeting properties: anti-digoxin 26-10 (sFv')$_2$ and anti-c-erbB-2 741F8 (sFv')$_2$ made by protein folding and bonded through C-terminal cysteinyl peptides. *Protein Engng*, **8**, 301–314.

McGregor, D.P., Molloy, P.E., Cunningham, C. and Harris, W.J. (1994) Spontaneous assembly of bivalent single chain antibody fragments in *Escherichia coli*. *Mol. Immunol.*, **31**, 219–226.

Meulemans, E.V., Slobbe, R., Wasterval, P., Ramaekers, F.C.S. and Van Eys, G.J.J.M. (1994) Selection of phage-displayed antibodies specific for a cytoskeletal antigen by competitive elution with a monoclonal antibody. *J. Mol. Biol.*, **244**, 353–360.

Milenic, D. E., Yokota, T., Filpula, D. R., Finkelman, M. A. J., Dodd, S. W., Wood, J. F., Whitlow, M. L., Snoy, P. and Schlom, J. (1991) Construction, binding properties, metabolism and tumor targeting of a single-chain Fv derived from the pancarcinoma monoclonal antibody CC49. *Cancer Res.*, **51**, 6363–6371.

Mullinax, R.L., Gross, E.A., Amberg, J.R., Hay, B.N., Hogrefe, H.H., Kubitz, M.M., Greener, A., Alting-Mees, M., Ardourel, D., Short, J.M., Sorge, J.A. and Shopes, B. (1990) Identification of human antibody clones specific for tetanus toxoid in a bacteriophage λ immunoexpression library. *Proc. Natl. Acad. Sci. USA* , **87**, 8095–8099.

Muyldermans, S., Atarhouch, T., Saldanha, J., Barbosa, J.A.R.G. and Hamers, R. (1994) Sequence and structure of V_H domain from naturally occurring camel heavy chain immunoglobulins lacking light chains. *Protein Engng*, **7**, 1129–1135.

Neri, D., Momo, M., Prospero, T. and Winter, G. (1995a) High-affinity antigen binding by chelating recombinant antibodies (CRAbs). *J. Mol. Biol.*, **246**, 367–373.

Neri, D., de Lalla, C., Petrul, H., Neri, P. and Winter, G. (1995b) Calmodulin as a versatile tag for antibody fragments. *Bio/Technology*, **13**, 373–377.

Newton, D.L., Nichols, P.J., Rybak, S.M. and Youle, R.J. (1994) Expression and characterization of recombinant human eosinophil-derived neurotoxin and eosinophil-derived neurotoxin-anti-transferrin receptor sFv. *J. Biol. Chem.*, **269**, 26739–26745.

Newton, D.L., Xue, Y., Olson, K.A., Fett, J.W. and Rybak, S.M. (1996) Angiogenin single-chain immunofusions: influence of peptide linkers and spacers between fusion protein domains. *Biochemistry*, **35**, 545–553.

Nicholls, P.J., Johnson, V.G., Andrew, S.M., Hoogenboom, H.R., Raus, J.C.M. and Youle, R.J. (1993a) Characterization of single-chain antibody (sFv)-toxin fusion proteins produced *in vitro* in rabbit reticulocyte lysate. *J. Biol. Chem.*, **268**, 5302–5308.

Nicholls, P.J., Johnson, V.G., Blanford, M.D. and Andrew, S.M. (1993b) An improved method for generating single-chain antibodies form hybridomas. *J. Immunol. Meth.*, **165**, 81–91.

Nissim, A., Hoogenboom, H.R., Tomlinson, I.M., Flynn, G., Midgley, C., Lane, D. and Winter, G. (1994) Antibody fragments from a "single pot" phage display library as immunochemical reagents. *EMBO J.*, **13**, 692–698.

Nyyssönen, E., Penttilä, M., Harkki, A., Saloheimo, A., Knowles, J.K.C. and Keränen, S. (1993) Efficient production of antibody fragments by the filamentous fungus *Trichoderma reesei*. *Bio/Technology*, **11**, 591–595.

Ohlin, M., Owman, H., Mach, M. and Borrebaeck, C.A.K. (1996) Light chain shuffling of a high affinity antibody resuts in a drift in epitope recognition. *Mol. Immunol.*, **33**, 47–56.

Ohtomo, T., Tsuchiya, M., Sato, K., Shimizu, K., Moriuchi, S., Miyao, Y., Akimoto, T., Akamatsu, K.-I., Hayakawa, T. and Ohsugi, Y. (1995) Humanization of mouse ONS-M21 antibody with the aid of hybrid variable regions. *Mol. Immunol.*, **32**, 407–416.

Orlandi, R., Güssow, D.H., Jones, P.T. and Winter, G. (1989) Cloning immunoglobulin variable domains for expression by the polymerase chain reaction. *Proc. Natl. Acad. Sci. USA*, **86**, 3833–3837.

Ørum, H., Andersen, P.S., Øster, A., Johansen, L.K., Riise, E., Bjørnvad, M., Svendsen, I. and Engberg, J. (1993) Efficient method for constructing comprehensive murine fab antibody libraries displayed on phage. *Nucl. Acid Res.*, **21**, 4491–4498.

Pack, P. and Plückthun A. (1992) Miniantibodies: use of amphipatic helices to produce functional, flexibly linked dimeric Fv fragments with high avidity in *Escherichia coli*. *Biochemistry*, **31**, 1579–1584.

Pack, P., Kujau, M., Schroeckh, V., Knüpfer, U., Wenderoth, R., Riesenberg, D. and Plückthun, A. (1993) Improved bivalent miniantibodies, with identical avidity as whole antibodies, produced by high cell density fermentation of *Escherichia coli*. *Bio/Technology*, **11**, 1271–1277.

Pack, P., Müller, K., Zahn, R. and Plückthun. A. (1995) Tetravalent miniantibodies with high avidity assembling in *Escherichia coli*. *J. Mol. Biol.*, **246**, 28–34.

Padlan, E.A. (1991) A possible procedure for reducing the immunogenicity of antibody variable domains while preserving their ligand-binding properties. *Mol. Immunol.*, **28**, 489–498.

Padlan, E.A. (1994) Anatomy of the antibody molecule. *Mol. Immunol.*, **31**, 169–217.

Page, M.J. and Sydenham, M.A. (1991) High level expression of the humanized monoclonal antibody CAMPATH-1H in Chinese hamster ovary cells. *Bio/Technology*, **9**, 64–68.

Pantoliano, M.W., Bird, R.E., Johnson, S., Asel, E.D., Dodd, S.W., Wood, J.F. and Hardman, K.D. (1991) Conformational stability, folding, and ligand-binding affinity of single-chain Fv immunoglobulin fragments expressed in *Escherichia coli*. *Biochemistry*, **30**, 10117–10125.

Park, J.W., Hong, K., Carter, P., Asgari, H., Guo, L.Y., Keller, G.A., Wirth, C., Shalaby, R., Kotts, C., Wood, W.I., Papahadjopoulos, D. and Benz, C.C. (1995) Development of anti-p185^{HER2} immunoliposomes for cancer therapy. *Proc. Natl. Acad. Sci. USA*, **92**, 1327–1331.

Parmley, S.F. and Smith, G.P. (1988) Antibody-selectable filamentous fd phage vectors: affinity purification of target genes. *Gene* , **73**, 305–318.

Pedersen, J.T., Henry, A.H., Searle, S.J., Guild, B.C., Roguska, M. and Rees, A.R. (1994) Comparison of surface accessible residues in human and murine immunoglobulin Fv domains. *J. Mol. Biol.*, **235**, 959–973.

Perisic, O., Webb, P.A., Holliger, P., Winter, G. and Williams, R.L. (1994) Crystal structure of a diabody, a bivalent antibody fragment. *Structure*, **2**, 1217–1226.

Persson, M.A.A., Caothien, R.H. and Burton, D.R. (1991) Generation of diverse high-affinity human monoclonal antibodies by repertoire cloning. *Proc. Natl. Acad. Sci. USA*, **88**, 2432–2436.

Pessi, A., Bianchi, E., Crameri, A., Venturini, S., Tramontano, A. and Sollazzo, M. (1993) A designed metal-binding protein with a novel fold. *Nature*, **362**, 367–369.

Pilkington, G.R., Duan, L., Zhu, M., Keil, W. and Pomerantz, R.J. (1996) Recombinant human Fab antibody fragments to HIV-1 Rev and Tat regulatory proteins: direct selection from a combinatorial phage display library. *Mol. Immunol.*, **33**, 439–450.

Plückthun, A. (1994) Antibodies from *Escherichia coli*. In M. Rosenberg, and G.P. Moore, (eds.), *Handbook of Experimental Pharmacology*, Vol. 113, *The Pharmacology of Monoclonal Antibodies*, Springer-Verlag, Berlin, Heidelberg, pp. 269–315.

Poul, M.-A., Cerutti, M., Chaabihi, H., Ticchioni, M., Deramoudt, F.-X., Bernard, A., Devauchelle, G., Kaczorek, M. and Lefranc, M.-P. (1995) Cassette baculovirus vectors for the production of chimeric, humanized, or human antibodies in insect cells. *Eur. J. Immunol.*, **25**, 2005–2009.

Pugsley, A.P. (1993) The complete general secretory pathway in Gram-negative bacteria. *Microbiol. Rev.*, **57**, 50–108.

Pulito, V.L., Roberts, V.A., Adair, J.R., Rothermel, A.L., Collins, A.M., Varga, S.S., Martocello, C., Bodmer, M.W., Jolliffe, L.K. and Zivin, R.A. (1996) Humanization and molecular modeling of the anti-CD4 monoclonal antibody, OKT4A. *J. Immunol.*, **156**, 2840–2850.

Queen, C., Schneider, W.P., Selick, H.E., Payne, P.W., Landolfi, N.F., Duncan, J.F., Avdalovic, N.M., Levitt, M., Junghans, R.P. and Waldmann, T.A. (1989) A humanized antibody that binds to the interleukin 2 receptor. *Proc. Natl. Acad. Sci. USA*, **86**, 10029–10033.

Reiter, Y., Brinkmann, U., Webber, K.O., Jung, S.-H., Lee, B. and Pastan, I. (1994a) Engineering interchain disulfide bonds into conserved framework regions of Fv fragments: improved biochemical characteristics of recombinant immunotoxins containing disulfide-stabilized Fv. *Protein Engng*, **7**, 697–704.

Reiter, Y., Pai, L.H., Brinkmann, U., Wang, Q.-c. and Pastan, I. (1994b) Antitumor activity and pharmacokinetics in mice of a recombinant immunotoxin containing a disulfide-stabilized Fv fragment. *Cancer Res.*, **54**, 2714–2718.

Reiter, Y., Brinkmann, U., Kreitman, R.J., Jung, S.-H., Lee, B. and Pastan, I. (1994c) Stabilization of the Fv fragments in recombinant immunotoxins by disuldide bonds engineered into conserved framework regions. *Biochemistry*, **33**, 5451–5459.

Reiter, Y., Brinkmann, U., Jung, S.-H., Pastan, I. and Lee, B. (1995) Disulfide stabilization of antibody Fv: computer predictions and experimental evaluation. *Protein Engng*, **8**, 1323–1331.

Richardson, J.H., Sodroski, J.G., Waldmann, T.A. and Marasco, W.A. (1995) Phenotypic knockout of the high-affinity human interleukin 2 receptor by intracellular single-chain antibodies against the α subunit of the receptor. *Proc. Natl. Acad. Sci. USA*, **92**, 3137–3141.

Ridder, R., Schmitz, R., Legay, F. and Gram, H. (1995) Generation of rabbit monoclonal antibody fragments from a combinatorial phage display library and their production in the yeast *Pichia pastoris*. *Bio/Technology*, **13**, 255–260.

Ridgway, J.B.B., Presta, L.G. and Carter, P. (1996) "Knobs-into-holes" engineering of antibody C_H3 domains for heavy chain heterodimerization. *Protein Engng*, **9**, 617–621.

Riechmann, L., Clark, M., Waldmann, H. and Winter, G. (1988) Reshaping human antibodies for therapy. *Nature*, **332**, 323–327.

Riechmann, L. and Weill, M. (1993) Phage display and selection of a site-directed randomized single-chain antibody Fv fragment for its affinity improvement. *Biochemistry*, **32**, 8848–8855.

Riethmüller, G., Schneider-Gädicke, E. and Johnson, J.P. (1993) Monoclonal antibodies in cancer therapy. *Curr. Opin. Immunol.*, **5**, 732–739.

Rode, H.-J., Little, M., Fuchs, P., Dörsam, H., Schooltink, H., de Inés, C., Dübel, S. and Breitling, F. (1996) Cell surface display of a single-chain antibody for attaching polypeptides. *BioTechniques*, **21**, 650–658.

Rodrigues, M.L., Shalaby, M.R., Werther, W., Presta, L. and Carter, P. (1992) Engineering a humanized bispecific $F(ab')_2$ fragment for improved binding to T cells. *Int. J. Cancer*, Suppl. **7**, 45–50.

Rodrigues, M.L., Snedecor, B., Chen, C., Wong, W.L.T., Garg, S., Blank, G.S., Maneval, D. and Carter, P. (1993) Engineering Fab′ fragments for efficient $F(ab)_2$ formation in *Escherichia coli* and for improved *in vivo* stability. *J. Immunol.*, **151**, 6954–6961.

Rodrigues, M.L., Presta, L., Kotts, C.E., Wirth, C., Mordenti, J., Osaka, G., Wong, W.L.T., Nuijens, A., Blackburn, B. and Carter, P. (1995) Development of a humanized disulfide-stabilized anti-$p185^{HER2}$ Fv-β-lactamase fusion protein for activation of a cephalosporin doxorubicin prodrug. *Cancer Res.*, **55**, 63–70.

Roguska, M.A., Pedersen, J.T., Keddy, C.A., Henry, A.H., Searle, S.J., Lambert, J.M., Goldmacher, V.S., Blättler, W.A., Rees, A.R. and Guild, B.C. (1994) Humanization of murine monoclonal antibodies through variable domain resurfacing. *Proc. Natl. Acad. Sci. USA*, **91**, 969–973.

Roguska, M.A., Pedersen, J.T., Henry, A.H., Searle, S.M.J., Roja, C.M., Avery, B., Hoffee, M., Cook, S., Lambert, J.M., Blättler, W.A., Rees, A.R. and Guild, B.C. (1996) A comparison of two murine monoclonal antibodies humanized by CDR-grafting and variable domain resurfacing. *Protein Engng*, **9**, 895–904.

Routledge, E.G., Lloyd, I., Gorman, S.D., Clark, M. and Waldmann, H. (1991) A humanized monovalent CD3 antibody which can activate homologous complement. *Eur. J. Immunol.*, **21**, 2717–2725.

Russel, S.J., Hawkins, R.E. and Winter, G. (1993) Retroviral vectors displaying functional antibody fragments. *Nucl. Acids Res.*, **21**, 1081–1085.

Sanna, P.P., De Logu, A., Williamson, R.A., Samson, M.E., Altieri, D.C., Bloom, F.E. and Burton, D.R. (1995a) Rapid assay of phage-derived recombinant human Fabs as bispecific antibodies. *Bio/Technology*, **13**, 1221–1224.

Sanna, P.P., Williamson, R.A., De Logu, A., Bloom, F.E. and Burton, D. (1995b) Directed selection of recombinant human monoclonal antibodies to herpes simplex virus glycoproteins from phage display libraries. *Proc. Natl. Acad. Sci. USA*, **92**, 6439–6443.

Sarvetnick, N., Gurushanthaiah, D., Han, N., Prudent, J., Schultz, P. and Lerner, R. (1993) Increasing the chemical potential of the germ-line antibody repertoire. *Proc. Natl. Acad. Sci. USA*, **90**, 4008–4011.

Sastry, L., Alting-Mees, M., Huse, W.D., Short, J.M., Sorge, J.A, Hay, B.N., Janda, K.D., Benkovic, S.J. and Lerner, R.A. (1989) Cloning of the immunological repertoire in *Escherichia coli* for generation of monoclonal catalytic antibodies: construction of a heavy chain variable region-specific cDNA library. *Proc. Natl. Acad. Sci. USA*, **86**, 5728–5732.

Sawyer, J.R., Schlom, J. and Kashmiri, S.V.S. (1994) The effect of induction conditions on production of a soluble anti-tumor sFv in *Escherichia coli*. *Protein Engng*, **7**, 1401–1406.

Schier, R., Marks, J.D., Wolf, E.J., Appell, G., Huston, J.S., Weiner, L.M. and Adams, G.P. (1995) *In vitro* and *in vivo* characterization of a human anti-c-*erb*B-2 single-chain Fv isolated from a filamentous phage antibody library. *Immunotechnology*, **1**, 73–81.

Schier, R., Bye, J., Appell, G., McCall, A., Adams, G.P., Malmqvist, M., Weiner, L.M. and Marks, J.D. (1996) Isolation of high-affinity monomeric human anti-c-*erb*B-2 single chain Fv using affinity-driven selection. *J. Mol. Biol.*, **255**, 28–43.

Scott, J.K. and Smith, G.P. (1990) Searching for peptide ligands with an epitope library. *Science*, **249**, 386–390.

Shalaby, M.R., Shepard, H.M., Presta, L., Rodrigues, M.L., Beverley, P.C.L., Feldmann, M. and Carter, P. (1992) Development of humanized bispecific antibodies reactive with cytotoxic lymphocytes and tumor cells overexpressing the *HER2* protooncogene. *J. Exp. Med.*, **175**, 217–225.

Shitara, K., Nakamura, K., Tokutake-Tanaka, Y., Fukushima, M. and Hanai, N. (1994) A new vector for the high level expression of chimeric antibodies in myeloma cells. *J. Immunol. Methods*, **167**, 271–278.

Shu, L., Qi, C.-F., Schlom, J. and Kashmiri, S.V.S. (1993) Secretion of a single-gene-encoded immunoglobulin from myeloma cells. *Proc. Natl. Acad. Sci. USA*, **90**, 7995–7999.

Simons, G.F.M., Konings, R.N.H. and Schoenmakers, J.G.G. (1981) Genes VI, VII and IX of phage M13 code for minor capsid proteins of the virion. *Proc. Natl. Acad. Sci. USA*, **78**, 4194–4198.

Sizmann, D., Takkinen, K., Laukkanen, M.-L., Saloheimo, M., Candussio, A., Veijola-Bailey, P. and Teeri, T.T. (1993) High-level production of an active single-chain Fv fragment in the culture supernatant of *Escherichia coli*. In C. Terhorst, F. Malavasi, and A. Albertini, (eds.), *Generation of antibodies by cell and gene immortalization. Year Immunol.*, Karger, Basel, vol. 7, pp. 119–130.

Skerra, A. and Plückthun, A. (1988) Assembly of a functional immunoglobulin Fv fragment in *E. coli*. *Science*, **240**, 1038–1041.

Skerra, A. and Plückthun, A. (1991) Secretion and *in vivo* folding of the Fab fragment of the antibody McPC603 in *Escherichia coli*: influence of disulfides and cis-prolines. *Protein Engng*, **4**, 971–979.

Skerra, A. (1993) Bacterial expression of immunoglobulin fragments. *Curr. Opin. Immunol.*, **5**, 256–262.

Schmidt, M., Hynes, N.E., Groner, B. and Wels, W. (1996) A bivalent single-chain antibody-toxin specific for ErbB-2 and the EGF receptor. *Int. J. Cancer.*, **65**, 538–546.

Smith, G.P. (1985) Filamentous fusion phage: novel expression vectors that display cloned antigens on the virion surface. *Science* , **228**, 1315–1317.

Smith, G.P. (1988) Filamentous phages as cloning vectors. In Rodriguez, R.L. and Denhardt, D.T. (eds.), *Vectors*, Butterworths, Boston, pp. 61–83.

Smith, J.W., Hu, D., Satterthwait, A., Pinzsweeney, S. and Barbas, C.F. (1994) Building synthetic antibodies as adhesive ligands for integrins. *J.Biol.Chem.*, **269**, 32788–32795.

Smith, R.I.F. and Morrison, S.L. (1994) Recombinant polymeric IgG: an approach to engineering more potent antibodies. *Bio/Technology*, **12**, 683–688.

Söderlind, E., Lagerkvist, A.C.S., Duenas, M., Malmborg, A.-C., Ayala, M., Danielsson, L. and Borrebaeck, C.A.K. (1993) Chaperonin assisted phage display of antibody fragments on filamentous bacteriophages. *Bio/Technology*, **11**, 503–507.

Solar, I. and Gershoni, J.M. (1995) Linker modification introduces useful molecular instability in a single chain antibody. *Protein Engng*, **8**, 717–723.

Stemmer, W.P.C., Morris, S.K., Kautzer, C.K. and Wilson, B.S. (1993a) Increased antibody expression from *E. coli* through wobble-base library mutagenesis by enzymatic inverse PCR. *Gene*, **123**, 1–7.

Stemmer, W.P.C., Morris, S.K. and Wilson, B.S. (1993b) Selection of an active single chain Fv antibody from a protein linker library prepared by enzymatic inverse PCR. *BioTechniques*, **14**, 256–265.

Studnicka, G.M., Soares, S., Better, M., Williams, R.E., Nadell, R. and Horwitz, A.H. (1994) Human-engineered monoclonal antibodies retain full specific binding activity by preserving non-CDR complementarity-modulating residues. *Protein Engng*, **7**, 805–814.

Sun, L.K., Curtis, P., Rakowicz-Szulczynska, E., Ghrayeb, J., Chang, N., Morrison, S.L. and Koprowski, H. (1987) Chimeric antibody with human constant regions and mouse variable regions directed against carcinoma-associated antigen 17-1A. *Proc. Natl. Acad. Sci. USA*, **84**, 214–218.

Tai, M.-S., Mudgett-Hunter, M., Levinson, D., Wu, G.-M., Haber, E., Oppermann, H. and Huston, J.S. (1990) A bifunctional fusion protein containing Fc-binding fragment B of staphylococcal protein A amino terminal to antidigoxin single-chain Fv. *Biochemistry*, **29**, 8024–8030.

Takkinen, K., Laukkanen, M.-L., Sizmann, D., Alfthan, K., Immonen, T., Vanne, L., Kaartinen, M., Knowles, J.K.C. and Teeri, T.T. (1991) An active single-chain antibody containing a cellulase linker domain is secreted by *Escherichia coli*. *Protein Engng*, **4**, 837–841.

Tavladoraki, P., Benvenuto, E., Trinca, S., De Martinis, D., Cattaneo, A. and Galeffi, P. (1993) Transgenic plants expressing a functional single-chain Fv antibody are specifically protected from virus attack. *Nature*, **366**, 469–472.

Tempest, P.R., Bremner, P., Lambert, M., Taylor, G., Furze, J.M., Carr, F.J. and Harris, W.J. (1991) Reshaping a human monoclonal antibody to inhibit human respiratory syncytial virus infection *in vivo*. *Bio/Technology*, **9**, 266–271.

Thompson, J., Pope, T., Tung, J.-S., Chan, C., Hollis, G., Mark, G. and Johnson, K.S. (1996) Affinity maturation of a high-affinity human monoclonal antibody against the third hypervariable loop of human immunodeficiency virus: use of phage display to improve affinity and broaden strain reactivity. *J. Mol. Biol.*, **256**, 77–88.

Traunecker, A., Lanzavecchia, A. and Karjalainen, K. (1991) Bispecific single chain molecules (Janusins) target cytotoxic lymphocytes on HIV infected cells. *EMBO J.*, **10**, 3655–3659.

Vargas-Madrazo, E., Lara-Ochoa and F., Almagro, J.C. (1995) Canonical structure repertoire of the antigen-binding site of immunoglobulins suggests strong geometrical restrictions associated to the mechanism of immune recognition. *J. Mol. Biol.*, **254**, 497–504.

Vaughan, T.J., Williams, A.J., Pritchard, K., Osbourn, J.K., Pope, A.R., Earnshaw, J.C., McCafferty, J., Hodits, R.A., Wilton, J. and Johnson, K.S. (1996) Human antibodies with sub-nanomolar affinities isolated from a large non-immunized phage display library. *Nature Bio/Technology*, **14**, 309–314.

Verhoeyen, M., Milstein, C. and Winter, G. (1988) Reshaping human antibodies: grafting an anti-lysozyme activity. *Science*, **239**, 1534–1536.

Ward, E.S., Güssow, D., Griffiths, A.D., Jones, P.T. and Winter, G. (1989). Binding activities of a repertoire of single immunoglobulin variable domains secreted from *E. coli*. *Nature*, **341**, 544–546.

Ward, E.S. (1995) V_H shuffling can be used to convert an Fv fragment of anti-hen lysozyme specificity to one that recognizes a T cell receptor Vα. *Mol. Immunol.*, **32**, 147–156.

Ward, R.L., Clark, M.A., Lees, J. and Hawkins, N.J. (1996) Retrieval of human antibodies from phage-display libraries using enzymatic cleavage. *J. Immunol. Methods*, **189**, 73–82.

Waterhouse, P., Griffiths, A., Johnson, K. S. and Winter, G. (1993) Combinatorial infection and *in vivo* recombination: a strategy for making large phage antibody repertoires. *Nucl. Acids Res.*, **21**, 2265–2266.

Webber, K.O., Reiter, Y., Brinkmann, U., Kreitman, R. and Pastan, I. (1995) Preparation and characterization of a disulfide-stabilized Fv fragment of the anti-Tac antibody: comparison with its single-chain analog. *Mol. Immunol.*, **32**, 249–258.

Webster, R.E. and Lopez, J. (1985) Structure and assembly of the class I filamentous bacteriophage. In Casjens, S. (ed.), *Virus Structure and Assembly*, Jones and Bartlett Inc., Boston/Portala Valley, USA, pp. 235–267.

Weiss, E., Chatellier, J. and Orfanoudakis, G. (1994) *In vivo* biotinylated recombinant antibodies: Construction, characterization, and application of a bifunctional Fab-BCCP fusion protein produced *in Escherichia coli. Protein Express. Purif.*, **5**, 509–517.

Welschof, M., Terness, P., Kolbinger, F., Zewe, M., Dübel, S., Dörsam, H., Hain, C., Finger, M., Jung, M., Moldenhauer, G., Hayashi, N., Little, M. and Opelz, G. (1995) Amino acid sequence based PCR primers for amplification of rearranged human heavy and light chain immunoglobulin variable region genes. *J. Immunol. Methods*, **179**, 203–214.

Whitlow, M. and Filpula, D. (1991) Single-chain Fv proteins and their fusion proteins. Methods: *A Companion to Meth. Enzymol.*, **2**, 97–105.

Whitlow, M., Bell, B.A., Feng, S.-L., Filpula, D., Hardman, K.D., Hubert, S.L., Rollence, M.L., Wood, J.F., Schott, M.E., Milenic, D.E., Yokota, T. and Schlom, J. (1993) An improved linker for single-chain Fv with reduced aggregation and enhanced proteolytic stability. *Protein Engng*, **6**, 989–995.

Whitlow, M., Filpula, D., Rollence, M.L., Feng, S.-L. and Wood, J.F. (1994) Multivalent Fvs: characterization of single-chain Fv oligomers and preparation of a bispecific Fv. *Protein Engng*, **7**, 1017–1026.

Williamson, R. A., Burioni, R., Sanna, P., Partridge, L., Barbas, C.F. and Burton, D.R. (1993) Human monoclonal antibodies against a plethora of viral pathogens from single combinatorial libraries. *Proc. Natl. Acad. Sci. USA*, **90**, 4141–4145.

Wood, C.R., Boss, M.A., Kenten, J.H., Calvert, J.E., Roberts, N.A. and Emtage, J.S. (1985) The synthesis and *in vivo* assembly of functional antibodies in yeast. *Nature*, **314**, 446–449.

Woodle, E.S., Thistlethwaite, J.R., Jolliffe, L.K., Zivin, R.A., Collins, A., Adair, J.R., Bodmer, M., Athwal, D., Alegre, M.-L. and Bluestone, J.A. (1992) Humanized OKT3 antibodies: successful transfer of immune modulating properties and idiotype expression. *J. Immunol.*, **148**, 2756–2763.

Wu, T.T. and Kabat, E.A. (1970) An analysis of the sequences of the variable regions of Bence-Jones proteins and myeloma light chains and their implication for antibody complementarity. *J. Exp. Med.*, **132**, 211–249.

Wu, X.-C., Ng, S.-C., Near, R.I. and Wong, S.-L. (1993) Efficient production of a functional single-chain antidigoxin antibody via an engineered *Bacillus subtilis* expression-secretion system. *Bio/Technology*, **11**, 71–76.

Yang, W.-P., Green, K., Pinz-Sweeney, S., Briones, A.T., Burton, D.R. and Barbas, C.F. (1995) CDR walking mutagenesis for the affinity maturation of a potent human anti-HIV-1 antibody into the picomolar range. *J. Mol. Biol.*, **254**, 392–403.

Yokota, T., Milenic, D.E., Whitlow, D.E., Whitlow and M., Schlom, J. (1992) Rapid tumor penetration of a single-chain Fv and comparison with other immunoglobulin forms. *Cancer Res.*, **52**, 3402–3408.

Young, R.J., Owens, R.J., Mackay, G.A., Chan, C.M.W., Shi, J., Hide, M., Francis, D.M., Henry, A.J., Sutton, B.J. and Gould, H.J. (1995) Secretion of recombinant human IgE-Fc by mammalian cells and biological activity of glycosilation site mutants. *Protein Engng*, **8**, 193–199.

Zapata, G., Ridgway, J.B.B., Mordenti, J., Osaka, G., Wong, W.L.T., Bennett, G.L. and Carter, P. (1995) Engineering linear $(Fab')_2$ fragments for efficient production in *Escherichia coli* and enhanced antiproliferative activity. *Protein Engng*, **8**, 1057–1062.

Zebedee, S.L., Barbas, C.F., Hom, Y.-L., Caothien, R.H., Graff, R., DeGraw, J., Pyati, J., LaPolla, R., Burton, D.R., Lerner, R.A. and Thornton, G.B. (1992) Human combinatorial antibody libraries to hepatitis B surface antigen. *Proc. Natl. Acad. Sci. USA*, **89**, 3175–3179.

Zhu, Z., Zapata, G., Shalaby, R., Snedecor, B., Chen, H. and Carter, P. (1996) High level secretion of a humanized bispecific diabody from *Escherichia coli. Bio/Technology*, **14**, 192–196.

2. BISPECIFIC ANTIBODIES

DAVID M. SEGAL, BARBARA A. VANCE, and GIUSEPPE SCONOCCHIA

Experimental Immunology Branch, National Cancer Institute,
Bethesda, MD 20892-1360, USA

INTRODUCTION

Bispecific antibodies (bsAbs) are immunoglobulin-based molecules that contain two different binding specificities. They have been prepared using classical protein chemistry, cell fusions, or molecular engineering, and can be used for any application in which two different molecules are brought together within a distance of 5–10 nm. Of particular interest is the ability of bsAbs to redirect immune effector cells against experimental targets *in vitro* and against tumors and other unwanted cells in a clinical setting.

Normally, cellular cytotoxicity procedes by a mechanism in which the cytotoxic cell binds the target cell, forming a bicellular conjugate with a tight, synaptic-like interface, followed by the delivery of a "lethal hit" that eventually kills the target cell (Figure 1A). Cytotoxic cells express several types of receptors specific for target

A. Normal Mechanism

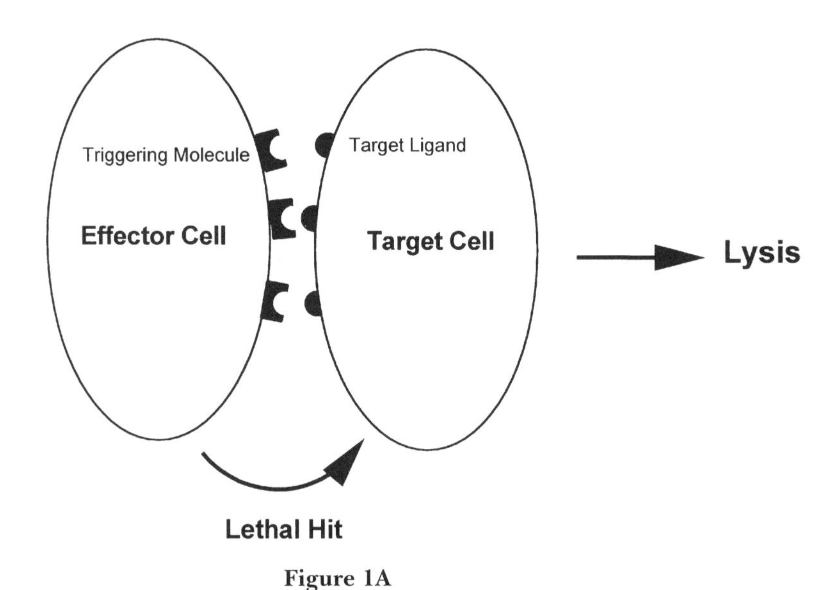

Figure 1A

B. Redirected Hybridoma Cell Lysis

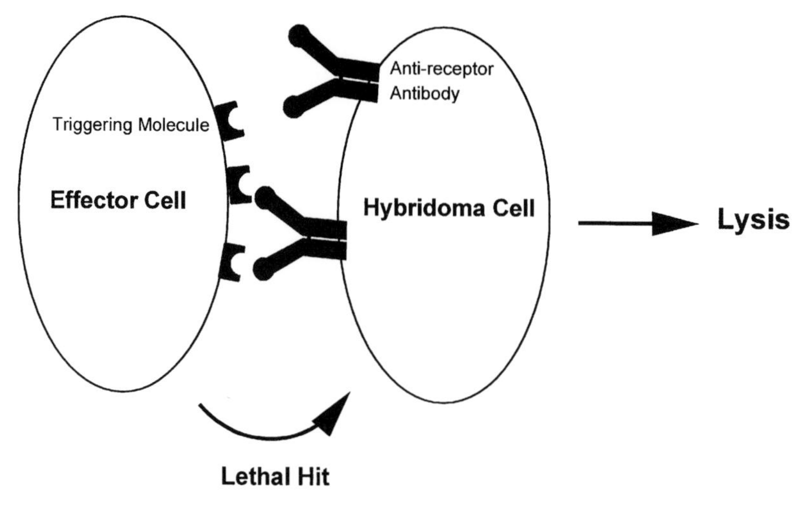

Figure 1B

C. Reverse ADCC

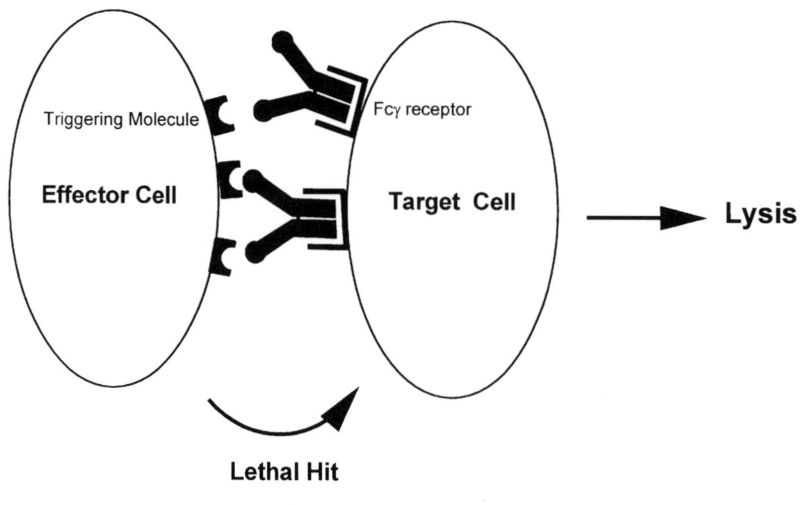

Figure 1C

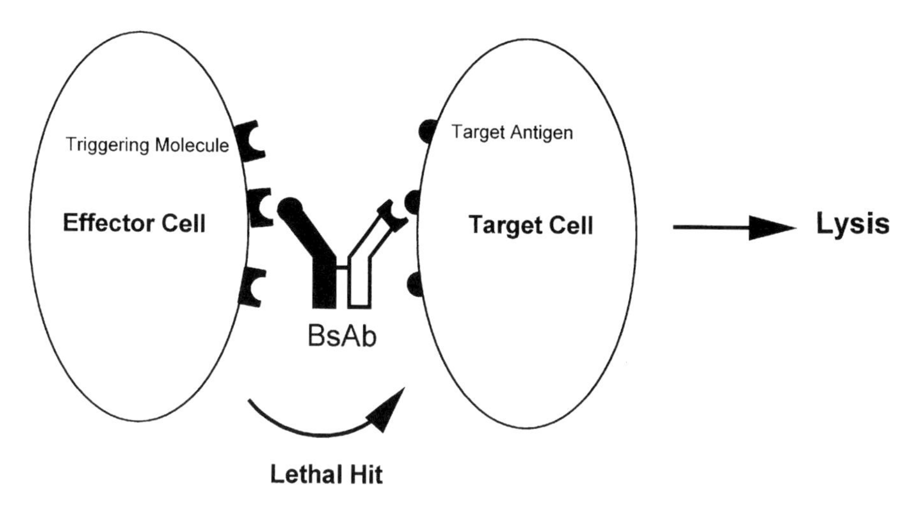

Figure 1D

Figure 1 Types of cell-mediated cytolysis. A. The normal mechanism. An effector cell binds ligands on a target cell through triggering molecules, thus forming an effector-target conjugate. This is followed by delivery of the lethal hit. B. Redirected lysis of hybridoma cells that produce mAbs against a triggering molecule. Cell surface antibodies on the hybridoma cell bind to triggering molecules on the effector cell, leading to hybridoma cell lysis. C. Reverse ADCC. An IgG antibody binds to cytotoxic triggering molecules on the effector cell and to FcγR on the target cell, leading to conjugate formation and lysis. D. Bispecific antibody mediated redirected lysis. A bsAb binds to both a triggering molecule on the cytotoxic cell and an antigen on the target cell, promoting conjugate formation and target cell lysis.

cell antigens that can participate in conjugate formation, but only a limited number of receptors, known as "cytotoxic triggering molecules" induce the effector cell to deliver the lethal hit. Non-antigen-specific adhesion molecules such as the integrins and selectins are examples of receptors that initiate and stabilize intercellular conjugate formation but do not normally trigger lysis. By contrast, cytotoxic triggering molecules such as the T cell receptor (TcR) or Fcγ receptors (FcγR) participate in both conjugate formation and the triggering of the lethal hit process.

Studies beginning more than 10 years ago (Staerz *et al.*, 1985; Perez *et al.*, 1985; Hoffman *et al.*, 1985; Lancki *et al.*, 1984; Kranz *et al.*, 1984; Karpovsky *et al.*, 1984) showed that the normal specificities of effector cells could be overridden by using antibodies against triggering molecules to bind target cells to cytotoxic cells, a process termed "redirected lysis" or "targeted cellular cytotoxicity". In fact, these studies defined triggering molecules and showed that ligand recognition was not required for eliciting cytotoxic responses. Redirected lysis can be brought about in several ways: (1) Hybridoma cells that produce antibodies against a cytotoxic triggering molecule are killed by effector cells that express that triggering molecule (Hoffman *et al.*, 1985; Graziano and Fanger, 1987; Lancki *et al.*, 1984). This is because hybridoma cells typically express a cell-surface form of the antibody that binds the hybridoma cell to triggering molecules on the effector cell (Figure 1B). Redirected hybridoma cell lysis has provided a convenient means for identifying hybridoma cells with specificity for triggering molecules. (2) The addition of antibodies against triggering molecules to cultures containing cytotoxic cells and FcγR positive target cells induces target cell lysis by bridging the FcγR on the targets to triggering molecules on the effector cells (Leeuwenberg *et al.*, 1985) (Figure 1C). This process, known as "reverse ADCC" is a very convenient method for determining if a particular receptor is a triggering molecule, especially when only small amounts of antibody are available. (3) BsAbs that simultaneously recognize triggering molecules on killer cells and cell surface determinants on target cells mediate lysis with the specificity determined by the anti-target arm of the bsAb (Figure 1D) (Titus *et al.*, 1987b; Perez *et al.*, 1985; Staerz *et al.*, 1985; Karpovsky *et al.*, 1984).

Of the different ways of inducing redirected lysis, bsAbs have the advantage that they can target cytotoxic cells against any cell containing a recognizable epitope, for example tumor or virally infected cells. It is this ability to specifically induce a variety of cytotoxic responses against unwanted cells that has made bsAbs the subject of a multitude of *in vitro*, pre-clinical and clinical studies. In this chapter we review the known triggering molecules and cytotoxic processes they evoke, the types of bsAbs currently available, and the results of pre-clinical and clinical studies using bsAbs. Other recent publications dealing with bsAbs include a book (Fanger, 1995) and several reviews (Weiner and De Gast, 1995; Clark and Weiner, 1995; Beun *et al.*, 1994; Segal *et al.*, 1993).

TRIGGERING MOLECULES

Over the past several years, a number of cytotoxic triggering molecules have been identified in leukocyte populations. The classic examples of triggering molecules are TcR on several subsets of T cells, and FcγR on NK and myeloid populations, but recently some adhesion molecules, such as CD2, CD44, and CD69 have also been found to trigger cytotoxic responses under certain conditions. The functions these molecules trigger include cellular cytotoxicity, cytokine release, production of active oxygen species and phagocytosis. In addition, many triggering molecules on MHC class II positive effector cells rapidly internalize and process substances bound to them (for example, by a bsAb), leading to enhanced antigen presentation. Table 1 lists the known triggering molecules.

T Cell Receptor

The TcR is a multi-chain protein consisting of antigen recognition and signal transducing portions (Malissen and Schmitt-Verhulst, 1993; Frank *et al.*, 1990).

Table 1 Known cytotoxic triggering molecules

Cell Type	*Triggering Molecules*	*References*
T cells	TCR/CD3	(Perez *et al.*, 1985; Staerz *et al.*, 1985)
	CD2	(Scott *et al.*, 1988; Bolhuis *et al.*, 1986)
	CD44	(Sconocchia *et al.*, 1994; Seth *et al.*, 1991; Galandrini *et al.*, 1993)
	CD69	(Moretta *et al.*, 1991)
	L-selectin (Mel 14)	(Seth *et al.*, 1991)
	Ly-6.2C	(Leo *et al.*, 1987b)
NK cells	CD16 (FcγRIII)	(Clark *et al.*, 1995; Titus *et al.*, 1987)
	CD2	(Siliciano *et al.*, 1985)
Monocytes/Macrophages	CD16 (FcγRIII)	(Wallace *et al.*, 1995)
	CD32 (FcγRII)	(Graziano and Fanger, 1987a; Shen *et al.*, 1986a)
	CD64 (FcγRI)	(Graziano and Fanger, 1987a; Shen *et al.*, 1986a)
PMN	CD32 (FcγRII)	(Wallace *et al.*, 1995; Graziano and Fanger, 1987)
	CD64 (FcγRI)	(Wallace *et al.*, 1995; Graziano and Fanger, 1987)

Antigen recognition is mediated by the α/β or γ/δ chains, antibody-like molecules that are mainly extracellular and consist of both variable and constant domains. These molecules associate non-covalently with five other polypeptides that transduce an intracellular signal upon TcR crosslinking. The signal transducing molecules consist of two CD3 heterodimers (CD3 γ/ε and δ/ε) and ζ–ζ or ζ–η disulfide-linked dimers. Of these, the CD3 components have extracellular portions that are non-variable in sequence, and therefore serve as a marker for all TcR. To date, every antibody that binds to the extracellular portion of the TcR, including those binding to α/β or γ/δ (Borst *et al.*, 1988; Staerz and Bevan, 1985), CD3ε or CD3δ/ε (Salmeron *et al.*, 1991; Shalaby *et al.*, 1992; Perez *et al.*, 1985; Leo *et al.*, 1987) has been able to induce redirected lysis in CTL. In addition, transmembrane fusion proteins that contain ζ on their cytoplasmic sides are also able to redirect lysis (Stancovski *et al.*, 1993; Hwu *et al.*, 1993; Eshhar *et al.*, 1993; Romeo and Seed, 1991; Irving and Weiss, 1991), indicating that ζ crosslinking *per se* induces a lytic signal.

Both CD4 and CD8 positive CTL clones mediate redirected lysis (Segal and Snider, 1989; Perez *et al.*, 1985; Liu *et al.*, 1985). In freshly isolated human T lymphocytes, a small subset of cells (approximately 2% of PBL) which is exclusively CD8+ and CD56+, mediate all CD3 redirected cytolysis, and these cells require IL-2 to maintain their optimal activity (Garrido *et al.*, 1990a). The great majority of freshly isolated PBL, however, are incapable of mediating lysis, and therefore require activation in order to generate targetable lytic activity. This has been achieved by several approaches, for example by crosslinking the TcR with immobilized anti-CD3 antibody and adding IL-2 (Garrido *et al.*, 1990a) or by stimulating with lectins plus IL-2 (Lamers *et al.*, 1992; Staerz *et al.*, 1985). In addition, T cells have been activated by using two different bsAbs, anti-target × anti-CD3 plus anti-target × anti-CD28 (Renner *et al.*, 1994; Jung *et al.*, 1991; Jung *et al.*, 1987), or by using an anti-CD3 mAb with FcγR+ targets transfected with B7.1 (the ligand for CD28) (Azuma *et al.*, 1992), thus making use of the known costimulatory property of CD28 on T cells. T cells have also been activated by using trispecific antibodies that crosslink CD3 to CD2 on the surface of T cells while simultaneously binding the target cell through the third arm of the trispecific antibody (Jung *et al.*, 1991; Tutt *et al.*, 1991).

Mechanisms of TcR Mediated Cytotoxicity

There are two general mechanisms by which T cells kill targets in redirected lysis assays. In the commonly used 4 hr ^{51}Cr release assay, the effector cell binds and lyses the target cell, but spares bystander cells (Perez *et al.*, 1986). Alternatively, T cells, resting or activated, secrete toxic substances that block the growth and eventually kill both target and bystander cells (Qian *et al.*, 1991). Using anti-cytokine mAbs, it has been demonstrated that IFN-γ and TNF-α are two mediators of this type of cytotoxic response, both *in vitro* (Qian *et al.*, 1991) and in animal studies (Demanet *et al.*, 1994).

The rapid, target specific process, is frequently mediated by a Ca^{2+} dependent, granule exocytosis mechanism (Podack *et al.*, 1991; Tschopp and Nabholz, 1990) in which the contents of specialized cytolytic granules are released into the narrow gap between the effector cell-target cell interface. Perforin, a component of the granules which has sequence homology with complement component c9 (Shinkai *et al.*, 1988; Lichtenheld *et al.*, 1988), aggregates into cylindrical structures and inserts into the lipid bilayer of the target cell. Pores are formed and lysis ensues (Dennert and Podack, 1983; Ishiura *et al.*, 1990). CTL granule release-mediated lysis of bystander cells does not occur presumably because the tight junction between killer and target cells prevents diffusion of cytotoxic substance into the medium. In addition, perforin is unstable in serum. The importance of perforin in mediating cellular cytotoxicity was shown by Shiver and Henkart (1991). They cloned the perforin gene into RBL cells, a non-cytotoxic rat mast cell line. When target cells were coated with TNP and anti-TNP IgE was added, the transfected RBL cells lysed the targets. This experiment suggested that perforin became a component of basophilic granules that were released upon IgE receptor cross-linking. Recently it was shown that perforin-deficient mice were unable to clear a lymphocytic choriomeningitis virus infection, presumably because of their failure to generate a competent CD8+ CTL response (Kagi *et al.*, 1994a). Granules also release proteases such as granzymes A and B into the cell interface, and it is thought that these trigger an apoptotic response in the target cell (Shiver *et al.*, 1992).

A second mechanism of rapid cellular cytotoxicity involves Fas and Fas-ligand interactions and has received much attention in recent years. Fas-ligand, a type II integral membrane protein, belongs to the TNF superfamily and is rapidly induced on T cells activated by TcR crosslinking or by PMA and ionomycin (Suda *et al.*, 1995; Vignaux *et al.*, 1995). Fas, a type I integral membrane protein, is a member of the TNF receptor superfamily and is expressed on a number of tumor cell lines. Fas bearing cells can be killed by crosslinking Fas with anti-Fas antibodies (Trauth *et al.*, 1989; Dhein *et al.*, 1992), or with purified Fas-ligand protein (Tanaka *et al.*, 1995), or with cells expressing Fas-ligand (Lynch *et al.*, 1994; Suda and Nagata, 1994). Death occurs by activation of the apoptotic pathway via a mechanism that is independent of extracellular Ca^{2+} (Dhein *et al.*, 1992; Rouvier *et al.*, 1993). The Fas-dependent component of cell mediated cytotoxicity was studied in perforin-deficient mice (Kagi *et al.*, 1994b) and in perforin-negative mutant cytotoxic cells (Kojima *et al.*, 1994). Both studies revealed that in the absence of perforin only Fas positive targets could be lysed in a standard ^{51}Cr release assay. Fas dependent lysis was antigen specific, and did not result in lysis of bystander cells.

Fcγ Receptors (FcγR)

FcγR trigger a variety of functions in different leukocyte populations. Three major classes of Fcγ receptors, FcγRI (CD64), FcγRII (CD32), and FcγRIII (CD16) have been defined by mAb binding, and by their DNA and protein sequences (Van de Winkel and Anderson, 1991; Ravetch and Kinet, 1991; Ravetch and Anderson, 1990). The three Fcγ receptors all contain C2 type immunoglobulin domains and are highly homologous to each other in their extracellular portions.

FcγRII and III each contain two immunoglobulin domains, and bind IgG with relatively low affinity. Both exist in multiple forms. FcγRII was originally identified as a 40 kD cell surface glycoprotein that bound mAb IV.3 (Looney *et al.*, 1986). cDNA sequencing (Warmerdam *et al.*, 1991; Warmerdam *et al.*, 1990; Stuart *et al.*, 1987; Stengelin *et al.*, 1988; Hibbs *et al.*, 1988; Brooks *et al.*, 1989), revealed that several structurally distinct forms of FcγRII exist. These forms of the receptor are produced by three genes on the long arm of chromosome 1 (Sammartino *et al.*, 1988; Qiu *et al.*, 1990). Further diversity arises from alternative splicing, which could potentially give rise to many structurally and functionally distinct receptors. FcγRIII is a heavily glycosylated 50–80 kD protein, originally identified by mAb 3G8 (Fleit *et al.*, 1982) on monocytes, lymphocytes, and PMN. Two genes for this receptor, also located on the long arm of chromosome 1 (Peltz *et al.*, 1989), code for proteins that differ only in the extracellular portion at residue 203, proximal to the plasma membrane. As a result of this single difference, FcγRIIIB, which has a Ser at position 203 and is expressed primarily on neutrophils, is anchored to the plasma membrane by a phosphatidylinositol glycan linkage, whereas FcγRIIIA, which has a Phe at position 203 and is expressed on NK cells and macrophages, is a transmembrane receptor (Ravetch and Perussia, 1989; Lanier *et al.*, 1989; Edberg *et al.*, 1989). The structural linkage to the membrane correlates with the ability of FcγRIII+ cells to lyse anti-CD16 hybridoma cell targets or antibody-coated human tumor cells; FcγRIIIA+ cells lyse tumor targets whereas cells expressing FcγRIIIB do not (Lanier *et al.*, 1988b).

FcγRI (CD64) contains three extracellular immunoglobulin domains and binds monomeric IgG with relatively high affinity (K_d=1–10 nM) FcγRI is a 72 kD transmembrane protein encoded by two genes (Van de Winkel *et al.*, 1991), again on the long arm of chromosome 1 (Osman *et al.*, 1992), that are expressed in myeloid cells (Van de Winkel *et al.*, 1991; Allen and Seed, 1989). FcγRI is recognized by the mAbs 22, 197 (Vance *et al.*, 1989; Guyre *et al.*, 1989), 32 (Anderson *et al.*, 1986), and others (Van de Winkel and Anderson, 1991). While it is known that cytokine release, ADCC, and phagocytosis can be triggered by the FcγRI recognized by these mAbs, little is known about possible differences in function mediated by the two different FcγRI gene products.

FcγR Expression

The expression and modulation by cytokines of FcγR on human leukocytes, tissues and myeloid cell lines (Van de Winkel and Anderson, 1991; Ravetch and Kinet, 1991; Fanger *et al.*, 1989) determines which cells, bsAbs, and cytokines can be used to induce FcγR-dependent cytotoxicity in laboratory experiments and clinical trials. FcγRI is expressed predominantly on monocytes, macrophages, and monocyte-like cell lines (U937, HL-60, and THP-1 cells). IFN-γ, IL-10, and IL-13 up-regulate the expression of FcγRI on monocytes and macrophages whereas IL-4 decreases its expression. IFN-γ induces expression of FcγRI on PMN and eosinophils (Anderson *et al.*, 1986; Petroni *et al.*, 1988; Hartnell *et al.*, 1992) and *in vivo* administration of G-CSF also upregulates FcγRI on neutrophils by a mechanism that is unclear (Repp *et al.*, 1991).

FcγRII is expressed on a wide variety of cell types including monocytes, macrophages, B cells, granulocytes, platelets, Langerhans cells, and placental endothelial cells. It is also expressed on U937, HL-60, THP-1, K562, Raji, and Daudi cell lines. This receptor is constitutively expressed on monocytes, B cells, and neutrophils, and its level of expression on these cells is only modestly affected by cytokines. On other cell types, however, FcγRII expression is strongly dependent on cytokines and other activation signals. For example, FcγRII is upregulated on platelets by thrombin, PMA, and ADP, while on eosinophils, FcγRII is upregulated by IFN-γ and IL-3 (Hartnell *et al.*, 1992). In one study (Van de Winkel *et al.*, 1989) protease treatment of FcγRII positive cells increased their affinity for human IgG complexes suggesting that proteases produced by infiltrating cells at sites of inflammation could modulate FcγRII function.

The transmembrane form of FcγRIII, FcγRIIIA, is expressed on NK cells, a small subset of T cells, macrophages, and 5–15% of circulating blood monocytes, while the GPI anchored form, FcγRIIIB, is expressed on neutrophils, and eosinophils. In addition, FcγRIII has been observed in kidney mesangial cells, placental trophoblasts, and PMA- or DMSO-treated HL-60 cells. Thus far, FcγRIII expression appears to be relatively unaffected by cytokines. Two possible exceptions are reports that IFN-γ may upregulate FcγRIIIB on eosinophils

(Hartnell *et al.*, 1992) and TGF-β may increase expression of FcγRIIIA on monocytes (Wong *et al.*, 1991; Welch *et al.*, 1990).

FcγR Triggered Functions

Like T cells, FcγR crosslinking can induce cytokine release and a cell mediated cytolytic response. In addition, FcγR initiate functions normally mediated by myeloid cells, but not by lymphocytes. These include phagocytosis, superoxide and nitric oxide production, and enhanced antigen presentation. Since FcγR positive effector cells often express more than one type of receptor, bsAbs, made from Fab fragments directed against one type of FcγR have proved essential in defining the functions of FcγR on different cell types. FcγR triggered responses have been reviewed to varying extents in (Van de Winkel and Anderson, 1991; Lanzavecchia, 1990; Fanger *et al.*, 1989; Fanger *et al.*, 1989).

Phagocytosis

It has been established that the binding of erythrocytes (E) to FcγRI and FcγRII on freshly isolated or cultured monocytes, alveolar macrophages, and IFN-γ treated PMN leads to their phagocytosis (Van Schie *et al.*, 1992; Munn *et al.*, 1991; Van de Winkel and Anderson, 1991; Anderson *et al.*, 1990). FcγR-mediated phagocytosis of tumor cells by macrophages appears to be inhibited by IFN-γ (Munn *et al.*, 1991; Backman and Guyre, 1994). Whether or not FcγRIIIB on PMN mediate phagocytosis is controversial (Salmon *et al.*, 1990; Anderson *et al.*, 1990). PMN do not phagocytose E coated with anti-FcγRIII $\times$ anti-E bsAb but do phagocytose E opsonized with anti-E IgG antibody. A role for FcγRIIIB in phagocytosing antibody-coated E was indicated by the observation that the NA1–NA2 polymorphism in the FcγRIIIB molecule affected the efficiency of phagocytosis; thus neutrophils from a heterozygous donor were less efficient than those from a NA1 homozygous donor in mediating phagocytosis (Salmon *et al.*, 1990).

Release of cytokines and superoxide

Induction of cytokine release by FcγR crosslinking has been reported for the three classes of human FcγR on monocytes, macrophages, NK cells and neutrophils (Van de Winkel and Anderson, 1991; Wallace *et al.*, 1995; Debets *et al.*, 1990). Depending upon the receptor type, cell type and activation state, cytokine profiles are likely to vary considerably in response to FcγR crosslinking. Some of the cytokines known to be released include TNF-α, IFN-γ, IL-1, and IL-6. Several studies show that FcγRI on PMN (Akerley *et al.*, 1991) and monocytes (Pfefferkorn and Fanger, 1989a, b) is capable of triggering superoxide production. Superoxide generation through FcγRII on PMN does not appear to be very efficient, but does show some synergy if both FcγRII and FcγRIII are triggered (Hundt and Schmidt, 1992). Moreover, it has been reported (Tosi and Berger, 1988) that FcγRII on PMN was effective for triggering superoxide production if cells were first treated with elastase. FcγRIIIB can trigger superoxide production when crosslinked with IgG immune complexes (Crockett-Torabi and Fantone, 1990), but not with mAbs (Tosi and Berger, 1988; Huizinga *et al.*, 1990).

Antigen presentation

MHC class II positive cells such as monocytes, dendritic cells and B cells present processed antigens to CD4 positive T cells during the initial stages of antigen-specific T cell responses. A critical step in the presentation of protein antigens is their uptake and degradation by the antigen presenting cell (APC), followed by the loading of peptide onto the MHC class II molecules. In the absence of a specific binding mechanism, soluble protein is taken up by the APC by fluid phase pinocytosis, an inefficient process requiring high concentrations of antigen. The efficiency of antigen presentation can be improved several hundred fold or more by specifically binding the antigen to a surface protein that frequently undergoes endocytosis. For example, antigen receptor on B-cells binds and endocytoses low concentrations of antigen, leading to a great enhancement in the efficiency of processing and presentation (Lanzavecchia, 1990). BsAbs that bind antigen to a number of different cell surface components can also lead to remarkable improvements in antigen presentation (Snider, 1992; Snider *et al.*, 1990; Snider and Segal, 1989; 1987). Because most APC express FcγR, and because these receptors are normally involved in endocytic processes, it might be predicted that the binding of antigen to FcγR would lead to enhanced presentation. Indeed, recent studies suggest that antigen directed with bsAbs to each of the classes of FcγRs expressed on the surface of human monocytes was presented at least 100-fold more efficiently than antigen in the absence of bsAb (Gosselin *et al.*, 1992; Liu and Guyre, 1995). Early studies showed that the targeting of antigen to FcγRII in mice with bsAbs greatly improved IgG antibody responses against the antigen, suggesting an enhancement of T helper responses (Snider *et al.*, 1990). More recently, transgenic mice expressing human FcγRI produced high-titers of IgG antigen-specific antibody against antigenic determinants on an anti-huFcγRI antibody used for immunization (Heijnen *et al.*, 1996). These data suggest that huFcγRI is capable of mediating enhanced antigen presentation *in vivo*. The ability of the other FcγRs to enhance antigen presentation *in vivo* have yet to be determined.

Antibody-dependent cellular cytotoxicity

NK cells are the most potent mediators of ADCC. These cells contain cytolytic granules and mediate short term lysis by the same mechanisms as described above for CTL. FcγRI, FcγRII, and FcγRIII on monocytes and PMN (IFN-γ treated for FcγRI) are each independently able to trigger killing of ox or chick erythrocytes in a 4 hr assay (Van de Winkel and Anderson, 1991; Fanger *et al.*, 1989; Fanger *et al.*, 1989). Whereas monocytes or macrophages are able to mediate lysis of tumor cells through all three FcγR, PMN are unable to lyse tumors through FcγRIIIB (Van de Winkel and Anderson, 1991; Fanger *et al.*, 1989). PMN are able to kill tumor targets when triggered through FcγRII, but only after IFN-γ treatment. IFN-γ must induce an intracellular killing mechanism in PMN, since it does not affect FcγRII expression.

Adhesion Molecules (CD2, CD44 and CD69)

Recently, a number of adhesion molecules have been shown to act as cytotoxic triggers. CD2 (LFA-2, the sheep red blood cell receptor) is expressed on T lymphocytes, NK cells and thymocytes in humans (Davis *et al.*, 1993) and, in addition, on B cells (Guckel *et al.*, 1991) and tissue macrophages (Beyers *et al.*, 1989) in rodents. CD2 is a glycosylated type 1 transmembrane receptor with two immunoglobulin superfamily domains on its extracellular portion and a conserved cytoplasmic domain with proline rich sequences that interact with SH3 domains of Src-like kinases $p56^{lck}$ and $p59^{src}$ (Bell and Imboden, 1995; Bell *et al.*, 1996; Davis *et al.*, 1993). Three known ligands for CD2 include CD58 (Shaw *et al.*, 1986; Dustin *et al.*, 1987), CD59 (Hahn *et al.*, 1992), and CD48 (Hahn *et al.*, 1992), and CD2 interacts with CD48 and CD58 through a highly charged binding surface on the N-terminal domain of the extracellular region (Arulanandam *et al.*, 1993; 1994). CD2 was first shown to be an activator of T cells in experiments where mAbs against two distinct epitopes outside of the ligand binding site on CD2, T11(2) and T11(3) induced T cell proliferation (Meuer *et al.*, 1984). Interestingly, the T11(3) epitope was exposed only upon binding of the anti-T11(2) mAb, and neither mAb alone, nor a mAb against a ligand binding-site epitope induced proliferation. Based upon these studies, it was shown that mAbs against T11(2) and T11(3) could redirect lysis mediated by either CTL or NK cells in reverse ADCC assays, or when conjugated to anti-target mAbs in bsAbs (Scott *et al.*, 1988; Siliciano *et al.*, 1985). As expected both antibodies together were required for lysis. More recently, it was reported that a trispecific antibodies with anti-CD2, CD3, and CD37 specificities retargeted resting, unprimed T cells from fresh human PBMC to lysis CD37+ tumor cells with much greater activity than a bsAb with anti-CD3 and anti-CD37 specificities (Tutt *et al.*, 1991). In this case, the anti-CD2 portion of the trispecific antibody was directed against the T11(1) epitope in the ligand binding region which had previously been found to be ineffective by itself in redirecting lysis (Meuer *et al.*, 1984). The trispecific antibody results suggest that the crosslinking of CD2 with the TcR may produce an enhancement of the lytic signal in T cells. In fact, it has been shown that T or NK cell activation resulting from CD2 crosslinking with the anti-T11(2) +T11(3) mAbs requires the presence of the TcR or FcγR (Spruyt *et al.*, 1991; Howard *et al.*, 1992), suggesting that theses antibodies may also promote an interaction of CD2 with the TcR or FcγR.

CD44, a type I integral membrane protein is expressed on most leukocytes. It recognizes extracellular matrix components, mainly glycosaminoglycans such as hyaluronan and chondroitin sulfate (Culty *et al.*, 1990; Lesley *et al.*, 1993) but also binds the chemokine, osteopontin (Weber *et al.*, 1996). Anti-CD44 mAbs trigger lysis in reverse ADCC assays in both murine and human T cell clones

(Galandrini *et al.*, 1993; Seth *et al.*, 1991), and MEL-14, a mAb against an antigenically related epitope on L-selectin is also cytotoxic in murine T cell clones (Seth *et al.*, 1991). In human PBL, CD44 is a triggering molecule on IL-2 activated NK, and to a lesser degree on CD56+ T cells (Sconocchia *et al.*, 1994).

CD69, a type II integral membrane protein with a C-lectin extracellular domain, is one of the first activation antigens to appear on the surface of T cells after stimulation with immobilized anti-CD2, anti-CD3, anti-CD5 or anti-CD28 mAbs (Cosulich *et al.*, 1992; Vandenberghe *et al.*, 1993). Furthermore, a dose-related increase in the number of PBL expressing CD69 was observed in patients treated with anti-CD3 mAbs (Urba *et al.*, 1992). CD69 crosslinking on T cells induces release of TNF-α (Santis *et al.*, 1992) and expression of genes for IL-2 and IFN-γ (Nakamura *et al.*, 1989; Testi *et al.*, 1989). IL-2, IL-12, IL-6 and IFN-α induce CD69 expression on NK cells which closely parallels the acquisition of NK activity (Lanier *et al.*, 1988a). In addition, the expression of CD69 on NK cells increases after they interact with NK targets or are stimulated by FcγRIII crosslinking (Borrego *et al.*, 1993). Using anti-CD69 mAbs in reverse ADCC assays it was demonstrated that CD69 is a cytotoxic triggering molecule on NK and γ/δ (but not α/β) T cells clones (Moretta *et al.*, 1991).

Monocytes and platelets constitutively express CD69; crosslinking of CD69 on monocytes leads to activation of both cyclooxygenase and lipoxygenase pathways (De Maria *et al.*, 1994) whereas CD69 crosslinking on platelets induces aggregation (Testi *et al.*, 1990). Eosinophils isolated from asthmatics express CD69 (Hartnell *et al.*, 1993) and GM-CSF, IL-3, IL-5, IFN-γ, but not PAF can induce CD69 expression on eosinophils *in vitro* (Hartnell *et al.*, 1993). CD69 is stored in neutrophils, probably in trans-Golgi structures, and is rapidly translocated to the surface upon activation with phorbol myristal acetate (PMA) or N-formyl-met-leu-phe (fMLP; Gavioli *et al.*, 1992). CD69 crosslinking in PMA-stimulated neutrophils caused an increase in the release of lysozyme but had no effect upon superoxide anion production or cytotoxic activity against K562 cells (Gavioli *et al.*, 1992).

TYPES OF BISPECIFIC ANTIBODIES

The simplest way of preparing a bispecific antibody is to chemically crosslink two antibodies using reagents that randomly link the antibodies by amino acid side chain groups, usually ε-amino groups on lysine residues (Segal, 1993; Carlson *et al.*, 1978; Karpovsky *et al.*, 1984; Segal and Bast, 1995) (Figure 2A). The most commonly used reagents for preparing such bsAbs use dithiol exchange reactions, which greatly favor the formation of hetero- over homoconjugates. Randomly crosslinked heteroconjugates work well for most *in vitro* applications, where, for example, a receptor on a cytotoxic cell needs to be linked to a cell surface component on a target cell. However, because of their heterogeneity in size and chemical composition, and because of variability between batches, heteroconjugates would not give consistent results *in vivo*, where size and composition greatly effect biodistribution and stability. Instead, homogeneous bispecific molecules are

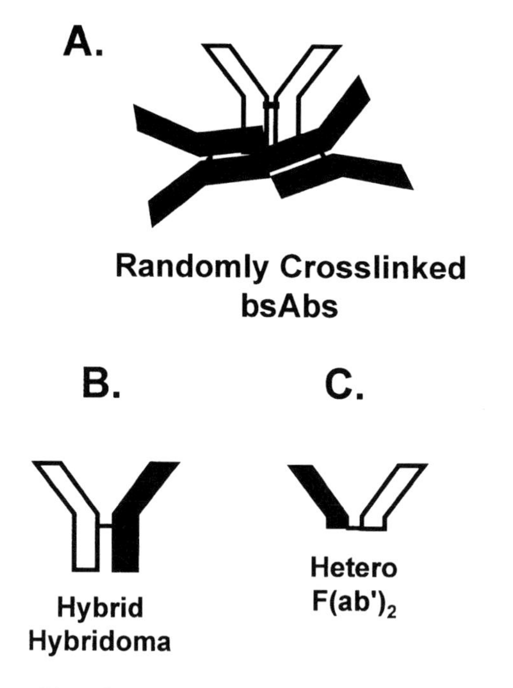

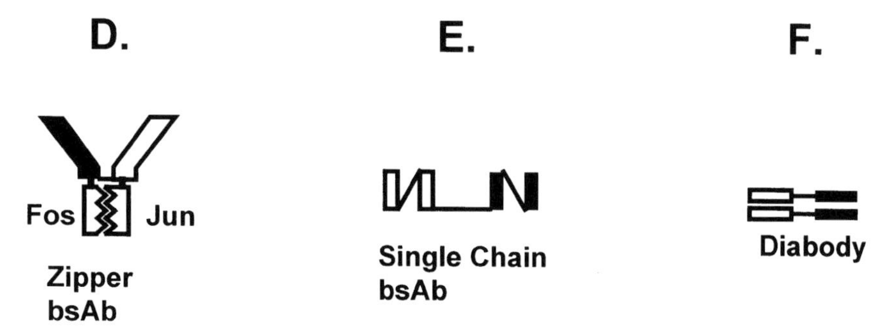

Figure 2 Different types of bispecific antibodies. A. Randomly crosslinked bsAbs, produced by crosslinking agents that bind to amino acid side chains on the parental antibodies. B. Hybrid hybridoma antibodies, produced by fusing two different hybridoma cells, and isolating molecules with the correct chain pairings from the secreted product. C. Hetero $F(ab')_2$ bsAbs, prepared by heterocrosslinking two F(ab′) fragments by their hinge sulfhydryls. D. Zipper bsAbs, genetically engineered F(ab′) fusion proteins containing Fos or Jun leucine zipper peptides at their C-terminal ends. On mixing, the two F(ab′) fusion proteins preferentially form heterodimers. E. Single chain bsAbs, formed by concatenating two single-chain Fvs. F. Diabodies, formed by mixing two different single-chain Fvs with shortened V_L–V_H linkers.

greatly preferred for *in vivo* animal and clinical studies. Two types of homogeneous bsAbs, the hybrid-hybridoma and the hinge-linked hetero-F(ab')$_2$ are currently in use, but a number of genetically engineered constructs have recently been described that hold great promise for simplifying the preparation of bispecific molecules. The different types of bsAbs are reviewed in this section.

Hybrid-hybridomas

The fusion of two different hybridomas can result in the formation of a stable "hybrid-hybridoma" or "quadroma" line that produces as one of its products a homogeneous bsAb (Figure 2B) (Milstein and Cuello, 1984; 1983). While hybrid-hybridoma lines have the advantage of providing a stable source of bsAb, their major drawback is that the combination of the two heavy and two light chains produced by the hybrid-hybridoma line can lead to the assembly of up to ten different immunoglobulin species, only one of which has both antibody activities in a single molecule. This leads to reduced yields of bsAb, and often necessitates complicated purification protocols.

Several different procedures have been described for producing hybrid-hybridomas (Segal *et al.*, 1993; Urnovitz *et al.*, 1988; Wong and Colvin, 1987; Karawajew *et al.*, 1990; Clark *et al.*, 1988; Karawajew *et al.*, 1987; Lanzavecchia and Scheidegger, 1987; Segal and Bast, 1995), but in general, two hybridoma lines are fused using polyethylene glycol. The resulting hybrid-hybridoma lines are then selected from the unfused parental lines using one of a number of different selectable drug markers. Alternatively, the fused lines can be selected using flow cytometry after labeling the parental lines with dyes that are distinguishable from one another. Once stable hybrid-hybridoma lines have been established, they are screened for bsAb production. The amount of bsAb secreted is highly variable between lines and depends upon the relative rates of synthesis of the heavy (H) and light (L) chains and upon preferential associations that might occur between chains. When the synthesis of the four chains is equal and when they associate randomly, 12.5% of the total immunoglobulin will be bsAb (Staerz and Bevan, 1986a). While the H–H interaction is usually near random, particularly when both parental Abs are of the same IgG isotypes, H–L pairings, because they involve interactions between variable residues in the V regions, often show preferences; in some cases homologous H–L chain pairing is preferred, but in other cases random, or even heterologous pairing occurs preferentially (De Lau *et al.*, 1991). Thus, depending upon the nature of the H–L interaction, the actual yield of bsAb can vary between 0% and 50%. If the H chains from the parental antibodies are of different IgG isotype, then the yield of mixed H–H chain combinations may drop (Milstein and Cuello, 1984) and, if two different classes of H chain are used, association between the heterologous H chains may not occur (Urnovitz *et al.*, 1988). The lower yield of mixed isotype bsAbs may, however, be offset by an increase in efficiency of purification (Segal and Bast, 1995; Milstein and Cuello, 1984); for example, IgG molecules consisting of two different H-chain isotype may be purified by Protein A chromatography by eluting at different pH values

(Koolwijk *et al.*, 1989). Mixed isotype bsAbs are often incapable of mediating ADCC, which may be advantageous in cases where effector cells bind the bsAb (Clark and Waldmann, 1987; Clark *et al.*, 1988). In spite of their difficulty in production and isolation, clinical grade F(ab′)$_2$ fragments of hybrid-hybridoma antibodies have been prepared in gram amounts and have been tested in patients in two different phase I–II trials (Weiner *et al.*, 1995; Canevari *et al.*, 1995).

Hetero-F(ab′)$_2$ bsAbs

Homogeneous bsAbs can be formed by crosslinking two different F(ab′) fragments via their hinge sulfhydryl residues (Figure 2C). F(ab′)$_2$ fragments are normally produced from mAbs by pepsin digestion and these are then reduced to the F(ab′) fragment (Andrew and Titus, 1992). Two different F(ab’) fragments can be crosslinked by either disulfide exchange or by forming a stable thioether link between the two F(ab′)s. For disulfide exchange, disulfide groups are formed on one of the F(ab′)s using a dithiol compound such as Ellman’s reagent. Mixing the reduced F(ab′) with the dithiol derivative produces only the bispecific molecule, uncontaminated by parental type F(ab′)$_2$ (Brennan *et al.*, 1985). Stable thioether bonds between two different F(ab′)s are produced using a dimaleimide reagent such as o-phenylenedimaleimide (Glennie *et al.*, 1987); the hinge region SH groups on one of the F(ab′) molecules are alkylated with the dimaleimide compound, thus providing free maleimide groups that are then reacted with the hinge region SH groups from the other F(ab′) (Glennie *et al.*, 1987). A thioether linked hetero-F(ab′)$_2$ has been prepared in large amounts under GMP conditions and is being studied in several clinical trials (Valone *et al.*, 1995). While hetero-F(ab′)$_2$ bsAbs are well-defined with respect to size and composition, their preparation requires multi-step procedures, each of which is accompanied by loss of protein.

Genetically Engineered bsAbs

Because of the difficulties involved in producing large amounts of homogeneous bsAbs by the “conventional” methods described above, several genetically engineered constructs have been described. To avoid losses incurred during proteolytic digestion steps, F(ab′) constructs have been expressed in both bacteria (Carter *et al.*, 1992; Shalaby *et al.*, 1992) and mammalian cells (Weiner *et al.*, 1994; Kostelny *et al.*, 1992). For the preparation of bsAbs, two different F(ab′) molecules were expressed and isolated separately, and then linked via their hinge sulfhydryl groups to give the hetero-F(ab′)$_2$. In one study F(ab′)s were crosslinked chemically using a dimaleimide reagent (Shalaby *et al.*, 1992), but this step was removed in the other study by inclusion of Fos or Jun leucine zippers at the C-terminal ends of the F(ab′) heavy chains (Figure 2D) (Kostelny *et al.*, 1992). Fos and Jun leucine zippers have a strong tendency to form heterodimers, and when F(ab′)s with Fos and Jun zippers are mixed under mildly oxidizing conditions, bispecific hetero-F(ab′)$_2$ molecules are obtained in high yields. Animal studies have shown that these bsAbs are stable in serum, and are cleared at about the same rate as conventional $F(ab')_2$ fragments (Weiner *et al.*, 1994; Bakacs *et al.*, 1995).

Bispecific constructs have also been produced using only the Fv portions of antibodies (Hayden *et al.*, 1994; Gruber *et al.*, 1994; Jost *et al.*, 1996; Kurucz *et al.*, 1995; Traunecker *et al.*, 1991; Dorai *et al.*, 1994; Pack and Pluckthun, 1992; Pluckthun, 1992; Mallender and Voss, Jr., 1994; Holliger *et al.*, 1993; Winter and Milstein, 1991). The Fv portion of an antibody is a 26 kD heterodimer consisting of the variable (V) domains of the heavy (H) and light (L) chains, and is the smallest fragment to bear the antigen-binding site (Hochman *et al.*, 1973). One construct known as a "diabody" (Figure 2F) uses two different polypeptide chains, V_{H1}–V_{H2} and V_{L1}–V_{L2}, that associate laterally and non-covalently to form a bsAb (Holliger *et al.*, 1993). Other constructs are composed of single-chain Fvs (sFvs). sFvs are antigen-binding constructs that are identical to Fvs except that a polypeptide linker joins the C-terminus of one V domain to the N-terminus of the other (Huston *et al.*, 1993; 1988; Bird *et al.*, 1988). Single-chain bsAbs have been produced by concatenating two different sFvs with a polypeptide linker (Figure 2E) (Jost *et al.*, 1996; Kurucz *et al.*, 1995; Hayden *et al.*, 1994; Gruber *et al.*, 1994; Mallender and Voss, Jr., 1994). When expressed in mammalian cells, single chain bsAbs have the same affinities for antigen and are secreted as rapidly as the parental sFvs (Hayden *et al.*, 1994; Jost *et al.*, 1996). Single chain bsAbs have also been produced in high amounts in bacteria and after *in vitro* refolding and purification, have given reasonably high yield of fully active single chain bsAb, capable of redirecting lysis in the ng/ml range (Kurucz *et al.*, 1995). It is not known how stable a single chain bsAb would be in serum, or how rapidly it would be cleared relative to its rate of tissue penetration. sFvs vary in their stabilities, so it might be expected that some single chain bsAbs would be more stable than others, both *in vivo* and *in vitro*. Also, because of their relatively low molecular weight, one would expect single chain bsAbs to be cleared more rapidly and to have greater tissue penetration than intact bsAbs. However, single chain bsAbs have a definite advantage over other constructs in that they are produced as a single bispecific molecule with no assembly required.

ANIMAL MODELS

Because bsAbs have the potential of targeting a wide variety of effector cells with a panoply of responses against tumor cells in cancer patients, much effort has been expended to learn how bsAbs function in experimental animals, mainly mice and rats. The animal models used in these studies fall into two categories: (1) xenograft models where human tumors are given to immunodeficient mice and the targeted cytotoxic cells used for treatment are derived from human sources, and (2) syngeneic models, where effector cells are provided by immunocompetent mice bearing syngeneic tumors.

Initial xenograft studies showed that human T or NK cells targeted with a bsAb hindered growth of human colon adenocarcinoma transplanted subcutaneously into nude mice (Titus *et al.*, 1987a, b). One mechanism used in blocking tumor growth entailed the release of cytokines following TcR crosslinking, as suggested by the inhibition of bystander tumor growth (Qian *et al.*, 1991). *In vitro*, IFN-γ and TNF-α were shown to be involved in the blockage of growth of this same tumor by targeted human T cells (Qian *et al.*, 1991). In later studies (Mezzanzanica *et al.*, 1991; Garrido *et al.*, 1990), targeted human T cells were used to treat nude mice bearing established intraperitoneal (ip) human ovarian tumors. Mice were treated with ip injections of activated T cells targeted with bsAb, and anti-tumor effects were assessed by counting the number of tumor cells recovered from the peritoneal cavity 11 days after treatment, or by measuring long term survival times. By both criteria, the targeted T cells strongly inhibited tumor growth. These studies formed the pre-clinical basis for clinical trials with ovarian cancer patients, described in the next section.

In a recent report (Renner *et al.*, 1994) a combination of two different bsAbs was used to cure SCID mice of subcutaneously growing, established Hodgkin's lymphoma. The two bsAbs both bound to CD30 on the tumor cells. In addition, one bsAb bound to CD3 and one to CD28 on T cells. The concept behind the use of these two bsAbs was that upon binding to tumor cells, the bsAbs would generate a potent activation signal on the effector cells by co-crosslinking CD28 and CD3. Mice bearing 6–8 mm tumors were given iv injections of bsAbs followed 24 hr later with iv injections of activated human PBL. Biodistribution studies and histological examination established that each bsAb individually reached the tumor and promoted an infiltration of human CD4+ and CD8+ T cells. However, eradication of the tumor occurred only when both bsAbs were given together, suggesting that both CD28 and CD3 crosslinking were required to produce competent effector cells.

Although *in vivo* studies using immunodeficient mice showed that bsAbs could cause substantial anti-tumor responses, such studies give only limited information on how an intact immune system would react against syngeneic tumor in the presence of bsAb, a situation that would apply in most clinical trials. Therefore it would be highly advantageous to study bsAbs in a totally syngeneic setting. Unfortunately, this has proven difficult due to the lack of mAbs specific for murine tumor antigens. Nevertheless, bsAb targeting studies employing syngeneic systems have been possible in a few cases. In the first such study (Staerz and Bevan, 1986b), Thy 1.2+ mice bearing a Thy 1.1+ lymphoma were treated with activated host T cells and a bsAb against Thy 1.1 and Vβ8$^+$ TcR. Treated mice showed a longer mean survival time than the untreated group.

BsAbs reacting with CD3 and the idiotypes of the surface immunoglobulin on mouse B cell lymphomas have formed the basis of a number of studies designed to treat hematological malignancies (Demanet *et al.*, 1992; 1994; 1991; Brissinck *et al.*, 1993; 1991; Weiner *et al.*, 1994; Weiner and Hillstrom, 1991). Mice bearing either BCL$_1$ or 38C13 lymphomas responded remarkably well to bsAb therapy. For example most mice treated with 5 µg of anti-CD3 × anti-idiotype bsAb 9 days after receiving BCL$_1$ lymphoma were tumor free 150 days later, whereas all mice were dead by about 80 days in the untreated controls (Brissinck *et al.*, 1991). Activation of T cells by allo-stimulation (Demanet *et al.*, 1991) or TcR crosslinking

(Weiner *et al.*, 1994) resulted in increased anti-tumor effects of the bsAbs. Both CD4 and CD8 positive T cells infiltrated the tumor in bsAb-treated mice, and blockage of tumor growth was at least in part mediated by cytokines such as IFN-γ and TNF-α that were secreted by the targeted T cells (Demanet *et al.*, 1994).

Solid tumors have been more refractive to bsAb treatment. Mouse mammary tumors often express mouse mammary tumor virus proteins such as gp52. BsAbs that bind gp52 and mouse CD3 induce activated mouse T cells to specifically kill gp52+ tumor cells *in vitro*, and to block subcutaneous tumor growth in Winn-type assays (Moreno *et al.*, 1995). However, bsAb treatment was unable to block the growth of established, subcutaneously growing mammary tumors, and only delayed the growth of lung metastases of mammary tumors (Bakacs *et al.*, 1995). In a second syngeneic mouse lung metastasis model (Penna *et al.*, 1994), melanoma cells which were transfected with the human melanoma antigen p97 were injected into mice, and 10 min later mice were given staphylococcal enterotoxin B (SE-B) with or without anti-CD3 × anti-p97 bsAb. Pulmonary metastases were counted on day 14. Combination of SE-B and bsAb significantly reduced the number of metastasis compared to SE-B alone or to a mixture of the unconjugated parental antibodies.

In a syngeneic rat colon carcinoma solid tumor model, bsAb treatment had very little effect on tumor growth unless the rats were given high amounts of IL-2 along with the bsAb treatment (Beun *et al.*, 1992). In a second syngeneic rat system (Kroesen *et al.*, 1995) immunocompetent rats were given squamous cell carcinoma cells transfected with the human pan-carcinoma-associated antigen EGP-2, resulting in the growth of tumor nodules in the lungs. Intravenous treatment of the tumor bearing rats with anti-CD3 × anti-EGP-2 bsAb and rat rIL-2 resulted in profound reduction of the tumor, whereas bsAb alone, IL-2 alone, or a mixture of anti-CD3, anti-EGP-2, and IL-2 were ineffective.

In summary animal models have shown that targeted effector cells given locally at the site of tumors can deliver a strong anti-tumor response. Systemic treatment of lymphomas also has been highly effective in both syngeneic and xenograft models. Solid tumors, however, are more resistant to bsAb treatment and require improved methods for activating effector cells *in vivo* and for delivering the targeted cells to the tumors.

CLINICAL STUDIES

The generation of anti-tumor immune responses in cancer patients remains a fundamental goal of cancer therapy. Standard therapies such as surgery, chemotherapy, radiotherapy and hormonal therapy, whose efficacies are limited to hematologic malignancies and to some solid tumors, frequently fail to completely remove malignant cells from patients with advanced cancers. During the past several years a variety of immunotherapeutic strategies have been tested in cancer patients (also reviewed in Chapter “Clinical Studies in Oncology”, see below). Adoptive cellular immunotherapies in which activated NK or TIL cells were passively transferred into cancer patients showed some anti-tumor effects, mainly in renal cell carcinomas and melanomas (Ettinghausen and Rosenberg, 1993).

Treatment of cancers with unconjugated antibodies against tumor associated antigens (TAA) gave only occasional anti-tumor responses, and was not accompanied by the infiltration of the tumor by FcγR+ cells (Dillman, 1989). Immunoconjugates containing toxins or radioisotopes have also been used but frequently without dramatic anti-tumor effects (Vitetta *et al.*, 1993; Epenetos *et al.*, 1986). Several reasons may account for the failure of immunotherapeutic approaches to eradicated tumors. Immunocompetent cells or antibodies may have restricted movement within tumor tissue, thus limiting their accessability to cancer cells (Jain, 1990; Epenetos *et al.*, 1986). In addition malignant cells can produce immunosuppressive factors such as prostaglandin E_2 and TGF-β that could profoundly impair the functions of immunocompetent cells, in particular, T cells (Tada *et al.*, 1991; Smyth *et al.*, 1991). Finally, the high concentrations of IgG in serum and interstitial fluids could block the interaction of tumor-bound IgG antibodies with complement and with FcγR on cellular effectors (NK cells, monocytes, and granulocytes) (Wallace *et al.*, 1995; Karpovsky *et al.*, 1984; Segal *et al.*, 1983).

BsAbs with specificity for TAA and cytotoxic triggering molecules on immune effector cells could provide an alternative approach for treating malignancies. Such bsAbs could redirect all T cells against tumors, regardless of their natural specificities, and could overcome the inhibition of FcγR-dependent cellular cytotoxic responses by non-immune IgG. In spite of problems in obtaining sufficient amounts of clinical grade bsAb, a number of phase I/II clinical trials have been initiated. The goals of these studies were to establish the maximum tolerated doses, to follow the pharmacokinetics, and to look for early signs of biological or clinical effects of bsAbs. In this section we review the current status of bsAb based clinical trials.

Retargeting of T Cells by Bispecific Antibodies

A hetero-F(ab')$_2$ with anti-CD3 and anti-glioma specificities was the first bsAb used to treat cancer patients (Nitta *et al.*, 1990). The study was carried out with 20 patients having grade III/IV malignant glioma. Taking advantage of the local growth of glioma, the investigators infused autologous PBMC, previously activated for 7–10 days with IL-2 in the presence (ten patients) or absence (ten patients) of bsAb, into the tumor site by an Ommaya reservoir that had been positioned during surgical removal of the glioma. The treatment regimen consisted of three bi-weekly infusions over a three week period. In the group of patients treated with the bsAb none relapsed in a 10–18 month follow-up period. CT scans and histological examination confirmed the clinical outcomes, showing eradication of residual glioma in four patients and tumor regression in four others. By contrast, in the group treated with autologous activated PBMC in the absence of bsAb, only two patients were alive after 2 years.

In a second study (Kroesen *et al.*, 1993), patients suffering from malignant ascites or pleural effusions resulting from a variety of solid tumors (colon, breast, and gastric carcinoma) were treated locally with *ex-vivo* activated autologous lymphocytes and a bsAb with specificity for CD3 and a pan-TAA. Local anti-tumor effects were observed early after treatment. Analyses of effusion samples showed the presence of conjugates between tumors cells and lymphocytes, which was followed by decreased numbers of the tumor cells and the normalization of carcinoembryonic antigen levels in the neoplastic effusions.

In a third study, 28 ovarian cancer patients were treated in a phase I/II clinical trial with intraperitoneal infusions of autologous T lymphocytes retargeted by a bsAb (Canevari *et al.*, 1995). The bsAb was a $F(ab')_2$ fragment from a mouse hybrid-hybridoma having specificity for CD3 on T cells and for a folate binding protein, which is over expressed on ovarian carcinoma cells. The patients were treated with two cycles of five intraperitoneal infusions of autologous peripheral T cells previously activated *in vitro* with IL-2 and retargeted with bsAb. The treatment gave only mild toxic effects consisting of fever, nausea, emesis, and fatigue, but anti-bsAb antibodies were detected at the end of treatment in twenty-two out of twenty-five patients. Of 26 patients evaluated, 27% responded to therapy, with a mean survival time of greater than 22 months. Nineteen patients were evaluated by surgery and histology: three patients achieved complete responses, one showed a complete intraperitoneal response with progressive disease in retroperitoneal lymph nodes, three gave partial responses, seven had stable disease, and five had progressive disease. This study clearly established that locoregional treatment of at least one type of solid tumor with activated autologous T cells targeted with a bsAb could lead to the eradication of tumors. However, tumors that were inaccessible to treatment showed progressive growth. An initial attempt to administer the anti-CD3 × anti-folate receptor bsAb systemically resulted in severe toxic symptoms in one patient (Tibben *et al.*, 1993). Thirty minutes after intravenous infusion of 1 mg of bsAb, the patient developed fever, chills, headache, hypotension, and fatigue, all of which disappeared after the treatment was stopped. The adverse effect of the treatment paralleled an increase in the serum concentration of TNF. The authors suggested that the toxicity was an effect of the monovalent binding of the bsAb to CD3 which triggered the systemic activation of T cells. Alternatively it is also possible that TNF release was induced by the bsAb binding to low levels of folate binding protein on normal cells, with concomitant crosslinking of TcR on all T cells.

Targeting Through Fcγ Receptors

Clinical studies that target effector cells through FcγR have been made possible by the development of two different bsAbs, MDX-210 (Wallace *et al.*, 1995) and 2B1 (Hsieh-Ma *et al.*, 1992). Both bsAbs recognize the HER-2/*neu* oncogene product that is over expressed on breast, ovarian, lung, and colon cancer cells. MDX-210 is a chemically crosslinked hetero-$F(ab')_2$ that has specificity for FcγRI (CD64) present on monocytes/macrophages and activated neutrophils, whereas 2B1 is a hybrid-hybridoma bsAb that binds FcγRIII (CD16) on NK cells, macrophages, and neutrophils.

In a phase I clinical study (Valone *et al.*, 1995) fifteen patients were treated with a single infusion of MDX-210 at doses ranging from 0.35 to 10 mg/m^2. The patients, who were suffering from stage III and IV metastatic breast and ovarian cancers that over expressed the HER-2/neu product, tolerated the treatment with only 1–2 grade toxicity, including fever, malaise, and hypotension. Transient monocytopenia and lymphopenia developed at 1–2 hours after treatment with MDX-210. Only two patients developed grade 3 hypotension, which resolved in a few hours after fluid infusion. Doses of MDX-210 at 7 mg/m^2 and higher induced elevated plasma doses of TNF-α, IL-6, G-CSF, and neopterin. Localization of MDX-210 in the tumor was observed in two patients. Among 10 patients assessed for clinical responses, two showed some effects of treatment; a breast cancer patient resolved subcutaneous nodules, and an ovarian cancer patient showed a substantial decrease in cervical adenopathy. In view of these early promising results, MDX-210 based studies have been expanded to several centers to study the effects of different dosing regimens and co-administration of cytokines on the efficacy of treatment.

Recently, the results of a phase I clinical trial in which 15 patients with advanced HER-2/neu positive tumors were treated with the 2B1 bsAb have been reported (Weiner *et al.*, 1995). Patients received infusions of 2B1 on days 1, 4, 5, 6, 7, and 8 at doses ranging from 1 to 5 mg/m^2. Thrombocytopenia was dose limiting in two patients at 5 mg/m^2 who had received extensive previous chemotherapy. The 2B1 antibody retained activity in serum and was cleared with a half life of about 20 hr. The initial treatment was accompanied by increases in serum levels of a number of cytokines, and several minor clinical responses such as reductions in tumor mass and resolution of pleural effusions and ascites.

Conclusions and Future Perspectives

Although all clinical trials using bsAbs to target cellular cytotoxicity have, to date, given biological responses, it is safe to say that we still have many obstacles to overcome before bsAbs can be used effectively. Obtaining sufficient quantities of bsAb suitable for clinical trials remains a major problem. What type of bsAb would be most effective – a small construct that would have improved tumor penetration, or a large construct, that would have a prolonged serum half life? All clinical trials have encountered anti-bsAb responses, limiting the efficacy of treatment to short periods of time. Would a totally "humanized" construct allow more prolonged treatment? Once a bsAb has become available, then the optimal routes of administration, doses, and schedules must be established. Would local or systemic administration or bsAb be superior in producing an anti-tumor response? And would the effector cells require activation? The answers to these and other questions will require extensive experimentation, but the mere fact that so many options exist gives one hope that ways will be found to improve the effectiveness of bsAb therapy in treating cancers, and perhaps other diseases as well.

REFERENCES

Akerley, W. L., Guyre, P. M. and Davis, B. H. (1991) Neutrophil activation through high-affinity Fc gamma receptor using a monomeric antibody with unique properties. *Blood* **77**, 607–615.

Allen, J. M. and Seed, B. (1989) Isolation and expression of functional high-affinity Fc receptor complementary DNAs. *Science* **243**, 378–381.

Anderson, C. L., Guyre, P. M., Whitin, J. C., Ryan, D. H., Looney, R. J. and Fanger, M. W. (1986) Monoclonal antibodies to Fc receptors for IgG on human mononuclear phagocytes. Antibody characterization and induction of superoxide production in a monocyte cell line. *J. Biol. Chem.* **261**, 12856–12864.

Anderson, C. L., Shen, L., Eicher, D. M., Wewers, M. D. and Gill, J. K. (1990) Phagocytosis mediated by three distinct Fc gamma receptor classes on human leukocytes. *J. Exp. Med.* **171**, 1333–1345.

Andrew, S. M. and Titus, J. A. (1992) Purification and fragmentation of antibodies. In *Current Protocols in Immunology* (Edited by Coligan, J. E., Kruisbeek, A. M., Margulies, D. H., Shevach, E. M. and Strober, W.), p. 2.7.1. Greene Publishing Associates and John Wiley & Sons, New York.

Arulanandam, A. R., Withka, J. M., Wyss, D. F., Wagner, G., Kister, A., Pallai, P., Recny, M. A. and Reinherz, E. L. (1993) The CD58 (LFA-3) binding site is a localized and highly charged surface area on the AGFCC'C face of the human CD2 adhesion domain. *Proc. Natl. Acad. Sci. USA* **90**, 11613–11617.

Arulanandam, A. R., Kister, A., McGregor, M. J., Wyss, D. F., Wagner, G. and Reinherz, E. L. (1994) Interaction between human CD2 and CD58 involves the major beta sheet surface of each of their respective adhesion domains. *J. Exp. Med.* **180**, 1861–1871.

Azuma, M., Cayabyab, M., Buck, D., Phillips, J. H. and Lanier, L. L. (1992) CD28 interaction with B7 costimulates primary allogeneic proliferative responses and cytotoxicity mediated by small, resting T lymphocytes. *J. Exp. Med.* **175**, 353–360.

Backman, K. A. and Guyre, P. M. (1994) Gamma-interferon inhibits Fc receptor II-mediated phagocytosis of tumor cells by human macrophages. *Cancer Res.* **54**, 2456–2461.

Bakacs, T., Lee, J., Moreno, M. B., Zacharchuk, C. M., Cole, M. S., Tso, J. Y., Paik, C. H., Ward, J. M. and Segal, D. M. (1995) A bispecific antibody prolongs survival in mice bearing lung metastases of syngeneic mammary adenocarcinoma. *Int. Immunol.* **7**, 947–955.

Bell, G. M., Fargnoli, J., Bolen, J. B., Kish, L. and Imboden, J. B. (1996) The SH3 domain of p56lck binds to proline-rich sequences in the cytoplasmic domain of CD2. *J. Exp. Med.* **183**, 169–178.

Bell, G. M. and Imboden, J. B. (1995) CD2 and the regulation of T cell anergy. *J. Immunol.* **155**, 2805–2807.

Beun, G. D., van Eendenburg, D. H., Corver, W. E., van de Velde, C. J. and Fleuren, G. J. (1992) T-cell retargeting using bispecific monoclonal antibodies in a rat colon carcinoma model. I. Significant bispecific lysis of syngeneic colon carcinoma CC531 is critically dependent on prolonged preactivation of effector T-lymphocytes by immobilized anti-T-cell receptor antibody. *J. Immunother.* **11**, 238–248.

Beun, G. D., van de Velde, C. J. and Fleuren, G. J. (1994) T-cell based cancer immunotherapy: direct or redirected tumor-cell recognition? *Immunol. Today* **15**, 11–15.

Beyers, A. D., Barclay, A. N., Law, D. A., He, Q. and Williams, A. F. (1989) Activation of T lymphocytes via monoclonal antibodies against rat cell surface antigens with particular reference to CD2 antigen. *Immunol. Rev.* **111**, 59–77.

Bird, R. E., Hardman, K. D., Jacobson, J. W., Johnson, S., Kaufman, B. M., Lee, S. M., Lee, T., Pope, S. H., Riordan, G. S. and Whitlow, M. (1988) Single-chain antigen-binding proteins. *Science* **242**, 423–426.

Bolhuis, R. L. H., Roozemond, R. C. and van de Griend, R. J. (1986) Induction and blocking of cytolysis in CD2+, CD3− NK and CD2+, CD3+ cytotoxic lymphocytes via CD2 50 KD sheep erythrocyte receptor. *J. Immunol.* **136**, 3939–3944.

Borrego, F., Pena, J. and Solana, R. (1993) Regulation of CD69 expression on human natural killer cells: differential involvement of protein kinase C and protein tyrosine kinases. *Eur. J. Immunol.* **23**, 1039–1043.

Borst, J., van Dongen, J. J. M., Bolhuis, R. L. H., Peters, P. J., Hafler, D. A., de Vries, E. and van de Griend, R. J. (1988) Distinct molecular forms of human T cell receptor gamma/delta detected on viable T cells by a monoclonal antibody. *J. Exp. Med.* **167**, 1625–1644.

Brennan, M., Davison, P. F. and Paulus, H. (1985) Preparation of bispecific antibodies by chemical recombination of monoclonal immunoglobulin G1 fragments. *Science* **229**, 81–83.

Brissinck, J., Demanet, C., Moser, M., Leo, O. and Thielemans, K. (1991) Treatment of mice bearing BCL_1 lymphoma with bispecific antibodies. *J. Immunol.* **147**, 4019–4026.

Brissinck, J., Demanet, C., Moser, M., Leo, O. and Thielemans, K. (1993) Bispecific antibodies in lymphoma. *Int. Rev. Immunol.* **10**, 187–194.

Brooks, D. G., Qiu, W. Q., Luster, A. D. and Ravetch, J. V. (1989) Structure and expression of human IgG FcRII(CD32). Functional heterogeneity is encoded by the alternatively spliced products of multiple genes. *J. Exp. Med.* **170**, 1369–1385.

Canevari, S., Stoter, G., Arienti, F., Bolis, G., Colnaghi, M. I., Di Re, E. M., Eggermont, A. M. M., Goey, S. H., Gratama, J. W., Lamers, C. H. J., Nooy, M. A., Parmiani, G., Raspagliesi, F., Ravagnani, F., Scarfone, G., Trimbos, J. B., Warnaar, S. O. and Bolhuis, R. L. H. (1995) Regression of advanced ovarian carcinoma by intraperitoneal treatment with autologous T lymphocytes retargeted by a bispecific monoclonal antibody. *J. Natl. Cancer Inst.* **87**, 1463–1469.

Carlson, J., Drevin, H. and Aven, R. (1978) Protein thiolation and reversible protein–protein conjugation. N-succinimidyl 3-(2-pyridyldithiol)propionate, a new heterobifunctional reagent. *Biochem. J.* **173**, 723–737.

Carter, P., Kelley, R. F., Rodrigues, M. L., Snedecor, B., Covarrubias, M., Velligan, M. D., Wong, W. L. T., Rowland, A. M., Kotts, C. E., Carver, M. E., Yang, M., Bourell, J. H., Shepard, H. M. and Henner, D. (1992) High level escherichia coli expression and production of a bivalent humanized antibody fragment. *Bio/Technology* **10**, 163–167.

Clark, J. I., Alpaugh, R. K. and Weiner, L. M. (1995) Natural killer cell-directed bispecific antibodies. In *Bispecific antibodies* (Edited by Fanger, M. W.), p. 77. R. G. Landes Company, Austin.

Clark, J. I. and Weiner, L. M. (1995) Biologic treatment of human cancer. *Curr. Probl. Cancer* **19**, 185–262.

Clark, M., Gilliland, L. and Waldmann, H. (1988) Hybrid antibodies for therapy. *Prog. Allergy.* **45**, 31–49.

Clark, M. R. and Waldmann, H. (1987) T-cell killing of target cells induced by hybrid antibodies: comparison of two bispecific monoclonal antibodies. *J. Natl. Cancer Inst.* **79**, 1393–1401.

Cosulich, M. E., Risso, A., Smilovich, D., Baldissarro, I., Capra, M. C. and Bargellesi, A. (1992) CD69 activation molecule: requirements for its expression on T cells. *Pharmacol. Res.* **26 Suppl 2**, 136–138.

Crockett-Torabi, E. and Fantone, J. C. (1990) Soluble and insoluble immune complexes activate human neutrophil NADPH oxidase by distinct Fc gamma receptor-specific mechanisms. *J. Immunol.* **145**, 3026–3032.

Culty, M., Miyake, K., Kincade, P. W., Sikorski, E., Butcher, E. C., Underhill, C. and Silorski, E. S. (1990) The hyaluronate receptor is a member of the CD44 (H-CAM) family of cell surface glycoproteins. *J. Cell Biol.* **111**, 2765–2774.

Davis, S. J., Jones, E. Y., Bodian, D. L., Barclay, A. N. and van der Merwe, P. A. (1993) Analysis of the structure and interactions of CD2. *Biochem. Soc. Trans.* **21**, 952–958.

De Lau, W. B. M., Heije, K., Neefjes, J. J., Oosterwegel, M., Rozemuller, E. and Bast, B. J. E. G. (1991) Absence of preferential homologous H/L chain association in hybrid hybridomas. *J. Immunol.* **146**, 906–914.

De Maria, R., Cifone, M. G., Trotta, R., Rippo, M. R., Festuccia, C., Santoni, A. and Testi, R. (1994) Triggering of human monocyte activation through CD69, a member of the natural killer cell gene complex family of signal transducing receptors. *J. Exp. Med.* **180**, 1999–2004.

Debets, J. M., Van de Winkel, J. G., Ceuppens, J. L., Dieteren, I. E. and Buurman, W. A. (1990) Cross-linking of both Fc gamma RI and Fc gamma RII induces secretion of tumor necrosis factor by human monocytes, requiring high affinity Fc-Fc gamma R interactions. Functional activation of Fc gamma RII by treatment with proteases or neuraminidase. *J. Immunol.* **144**, 1304–1310.

Demanet, C., Brissinck, J., Van Mechelen, M., Leo, O. and Thielemans, K. (1991) Treatment of murine B cell lymphoma with bispecific monoclonal antibodies (anti-idiotype × anti-CD3). *J. Immunol.* **147**, 1091–1097.

Demanet, C., Brissinck, J., Moser, M., Leo, O. and Thielemans, K. (1992) Bispecific antibody therapy of two murine B-cell lymphomas. *Int. J. Cancer Suppl.* **7**, 67–68.

Demanet, C., Brissinck, J., Leo, O., Moser, M. and Thielemans, K. (1994) Role of T-cell subsets in the bispecific antibody (anti-idiotype × anti-CD3) treatment of the BCL_1 lymphoma. *Cancer Res.* **54**, 2973–2978.

Dennert, G. and Podack, E. R. (1983) Cytolysis by H-2-specific T killer cells. Assembly of tubular complexes on target membranes. *J. Exp. Med.* **157**, 1483–1495.

Dhein, J., Daniel, P. T., Trauth, B. C., Oehm, A., Moller, P. and Krammer, P. H. (1992) Induction of apoptosis by monoclonal antibody anti-APO-1 class switch variants is dependent on cross-linking of APO-1 cell surface antigens. *J. Immunol.* **149**, 3166–3173.

Dillman, R. O. (1989) Monoclonal antibodies for treating cancer. *Ann. Int. Med.* **111**, 592–603.

Dorai, H., McCartney, J. E., Hudziak, R. M., Tai, M.-S., Laminer, A., Houston, L. L., Huston, J. S. and Oppermann, H. (1994) Mammalian cell expression of single-chain Fv (sFv) antibody proteins and their C-terminal fusions with interleukin-2 and other effector domains. *Bio/Technology* **12**, 890–897.

Dustin, M. L., Sanders, M. E., Shaw, S. and Springer, T. A. (1987) Purified lymphocyte function-associated antigen 3 binds to CD2 and mediates T lymphocyte adhesion. *J. Exp. Med.* **165**, 677–692.

Edberg, J. C., Redecha, P. B., Salmon, J. E. and Kimberly, R. P. (1989) Human Fc gamma RIII (CD16). Isoforms with distinct allelic expression, extracellular domains, and membrane linkages on polymorphonuclear and natural killer cells. *J. Immunol.* **143**, 1642–1649.

Epenetos, A. A., Snook, D., Durbin, H., Johnson, P. M. and Taylor-Papadimitriou, J. (1986) Limitations of radiolabeled monoclonal antibodies for localization of human neoplasms. *Cancer Res.* **46**, 3183–3191.

Eshhar, Z., Waks, T., Gross, G. and Schindler, D. G. (1993) Specific activation and targeting of cytotoxic lymphocytes through chimeric single chains consisting of antibody-binding domains and the gamma or zeta subunits of the immunoglobulin and T-cell receptors. *Proc. Natl. Acad. Sci. USA* **90**, 720–724.

Ettinghausen, S. E. and Rosenberg, S. A. (1993) Clinical trials of immunotherapy of cancer utilizing cytotoxic cells. In *Cytotoxic cells, recognition, effector function, generation, and methods* (Edited by Sitkovsky, M. V. and Henkart, P. A.), p. 407. Birkhauser, Boston.

Fanger, M. W., Graziano, R. F., Shen, L. and Guyre, P. M. (1989a) Fc gamma R cytotoxicity exerted by mononuclear cells. *Chem. Immunol.* **47**, 214–253.

Fanger, M. W., Shen, L., Graziano, R. F. and Guyre, P. M. (1989b) Cytotoxicity mediated by human Fc receptors for IgG. *Immunol. Today.* **10**, 92–99.

Fanger, M. W. (1995) *Bispecific antibodies.* R. G. Landes Co. Austin.

Fleit, H. B., Wright, S. D. and Unkeless, J. C. (1982) Human neutrophil Fc gamma receptor distribution and structure. *Proc. Natl. Acad. Sci. USA* **79**, 3275–3279.

Frank, S. J., Samelson, L. E. and Klausner, R. D. (1990) The structure and signalling functions of the invariant T cell receptor components. *Semin. Immunol.* **2**, 89–97.

Galandrini, R., Albi, N., Tripodi, G., Zarcone, D., Terenzi, A., Moretta, A., Grossi, C. E. and Velardi, A. (1993) Antibodies to CD44 trigger effector functions of human T cell clones. *J. Immunol.* **150**, 4225–4235.

Garrido, M. A., Perez, P., Titus, J. A., Valdayo, M. J., Winkler, D. A., Barbieri, S. A., Wunderlich, J. R. and Segal, D. M. (1990a) Targeted cytotoxic cells in human peripheral blood lymphocytes. *J. Immunol.* **144**, 2891–2898.

Garrido, M. A., Valdayo, M. J., Winkler, D. F., Titus, J. A., Hecht, T. T., Perez, P., Segal, D. M. and Wunderlich, J. R. (1990b) Targeting human T lymphocytes with bispecific antibodies to react against human ovarian carcinoma cells in nu/nu mice. *Cancer Res.* **50**, 4227–4232.

Gavioli, R., Risso, A., Smilovich, D., Baldissarro, I., Capra, M. C., Bargellesi, A. and Cosulich, M. E. (1992) CD69 molecule in human neutrophils: its expression and role in signal-transducing mechanisms. *Cell Immunol.* **142**, 186–196.

Glennie, M. J., McBride, H. M., Worth, A. T. and Stevenson, G. T. (1987) Preparation and performance of bispecific $F(ab')_2$ antibody containing thioether-linked F(ab') gamma fragments. *J. Immunol.* **139**, 2367–2375.

Gosselin, E. J., Wardwell, K., Gosselin, D. R., Alter, N., Fisher, J. L. and Guyre, P. M. (1992) Enhanced antigen presentation using human Fc gamma receptor (monocyte/macrophage)-specific immunogens. *J. Immunol.* **149**, 3477–3481.

Graziano, R. F. and Fanger, M. W. (1987a) Fc gamma RI and Fc gammaRII on monocytes and granulocytes are cytotoxic trigger molecules for tumor cells. *J. Immunol.* **139**, 3536–3541.

Graziano, R. F. and Fanger, M. W. (1987b) Human monocyte-mediated cytotoxicity: the use of Ig-bearing hybridomas as target cells to detect trigger molecules on the monocyte cell surface. *J. Immunol.* **138**, 945–950.

Gruber, M., Schodin, B. A., Wilson, E. R. and Kranz, D. M. (1994) Efficient tumor cell lysis mediated by a bispecific single chain antibody expressed in *Escherichia coli*. *J. Immunol.* **152**, 5368–5374.

Guckel, B., Berek, C., Lutz, M., Altevogt, P., Schirrmacher, V. and Kyewski, B. A. (1991) Anti-CD2 antibodies induce T cell unresponsiveness *in vivo*. *J. Exp. Med.* **174**, 957–967.

Guyre, P. M., Graziano, R. F., Vance, B. A., Morganelli, P. M. and Fanger, M. W. (1989) Monoclonal antibodies that bind to distinct epitopes on Fc gamma RI are able to trigger receptor function. *J. Immunol.* **143**, 1650–1655.

Hahn, W. C., Menu, E., Bothwell, A. L., Sims, P. J. and Bierer, B. E. (1992) Overlapping but nonidentical binding sites on CD2 for CD58 and a second ligand CD59. *Science* **256**, 1805–1807.

Hartnell, A., Kay, A. B. and Wardlaw, A. J. (1992) IFN-gamma induces expression of Fc gamma RIII (CD16) on human eosinophils. *J. Immunol.* **148**, 1471–1478.

Hartnell, A., Robinson, D. S., Kay, A. B. and Wardlaw, A. J. (1993) CD69 is expressed by human eosinophils activated *in vivo* in asthma and *in vitro* by cytokines. *Immunology* **80**, 281–286.

Hayden, M. S., Linsley, P. S., Gayle, M. A., Bajorath, J., Brady, W. A., Norris, N. A., Fell, H. P., Ledbetter, J. A. and Gilliland, L. K. (1994) Single-chain mono- and bispecific antibody derivatives with novel biological properties and antitumour activity from a COS cell transient expression system. *Therapeutic Immunol.* **1**, 3–15.

Heijnen, I. A., van Vugt, M. J., Fanger, N. A., Graziano, R. F., de Wit, T. P., Hofhuis, F. M., Guyre, P. M., Capel, P. J., Verbeek, J. S. and Van de Winkel, J. G. (1996) Antigen targeting to myeloid-specific human Fc gamma RI/CD64 triggers enhanced antibody responses in transgenic mice. *J. Clin. Invest.* **97**, 331–338.

Hibbs, M. L., Bonadonna, L., Scott, B. M., McKenzie, I. F. C. and Hogarth, P. M. (1988) Molecular cloning of a human immunoglobulin G Fc receptor. *Proc. Natl. Acad. Sci. USA* **85**, 2240–2244.

Hochman, J., Inbar, D. and Givol, D. (1973) An active antibody fragment (Fv) composed of the variable portions of heavy and light chains. *Biochemistry* **12**, 1130–1135.

Hoffman, R. W., Bluestone, J. A., Leo, O. and Shaw, S. (1985) Lysis of anti-T3-bearing murine hybridoma cells by human allospecific cytotoxic T cell clones and inhibition of that lysis by anti-T3 and anti-LFA-1 antibodies. *J. Immunol.* **135**, 5–8.

Holliger, P., Prospero, T. and Winter, G. (1993) "Diabodies": Small bivalent and bispecific antibody fragments. *Proc. Natl. Acad. Sci. USA* **90**, 6444–6448.

Howard, F. D., Moingeon, P., Moebius, U., McConkey, D. J., Yandava, B., Gennert, T. E. and Reinherz, E. L. (1992) The CD3 zeta cytoplasmic domain mediates CD2-induced T cell activation. *J. Exp. Med.* **176**, 139–145.

Hsieh-Ma, S. T., Eaton, A. M., Shi, T. and Ring, D. B. (1992) *In vitro* cytotoxic targeting by human mononuclear cells and bispecific antibody 2B1, recognizing c-*erb*B-2 protooncogene product and Fcgamma receptor III. *Cancer Res.* **52**, 6832–6839.

Huizinga, T. W., Dolman, K. M., van der Linden, N. J., Kleijer, M., Nuijens, J. H., von dem Borne, A. E. and Roos, D. (1990) Phosphatidylinositol-linked FcRIII mediates exocytosis of neutrophil granule proteins, but does not mediate initiation of the respiratory burst. *J. Immunol.* **144**, 1432–1437.

Hundt, M. and Schmidt, R. E. (1992) The glycosylphosphatidylinositol-linked Fc gamma receptor III represents the dominant receptor structure for immune complex activation of neutrophils. *Eur. J. Immunol.* **22**, 811–816.

Huston, J. S., Levinson, D., Mudgett-Hunter, M., Tai, M. S., Novotny, J., Margolies, M. N., Ridge, R. J., Bruccoleri, R. E., Haber, E., Crea, R. and Oppermann, H. (1988) Protein engineering of antibody binding sites: recovery of specific activity in an anti-digoxin single-chain Fv analogue produced in *Escherichia coli*. *Proc. Natl. Acad. Sci. USA* **85**, 5879–5883.

Huston, J. S., McCartney, J., Tai, M.-S., Mottola-Hartshorn, C., Jin, D., Warren, F., Keck, P. and Oppermann, H. (1993) Medical applications of single-chain antibodies. *Intern. Rev. Immunol.* **10**, 195–217.

Hwu, P., Shafer, G. E., Treisman, J., Schindler, D. G., Gross, G., Cowherd, R., Rosenberg, S. A. and Eshhar, Z. (1993) Lysis of ovarian cancer cells by human lymphocytes redirected with a chimeric gene composed of an antibody variable region and the Fc receptor gamma chain. *J. Exp. Med.* **178**, 361–366.

Irving, B. A. and Weiss, A. (1991) The cytoplasmic domain of the T cell receptor zeta chain is sufficient to couple to receptor-associated signal transduction pathways. *Cell* **64**, 891–901.

Ishiura, S., Matsuda, K., Koizumi, H., Tsukahara, T., Arahata, K. and Sugita, H. (1990) Calcium is essential for both the membrane binding and lytic activity of pore-forming protein (perforin) from cytotoxic T-lymphocyte. *Mol. Immunol.* **27**, 803–807.

Jain, R. K. (1990) Physiological barriers to delivery of monoclonal antibodies and other macromolecules in tumors. *Cancer Res.* **50**, 814s–819s.

Jost, C. R., Titus, J. A., Kurucz, I. and Segal, D. M. (1996) A single-chain bispecific Fv2 molecule produced in mammalian cells redirects lysis by activated CTL. *Mol. Immunol.* **33**, 211–219.

Jung, G., Ledbetter, J. A. and Muller-Eberhard, H. J. (1987) Induction of cytotoxicity in resting human T lymphocytes bound to tumor cells by antibody heteroconjugates. *Proc. Natl. Acad. Sci. USA* **84**, 4611–4615.

Jung, G., Freimann, U., Von Marschall, Z., Reisfeld, R. A. and Wilmanns, W. (1991) Target cell-induced T cell activation with bi- and trispecific antibody fragments. *Eur. J. Immunol.* **21**, 2431–2435.

Kagi, D., Ledermann, B., Burki, K., Seiler, P., Odermatt, B., Olsen, K. J., Podack, E. R., Zinkernagel, R. M. and Hengartner, H. (1994a) Cytotoxicity mediated by T cells and natural killer cells is greatly impaired in perforin-deficient mice. *Nature* **369**, 31–37.

Kagi, D., Vignaux, F., Ledermann, B., Burki, K., Depraetere, V., Nagata, S., Hengartner, H. and Golstein, P. (1994b) Fas and perforin pathways as major mechanisms of T cell-mediated cytotoxicity. *Science* **265**, 528–530.

Karawajew, L., Micheel, B., Behrsing, O. and Gaestel, M. (1987) Bispecific antibody-producing hybrid hybridomas selected by a fluorescence activated cell sorter. *J. Immunol. Methods* **96**, 265–270.

Karawajew, L., Rudchenko, S., Wlasik, T., Trakht, I. and Rakitskaya, V. (1990) Flow sorting of hybrid hybridomas using the DNA stain Hoechst 33342. *J. Immunol. Methods* **129**, 277–282.

Karpovsky, B., Titus, J. A., Stephany, D. A. and Segal, D. M. (1984) Production of target-specific effector cells using hetero-cross-linked aggregates containing anti-target cell and anti-Fc gamma receptor antibodies. *J. Exp. Med.* **160**, 1686–1701.

Kojima, H., Shinohara, N., Hanaoka, S., Someya-Shirota, Y., Takagaki, Y., Ohno, H., Saito, T., Katayama, T., Yagita, H., Okumura, K., Shinkai, Y., Alt, F. A., Matsuzawa, A., Yonehara, S. and Takayama, H. (1994) Two distinct pathways of specific killing revealed by perforin mutant cytotoxic T lymphocytes. *Immunity* **1**, 357–364.

Koolwijk, P., Spierenburg, G. T., Frasa, H., Boot, J. H., Van de Winkel, J. G. and Bast, B. J. (1989) Interaction between hybrid mouse monoclonal antibodies and the human high-affinity IgG FcR, huFc gamma RI, on U937. Involvement of only one of the mIgG heavy chains in receptor binding. *J. Immunol.* **143**, 1656–1662.

Kostelny, S. A., Cole, M. S. and Tso, J. Y. (1992) Formation of a bispecific antibody by the use of leucine zippers. *J. Immunol.* **148**, 1547–1553.

Kranz, D. M., Tonegawa, S. and Eisen, H. N. (1984) Attachment of an anti-receptor antibody to non-target cells renders them susceptible to lysis by a clone of cytotoxic T lymphocytes. *Proc. Natl. Acad. Sci. USA* **81**, 7922–7926.

Kroesen, B. J., Ter Haar, A., Spakman, H., Willemse, P., Sleijfer, D. T., De Vries, E. G. E., Mulder, N. H., Berendsen, H. H., Limburg, P. C., The, T. H. and De Leij, L. (1993) Local antitumour treatment in carcinoma patients with bispecific-monoclonal-antibody-redirected T cells. *Cancer Immunol. Immunother.* **37**, 400–407.

Kroesen, B. J., Helfrich, W., Bakker, A., Wubbena, A. S., Bakker, H., Kal, H. B., The, T. H. and De Leij, L. (1995) Reduction of EGP-2-positive pulmonary metastases by bispecific-antibody-redirected T cells in an immunocompetent rat model. *Int. J. Cancer* **61**, 812–818.

Kurucz, I., Titus, J. A., Jost, C. R., Jacobus, C. M. and Segal, D. M. (1995) Retargeting of CTL by an efficiently refolded bispecific single-chain Fv dimer produced in bacteria. *J. Immunol.* **154**, 4576–4582.

Lamers, C. H., van de Griend, R. J., Braakman, E., Ronteltap, C. P., Benard, J., Stoter, G., Gratama, J. W. and Bolhuis, R. L. (1992) Optimization of culture conditions for activation and large-scale expansion of human T lymphocytes for bispecific antibody-directed cellular immunotherapy. *Int. J. Cancer* **51**, 973–979.

Lancki, D. W., Ma, D. I., Havran, W. L. and Fitch, F. W. (1984) Cell surface structures involved in T cell activation. *Immunol. Rev.* **81**, 65–94.

Lanier, L. L., Buck, D. W., Rhodes, L., Ding, A., Evans, E., Barney, C. and Phillips, J. H. (1988a) Interleukin 2 activation of natural killer cells rapidly induces the expression and phosphorylation of the Leu-23 activation antigen. *J. Exp. Med.* **167**, 1572–1585.

Lanier, L. L., Ruitenberg, J. J. and Phillips, J. H. (1988b) Functional and biochemical analysis of CD16 antigen on natural killer cells and granulocytes. *J. Immunol.* **141**, 3478–3485.

Lanier, L. L., Cwirla, S., Yu, G., Testi, R. and Phillips, J. H. (1989) Membrane anchoring of a human IgG Fc receptor (CD16) determined by a single amino acid. *Science* **246**, 1611–1613.

Lanzavecchia, A. (1990) Receptor-mediated antigen uptake and its effect on antigen presentation to class II-restricted T lymphocytes. *Annu. Rev. Immunol.* **8**, 773–793.

Lanzavecchia, A. and Scheidegger, D. (1987) The use of hybrid hybridomas to target human cytotoxic T lymphocytes. *Eur. J. Immunol.* **17**, 105–111.

Leeuwenberg, J. T. M., Spits, H., Tax, W. J. M. and Capel, P. J. A. (1985) Induction of nonspecific cytotoxicity by monoclonal anti-T3 antibodies. *J. Immunol.* **134**, 3770–3775.

Leo, O., Foo, M., Sachs, D. H., Samelson, L. E. and Bluestone, J. A. (1987a) Identification of a monoclonal antibody specific for a murine T3 polypeptide. *Proc. Natl. Acad. Sci. USA* **84**, 1374–1378.

Leo, O., Foo, M., Segal, D. M., Shevach, E. and Bluestone, J. A. (1987b) Activation of murine T lymphocytes with monoclonal antibodies: Detection on Lyt2+ cells of an antigen not associated with the T cell receptor complex but involved in T cell activation. *J. Immunol.* **139**, 1214–1222.

Lesley, J., Hyman, R. and Kincade, P. W. (1993) CD44 and its interaction with extracellular matrix. *Adv. Immunol.* **54**, 271–335.

Lichtenheld, M. G., Olsen, K. J., Lu, P., Lowrey, D. M., Hameed, A., Hengartner, H. and Podack, E. R. (1988) Structure and function of human perforin. *Nature* **335**, 448–451.

Liu, C. and Guyre, P. M. (1995) Antigen targeting. In *Bispecific antibodies* (Edited by Fanger, M. W.), p. 133. R.G. Landes Company, Austin.

Liu, M. A., Kranz, D. M., Kurnick, J. T., Boyle, L. A., Levy, R. and Eisen, H. N. (1985) Heteroantibody duplexes target cells for lysis by cytotoxic T lymphocytes. *Proc. Natl. Acad. Sci. USA* **82**, 8648–8652.

Looney, R. J., Abraham, G. N. and Anderson, C. L. (1986) Human monocytes and U937 cells bear two distinct Fc receptors for IgG. *J. Immunol.* **136**, 1641–1647.

Lynch, D. H., Watson, M. L., Alderson, M. R., Baum, P. R., Miller, R. E., Tough, T., Gibson, M., Davis-Smith, T., Smith, C. A., Hunter, K., Bhat, D., Din, W., Goodwin, R. G. and Seldin, M. F. (1994) The mouse Fas-ligand gene is mutated in gld mice and is part of a TNF family gene cluster. *Immunity* **1**, 131–136.

Malissen, B. and Schmitt-Verhulst, A. M. (1993) Transmembrane signalling through the T-cell-receptor-CD3 complex. *Curr. Opin. Immunol.* **5**, 324–333.

Mallender, W. D. and Voss, E. W., Jr. (1994) Construction, expression, and activity of a bivalent bispecific single-chain antibody. *J. Biol. Chem.* **269**, 199–206.

Meuer, S. C., Hussey, R. E., Fabbi, M., Fox, D., Acuto, O., Fitzgerald, K. A., Hodgdon, J. C., Protentis, J. P., Schlossman, S. F. and Reinherz, E. L. (1984) An alternative pathway of T-cell activation: a functional role for the 50 kd T11 sheep erythrocyte receptor protein. *Cell* **36**, 897–906.

Mezzanzanica, D., Garrido, M. A., Neblock, D. S., Daddona, P. E., Andrew, S. M., Zurawski, V. R., Segal, D. M. and Wunderlich, J. R. (1991) Human T-lymphocytes targeted against an established ovarian carcinoma with bispecific $F(ab')_2$ antibody prolong host survival in a murine xenograft model. *Cancer Res.* **51**, 5716–5721.

Milstein, C. and Cuello, A. C. (1983) Hybrid hybridomas and their use in immunohistochemistry. *Nature* **305**, 537–540.

Milstein, C. and Cuello, A. C. (1984) Hybrid-hybridomas and production of bi-specific monoclonal antibodies. *Immunol. Today* **5**, 299–304.

Moreno, M. B., Titus, J. A., Cole, M. S., Tso, J. Y., Le, N., Paik, C. H., Bakács, T., Zacharchuk, C. M., Segal, D. M. and Wunderlich, J. R. (1995) Bispecific antibodies retarget murine T cell cytotoxicity against syngeneic breast cancer *in vitro* and *in vivo*. *Cancer Immunol. Immunother.* **40**, 182–190.

Moretta, A., Poggi, A., Pende, D., Tripodi, G., Orengo, A. M., Pella, N., Augugliaro, R., Bottino, C., Ciccone, E. and Moretta, L. (1991) CD69-mediated pathway of lymphocyte activation: anti-CD69 monoclonal antibodies trigger the cytolytic activity of different lymphoid effector cells with the exception of cytolytic T lymphocytes expressing T cell receptor α/β. *J. Exp. Med.* **174**, 1393–1398.

Munn, D. H., McBride, M. and Cheung, N. K. (1991) Role of low-affinity Fc receptors in antibody-dependent tumor cell phagocytosis by human monocyte-derived macrophages. *Cancer Res.* **51**, 1117–1123.
Nakamura, S., Sung, S. S., Bjorndahl, J. M. and Fu, S. M. (1989) Human T cell activation. IV. T cell activation and proliferation via the early activation antigen EA 1. *J. Exp. Med.* **169**, 677–689.
Nitta, T., Sato, K., Yagita, H., Okumura, K. and Ishii, S. (1990) Preliminary trial of specific targeting therapy against malignant glioma. *Lancet* **335**, 368–371.
Osman, N., Kozak, C. A., McKenzie, I. F. and Hogarth, P. M. (1992) Structure and mapping of the gene encoding mouse high affinity Fc gamma RI and chromosomal location of the human Fc gamma RI gene. *J. Immunol.* **148**, 1570–1575.
Pack, P. and Pluckthun, A. (1992) Miniantibodies: use of amphipathic helices to produce functional, flexibly linked dimeric FV fragments with high avidity in *Escherichia coli*. *Biochemistry* **31**, 1579–1584.
Peltz, G. A., Grundy, H. O., Lebo, R. V., Yssel, H., Barsh, G. S. and Moore, K. W. (1989) Human Fc gamma RIII: cloning, expression, and identification of the chromosomal locus of two Fc receptors for IgG. *Proc. Natl. Acad. Sci. USA* **86**, 1013–1017.
Penna, C., Dean, P. A. and Nelson, H. (1994) Antitumor × anti-CD3 bifunctional antibodies redirect T-cells activated *in vivo* with staphylococcal enterotoxin B to neutralize pulmonary metastases. *Cancer Res.* **54**, 2738–2743.
Perez, P., Hoffman, R. W., Shaw, S., Bluestone, J. A. and Segal, D. M. (1985) Specific targeting of cytotoxic T cells by anti-T3 linked to anti-target cell antibody. *Nature* **316**, 354–356.
Perez, P., Hoffman, R. W., Titus, J. A. and Segal, D. M. (1986) Specific targeting of human peripheral blood T cells by heteroaggregates containing anti-T3 crosslinked to anti-target cell antibodies. *J. Exp. Med.* **163**, 166–178.
Petroni, D. C., Li, S. and Guyre, P. M. (1988) Modulation of human polymorphonuclear leukocyte IgG Fc receptors and Fc receptor-mediated functions by IFN-gamma and glucocorticoids. *J. Immunol.* **140**, 3467–3472.
Pfefferkorn, L. C. and Fanger, M. W. (1989a) Transient activation of the NADPH oxidase through Fc gamma RI. Oxidase deactivation precedes internalization of cross-linked receptors. *J. Immunol.* **143**, 2640–2649.
Pfefferkorn, L. C. and Fanger, M. W. (1989b) Cross-linking of the high affinity Fc receptor for human immunoglobulin G1 triggers transient activation of NADPH oxidase activity. Continuous oxidase activation requires continuous de novo receptor cross-linking. *J. Biol. Chem.* **264**, 14112–14120.
Pluckthun, A. (1992) Mono- and bivalent antibody fragments produced in *Escherichia coli*: engineering, folding and antigen binding. *Immunol. Rev.* **130**, 151–188.
Podack, E. R., Hengartner, H. and Lichtenheld, M. G. (1991) A central role of perforin in cytolysis? *Annu. Rev. Immunol.* **9**, 129–157.
Qian, J., Titus, J. A., Andrew, S. M., Mezzanzanica, D., Garrido, M. A., Wunderlich, J. R. and Segal, D. M. (1991) Human PBL targeted with bispecific antibodies release cytokines that are essential for inhibiting tumor growth. *J. Immunol.* **146**, 3250–3256.
Qiu, W. Q., de Bruin, D., Brownstein, B. H., Pearse, R. and Ravetch, J. V. (1990) Organization of the human and mouse low-affinity Fc gamma R genes: duplication and recombination. *Science* **248**, 732–735.
Ravetch, J. V. and Anderson, C. L. (1990) Fc gamma R family: proteins, transcripts, and genes. In *Fc receptors and the action of antibodies* (Edited by Metzger, H.), p. 211. American Society for Microbiology, Washington, D.C.
Ravetch, J. V. and Kinet, J. P. (1991) Fc receptors. *Annu. Rev. Immunol.* **9**, 457–492.
Ravetch, J. V. and Perussia, B. (1989) Alternative membrane forms of Fc gamma RIII(CD16) on human natural killer cells and neutrophils. Cell type-specific expression of two genes that differ in single nucleotide substitutions. *J. Exp. Med.* **170**, 481–497.
Renner, C., Jung, W., Sahin, U., Denfeld, R., Pohl, C., Trumper, L., Hartmann, F., Diehl, V., Van Lier, R. and Pfreundschuh, M. (1994) Cure of xenografted human tumors by bispecific monoclonal antibodies and human T cells. *Science* **264**, 833–835.
Repp, R., Valerius, T., Sendler, A., Gramatzki, M., Iro, H., Kalden, J. R. and Platzer, E. (1991) Neutrophils express the high affinity receptor for IgG (Fc gamma RI, CD64) after *in vivo* application of recombinant human granulocyte colony-stimulating factor. *Blood* **78**, 885–889.

Romeo, C. and Seed, B. (1991) Cellular immunity to HIV activated by CD4 fused to T cell or Fc receptor polypeptides. *Cell* **64**, 1037–1046.

Rouvier, E., Luciani, M. F. and Golstein, P. (1993) Fas involvement in Ca(2+)-independent T cell-mediated cytotoxicity. *J. Exp. Med.* **177**, 195–200.

Salmeron, A., Sanchez-Madrid, F., Ursa, M. A., Fresno, M. and Alarcon, B. (1991) A conformational epitope expressed upon association of CD3-epsilon with either CD3-delta or CD3-gamma is the main target for recognition by anti-CD3 monoclonal antibodies. *J. Immunol.* **147**, 3047–3052.

Salmon, J. E., Edberg, J. C. and Kimberly, R. P. (1990) Fc gamma receptor III on human neutrophils. Allelic variants have functionally distinct capacities. *J. Clin. Invest.* **85**, 1287–1295.

Sammartino, L., Webber, L. M., Hogarth, P. M., McKenzie, I. F. and Garson, O. M. (1988) Assignment of the gene coding for human FcRII (CD32) to bands q23q24 on chromosome 1. *Immunogenetics* **28**, 380–381.

Santis, A. G., Campanero, M. R., Alonso, J. L., Tugores, A., Alonso, M. A., Yague, E., Pivel, J. P. and Sanchez-Madrid, F. (1992) Tumor necrosis factor-alpha production induced in T lymphocytes through the AIM/CD69 activation pathway. *Eur. J. Immunol.* **22**, 1253–1259.

Sconocchia, G., Titus, J. A. and Segal, D. M. (1994) CD44 is a cytotoxic triggering molecule in human peripheral blood NK cells. *J. Immunol.* **153**, 5473–5481.

Scott, C. F. J., Lambert, J. M., Kalish, R. S., Morimoto, C. and Schlossman, S. F. (1988) Human T cells can be directed to lyse tumor targets through the alternative activation/T11-E rosette receptor pathway. *J. Immunol.* **140**, 8–14.

Segal, D. M., Dower, S. K. and Titus, J. A. (1983) The role of non-immune IgG in controlling IgG-mediated effector functions. *Mol. Immunol.* **20**, 1177–1189.

Segal, D. M. (1993) SPDP crosslinking of antibodies to form heteroconjugates mediating redirected cytotoxicity. In *Cytotoxic cells; Recognition, effector function, generation, and methods* (Edited by Sitkovsky, M. V. and Henkart, P. A.), p. 485. Birkhauser, Boston.

Segal, D. M., Jost, C. R. and George, A. J. T. (1993) Targeted cellular cytotoxicity. In *Cytotoxic cells: generation, recognition, effector functions, methods* (Edited by Sitkovsky, M. V. and Henkart, P. A.), p. 96. Birkhäuser, Boston.

Segal, D. M. and Bast, B. J. E. G. (1995) Production of bispecific antibodies. In *Current protocols in immunology* (Edited by Coligan, J. E., Kruisbeek, A. M., Margulies, D. H., Shevach, E. M. and Strober, W.), p. 2.13.1. John Wiley & Sons.

Segal, D. M. and Snider, D. P. (1989) Targeting and activation of cytotoxic lymphocytes. *Chem.Immunol.* **47**, 179–213.

Seth, A., Gote, L., Nagarkatti, M. and Nagarkatti, P. S. (1991) T-cell-receptor-independent activation of cytolytic activity of cytotoxic T lymphocytes mediated through CD44 and gp90^{MEL-14}. *Proc. Natl. Acad. Sci. USA* **88**, 7877–7881.

Shalaby, M. R., Shepard, H. M., Presta, L., Rodrigues, M. L., Beverley, P. C. L., Feldmann, M. and Carter, P. (1992) Development of humanized bispecific antibodies reactive with cytotoxic lymphocytes and tumor cells overexpressing the HER2 protooncogene. *J. Exp. Med.* **175**, 217–225.

Shaw, S., Luce, G. E. G., Quinones, R., Gress, R. E., Springer, T. A. and Sanders, M. E. (1986) Two antigen-independent adhesion pathways used by human cytotoxic T-cell clones. *Nature* **323**, 262–264.

Shen, L., Guyre, P. M., Anderson, C. L. and Fanger, M. W. (1986) Heteroantibody-mediated cytotoxicity: antibody to the high affinity Fc receptor for IgG mediates cytotoxicity by human monocytes that is enhanced by interferon-gamma and is not blocked by human IgG. *J. Immunol.* **137**, 3378–3382.

Shinkai, Y., Takio, K. and Okumura, K. (1988) Homology of perforin to the ninth component of complement (C9). *Nature* **334**, 525–527.

Shiver, J. W., Su, L. and Henkart, P. A. (1992) Cytotoxicity with target DNA breakdown by rat basophilic leukemia cells expressing both cytolysin and granzyme A. *Cell* **71**, 315–322.

Shiver, J. W. and Henkart, P. A. (1991) A noncytotoxic mast cell tumor line exhibits potent IgE-dependent cytotoxicity after transfection with the cytolysin/perforin gene. *Cell* **64**, 1175–1181.

Siliciano, R. F., Pratt, J. C., Schmidt, R. E., Ritz, J. and Reinherz, E. L. (1985) Activation of cytotoxic T lymphocyte and natural killer cell function through the T11 sheep erythrocyte binding protein. *Nature* **317**, 428–429.

Smyth, M. J., Strobl, S. L., Young, H. A., Ortaldo, J. R. and Ochoa, A. C. (1991) Regulation of lymphokine-activated killer activity and pore-forming protein gene expression in human peripheral blood CD8+ T lymphocytes. Inhibition by transforming growth factor-beta. *J. Immunol.* **146**, 3289–3297.

Snider, D. P., Kaubisch, A. and Segal, D. M. (1990) Enhanced antigen immunogenicity induced by bispecific antibodies. *J. Exp. Med.* **171**, 1957–1963.

Snider, D. P. (1992) Immunization with antigen bound to bispecific antibody induces antibody that is restricted in epitope specificity and contains antiidiotype. *J. Immunol.* **148**, 1163–1170.

Snider, D. P. and Segal, D. M. (1987) Targeted antigen presentation using crosslinked antibody heteroaggregates. *J. Immunol.* **139**, 1609–1616.

Snider, D. P. and Segal, D. M. (1989) Efficiency of antigen presentation after antigen targeting to surface IgD, IgM, MHC, Fc gamma RII, and B220 molecules on murine splenic B cells. *J. Immunol.* **143**, 59–65.

Spruyt, L. L., Glennie, M. J., Beyers, A. D. and Williams, A. F. (1991) Signal transduction by the CD2 antigen in T cells and natural killer cells: requirement for expression of a functional T cell receptor or binding of antibody Fc to the Fc receptor, Fc gamma RIIIA (CD16). *J. Exp. Med.* **174**, 1407–1415.

Staerz, U. D., Kanagawa, O. and Bevan, M. J. (1985) Hybrid antibodies can target sites for attack by T cells. *Nature* **314**, 628–631.

Staerz, U. D. and Bevan, M. J. (1985) Cytotoxic T lymphocyte-mediated lysis via the Fc receptor of target cells. *Eur. J. Immunol.* **15**, 1172–1177.

Staerz, U. D. and Bevan, M. J. (1986a) Hybrid hybridoma producing a bispecific monoclonal antibody that can focus effector T-cell activity. *Proc. Natl. Acad. Sci. USA* **83**, 1453–1457.

Staerz, U. D. and Bevan, M. J. (1986b) Use of anti-receptor antibodies to focus T cell activity. *Immunol. Today* **7**, 241–245.

Stancovski, I., Schindler, D. G., Waks, T., Yarden, Y., Sela, M. and Eshhar, Z. (1993) Targeting of T lymphocytes to Neu/HER2-expressing cells using chimeric single chain Fv receptors. *J. Immunol.* **151**, 6577–6582.

Stengelin, S., Stamenkovic, I. and Seed, B. (1988) Isolation of cDNAs for two distinct human Fc receptors by ligand affinity cloning. *EMBO J.* **7**, 1053–1059.

Stuart, S. G., Trounstine, M. L., Vaux, D. J. L., Koch, T., Martens, C. L., Mellman, I. and Moore, K. W. (1987) Isolation and expression of cDNA clones encoding a human receptor for IgG (Fc gamma RII). *J. Exp. Med.* **166**, 1668–1684.

Suda, T., Okazaki, T., Naito, Y., Yokota, T., Arai, N., Ozaki, S., Nakao, K. and Nagata, S. (1995) Expression of the Fas ligand in cells of T cell lineage. *J. Immunol.* **154**, 3806–3813.

Suda, T. and Nagata, S. (1994) Purification and characterization of the Fas-ligand that induces apoptosis. *J. Exp. Med.* **179**, 873–879.

Tada, T., Ohzeki, S., Utsumi, K., Takiuchi, H., Muramatsu, M., Li, X. F., Shimizu, J., Fujiwara, H. and Hamaoka, T. (1991) Transforming growth factor-beta-induced inhibition of T cell function. Susceptibility difference in T cells of various phenotypes and functions and its relevance to immunosuppression in the tumor-bearing state. *J. Immunol.* **146**, 1077–1082.

Tanaka, M., Suda, T., Takahashi, T. and Nagata, S. (1995) Expression of the functional soluble form of human fas ligand in activated lymphocytes. *EMBO J.* **14**, 1129–1135.

Testi, R., Phillips, J. H. and Lanier, L. L. (1989) T cell activation via Leu-23 (CD69). *J. Immunol.* **143**, 1123–1128.

Testi, R., Pulcinelli, F., Frati, L., Gazzaniga, P. P. and Santoni, A. (1990) CD69 is expressed on platelets and mediates platelet activation and aggregation. *J. Exp. Med.* **172**, 701–707.

Tibben, J. G., Boerman, O. C., Claessens, R. A. M. J., Corstens, F. H. M., Van Deuren, M., De Mulder, P. H. M., Van der Meer, J. W. M., Keijser, K. G. G. and Massuger, L. F. A. G. (1993) Cytokine release in an ovarian cancer patient following intravenous administration of bispecific antibody OC/TR $F(ab')_2$. *J. Natl. Cancer Inst.* **85**, 1003–1004.

Titus, J. A., Garrido, M. A., Hecht, T. T., Winkler, D. F., Wunderlich, J. R. and Segal, D. M. (1987a) Human T cells targeted with anti-T3 crosslinked to anti-tumor antibody prevent tumor growth in nude mice. *J. Immunol.* **138**, 4018–4022.

Titus, J. A., Perez, P., Kaubisch, A., Garrido, M. A. and Segal, D. M. (1987b) Human K/NK cells targeted with heterocrosslinked antibodies specifically lyse tumor cells *in vitro* and prevent tumor growth *in vivo*. *J. Immunol.* **139**, 3153–3158.

Tosi, M. F. and Berger, M. (1988) Functional differences between the 40 kDa and 50 to 70 kDa IgG Fc receptors on human neutrophils revealed by elastase treatment and antireceptor antibodies. *J. Immunol.* **141**, 2097–2103.

Traunecker, A., Lanzavecchia, A. and Karjalainen, K. (1991) Bispecific single chain molecules (Janusins) target cytotoxic lymphocytes on HIV infected cells. *EMBO J.* **10**, 3655–3659.

Trauth, B. C., Klas, C., Peters, A. M., Matzku, S., Moller, P., Falk, W., Debatin, K. M. and Krammer, P. H. (1989) Monoclonal antibody-mediated tumor regression by induction of apoptosis. *Science* **245**, 301–305.

Tschopp, J. and Nabholz, M. (1990) Perforin-mediated target cell lysis by cytolytic T lymphocytes. *Annu. Rev. Immunol.* **8**, 279–302.

Tutt, A., Stevenson, G. T. and Glennie, M. J. (1991) Trispecific $F(ab')_3$ derivatives that use cooperative signaling via the TCR/CD3 complex and CD2 to activate and redirect resting cytotoxic T cells. *J. Immunol.* **147**, 60–69.

Urba, W. J., Ewel, C., Kopp, W., Smith, J. W.,2d, Steis, R. G., Ashwell, J. D., Creekmore, S. P., Rossio, J., Sznol, M. and Sharfman, W. (1992) Anti-CD3 monoclonal antibody treatment of patients with CD3-negative tumors: a phase IA/B study. *Cancer Res.* **52**, 2394–2401.

Urnovitz, H. B., Chang, Y., Scott, M., Fleischman, J. and Lynch, R. G. (1988) IgA:IgM and IgA:IgA hybrid hybridomas secrete heteropolymeric immunoglobulins that are polyvalent and bispecific. *J. Immunol.* **140**, 558–563.

Valone, F. H., Kaufman, P. A., Guyre, P. M., Lewis, L. D., Memoli, V., Deo, Y., Graziano, R., Fisher, J. L., Meyer, L. and Mrozek-Orlowski, M. (1995) Phase Ia/Ib trial of bispecific antibody MDX-210 in patients with advanced breast or ovarian cancer that overexpresses the proto-oncogene HER-2/neu. *J. Clin. Oncol.* **13**, 2281–2292.

Van de Winkel, J. G., Van Ommen, R., Huizinga, T. W., de Raad, M. A., Tuijnman, W. B., Groenen, P. J., Capel, P. J., Koene, R. A. and Tax, W. J. (1989) Proteolysis induces increased binding affinity of the monocyte type II FcR for human IgG. *J. Immunol.* **143**, 571–578.

Van de Winkel, J. G., Ernst, L. K., Anderson, C. L. and Chiu, I. M. (1991) Gene organization of the human high affinity receptor for IgG, Fc gamma RI (CD64). Characterization and evidence for a second gene. *J. Biol. Chem.* **266**, 13449–13455.

Van de Winkel, J. G. and Anderson, C. L. (1991) Biology of human immunoglobulin G Fc receptors. *J. Leukoc. Biol.* **49**, 511–524.

Van Schie, R. C., Verstraten, R. G., Van de Winkel, J. G., Tax, W. J. and de Mulder, P. H. (1992) Effect of recombinant IFN-gamma (rIFN-gamma) on the mechanism of human macrophage IgG FcRI-mediated cytotoxicity. rIFN-gamma decreases inhibition by cytophilic human IgG and changes the cytolytic mechanism. *J. Immunol.* **148**, 169–176.

Vance, B. A., Karlson, K. H., Morganelli, P. M. and Guyre, P. M. (1989) Single step screening of monoclonal antibodies against interferon-gamma-induced surface molecules on human monocytes. *J. Immunol. Methods* **118**, 287–296.

Vandenberghe, P., Verwilghen, J., Van Vaeck, F. and Ceuppens, J. L. (1993) Ligation of the CD5 or CD28 molecules on resting human T cells induces expression of the early activation antigen CD69 by a calcium- and tyrosine kinase-dependent mechanism. *Immunology* **78**, 210–217.

Vignaux, F., Vivier, E., Malissen, B., Depraetere, V., Nagata, S. and Golstein, P. (1995) TCR/CD3 coupling to Fas-based cytotoxicity. *J. Exp. Med.* **181**, 781–786.

Vitetta, E. S., Thorpe, P. E. and Uhr, J. W. (1993) Immunotoxins: magic bullets or misguided missiles? *Immunol. Today* **14**, 252–259.

Wallace, P. K., Valone, F. H. and Fanger, M. W. (1995) Myeloid cell-targeted cytotoxicity of tumor cells. In *Bispecific Antibodies* (Edited by Fanger, M. W.), p. 43. R.G. Landes Company, Austin.

Warmerdam, P. A., Van de Winkel, J. G., Gosselin, E. J. and Capel, P. J. (1990) Molecular basis for a polymorphism of human Fc gamma receptor II (CD32). *J. Exp. Med.* **172**, 19–25.

Warmerdam, P. A., Van de Winkel, J. G., Vlug, A., Westerdaal, N. A. and Capel, P. J. (1991) A single amino acid in the second Ig-like domain of the human Fc gamma receptor II is critical for human IgG2 binding. *J. Immunol.* **147**, 1338–1343.

Weber, G. F., Ashkar, S., Glimcher, M. J. and Cantor, H. (1996) Receptor-ligand interaction between CD44 and osteopontin (Eta-1). *Science* **271**, 509–512.

Weiner, G. J., Kostelny, S. A., Hillstrom, J. R., Cole, M. S., Link, B. K., Wang, S. L. and Tso, J. Y. (1994) The role of T cell activation in anti-CD3 × antitumor bispecific antibody therapy. *J. Immunol.* **152**, 2385–2392.

Weiner, G. J. and De Gast, G. C. (1995) Bispecific monoclonal antibody therapy of B-cell malignancy. *Leuk. Lymphoma.* **16**, 199–207.

Weiner, G. J. and Hillstrom, J. R. (1991) Bispecific anti-idiotype/anti-CD3 antibody therapy of murine B cell lymphoma. *J. Immunol.* **147**, 4035–4044.

Weiner, L. M., Clark, J. I., Davey, M., Li, W. S., Garcia de Palazzo, I., Ring, D. B. and Alpaugh, R. K. (1995) Phase I trial of 2B1, a bispecific monoclonal antibody targeting c-erbB-2 and Fc gamma RIII. *Cancer Res.* **55**, 4586–4593.

Welch, G. R., Wong, H. L. and Wahl, S. M. (1990) Selective induction of Fc gamma RIII on human monocytes by transforming growth factor-beta. *J. Immunol.* **144**, 3444–3448.

Winter, G. and Milstein, C. (1991) Man-made antibodies. *Nature* **349**, 293–299.

Wong, H. L., Welch, G. R., Brandes, M. E. and Wahl, S. M. (1991) IL-4 antagonizes induction of Fc gamma RIII (CD16) expression by transforming growth factor-beta on human monocytes. *J. Immunol.* **147**, 1843–1848.

Wong, J. T. and Colvin, R. B. (1987) Bi-specific monoclonal antibodies: selective binding and complement fixation to cells that express two different surface antigens. *J. Immunol.* **139**, 1369–1374.

3. TARGETED CYTOTOXICITY: ANTIBODY–DRUG AND ANTIBODY–TOXIN CONJUGATES

UWE ZANGEMEISTER-WITTKE [1] and WINFRIED WELS [2]

[1] *Division of Oncology, University Hospital Zürich, Häldeliweg 4, CH-8044 Zürich, Switzerland,* [2] *Institute for Experimental Cancer Research, Tumor Biology Center, Breisacher Strasse 117, D-79106 Freiburg, Germany*

INTRODUCTION

Conventional cytotoxic approaches such as chemotherapy could not prove curative for most prevalent tumors. Although chemotherapeutic drugs are designed to be selectively toxic to rapidly dividing cells, they lack tumor specificity and their effectiveness is limited by the emergence of drug resistant tumor cells. The inherent antigen specificity of antibodies and their ability to direct to tumor cells effector cells of the immune system, and a variety of exogenous antitumor effector molecules including drugs, toxins, and radioisotopes has been widely exploited for targeted tumor therapy. In order to be superior to conventional treatments, the antibodies must be directed towards antigens which are exclusively or at least preferentially expressed on tumor cells compared to normal tissues. Although many antibodies have shown potential to localize to tumors and to concentrate high doses of cytotoxic effector molecules on the tumor cell surface in preclinical tumor models, this has not yet been verified in clinical studies.

The failure of antibodies and cytotoxic antibody conjugates to efficiently localize to tumors in patients is mainly due to their rapid uptake by cells of the reticuloendothelial tissue, and to their immunogenicity which leads to the formation of immune complexes and precludes their repeated administration. A further hindrance especially in solid tumors is that the low percentage of the injected antibody dose which localizes does not evenly distribute in the tumor tissue, and a large fraction of tumor cells will not be exposed to the cytotoxic agent. The inaccessibility of solid tumors to antibodies and antibody conjugates is mainly due to the unfavorable large size of the macromolecules, but also to unfavorable intrinsic properties of the tumor, such as elevated interstitial pressure leading to radially outward convection, and heterogeneous antigen expression. In view of these obstacles it is conceivable that antibody conjugates could prove to be more promising for the treatment of minimal residual disease in combination with a conventional debulking therapy. However, for ethical reasons antibody-based therapy has been limited to the treatment of patients with advanced disease and high tumor loads, and it is thus not surprising that the few impressive clinical results were achieved in patients with hematopoietic tumors.

Recently, antibody-based therapy has been revolutionized by advances, such as the identification of novel tumor-associated antigens which are homogeneously expressed in the tumor and lack expression on normal vital tissues, and chemical and genetic engineering to produce tailor-made targeting vehicles with improved tumor localization properties. It can be expected that these improvements will provide more effective antitumor agents also for the treatment of solid tumors in carefully selected clinical settings.

THE CONCEPT OF TARGETED CYTOTOXICITY

The advantage of chemotherapy is that, in contrast to surgery and radiotherapy, it can be used for disseminated as well as localized cancer. Unfortunately, examples of long-term survival after chemotherapy represent the minority of cases, typically 5–10% with current treatments. The major limitation for systemic therapy is a lack of tumor specificity, so that agents effective in killing neoplastic cells usually also have detrimental effects on normal tissues. In addition, most metastatic cancers are either constitutive resistant to chemotherapy or respond to chemotherapy, but later recur as cancers that have acquired resistance (Gottesman and Pastan, 1993). Most frequently, the pattern of resistance includes a variety of cytotoxic drugs that do not have a common structure or a common intracellular target. This broad resistance to chemotherapy termed multidrug resistance is mediated by energy-dependent drug transport proteins located in the cell membrane. Although high doses of drugs can overcome multidrug resistance by overstraining the transport mechanisms, further effective treatment of refractory cancer is inevitably limited due to unspecific toxicity *in vivo*.

A rational strategy to overcome multidrug resistance of tumor cells and the limitations of unspecific toxicity to normal tissues is to attach the chemotherapeutic drug to a macromolecular carrier which concentrates the drug at the tumor site. Moreover, additional potential to overcome drug resistance is provided if the conjugate is able to by-pass the transport proteins in the cell membrane by entering cells via endocytosis. Another promising approach is to attach to the carrier cytotoxic effector molecules, e.g. protein toxins which act by different mechanisms than chemotherapeutic drugs and to which tumor cells are not cross-resistant. An ideal carrier to increase the antitumor efficacy of cytotoxic agents and to avoid damage to normal tissues would specifically recognize the tumor cell and increase the local concentration of the drug at the tumor site. This concept of targeted cytotoxicity, which was first proposed by Ehrlich nearly a century ago (Ehrlich, 1956), can be best realized using antibodies as ligands that bind to antigens on the surface of tumor cells. The monoclonal antibody technique (Köhler and Milstein, 1975) has made it possible to produce large amounts of uniform antibody molecules which can be further engineered chemically (Werlen *et al.*, 1995; Offord *et al.*, 1992) or by recombinant DNA technology (Plückthun, 1992; Pack and Plückthun, 1992) to produce tailor-made targeting vehicles in quantities useful for human therapy. To date cytotoxic antibody

Table 1 Toxins and drugs conjugated to antibodies

Toxins	Plant	Ricin, abrin, saporin, gelonin, modecin
	Bacterial	*Pseudomonas* exotoxin A, diphtheria toxin
Drugs	Antimetabolites	Aminopterin, methotrexate, cytosine arabinoside, 5-fluorouracil, 5-fluorodeoxyuridine
	Anthracyclines	Doxorubicin, idarubicin, daunomycin, morpholino doxorubicin
	Alkylating agents	Melphalan, chlorambucil, mitomycin C, Cis-platinum
	Anti-mitotic agents	Vinca alkaloids, colchicine, podophyllotoxin
	Miscellaneous	Neocarzinostain, bleomycin, calicheamycin

conjugates have emerged as a novel promising group of antitumor agents and a great variety of cytotoxic agents have been linked to antibodies including chemotherapeutic drugs and toxins (Table 1). Some of these agents are extremely potent or exhibit unspecific cell binding activity so that they must be harnessed by a tumor selective targeting moiety, and cannot be used in free form (reviewed by Poznansky and Juliano, 1984; Engert and Thorpe, 1990). Therefore, the targeting moiety, e.g. an antibody directed towards a tumor-associated antigen, must be carefully examined by immunohistochemistry on tissue sections and evaluated in patients using immunoscintigraphy before controlled phase I studies are performed.

Intact antibody molecules or recombinant and chemical fragments thereof as well as drugs and toxins, display a number of biological properties dictated by their normal functions. The key to making antibody–drug and antibody–toxin conjugates useful for cancer therapy is to eliminate unwanted functions of the components without adverse effect on other useful properties. Based on the great variety of drugs and toxins available with well-defined antitumor activity and owing to improved protein modification and genetic engineering techniques, various strategies have been devised to produce tailor-made chemical and recombinant antibody–drug, and antibody–toxin conjugates for clinical use. The rationale, the potential, and the limitation of the underlying approaches, which have traditionally focused independently on the two key elements: the targeting system comprising the antibody and the antigen on the target cell, and the effector molecule, will be discussed.

THE ANTIBODY: FACTORS INFLUENCING THERAPEUTIC EFFICACY

The monoclonal antibody technique allows the production of large amounts of uniform antibodies with defined tumor cell specificity (Köhler and Milstein, 1975). They can be linked to cytotoxic effector molecules to produce a novel class of antitumor agents. Such antibody conjugates combine in a single molecule the antigen specificity of the ligand and the cell killing potential of the cytotoxic effector. Most frequently antibodies of the IgG class have been used for this purpose. These are glycoproteins of 150 kDa possessing a number of potential sites to which effector molecules can be linked by chemical conjugation without perturbing the antigen binding property. Unfortunately, the benefit of intact antibodies for targeted toxin- and chemotherapy of solid tumors has been hampered by a number of shortcomings. Due to their relatively large size, intact antibodies and cytotoxic antibody conjugates cannot efficiently penetrate and distribute in larger tumor masses. This is a major limitation for the antitumor efficacy of cytotoxic agents which must get in direct contact with the target cell to kill it. As a consequence, the use of intact antibodies against solid tumors has been mainly limited to diagnostic applications, and to the delivery of radioisotopes and prodrug activating enzymes which employ a bystander effect to damage cells not directly targeted by the antibody conjugate. Another complication is caused by the immunogenicity of large antibodies produced in non-human species, usually in mice. Despite the fact that most patients are immunosuppressed secondary to extensive chemotherapy, with the exception of some hematopoietic malignancies human-anti-mouse antibody (HAMA) responses occur in the majority of patients treated (Durrant *et al.*, 1989; Schneck *et al.*, 1989; Oratz *et al.*, 1990). The development of neutralizing antibodies and the formation of immune complexes due to a patient's immune response is a major therapeutic hindrance. It precludes repeated cycles of therapy, reduces the amount of antibody available to localize to the tumor and, if cytotoxic antibody conjugates are used, may damage the clearing tissues. Several strategies including the use of polyethylene glycol for antibody modification (Kitamura *et al.*, 1990), or immunosuppressive drugs such as cyclophosphamide (Oratz *et al.*, 1990) and cyclosporin (Ledermann *et al.*, 1988) have been explored, but all failed to fully overcome the problem of antibody immunogenicity. Intact antibodies are currently being replaced by a new generation of antibodies with more favorable targeting properties *in vivo*, such as increased tumor localization, and rapid uptake and homogeneous distribution in solid tumor tissues.

To analyze the tumor localization potential of antibodies, imaging studies have been performed in animals and patients using radiolabelled antibodies. The low proportion of antibodies which localize to tumors in patients, usually less than 0.01% of the total injected dose, was disappointing in comparison to the high levels of tumor uptake achieved in some animal models (Pimm, 1988). Mathematical models and microdistribution studies have also provided theoretical descriptions of the variables important for the uptake and distribution of antibodies in tumors (Jain and Baxter, 1988; Sung *et al.*, 1994; Shockley *et al.*, 1992). These variables include the amount of antibody administered (Fenwick *et al.*, 1989), the size of the antibody (Colapinto *et al.*, 1988; Yokota *et al.*, 1992), its antigen binding affinity (Fujimori *et al.*, 1989), and the kinetics of cell binding and internalization (Matzku *et al.*, 1987, 1988). Although these determinants have been examined extensively in preclinical studies, there is still a substantial lack of data on the autonomous role and relevance of each of the variables and their complex interactions. Most studies have investigated either a single specific antibody in different tumor models or, conversely, various antibodies differing in more than one property in a single tumor model.

The reduction of the antibody size is a verified strategy to achieve rapid uptake and more even distribution of antibodies in solid tumors (Yokota *et al.*, 1992). Antibody fragments such as Fab′ and $F(ab)'_2$ produced by enzymatic cleavage, or Fv fragments produced by recombinant DNA technology have now been used frequently for targeted tumor therapy. Antibody fragments, especially those lacking the Fc portion, are more rapidly cleared from tissues and excreted. This increases the tumor to blood ratio of cytotoxic antibody conjugates and helps to reduce damage to non-target tissues. On the other hand, the shorter half-life of smaller antibody fragments and their conjugates in the circulation reduces the time available for binding to the antigens on the tumor cells and thus limits the dose of antibody that can accumulate in the tumor (Colcher *et al.*, 1990; Milenic *et al.*, 1991). Therefore, antibody fragments have been administered *in vivo* preferentially via continuous infusion protocols to achieve higher and more stable serum levels under controlled conditions. In addition to effects on the pharmacokinetics, the reduction of the antibody size also has the potential to reduce the immunogenicity of the macromolecule in patients, and antibody fragments which lack Fc regions will not bind to non-target cells bearing Fc receptors. Other antibody engineering techniques, such as the reshaping of antibodies by introducing the hypervariable regions from the heavy (V_H) and light chain (V_L) variable domains of antibodies raised in animals into a human antibody framework, have also been used to make less immunogenic targeting vehicles (Riechmann *et al.*, 1988).

The important role of the binding affinity of antibodies for tumor targeting has been recognized for many years and has been examined extensively *in vitro* and in animal models. Antibody–antigen interactions belong to the group of non-covalent biological binding reactions which depend on a structural complementarity between a ligand and a binding site on a macromolecule. In general, high affinity antibodies are superior to low affinity antibodies in a number of biological traits. They form more stable bonds with antigens and thus are more effective in neutralizing toxins or in protecting against bacterial or viral infections. In principle, this superiority might also apply to cytotoxic antibody conjugates directed towards antigens on the surface of tumor cells. The affinity of the antibody may affect the antitumor efficacy since it determines the number of cytotoxic molecules bound to a cell at a given antibody concentration. In addition, it is conceivable that the number of cytotoxic molecules that can bind to the cell surface is also determined by the number of antigens present on the cell. Thus, an ideal setting for effective antibody-mediated therapy appears to be the use of large amounts of a high affinity antibody directed towards an antigen which is abundantly expressed on the tumor cells. However, antibody–antigen interactions *in vivo* are much more complex and it is still a matter of debate whether high or low affinity antibodies are of greater therapeutic benefit. High affinity binding has shown advantage with regard to tumor uptake especially for small antibody fragments (Thomas *et al.*, 1989) and for the application of low doses of antibody (Sung *et al.*, 1992). On the other hand, there is experimental evidence that high affinity binding to antigens directly after extravasation in the proximity of the blood vessels significantly decreases the number of antibody molecules available for penetrating deeper into the tumor ("binding site barrier") (Juweid *et al.*, 1992).

The ability of an antibody to localize to and to evenly distribute in tumors *in vivo* cannot be predicted based on its whole body distribution, its uptake and distribution in the tumor, or its antigen binding properties alone pather, it also depends on a number of properties intrinsic to the tumor. For example, in tumors antigen expression is rarely homogeneous and antigens may exist on cells in different forms and in variable densities (Matzku *et al.*, 1987). Other factors which strongly influence the distribution of antibodies in tumors include the quantity and quality of blood vessels, the rate of blood flow, and the transport of interstitial fluid in the tumor (Fujimori *et al.*, 1989; Dvorak *et al.*, 1991; Shockley *et al.*, 1992). Fluid transport in solid tumors is determined by the elevated interstitial pressure which may reduce the driving force for extravasation, and leads to a radially outward convection opposing the inward diffusion of the macromolecule (Jain and Baxter, 1988). This partly explains why most frequently deposition of antibodies has been found to be maximal near blood vessels. Due to the important role of the vascular system and because it is the most frequently used route of antibody and drug administration for delivery to tumors, attempts to enhance blood flow such as the use of vasoconstricting agents have been examined in preclinical studies. These studies indicated that there was a differential response in tumors and normal tissues, as tumor blood vessels usually differ in structure and function from normal blood vessels (Jain, 1988), and that the combination of cytotoxic antibody conjugates with propanolol or Angiotensin II could enhance the antitumor efficacy (Smyth *et al.*, 1987; Zhou *et al.*, 1992).

A different approach to circumvent the problem of poor penetration of solid tumors by antibodies and antibody conjugates is to attack the endothelial cells lining the blood vessels of the tumor rather than the tumor cells themselves. The rationale behind this approach is that many tumor cells rely on each capillary for oxygen and nutrients and that vascular endothelial cells are directly accessible to cytotoxic agents administered into the blood stream. Using immunotoxins directed towards highly selective antigens on the surface of proliferating endothelial cells, this approach has shown to cause thrombosis, wide-spread infarction and regression of tumors in animal models (Burrows and Thorpe, 1994; Burrows *et al.*, 1994).

THE IMPORTANCE OF THE TUMOR ANTIGEN

The therapeutic efficacy of cytotoxic antibody conjugates not only depends on the targeting potential of the antibody, but also on the properties of the antigen towards which the antibody is directed. One requirement for effective targeting is that the antigen is differentially expressed in the tumor and in normal tissues. Ideally it is abundantly expressed on tumor cells and absent on normal cells. Moreover, it should be uniformly expressed on the malignant cell population, not be shed from the cell surface, and not downmodulated upon binding of the antibody. Malignant transformation of cells is often associated with changes in their antigenic profiles. The newly expressed antigens include differentiation antigens, the products of oncogenes, and mutated tumor suppressor gene products (Ben-Mahrez *et al.*, 1990; Crawford *et al.*, 1982). The use of monoclonal antibodies has allowed the identification of a large number of these tumor-associated antigens, some of which have also been cloned and functionally defined. However, apart from a few truly tumor-specific antigens like immunoglobulin idiotypes identified on B cells (Thielemans *et al.*, 1984) these antibodies generally cross-react to some extent also with antigens on normal tissues. This may be acceptable as long as the normal cells are not of vital interest or the level of antigen expression on these cells is significantly below that found on the tumor cells. Nevertheless, it should be emphasized that even minor cross-reactivities with normal tissues, which might be overlooked when an unconjugated antibody is screened, can lead to major toxicity when the antibody is armed with extremely potent effector molecules. In Table 2, examples of tumor-associated antigens which have been used as targets for antibodies in tumor diagnosis and therapy are listed. Recently, Burrows and Thorpe (1994) have described an approach for the therapy of solid tumors which is based on targeting the vasculature of solid tumors. This approach opens a new spectrum of antigens selectively expressed on proliferating endothelial cells as potential targets for antibody-based cancer therapy.

Most of the preclinical studies with antibodies have been completed in athymic mice bearing human tumor xenografts. These models are useful to estimate tumor localization and therapeutic potential of cytotoxic antibody conjugates under complex physiological conditions. However, since the antibodies used might not recognize the murine homologues of the antigens or the respective antigens might not be expressed on normal murine tissues, it is generally not possible to predict from these studies the potential cross-reactivity of the antibodies in patients. For example, a CD24-specific antibody which demonstrated efficient localization to lung cancer xenografts in mice (Smith *et al.*, 1989; Zangemeister-Wittke *et al.*, 1993)

Table 2 Tumor-associated antigens as potential targets for antibody-based therapy

Tumor	*Antigen*
Pancarcinoma	Le^y, GA733-2 (EGP-2), TAG-72, mucin, HMV mucin, CEA
Breast	EGF receptor, erbB-2, gp55
Lung	EGF receptor, erbB-2, NCAM, CD24
Ovary	38–40 kDa glycoprotein, CA125
Prostate	PSA
Gastrointestinal tract	HMW glycoprotein, gp72
Melanoma	HMW-MAA, gp100, p97, GD_2 ganglioside, GD_3 ganglioside
Lymphoma, Leukemia	CD19, CD22, CD25, CD5, CD7

was rapidly entrapped by the reticuloendothelial system in patients, possibly due to species specific cross-reactivity with human leukocytes (Ledermann *et al.*, 1993). Therefore, the limited tumor specificity of available antibodies requires their extensive screening on human tissue sections and in imaging studies in patients to assess the risk of damage to normal tissues by cytotoxic antibody conjugates.

With few exceptions chemotherapeutic drugs and toxins act on intracellular targets. Following coupling to antibodies these molecules must therefore enter cells after binding of the antibody to its antigen on the cell surface. Thus, the antitumor efficacy of antibody conjugates not only relies on the quantitative delivery of the cytotoxic effector portion to the tumor cell, but also on the efficient transfer from the initial binding site on the cell surface to the intracellular site of action in the cytoplasm or the nucleus. To fulfill this requirement antibody conjugates should ideally bind to antigens with high affinity (Ramakrishnan and Houston, 1984) and the antigen should be abundantly expressed on the tumor cell and internalized at high rates (Embleton *et al.*, 1986; Preijers *et al.*, 1988; Wargalla and Reisfeld, 1989; Froesch *et al.*, 1996).

The ability of different surface antigens to internalize following antibody binding has been addressed by several studies and different rates have been observed (Garnett and Baldwin, 1986; Starling *et al.*, 1988; Press *et al.*, 1989). Different rates of internalization have also been observed for an individual antigen on different cell lines (Wargalla and Reisfeld, 1989) and even different epitopes on an individual antigen may differ in their potential to become internalized subsequent to antibody binding (Matzku *et al.*, 1990). Another important issue is that some antigens require cross-linking on the cell surface to become internalized, whereas other antigens have intrinsic internalization properties also in their monomeric form. This knowledge may help to predict whether the targeting antibody must be bivalent, or whether monovalent antibody fragments can be used to achieve intracellular delivery of the cytotoxic effector molecule. Following internalization, the antibody–antigen complex is transported to the endosome where cleavage of pH sensitive linkers, reduction of disulfide bonds, and dissociation of the macromolecular complex may occur. Most of the complexes are further transported to lysosomes where they are degraded. Alternatively, dependent on the antigen, retrograde transport via the trans-Golgi network and the Golgi cisterns to the endoplasmic reticulum can occur (Garred *et al.*, 1995). In order to be cytotoxic, drugs and toxins must leave the vesicular compartment to enter the cytoplasm and to escape lysosomal degradation. If a small chemotherapeutic drug can permeate through the vesicular membrane, intoxication of cells only requires that the drug is released from the antibody in its active form. This may occur at any stage following endocytosis. In the case of other effector molecules, such as toxins, the process of translocation into the cytoplasm is much more complex and requires the interaction with specialized receptors in an appropriate compartment. For example, with the exception of diphtheria toxin from *Corneybacterium diphtheriae* which translocates from acidified endosomes, most commonly used toxins must be sequestered into the neutral compartment of the endoplasmic reticulum to translocate into the cytoplasm

(Press *et al.*, 1986). Different antigens and even different epitopes on the same antigen may use different pathways of intracellular routing and processing, and thus may differ in their potential to mediate the cytotoxic activity of these immunotoxins (Press *et al.*, 1988; May *et al.*, 1991). In view of the complexity of the intoxication process which makes it difficult to predict whether a given antibody will result in an effective immunotoxin, Weltman *et al.* (1987) and Till *et al.* (1988) have developed indirect screening assays to analyze the potential of tumor-associated antigens to serve as targets for immunotoxin therapy.

ANTIBODY–DRUG CONJUGATES

Since the discovery of antifolates and alkylating agents and their introduction as effective antileukemic agents, drug therapy has revolutionized the treatment of malignant disease. However, with the exception of leukemias and lymphomas, testicular carcinoma and a few others, curative effects have been the exception (Spitler *et al.*, 1988). Chemotherapeutic drugs are of many different types and include alkylating agents, intercalating drugs, microtubule inhibiting substances, and antimetabolites. All of these agents are complex chemical structures and can be used to interfere in a particular biological function in the cell leading to cell death (Cassidy and Douros, 1980; Chabner and Collins, 1990).

In addition to efforts to identify more effective agents or combinations thereof, the need to limit unwanted side effects and damage to normal tissues, and to overcome multidrug resistance also demands the modification of known drugs with clinically well-defined activities. One possible modification is to target chemotherapeutic drugs to tumor cells using antibody–drug conjugates directed towards antigens selectively expressed on tumor cells. This has the potential to concentrate high doses of drug at the tumor site and to reduce unwanted damage to normal tissues. Some antitumor antibiotics, such as the calicheamicins or the CC-1065 analogues are so extremely potent that they can be used *in vivo* only if their unspecific toxicity is harnessed by a powerful tumor-selective targeting moiety (Hinman *et al.*, 1993; Chari *et al.*, 1995). Despite the well-defined activities and side effects of conventional chemotherapeutic drugs in patients, it should be emphasized that antibody–drug conjugates differ in many respects from the unconjugated low molecular weight drugs. For example, they are less rapidly excreted from tissues, have longer half-lives in the circulation, and display a different whole body distribution. Consequently, the side effects induced by these macromolecular conjugates may also be different. Moreover, coupling of drugs to antibodies may protect them from degradation and rapid excretion and facilitate their controlled release in the target tissue (Arnon and Sela, 1982).

Antibody–drug conjugates have the potential to overcome multidrug resistance of tumor cells by increasing the concentration of the cytotoxic activity in the immediate vicinity of the tumor cells, and by promoting drug internalization into cells via receptor-mediated endocytosis (Mellman *et al.*, 1986). This may result in decreased efflux of the drug by transport proteins located in the cell membrane and has been demonstrated for anthracyclines and methothrexate which were stably linked to antibodies (Sheldon *et al.*, 1989; Sivam *et al.*, 1995; Dillman *et al.*, 1989). However, the ability to overcome drug resistance also depends on other variables, e.g. the property of the linker between drug and antibody, and the subcellular distribution and the functional requirement of the various drug transport proteins. Endocytosis of the drug may be sufficient to overcome drug resistance mediated by efflux pumps located in the plasma membrane, such as the p-glycoprotein and the multi-drug resistance associated protein (MRP) (Beck, 1987; Gottesman and Pastan, 1993; Cole *et al.*, 1992). It may, however, not be sufficient if the transport protein is located in the membrane of intracellular vesicles, such as the lung resistance related protein (LRP) (Scheper *et al.*, 1993). As shown recently, a doxorubicin–antibody conjugate employing an acid sensitive linker could not overcome drug resistance of lung cancer cells (Froesch *et al.*, 1996).

Chemical Cross-Linking of Antibodies and Drugs

In contrast to toxins, conventional chemotherapeutic drugs are not endowed with catalytic activity. As a consequence they are up to 1000 times less potent than toxins at equimolar doses and therefore maximum loading should be achieved. However, most drugs are hydrophobic molecules which are difficult to conjugate to hydrophilic antibodies and there is a limit in the number of drug molecules which can be conjugated to an antibody without perturbing its structural and functional integrity. The maximal amount of drug which can be conjugated may differ between antibodies, and there is a point at which antigen binding activity is lost upon further loading. The possible number of drug molecules coupled is further reduced when smaller antibody fragments are used as carriers. Another requisite for the preparation of effective antibody–drug conjugates is that the coupling procedure preserves the cytotoxic function of the drug. Because of the small size of such molecules this is sometimes difficult to achieve.

Conventional chemotherapeutic drugs act on intracellular targets. Compared to antibodies they are very small molecules and with rare exceptions (Starling *et al.*, 1988) must be liberated from the antibody to be fully active. The linkage between antibody and drug thus has major implication on the selectivity and the antitumor efficacy of these conjugates since it determines the site and the rate at which liberation occurs. Based on the information available on antibody and drug activity various cross-linking strategies have been devised, which can fulfill the different requirements. However, in each case the cleavable site must be stable in the circulation so that the drug is not released prematurely. If the drug is hydrophobic enough to permeate through the cell membrane its release from the antibody can occur at any stage following binding to the tumor cell. On the cell surface this would require a linker that prefers conditions specifically found in the microenvironment of the tumor, e.g. one that is cleaved under the mild acidic conditions found in many solid tumors. Other linkers must be designed to be cleaved under conditions found in intracellular compartments. This may occur by random proteolytic fragmentation of the conjugate in endosomes if the drug is conjugated to the antibody via a stable covalent bond (Sheldon *et al.*, 1989; Sivam *et al.*, 1995; Dillman *et al.*, 1989). Since this random fragmentation may result in a significant number of cytotoxically inactive antibody–drug debris, acid-labile linkers have been used which also allow the controlled release of the drugs inside cells (Trail *et al.*, 1993; Froesch *et al.*, 1996; Greenfield *et al.*, 1990).

For the coupling of drugs to antibodies use can be made of the various reactive groups present on each moiety. However, not all groups are amenable to chemical modification due to their location close to active sites. In addition, the chemistry of conjugating hydrophobic drugs to hydrophilic antibodies is difficult and the number of reactive sites or drug molecules linked to the antibody is limited. For example, modification of primary amino groups using 2-iminothiolane (2-IT) is a verified and reproducible method to introduce sulfhydryl groups into antibodies as long as low numbers are sufficient for conjugation, e.g. for the preparation of immunotoxins. In contrast, greater numbers of 2-IT residues, which would be needed for the preparation of effective antibody–drug conjugates, will cause aggregation and precipitation of the antibody. As demonstrated for antibody–doxorubicin conjugates use of the sulfhydryl groups, obtained by mild reduction of the antibody, was more effective in providing reactive sites and did not result in disintegration of the antibody (Willner *et al.*, 1993; Froesch *et al.*, 1996). Another requirement is that the coupling procedure also preserves the active site of the drug. For example, in the case of the alkylating agent chlorambucil modification of the chloroethyl group yields inactive drugs (Smyth *et al.*, 1986). With methothrexate, modification of the pteridine nucleus results in drug inactivation, whereas the carboxyl group of the glutamic acid can be linked to antibodies (Kanellos *et al.*, 1985). In view of these difficulties indirect conjugation methods based on the use of intermediate carriers, such as human serum albumin, which contains numerous amino groups for amide formation and provides a cysteine at residue 34 for coupling to antibodies, have been used (Fitzpatrick and Garnett, 1995). The various conjugation strategies which have been developed for the direct and indirect coupling of drugs to antibodies are described in more detail by Upeslacis and Hinman (1988).

One of the best characterized anti-cancer drugs which is frequently used for the treatment of solid tumors is the anthracycline doxorubicin. Employing its different functional groups, doxorubicin has been covalently linked to antibodies using different strategies. Cross-linking was accomplished either directly (Hurwitz *et al.*, 1975), or indirectly using small spacer molecules (Dillman *et al.*, 1988; Greenfield *et al.*, 1990), or macromolecular carriers which facilitate the linkage of a greater number of drug molecules to the antibody (Galun *et al.*, 1990; Shih *et al.*, 1994). In functional studies the most active and antigen-specific conjugates employed acid sensitive linkers (Trail *et al.*, 1993; Zhu *et al.*, 1995; Froesch *et al.*, 1996). These are cleaved in the acidic environment of endosomes to release the active drug from the antibody carrier. Cleavage of acid-sensitive linkers may also occur immediately after extravasation of the conjugate under the mild acidic conditions found in solid tumors. If the free drug can enter cells independent of the antibody, e.g. by permeation through the cell membrane, this will be of advantage if the homogeneous distribution of the macromolecular conjugate in the tumor tissue is not ensured. Acid-sensitive linkers can be produced by coupling a *cis*-aconityl linker to the amino sugar of doxorubicin (Dillman *et al.*, 1988). A different approach is the use of the 13-keto group of the anthracycline as an attachment site to produce a hydrazone bond (Greenfield *et al.*, 1990). We have recently described a carcinoma reactive antibody–doxorubicin conjugate generated by reacting the hydrazide derivative of the cross-linking reagent succinimidyl 4-(N-maleimidomethyl)-cyclohexane-1-carboxylate (SMCC) with the C13 carbonyl group of the anthracycline (Froesch *et al.*, 1996). The molecular organization of an antibody–doxorubicin conjugate containing an acid sensitive hydrazone linker is schematically shown in Figure 1. Compared with the linker strategy described by Greenfield *et al.* (1990), this protocol takes advantage of commercially available reagents and a reduced number of reaction steps.

Clinical Trials with Antibody–Drug Conjugates

Compared to radioimmunoconjugates and immunotoxins there is only limited experience with antibody–drug conjugates in the treatment of cancer. The preclinical animal studies and clinical trials with patients completed over the past twenty years have demonstrated the potential of these agents for cancer therapy, but they have also highlighted the problems associated with insufficient tumor localization, uptake and distribution in the tumor, and low potency of the drug. The published clinical studies with antibody–drug conjugates are listed in Table 3. They include the use of doxorubicin, daunomycin, methotrexate, N-acetyl-melphalan, mitomycin C, vinca alkaloid, and vindesine, and were mainly phase I studies carried out in patients with advanced disease. Although gram quantities of antibody linked to up to several hundred milligrams of drug have been administered to patients,

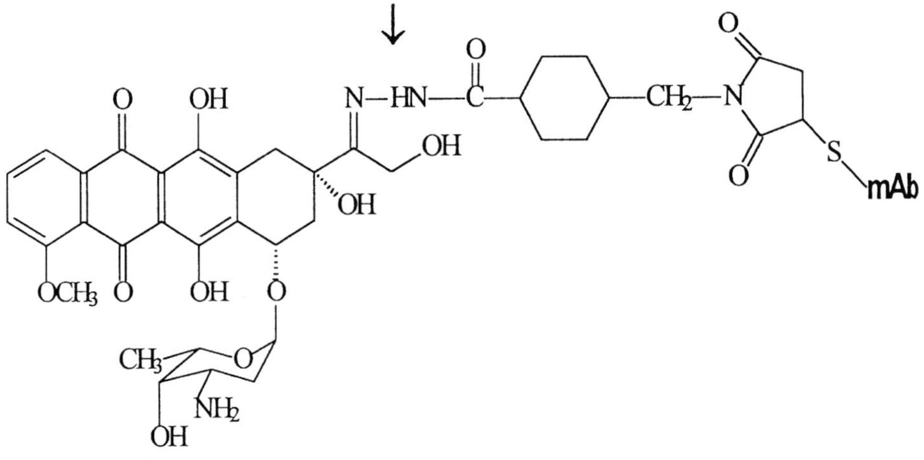

Figure 1 Molecular organization of an antibody–doxorubicin conjugate containing an acid sensitive hydrazone linker (arrow) and a stable thioether bond.

Table 3 Clinical trials of cancer therapy with antibody–drug conjugates

Tumor	*Antibody*	*Drug*	*Response*	*Toxicity*	*Reference*
Colon	Anti-colon	N-acetyl-melphalan	3/9	Serum sickness	Tjandra *et al.*, 1989
	Anti-colon	Neocarzinostatin	3/8	Leukocytosis	Takahashi *et al.*, 1988
Lung	Anti-GA733-2	Methotrexate	1/11	Fever, anorexia, diarrhoea, abdominal pain	Elias *et al.*, 1994
Lung and Colon	Anti-GA733-2	Vinca alkaloid	0/13	Gastrointestinal toxicity, abdominal pain	Schneck *et al.*, 1989
Ovary and Colon	Anti-CEA	Vindesine	0/4	None	Ford *et al.*, 1982
Neuroblastoma	Anti-neuroblastoma (polyclonal)	Daunomycin	3/7	Marrow depression	Melino *et al.*, 1984
		Chlorambucil	2/7	Marrow depression	
Various	Combination	Doxorubicin	5/23	Fever, rash, chills, marrow depression	Oldham *et al.*, 1988
	Combination	Mitomycin C	0/19	Thrombocytopenia, diarrhoea, gastrointestinal toxicity	Orr *et al.*, 1989
	Combination	Doxorubicin	5/43	Fever, chills, rash, marrow depression	Oldham *et al.*, 1989

compared to preclinical studies the results are less than impressive. Some partial and mixed responses were observed in patients with colon and lung tumors, and also in patients with neuroblastoma. In addition to the low tumor localization of the antibodies and the low potency of the drugs, in nearly all patients HAMA responses have precluded the administration of more effective doses. Nevertheless, despite the high failure rate it is noteworthy that toxicity due to the release of free drug from the conjugate was reported in only one study (Oldham *et al.*, 1988). Thus, with the reduction of the toxicity of the free drug at least one of the major aims of targeted cytotoxicity has been achieved. Heterogeneous antigen expression in patient's tumors has also been recognized as an unfavorable property, enabling tumor cells to escape from cytotoxic damage. This problem has been addressed by some investigators using cocktails of different tumor-reactive antibodies with the result of slightly increased response rates (Orr *et al.*, 1989; Oldham *et al.*, 1988; Tjandra *et al.*, 1989).

CHEMICAL IMMUNOTOXINS

An immunotoxin is a macromolecular complex composed of an antibody and an enzymatically active toxin of plant or bacterial origin. Due to their protein nature antibodies and toxins can be linked either chemically using cross-linking reagents or by fusion of their genes to produce single-chain fusion toxins. Whereas the antibody moiety directs the immunotoxin to antigens on the cell surface, the toxin arrests protein synthesis in eukaryotic cells in an irreversible manner. Since this mechanism of action greatly differs from that exerted by conventional chemotherapeutic drugs immunotoxins might prove particularly promising for the treatment of cancers resistant to chemotherapy. Due to the high potency of toxins, a major concern of immunotoxin therapy is that the toxin must be harnessed using a carefully selected targeting system to limit unwanted side effects in patients. In comparison to their recombinant counterparts, chemical immunotoxins are heterogeneous with respect to the number of linkers and the sites of amino acid side chain substitutions introduced into the antibody. Consequently, chemical immunotoxins consist of mixtures of unconjugated, singly- or multiply-substituted antibodies, and further structural heterogeneity of chemical immunotoxins is caused by the attachment of toxins to different sites of the antibodies. Nevertheless, owing to improved protein modification, cross-linking, and purification techniques highly selective and potent immunotoxins can be prepared directly from natural sources and in quantities sufficient for human therapy. Chemical immunotoxins are usually larger in size than recombinant immunotoxins and therefore penetrate solid tumors less efficiently. On the other hand, their longer half-life in the circulation may increase the total dose that localizes to the tumor and is of advantage for the treatment of hematopoietic malignancies.

Several extremely toxic proteins of bacterial and plant origin have been employed for the preparation of chemical immunotoxins. The most commonly used are summarized in Table 4. Intoxication of cells by these proteins results from entry of the catalytic fragments into the cytoplasm of mammalian cells where they

Table 4 Structure and mode of action of toxins used for the preparation of immunotoxins

Toxin	*Source*	*M.W.*	*Structure*	*Mode of action*
Plant				
Ricin	Seeds of castor bean *Ricinus communis*	62 kDa	2 chains (A-ss-B)	Inactivates 28S rRNA
Abrin	Seeds of *Abrus precatorius*	62 kDa	2 chains (A-ss-B)	Inactivates 28S rRNA
Gelonin	Seeds of *Gelonium multiflorum*	30 kDa	1 chain (A)	Inactivates 28S rRNA
Saporin	*Saponaria officinalis*	30 kDa	1 chain (A)	Inactivates 28S rRNA
Bacterial				
Pseudomonas exotoxin A	*Pseudomonas aeruginosa*	66 kDa	1 chain (I-II-III)	ADP-ribosylation of EF-2
Diphtheria toxin	*Corneybacterium diphtheria*	58 kDa	2 chains (A-ss-B)	ADP-ribosylation of EF-2

covalently modify components of the protein synthesis machinery. The bacterial toxins *Pseudomonas aeruginosa* exotoxin A (ETA) and diphtheria toxin (DT) secreted by toxigenic *Corynebacterium diphtheriae* are both synthesized as single-chain polypeptides of 59 and 67 kDa, respectively. Both toxins consist of several functional domains which were defined by crystal structure and deletion analysis (Allured *et al.*, 1986; Hwang *et al.*, 1987; Choe *et al.*, 1992). Distinct domains are responsible for the binding to receptors on the surface of mammalian cells, intracellular processing, and catalytic activity. ETA and DT differ in their molecular organization, but possess identical reaction mechanisms. The catalytic domain ADP-ribosylates and inactivates eukaryotic elongation factor 2 (EF-2) which is an essential component in protein synthesis (Iglewski *et al.*, 1977; Collier *et al.*, 1975). This modification occurs at a post-translationally modified histidine derivative, diphthamide, which is located in the ribsomal binding site of EF-2 (described in detail by Perentesis *et al.*, 1992). Processing of DT by extracellular proteases yields two fragments which are held together by a disulfide bond. The N-terminal A fragment of 21 kDa harbors the catalytic activity; the C-terminal B fragment of 38 kDa mediates cell binding and uptake into the cytoplasm. Due to the ability to manipulate its structure, most progress in the understanding of the mechanism of intoxication has been made with ETA. In contrast to DT the catalytic domain of ETA is located at the C-terminus (Hwang *et al.*, 1987). After cell binding and internalization ETA is cleaved by a cellular protease and a N-terminal 28 kDa and a C-terminal 37 kDa fragment are generated. After reduction of a disulfide bond the C-terminal fragment which contains the enzymatic activity translocates to the cytoplasm (Ogata *et al.*, 1992).

The plant toxins, ricin and abrin consist of two 30 kDa chains, the A and B-chains, which are generated by proteolytic cleavage of a precursor molecule and remain linked via a single disulfide bond (Harley and Lord, 1985). In direct analogy to DT the A-chain harbors the enzymatic activity whereas the B-chain is responsible for cell binding. The A-chain of ricin and abrin catalytically inactivate the 60S ribosomal subunit by a specific N-glycosidic cleavage of the 28S rRNA (Endo *et al.*, 1987). In contrast to the A-chain of ricin, abrin A-chain occurs naturally in a form devoid of side-chain glycosylation and consequently is subject to lower levels of hepatic uptake *in vivo* (Skilleter *et al.*, 1989). A large family of other plant derived 30 kDa ribosome inactivating proteins (RIPs), which occur naturally as single-chain proteins, has also been identified (Stirpe *et al.*, 1986). These RIPs resemble ricin A-chain in size and mode of action, but differ from ricin A-chain and from one another in primary structure, and the degree and type of glycosylation.

To gain access to the protein synthesis machinery, toxins must become internalized via receptor-mediated endocytosis and the catalytic domain must translocate across the membrane of an appropriate intracellular compartment. The processes involved include proteolysis, disulfide bond reduction, binding to intracellular receptors and interaction with the membrane bilayer in a precise temporal and spatial order. The details, however, differ among the various toxins. With the exception of DT, which enters the cytosol directly from acidified endosomes (Hudson *et al.*, 1988), the other toxins prefer a neutral environment for translocation to the cytoplasm, the location of which, however, is still unknown. At least for ricin and ETA there is convincing evidence that these toxins must undergo retrograde transport through the Golgi stack to the endoplasmic reticulum before translocation to the cytoplasm occurs (Wales *et al.*, 1993; Simpson *et al.*, 1995). At the C-terminus of the catalytic domain III of ETA an amino acid sequence has been identified which resembles the consensus sequence KDEL (Chaudhary *et al.*, 1990). This sequence is responsible for signaling the retention of soluble proteins within the endoplasmic reticulum. Binding of the catalytic domain to the KDEL receptor may be an essential step leading to its translocation to the cytosol. A schematic model of the internalization pathway of ETA is shown in Figure 2. Although, as demonstrated with Shiga toxin, a KDEL-like sorting signal is not required for efficient translocation of all toxins (Garred *et al.*, 1995), ricin A-chain modified by adding the KDEL sequence to the C-terminus demonstrates enhanced entry into the cytosol (Wales *et al.*, 1993).

The cytotoxic potency of an immunotoxin is dictated by the extent to which it can undergo the series of steps comprising the intoxication mechanism exploited by the parental toxin. With the exception of the single-chain RIPs specialized domains of native toxins mediate binding and entry into cells via receptors ubiquitously expressed on mammalian cells. Ricin and abrin are lectins containing two binding sites which recognize galactose-terminating glycoproteins and glycolipids (Blakey *et al.*, 1988). ETA and DT recognize the α2-macroglobulin receptor and a heparin-binding EGF-like growth factor precursor, respectively (Kounnas *et al.*, 1992; Naglich *et al.*, 1992). Although potent immunotoxins can be made by linking native toxins to antibodies these conjugates cross-react with normal cells by means of the toxins' cell-binding sites. The most effective way to avoid the serious drawback of non-specific binding and to prepare more tumor-specific immunotoxins is to replace the cell binding function of the toxin with the antibody. This can be achieved by completely eliminating the binding domain of the toxin by chemical modification or genetic engineering and to target the catalytic domain alone. This strategy proved to be successful in the case of immunotoxins made with the A-chain of ricin and abrin (Cumber *et al.*, 1982; Blakey and Thorpe, 1988; Wawrzynczak *et al.*, 1992), and with ETA and DT (FitzGerald and Pastan, 1989; Pastan *et al.*, 1995). However, in contrast to immunotoxins made with native toxins which exploit the natural pathway of toxin entry, immunotoxins made with modified toxins devoid of the cell binding domain rely on the target antigen to mediate entry into the cell. Moreover, their potency depends on the efficiency with which the respective pathway of internalization delivers the enzyme to the translocation compartment. The use of lysosomotropic amines such as ammonium chloride, chloroquine, and the carboxylic ionophore monensin has been recognized as a strategy to potentiate the cytotoxicity of A-chain immunotoxins. These agents increase lysosomal pH and thus may retard toxin degradation (Ohkuma and Poole, 1978). Monensin also blocks the intracellular transport of vesicles from the Golgi apparatus (Tartakoff and Vassalli, 1978). Although these agents are active *in vitro*, their efficacy *in vivo* has not been demonstrated and additional modifications are necessary to make them clinically useful (Weil-Hillman *et al.*, 1985; Roth *et al.*, 1988; Derbyshire *et al.*, 1992).

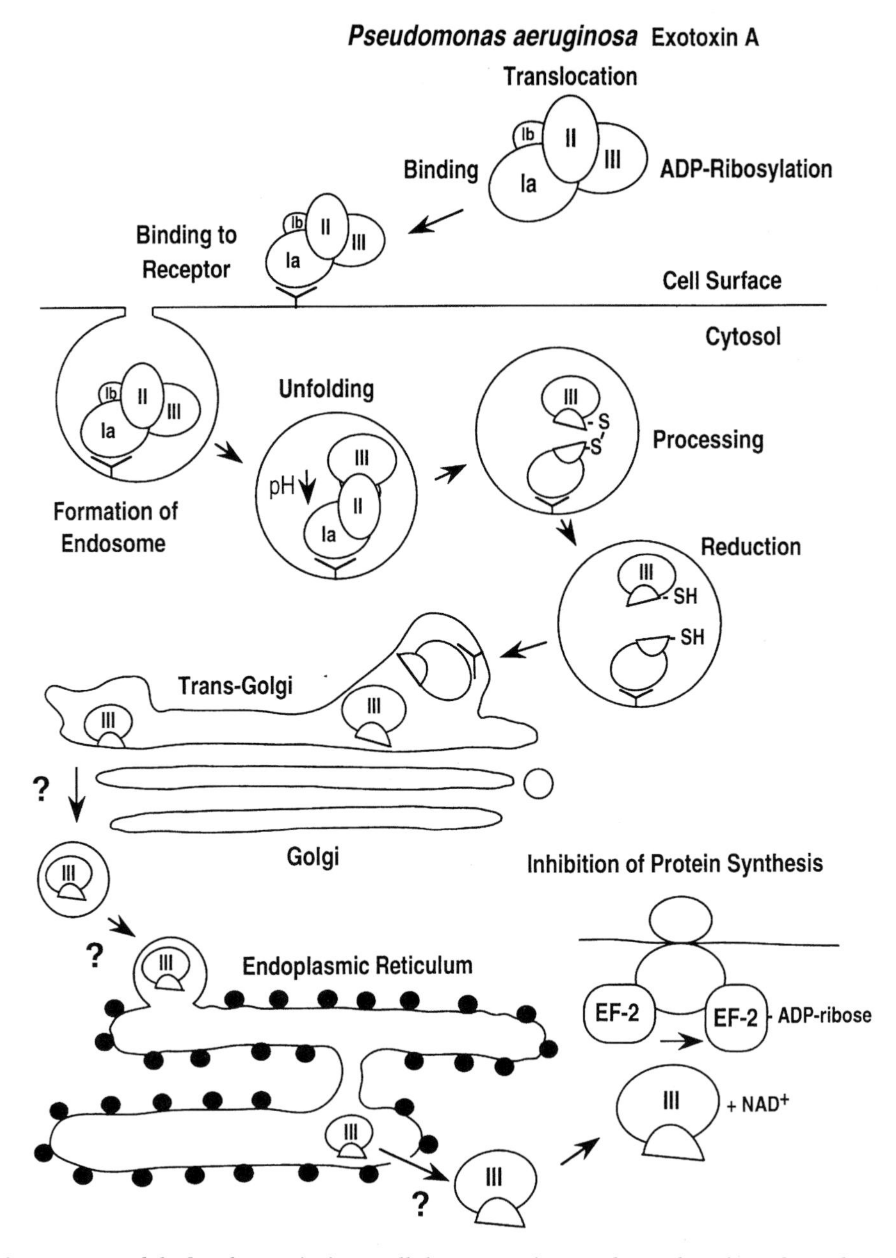

Figure 2 Model of endocytosis, intracellular processing, and translocation of *Pseudomonas aeruginosa* exotoxin A (ETA). After binding to receptors on the cell surface via the N-terminal binding domain Ia of ETA, the toxin-receptor complex is internalized via the endosomal pathway. Acidification of the endosome facilitates unfolding of the toxin which is then cleaved between Arg^{279} and Gly^{280} residues by a cellular protease. After reduction of a disulfide bond the 37 kDa C-terminal fragment is transported through the Golgi network and most likely reaches the endoplasmic reticulum. Translocation into the cytoplasm may occur accross the endoplasmic reticulum membrane (modified after Pastan *et al*., 1992).

A different way to address the problem of non-specific binding of intact toxins is to chemically modify their binding sites. With ricin the galactose binding sites of the B-chain has been "blocked" by the covalent reaction with a glycopeptide containing a triantennary N-linked oligosaccharide from fetuin (Lambert *et al.*, 1991a). Immunotoxins made with blocked ricin reach the cytotoxic activity of native ricin with antibody dependent specificity (Lambert *et al.*, 1991b). However, the two functions of the B-chain, cell binding and translocation, cannot be separated. Therefore, the high potency of these immunotoxins, which is attributed to the preserved translocation function of the B-chain, is dependent on a minimal residual lectin activity of the blocked ricin molecule (Goldmacher *et al.*, 1992). As an alternative approach, the B-chain of ricin, like domain I of ETA, can be obstructed by reacting it with thiolated antibody to form a thioether bond resulting in the sterical hindrance of the toxin's cell binding function (Brusa *et al.*, 1989; Pai *et al.*, 1991; Zangemeister-Wittke *et al.*, 1994; Zimmermann *et al.*, 1997). Other chemical modifications of the toxin moiety have also been employed to enhance the antitumor efficacy of immunotoxins. For example, the carbohydrates present on ricin A-chain, which *in vivo* cause the rapid clearance of the respective immunotoxins by the reticuloendothelial system, can be destroyed producing deglycosylated ricin A-chain (Blakey *et al.*, 1987; Thorpe *et al.*, 1988). Protein engineering techniques now allow the production of these second generation toxins in recombinant form for the preparation of chemical immunotoxins in quantities sufficient for clinical applications.

The Chemistry of Antibody–Toxin Cross-Linking

The covalent linkage of two different proteins by chemical means requires the modification of each protein prior to the cross-linking reaction. Immunotoxins must retain the antigen-binding activity of the antibody as well as the ability of the toxin to exert its cytotoxic effect, a requisite which limits the number of methods applicable for protein modification. Moreover, immunotoxins are most active if composed of a one to one molar ratio of antibody and toxin. Therefore, the method of cross-linking must be controllable and reproducible so that conjugates of defined structure can be routinely constructed in quantities sufficient for use in patients. This, however, is difficult to achieve since chemical modifications of antibodies are heterogeneous with respect to the number of linkers introduced and the sites of amino acid residue side chain substitution. Consequently, chemical immunotoxins usually consist of a mixture of unconjugated, singly or multiply-substituted antibodies requiring efficient purification. Further structural heterogeneity is caused by attachment of toxins to different sites of the antibody. Consequently, only a fraction of the immunotoxin product is endowed with the maximum cell binding and cytotoxic activity of the respective unconjugated components.

The antibody is usually modified with a heterobifunctional cross-linking reagent to introduce a selectively reactive group useful for conjugation. Most often an alkylating function or a sulfhydryl group have served this purpose. Dependent on its structure, the toxin may also be modified. In proteins, several amino acid side chains contain reactive groups that can serve as sites of attachment for cross-linking reagents (described in detail by Wawrzynczak and Thorpe, 1987). The ε-amino group of lysine, which is present on the surface of most proteins, is especially suitable for cross-linking since it reacts with a number of different reagents under conditions that do not affect other chemical groups in the protein. 2-iminothiolane (2-IT) has been frequently used to introduce sulfhydryl groups into primary amines for the reaction with free thiols or maleimide residues provided by cross-linking reagents, to form reducible disulfide bonds or stable thioether bonds, respectively. The native thiol groups of cysteine residues represent other suitable target sites for cross-linkers containing alkylating groups. Alternatively, the cysteinyl thiol group can be reacted with thiol-containing cross-linkers to form disulfide bonds.

Another important requirement for the construction of therapeutically effective immunotoxins is that the linkage is stable in the circulation *in vivo*, but enables release of the toxin from the antibody carrier following binding and internalization into cells. A-chain immunotoxins containing non-reducible linkers are consistently less cytotoxic than their disulfide-bonded counterparts indicating that for the cytotoxic process the reductive cleavage of the disulfide bond is important to release the catalytic moiety into the cytosol (Thorpe and Ross, 1982). For this reason cross-linking reagents have been developed for the preparation of A-chain immunotoxins which contain hindered linkers to protect the disulfide bond from attack by thiolate anions. These linkers are more stable in the circulation and may help to reduce the premature release of the toxin. The cytotoxic activity of the resulting immunotoxins is comparable to that of immunotoxins prepared with conventional cross-linking reagents (Thorpe *et al.*, 1987, 1988).

Clinical Trials with Chemical Immunotoxins

The antitumor efficacy of chemical immunotoxins has been extensively evaluated *in vitro* and in preclinical animal models. Most often the treatment delayed the growth of tumors or prolonged the lifespan of the animals (Byers *et al.*, 1987; Zangemeister-Wittke *et al.*, 1993; Zimmermann *et al.*, 1997; Engert *et al.*, 1990). In some cases complete regressions were also observed (Brinkmann *et al.*, 1991; Winkler *et al.*, 1994). Because these preclinical studies indicated a sufficient therapeutic window between the maximum tolerated dose and immunotoxin concentrations required to achieve antitumor effects, a number of clinical trials with immunotoxins for cancer therapy were initiated. Unfortunately, the hopes raised by the preclinical results have not yet been fulfilled in patients. Despite great efforts therapeutic effects have been prevented by the difficulties associated with conjugation chemistry, large scale production, and by safety aspects. Moreover, the limitations of available animal models, e.g. human tumors growing in athymic mice, became obvious. For example, in patients toxic side effects not predicted by animal studies have been reported (Gould *et al.*, 1989). The results of some published trials are summarized in Table 5. Only limited clinical activity, resulting in few partial responses was seen against solid tumors including melanoma, breast and colon cancer. A phase II study with an NCAM-specific blocked ricin immunotoxin in refractory small cell lung cancer is still in progress (Lynch *et al.*, 1995).

Table 5 Clinical trials of cancer therapy with immunotoxins

Tumor	*Antibody*	*Toxin*	*Response*	*Toxicity*	*Reference*
Colon	Anti-gp72	Ricin A-chain	0/17	Vascular leak syndrom, renal and neuro toxicity	Byers *et al.*, 1989
Small cell lung cancer	Anti-NCAM	Blocked ricin	1/19	Vascular leak syndrom	Lynch *et al.*, 1993
Breast	Anti-gp55	Ricin A-chain (recombinant)	0/4	Vascular leak syndrom, neuropathy	Weiner *et al.*, 1989
Melanoma	Anti-gp240	Ricin A-chain	1/22	Vascular leak syndrom	Spitler *et al.*, 1987
	Anti-gp240	Ricin A-chain	4/20	Vascular leak syndrom	Oratz *et al.*, 1990
Ovary	OVB3	*Pseudomonas* exotoxin A	0/23	CNS toxicity	Pai *et al.*, 1991
Brain	Anti-transferrin rec.	Diphtheria toxin	8/15	None	Laske *et al.*, 1995
Various solid tumors	Anti-Le^y	*Pseudomonas* exotoxin A (recombinant truncated form)	2/32	Vascular leak syndrom, hypoalbuminemia, weight gain	Pai *et al.*, 1995
CLL	Anti-CD5	Ricin A-chain	0/8	Fever	Hertler *et al.*, 1989
B-cell lymphoma	Anti-CD22	Ricin A-chain (deglycosylated)	6/26	Vascular leak syndrom, rhabdomyolysis	Amlot *et al.*, 1993
	Anti-CD19	Blocked ricin	3/25	Vascular leak syndrom, hepatic- and hematologic toxicity	Grossbard *et al.*, 1992
Hodgkin's lymphoma	Anti-CD25	Ricin A-chain (deglycosylated)	1/12	Vascular leak syndrom	Schnell *et al.*, 1995
Cutaneous TCL	Anti-CD5	Ricin A-chain	4/14	Vascular leak syndrom	LeMaistre *et al.*, 1991

The most common dose limiting toxicity associated with immunotoxins is vascular leak syndrome which is likely secondary to non-specific endothelial damage. Moreover, hypoalbuminemia, fluid retention, and weight gain are commonly observed during treatment of patients with immunotoxins. Similar to antibodies, toxins are highly immunogenic and most patients develop an immune response to the toxin which cannot be completely abrogated by immunosuppressive treatment (Oratz *et al.*, 1990). The rapid appearance of host immune responses to antibody and toxin has limited the delivery of multiple courses of therapy (Durrant *et al.*, 1989). Unfortunately, although it is possible to reduce the immunogenicity of antibodies by various engineering techniques, toxins cannot be modified in this sense. Other major side effects in patients treated with immunotoxins were central nervous toxicity, hepatocytotoxicity and encephalopathy (Gould *et al.*, 1989; Spitler *et al.*, 1987; Pai *et al.*, 1991). Recently, Pai *et al.* (1996) have reported results from a phase I trial using the antibody B3, which is directed against a Le^y carbohydrate epitope, conjugated with a recombinant form of ETA lacking the cell binding domain. This LMB-1 immunotoxin exerted antitumor activity in 2 patients, but also produced the expected unwanted side effects. It is hoped that fully recombinant immunotoxins can overcome the limitations of chemical immunotoxins and prove more effective against solid tumors in the clinic.

Chemical immunotoxins are relatively large molecules with low potential to penetrate solid tumors. On the other hand, their long half-life in the circulation makes them ideal for the treatment of hematopoietic malignancies. Thus, not surprisingly, more encouraging clinical activity resulting in some complete responses has been reported in patients with non-Hodgkin's and Hodgkin's lymphomas receiving immunotoxins containing recombinant ricin A-chain or blocked ricin (Grossbard *et al.*, 1992; Amlot *et al.*, 1993; Sausville *et al.*, 1995; Schnell *et al.*, 1995). Moreover, chemical immunotoxins also proved to be highly effective in *ex vivo* purging of bone marrow from contaminating tumor cells (Roy *et al.*, 1991; Myklebust *et al.*, 1993).

RECOMBINANT IMMUNOTOXINS

The elucidation of the molecular organization of bacterial toxins, such as diphtheria toxin from *Corynebacterium diphtheriae* (DT) and exotoxin A from *Pseudomonas aeruginosa* (ETA), and the elimination of the domains responsible for target cell recognition has made it possible to derive recombinant toxin variants which lack the ability to autonomously bind to and enter mammalian cells. Such truncated toxins, produced in high amounts in bacterial expression systems, have been useful for the construction of improved chemical antibody–toxin conjugates (Batra *et al.*, 1989; Theuer *et al.*, 1993). More importantly, they also provide the basis for the construction of a novel class of biotechnologically produced molecules where the native binding domain of the toxin is replaced with a heterologous ligand domain directly linked to the toxin moiety via gene fusion. Their expression in *E. coli* has resulted in the production of cytotoxic proteins with novel target cell specificty with a precise structure and homogeneous composition. Initially cDNA sequences encoding peptide hormones, cytokines, or growth factors have been fused to truncated DT and ETA genes (Murphy *et al.*, 1986; Williams *et al.*, 1987; Chaudhary *et al.*, 1987; Jeschke *et al.*, 1995). Since they irreversibly inactivate the protein synthesis machinery of eukaryotic cells, the production of recombinant DT and ETA and gene fusions derived thereof is limited to bacterial expression systems.

The smallest antibody fragment which can theoretically be generated by enzymatic cleavage of intact antibodies is the 25 kDa Fv molecule, consisting only of the immunoglobulin heavy (V_H) and light chain (V_L) variable domains. Since the two chains of such molecules are not covalently linked they dissociate rapidly and can therefore not be used for the coupling to effector molecules. With the advent of recombinant antibody technology it has become possible to produce Fv fragments which are stable and which can be directly linked to toxins. Recombinant single-chain (sc) Fv fragments consist of V_L and V_H domains covalently linked together by a flexible peptide linker or a disulfide bond (Reiter *et al.*, 1994). They can be routinely produced in bacteria in functional form by cloning and expression of cDNA derived from hybridoma cell mRNA using reverse transcription and polymerase chain reaction (Orlandi *et al.*, 1989; Plückthun, 1992).

Most recombinant antibody–toxins have been constructed using ETA. Single chain antibody–ETA fusion proteins have been derived by replacing the N-terminal binding domain Ia of ETA with the antibody sequence resulting in molecules very similar in size to the original 66 kDa toxin. The molecular organization of such chimeric toxins is schematically shown in Figure 3. Among others single chain antibody–toxins with specificity for the p55 subunit of the interleukin-2 receptor (Chaudhary *et al.*, 1989), the transferrin receptor (Batra *et al.*, 1991), carbohydrate antigens (Brinkmann *et al.*, 1991; Siegall *et al.*, 1994), and growth factor receptors which are frequently overexpressed in human cancers, such as the EGF receptor (Wels *et al.*, 1995) and ErbB-2 (Wels *et al.*, 1992; Batra *et al.*, 1992), have been constructed and shown potent antitumor activity *in vitro* and in animal models.

Due to their small size scFv-ETA fusion proteins generally display a very short half-life in the circulation of mice of 30 minutes or less (Wels *et al.*, 1992) requiring repeated injection of the molecules or continuous infusion in order to achieve potent therapeutic effects in animal experiments. Chemical toxin conjugates with intact antibodies carry two binding sites which allow cross-linking of the antigen on the target-cell surface, in some cases a prerequisite to induce efficient internalization. Due to their limitation to the single antigen binding site of the scFv antibody, recombinant immunotoxins are monovalent and therefore much more dependent on the internalization properties of the antigen and its potential to deliver the toxin to its intracellular target (Wels *et al.*, 1995). Care must also be taken with the choice of the recombinant scFv domain. The yield of biologically active single-chain immunotoxins produced in *E. coli* is highly dependent on the sequence of the individual scFv domain used, and is generally lower than that obtained with recombinant toxins carrying growth factors as a targeting domain. Therefore attempts have been made to improve the yield and

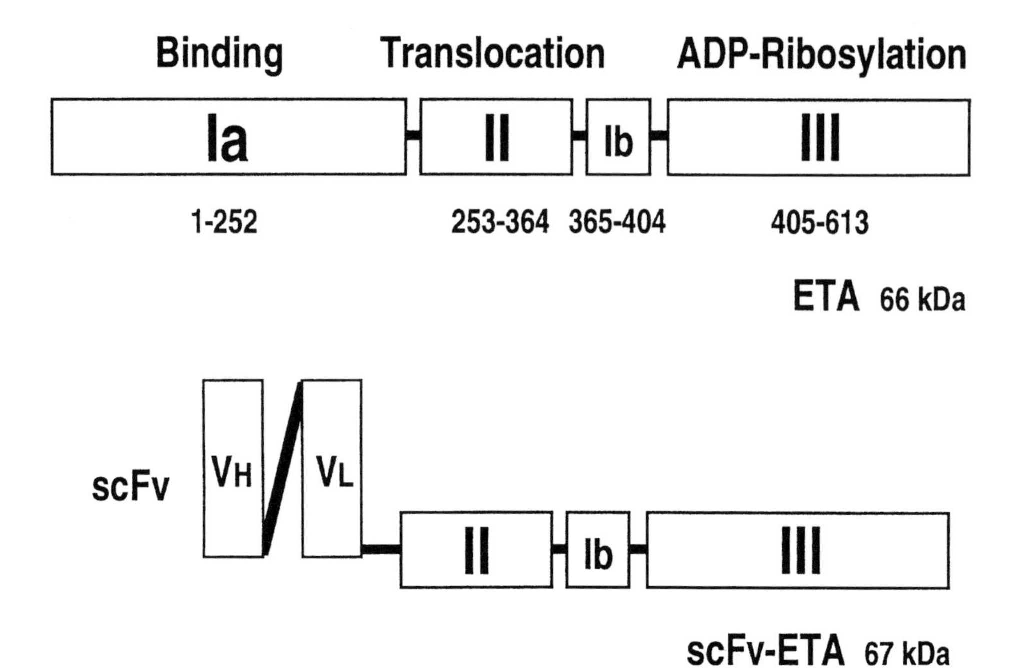

Figure 3 Domain structure of *Pseudomonas aeruginosa* exotoxin A and recombinant single-chain immunotoxins.

quality of recombinant antibody–toxins (Buchner *et al.*, 1992). Furthermore, some scFv-toxins have proven to be stable for a prolonged period of time (Wels *et al.*, 1992), whereas other single-chain immunotoxins, due to unfolding of the scFv domain at 37°C, rapidly lose binding activity.

As an alternative to the derivation of scFv domains, stabilized Fv fragments have also been generated by introducing a disulfide bond between the framework regions of the antibody variable domains (Reiter *et al.*, 1994). Immunotoxins can then be produced by genetically fusing the toxin sequences to either one of the variable domains. In contrast to single-chain immunotoxins, however, the two chains of the molecule have to be independently expressed as recombinant proteins and recombined *in vitro*. Single-chain immunotoxins with unique features can be obtained by combining in a single polypeptide chain two scFv domains of different specificity and the toxin moiety (Schmidt *et al.*, 1996). Such bispecific single-chain immunotoxins show enhanced cytotoxic activity on tumor cells expressing both target antigens and could be useful to improve tumor cell specificity and cellular uptake of the toxin.

Clinical Trials with Recombinant Immunotoxins

Up to now only few data are available on the activity of single-chain immunotoxins in tumor patients. It is generally expected that recombinant antibody–toxins will be safe and effective in the clinic, similar to the promising results obtained in some patients with recombinant single-chain toxins which employ natural ligands as targeting domains (Tepler *et al.*, 1994). Recently, a phase I clinical study has been initiated with LMB-7, a single-chain immunotoxin containing the scFv domain of antibody B3 which is directed against a Le^y carbohydrate epitope found on many solid tumors (Brinkmann *et al.*, 1991; Pai *et al.*, 1996). It will be of special interest to compare the *in vivo* activity of the recombinant LMB-7 directly with that of LMB-1, a chemical immunoconjugate composed of the parental B3 antibody and a truncated form of ETA. Using point mutated DT linked to transferrin for regional therapy of brain tumors, Laske *et al.* (1997) have reported the significant reduction in tumor volume in a phase I clinical trial.

CLOSING REMARKS AND PERSPECTIVES

The development of novel strategies for targeted cancer therapy is a compelling issue and motivated by the high failure rate and the lack of specificity of conventional treatments. The concept of the "magic bullet" against cancer cells is a rational approach to targeted cancer therapy and has been introduced in the clinical setting by the application of antibody–drug and antibody–toxin conjugates. Although the original concept of targeted cytotoxicity has lost some of its attractiveness due to the limitations and disappointing clinical results achieved with solid tumors, it has demonstrated promising therapeutic potential in the treatment of hematopoietic malignancies, localized disease in areas not accessible for surgery, and in *ex vivo* purging studies. Recently, the idea to combine in a single moiety cell recognition and antitumor function has been applied in other areas of ligand-mediated targeting and is the rationale also for the retargeting of effector cells of the immune system. Based on the progress and advances made in molecular oncology and in the understanding of antibody and drug activity, the preparation of a new generation of antibody conjugates has been addressed by many investigators. It is now possible to produce tailor-made antibodies with improved tumor localization properties, and better or more linkage sites for hydrophobic drugs in quantities sufficient for human therapy. In addition, novel cytotoxic drugs of defined activity are available, and genetic engineering techniques and improvements in linker technology have allowed the conjugation of these missiles to antibodies. In the future, major improvements can be expected from developments in three areas. First, identification of truly tumor-specific antigens or of novel tumor-associated antigens which are more selectively expressed on tumor cells compared to normal tissues. Second, the generation of cytotoxic antibody conjugates with defined structure which display improved potency and selectivity, improved tumor localization, lower toxicity and reduced immunogenicity. Third, the use of cytotoxic antibody conjugates in clinical settings which have been more carefully selected so that better responses or even cures become realistic. This could be in combination with agents which have complementary antitumor activity and different unspecific toxicity, or as second-line therapy in the adjuvant setting following debulking by conventional treatment. Since tailor-made antibodies and cytotoxic antibody conjugates produced by state-of-the-art technology, such as humanized antibodies and recombinant fusion toxins, are just on their way to the clinic, ongoing studies may hold promise for the application of Ehrlich's concept to help change the prognosis of metastatic cancer from fatal to curable.

REFERENCES

Allured, V.S., Collier, R.J., Carroll, S.F. and McKay, D.B. (1986) Structure of exotoxin A of Pseudomonas aeruginosa at 3.0-Ångstrom resolution. *Proc. Natl. Acad. Sci. USA*, **83**, 1320–1324.

Amlot, P.L., Stone, M.J., Cunningham, D., Fay, J., Newman, J., Collins, R., May, R., McCarthy, M., Richardson, J., Ghetie, V., Ramilo, O., Thorpe, P.E. and Vitetta, E.S. (1993) A phase I study of an anti-CD22-deglycosylated ricin A chain immunotoxin in the treatment of B-cell lymphomas resistant to conventional therapy *Blood*, **82**, 2624–2633.

Arnon, R. and Sela, M. (1982) *In vitro* and *in vivo* efficacy of conjugates of daunomycin with anti-tumor antibodies. *Immunol. Rev.*, **62**, 5–27.

Batra, J.K., FitzGerald, D.J., Chaudhary, V.K. and Pastan, I. (1991) Single-chain immunotoxins directed at the human transferrin receptor containing Pseudomonas exotoxin A or diphtheria toxin: anti-TFR(Fv)-PE40 and DT388-anti-TFR(Fv). *Mol. Cell. Biol.*, **11**, 2200–2205.

Batra, J.K., Jinno, Y., Chaudhary, V.K., Kondo, T., Willingham, M.C., FitzGerald, D.J. and Pastan, I. (1989) Antitumor activity in mice of an immunotoxin made with anti-transferrin receptor and a recombinant form of Pseudomonas exotoxin. *Proc. Natl. Acad. Sci. USA*, **86**, 8545–8549.

Batra, J.K., Kasprzyk, P.G., Bird, R.E., Pastan, I. and King, C.R. (1992) Recombinant anti-erbB-2 immunotoxins containing *Pseudomonas* exotoxin. *Proc. Natl. Acad. Sci. USA*, **89**, 5867–5871.

Ben-Mahrez, K., Sorokine, I., Thierry, D., Kawasumi, T., Ishii, S., Salmon, R. and Kohiyama, M. (1990) Circulating antibodies against c-myc oncogene products in sera of colorectal cancer patients. *Int. J. Cancer*, **46**, 35–38.

Blakey, D.C. and Thorpe, P.E. (1988) An overview of therapy with immunotoxins containing ricin or its A-chain. *Antibody, Immunoconjugates, and Radiopharmaceuticals*, **1**, 1–16.

Blakey, D.C., Wawrzynczak, W.J., Wallace, P.M., Thorpe, P.E. (1988) Antibody toxin conjugates: A perspective. In H. Waldmann, (ed.), *Monoclonal Antibody Therapy. Prog. Allergy*, Vol. 45, Karger, Basel, pp. 50–90.

Brinkmann, U., Pai, L.H., FitzGerald, D.J., Willingham, M. and Pastan, I. (1991) B3(Fv)-PE38KDEL, a single-chain immunotoxin that causes complete regression of a human carcinoma in mice. *Proc. Natl. Acad. Sci., USA*, **88**, 8616–8620.

Byers, V.S., Pimm, M.V., Scannon, P.J., Pawluczyk, I. and Baldwin, R.W. (1987) Inhibition of growth of human tumor xenografts in athymic mice treated with ricin toxin A chain-monoclonal antibody 791T/36 conjugates. *Cancer Res.*, **47**, 5042–5046.

Brusa, P., Pietribiasi, F., Bussolati, G., Dosio, F., Arione, R., Comoglio, P.M., Prat, M. and Cattel, L. (1989) Blocked and not blocked whole-ricin-antibody immunotoxins: intraperitoneal therapy of human tumour xenografted in nude mice. *Cancer Immunol. Immunother.*, **19**, 185–192.

Buchner, J., Brinkmann, U. and Pastan, I. (1992) Renaturation of single-chain immunotoxin faciliated by chaperones and protein disulfide isomerase. *BioTechnology*, **10**, 682–685.

Burrows, F.J. and Thorpe, P.E. (1994) Vascular targeting – A new approach to the therapy of solid tumors. *Pharmac. Ther.*, **64**, 155–174.

Burrows, F.J., Overholser, J.P. and Thorpe, P.E. (1994) Potent antitumor effects of an antitumor endothelial cell immunotoxin in a murine vascular targeting model. *Cell Biophys.*, **24/25**, 15–25.

Byers, V.S., Rodvien, R., Grant, K., Durrant, L.G., Hudson, K.H., Baldwin, R.W. and Scannon, P.J. (1989) Phase I study of monoclonal antibody ricin A chain immunotoxins XomaZyme 791 in patients with metastatic colon cancer. *Cancer Res.*, **49**, 6153–6160.

Cassidy, J.M. and Douros, J.D., (eds.), (1980) Anti-cancer agents based on natural product models. *Medicinal Chemistry*, **16**, Academic Press, N.Y.

Chabner, B.A. and Collins, J.M. (eds.) (1990) *Cancer chemotherapy: principles and practice*, J.B. Lippincott, PA.

Chari, R.V., Jackel, K.A., Bourret, L.A., Derr, S.M., Tadayoni, B.M., Mattocks, K.M., Shah, S.A., Liu, C., Blattler, W.A. and Goldmacher, V.S. (1995) Enhancement of the selectivity and antitumor efficacy of a CC-1065 analogue through immunoconjugate formation. *Cancer Res.*, **55**, 4079–4084.

Chaudhary, V.K., FitzGerald, D.J., Adhya, S. and Pastan, I. (1987) Activity of a recombiant fusion protein between transforming growth factor type α and *Pseudomonas* toxin. *Proc. Natl. Acad. Sci. USA*, **84**, 4538–4542.

Chaudhary, V.K., Jinno, Y., FitzGerald, D. and Pastan, I. (1990) Pseudomonas exotoxin contains a specific sequence at the carboxyl terminus that is required for cytotoxicity. *Proc. Natl. Acad. Sci. USA*, **87**, 308–312.

Chaudhary, V.K., Queen, C., Junghans, R.P., Waldman, T.A., FitzGerald, D.J. and Pastan, I. (1989) A recombinant immunotoxin consisting of two antibody variable domain fused to Pseudomonas exotoxin. *Nature*, **339**, 394–397.

Choe, S., Bennett, M.J., Fujii, G., Curmi, P.M.G., Kantardjieff, K.A., Collier, R.J. and Eisenberg, D. (1992) The crystal structure of diphtheria toxin. *Nature*, **357**, 216–222.

Colapinto, E., V., Humphrey, P.A., Zalutsky, M.R., Groothuis, D.R., Friedman, H.S., Tribolet, N., Carrel, S. and Bigner, D.D. (1988) Comparative localization of murine monoclonal antibody Mel-14 $F(ab')_2$ fragment and whole IgG2a in human glioma xenografts. *Cancer Res.*, **48**, 5701–5707.

Colcher, D., Bird, R., Roselli, M., Hardman, K.D., Johnson, S., Pope, S., Dodd, S.W., Pantoliano, W., Milenic, D.E. and Schlom, J. (1990) *In vivo* tumor targeting of a recombinant single-chain antigen binding protein. *J. Natl. Cancer Inst.*, **82**, 1191–1197.

Cole, S.P.C., Bhardwai, G., Gerlach, J.H., Mackie, J.E., Grant, C.E., Almquist, K.C., Stewart, A.J., Kurz, E.U., Duncan, A.M.V. and Deeley, R.G. (1992) Overexpression of a transporter gene in a multidrug-resistant human lung cancer cell line. *Science*, **258**, 1650–1654.

Collier, R.J. (1975) Diphtheria toxin: mode of action. *Bacteriol. Rev.*, **39**, 54–85.

Crawford, L.V., Pim, D.C. and Bulbrook, R.D. (1982) Detection of antibodies against the cellular protein p53 in sera from patients with breast cancer. *Int. J. Cancer*, **30**, 403–408.

Cumber, A.J., Forrester, J.A., Foxwell, B.M.J., Ross, W.C.J. and Thorpe, P.E. (1982) Preparation of antibody-toxin conjugates. *Methods in Enzymology*, **112**, 207–225.

Derbyshire, E.J., Stahel, R.A. and Wawrzynczak, E.J. (1992) Potentiation of a weakly active ricin A-chain immunotoxin recognizing the neural cell adhesion molecule. *Clin. exp. Immunol.*, **89**, 336–340.

Dillman, R.O., Johnson, D.E., Shawler, D.L., Koziol, J.A. (1988) Superiority of an acid-labile daunomycin-monoclonal antibody immunoconjugate compared to free drug. *Cancer Res.*, **48**, 6097–6102.

Dillman, R.O., Johnson, D.E., Ogden, J.R. and Beidler, D. (1989) Significance of antigen, drug and tumor cell targets in the preclinical evaluation of doxorubicin, daunorubicin, methotrexate, and mitomycin-C monoclonal antibody immunoconjugates. *Mol. Biother.*, **1**, 250–255.

Durrant, L.G., Byers, V.S., Scannon, P.J., Rodvien, R., Grant, K., Robins, A., Marksman, R.A. and Baldwin, R.W. (1989) Humoral immune responses to XMMCO-791-RTA immunotoxin in colorectal cancer patients. *Clin. exp. Immunol.*, **75**, 258–264.

Dvorak, H.F., Nagy, J.A. and Dvorak, A.M. (1991) Structure of solid tumors and their vasculature implication for therapy with monoclonal antibodies. *Cancer Cell*, **3**, 77–85.

Ehrlich, P. The relationship existing between chemical constitution, distribution and pharmacological action (1956). In F. Himmelweit, M. Marquardt and Sir H. Dale, (eds.), *The Collected Papers of Paul Ehrlich*, Vol. 1, John Wiley, N.Y., pp. 442–447.

Elias, D.J., Kline, L.E., Robbins, B.A., Johnson, H.C.L., Pekny, K., Benz, M., Robb, J.A., Walker, L.E., Kosty, M. and Dillman, R.O. (1994) Monoclonal antibody KS1/4-methotrexate immunoconjugate studies in non-small cell lung carcinoma. *Am. J. Respir. Crit. Care Med.*, **150**, 1114–1122.

Embleton, M.J., Byers, V.S., Lee, H.M., Scannon, P., Blackhall, N.W. and Baldwin, R.W. (1986) Sensitivity and selectivity of ricin toxin A chain monoclonal antibody 791T/36 conjugates against human tumor cell lines. *Cancer Res.*, **46**, 5524–5528.

Endo, Y., Mitsui, K., Motizuki, M. and Tsurugi, K. (1987) The mechanism of action of ricin and related toxin lectins on eukaryotic ribosomes. *J. Biol. Chem.*, **262**, 5908–5913.

Engert, A. and Thorpe, P. (1990) Immunotoxin: from the idea of "magic bullets" to clinical applications. *Med. Klin.*, **85**, 555–560.

Engert, A., Martin, G., Pfreundschuh, M., Amlot, P., Hsu, S.-M., Diehl, V. and Thorpe, P. (1990) Antitumor effects of ricin A chain immunotoxins prepared from intact antibodies and Fab' fragments on solid human hodgkin's disease tumors in mice. *Cancer Res.*, **50**, 2929–2935.

Fenwick, J.R., Philpott, G.W. and Connett, J.M. (1989) Biodistribution and histological localization of anti-human colon cancer monoclonal antibody (Mab) 1A3: the influence of administered antibody dose on tumor uptake. *Int. J. Cancer*, **44**, 1017–1027.

FitzGerald, D. and Pastan, I. (1989) Targeted toxin therapy for the treatment of cancer. *J. Natl. Cancer Inst.*, **81**, 1455–1463.

Fitzpatrick, J.J. and Garnett, M.C. (1995) Studies on the mechanism of action of an MTX-HSA-MoAb conjugate. *Anti-Cancer Drug Design*, **10**, 11–24.

Ford, C.H.J., Newman, C.E., Johnson, J.R., Woodhouse, C.S., Reeder, T.A., Roeland, C.F. and Simmons, R.G. (1982) Localisation and toxicity study of a vindesine-anti-CEA conjugate in patientswith advanced cancer. *Br. J. Cancer*, **47**, 35–42.

Froesch, B.A., Stahel R.A. and Zangemeister-Wittke, U. (1996) Preparation and functional evaluation of new doxorubicin immunoconjugates containing an acid sensitive linker on small cell lung cancer cell lines. *Cancer Immunol. Immunother.*, **42**, 55–63.

Fujimori, K., Covell, D.G., Fletcher, J.E. and Weinstein, J.N. (1989) Modeling analysis of the global and microscopic distribution of immunoglobulin G, $F(ab')_2$, and Fab in tumor. *Cancer Res.*, **49**, 5656–5663.

Galun, E., Shouval, D., Adler, R., Shahaar, M., Wilchek, M., Hurwitz, E. and Sela, M. (1990) The effect of anti-α-fetoprotein-adriamycin conjugate on a human hepatoma. *Hepatology*, **11**, 578–584.

Garred, O., Dubinina, E., Holm, P.K., Olsnes, S., van Deurs, B., Kozlov, J.V. and Sandvig, K. (1995) Role of processing and intracellular transport for optimal toxicity of shiga toxin and toxin mutants. *Exp. Cell Res.*, **218**, 39–49.

Goldmacher, V.S., Lambert, J.M. and Blättler, W.A. (1992) The specific cytotoxicity of immunoconjugates containing blocked ricin is dependent on the residual binding capacity of blocked ricin: evidence that the membrane binding and A-chain translocation activities of ricin cannot be separated. *Biochem. Biophys. Res. Comm.*, **183**, 758–766.

Gottesman, M.M. and Pastan, I. (1993) Biochemistry of multidrug resistance mediated by the multidrug transporter. *Annu. Rev. Biochem.*, **62**, 385–427.

Gould, B.J., Borowitz, M.J., Groves, E.S., Carter, P.W., Anthony, D., Weiner, L.M. and Frankel, A.E. (1989) Phase I study of an anti-breast cancer immunotoxin by continuous infusion: report of a targeted toxic effect not predicted by animal studies. *J. Natl. Cancer Inst.*, **81**, 775–781.

Greenfield, R.S., Kaneko, T., Daues, A., Edson, M., Fitzgerald, K.A., Olech, L.J., Grattan, J.A., Spitalny, G.L. and Braslawsky, G.R. (1990) Evaluation *in vitro* of adriamycin immunoconjugates synthesized using an acid sensitive hydrazone linker. *Cancer Res.*, **50**, 6600–6607.

Grossbard, M.L., Freedman, A.S., Ritz, J., Coral, F., Goldmacher, V.S., Eliseo, L., Spector, N., Dear, K., Lambert, J.M., Blättler, W.A., Taylor, J.A. and Nadler, L.M. (1992) Serotherypy of B-cell neoplasms with anti-B4-blocked ricin: A phase I trial of daily bolus infusion. *Blood*, **79**, 576–585.

Harley, S.M. and Lord, J.M. (1985) *In vitro* endoproteolytic claevage of castor bean lectin precursors *Plant Science*, **41**, 111–116.

Hertler, A.A., Schlossman, D.M. and Borowitz, M.J. (1988) A phase I study of T10-ricin A-chain immunotoxin in refractory chronic lymphocytic leukemia. *J. Biol. Response Mod.*, **7**, 97–113.

Hinman, L.M., Hamann, P.R., Wallace, R., Menendez, A.T., Durr, F.E. and Upeslacis, J. (1993) Preparation and characterization of monoclonal antibody conjugates of the calicheamicins: a novel and potent family of antitumor antibiotics. *Cancer Res.*, **53**, 3336–3342.

Hudson, T.H., Scharff, J., Kimak, M.A.G. and Neville, Jr., D.M. (1988) Energy requirements for diphtheria toxin translocation are coupled to the maintenance of a plasma membrane potential and a proton gradient. *J. Biol. Chem.*, **263**, 4773–4781.

Hurwitz, E., Maron, R., Arnon, R., Wilchek, M. and Sela, M. (1975) The covalent binding of daunomycin and adriamycin to antibodies with retention of both drug and antibody activities. *Cancer Res.*, **35**, 1175–1181.

Hwang, J., FitzGerald, D.J., Adhya, S. and Pastan, I. (1987) Functional domains of Pseudomonas exotoxin identified by deletion analysis of the gene expressed in *E. coli*. *Cell*, **48**, 129–136.

Iglewski, B.H., Liu, P.V. and Kabat, D. (1977) Mechanism of action of pseudomonas aeruginosa exotoxin A: adenosine diphoshate-ribosylation of mammalian elongation factor 2 *in vitro* and *in vivo*. *Infect. Immun.*, **15**, 138–144.

Jain, R.K. (1987) Transport of molecules in the tumor interstitium: A review. *Cancer Res.*, **47**, 3039–3051.

Jain, R.K. (1988) Determinants of tumor blood flow: A review. *Cancer Res.*, **48**, 2641–2658.

Jain, R.K. and Baxter, L.T. (1988) Mechanisms of heterogeneous distribution of monoclonal antibodies and other macromolecules in tumors: Significance of elevated interstitial pressure. *Cancer Res.*, **48**, 7022–7032.

Jeschke, M., Wels, W., Dengler, W., Imber, R., Stöcklin, E. and Groner, B. (1995) Targeted inhibition of tumor-cell growth by recombinant heregulin-toxin fusion proteins. *Int. J. Cancer*, **60**, 730–739.

Juweid, M., Neumann, R., Paik, C., Perez-Bacete, M.J., Sato, J., van Osdol, W. and Weinstein, J.N. (1992) Micropharmacology of monoclonal antibodies in solid tumors: Direct evidence for a binding site barrier. *Cancer Res.*, **52**, 5144–5153.

Kanellos, J., Pietersz, G.A. and McKenzie, I.F.C. (1985) Studies of methothrexate monoclonal antibody conjugates for immunotherapy. *J. Natl. Cancer Inst.*, **75**, 319–332.

Kitamura, K., Takahashi, T., Takashina, K., Yamaguchi, T., Tsurumi, H., Tojokuni, T. and Hakamori, S. (1990) Polyethylene glycol modification of the monoclonal antibody A7 enhances its tumor localization. *Biochem. Biophys. Res. Commun.*, **171**, 1387–1394.

Köhler, G. and Millstein, C. (1975) Continous cultures of fused cells secreting antibody of predetermined specificity. *Nature*, **256**, 495–497.

Kounnas, M.Z., Morris, R.E., Thompson, M.R., FitzGgerald, D.J. Strickland, D.K. and Saelinger, C.B. (1992) The α-macroglobulin receptor/low density lipoprotein receptor-related protein binds and internalizes pseudomonas exotoxin A. *J. Biol. Chem.*, **267**, 12420–12423.

Lambert, J.M., McIntyre, G., Gauthier, M.N., Zullo, D., Rao, V., Steeves, R.M., Goldmacher, V.S. and Blättler, W.A. (1991a) The galactose binding sites of the cytotoxic lectin ricin can be chemically blocked in high yield with reactive ligands prepared by chemical modification of glycopeptides containing triantennary N-linked oligosaccharides. *Biochemistry*, **30**, 3234–3247.

Lambert, J.M., Goldmacher, V.S., Collinson, A.R., Nadler, L.M. and Blättler, W.A. (1991b) An immunotoxin prepared with blocked ricin: A natural plant toxin adapted for therapeutic use. *Cancer Res.*, **51**, 6236–6242.

Laske, D.W., Muraszko, K.M., Oldfield, E.H., DeVroom, H.L., Sung, C., Dedrick, R.L., Simon, T.R., Colandrea, J. Copeland, C., Katz, D., Greenfield, L., Groves, E.S., Houston, L.L., Youle, R.J. (1997) Intraventricular immunotoxin therapy for leptomeningeal neoplasia. *Neurosurgery*, **41**, 1039–1049.

Ledermann, J.A., Begent, R.H. and Bagshawe, K.D. (1988) Cyclosporin A prevents the anti-murine antibody response to a monoclonal anti-tumor antibody in rabbits. *Br. J. Cancer*, **58**, 562–566.

Ledermann, J.A., Marston, N.J., Stahel, R.A., Waibel, R., Buscombe, J.R. and Ell, P.J. (1993) Biodistribution and tumor localisation of ^{131}I SWA11 recognising the cluster w4 antigen in patients with small cell lung cancer. *Br. J. Cancer*, **68**, 119–121.

LeMaistre, C.F., Rosen, S., Frankel, A., Kornfeld, S., Saria, E., Meneghetti, C., Drajesk, J., Fishwild, D., Scannon, P. and Byers, V. (1991) Phase I trial of H65-RTA immunoconjugate in patients with cutaneous T-cell lymphoma. *Blood*, **78**, 1173–1182.

Lynch, T.J., Grossbard, M., Fidias, P., Bartholomay, M., Coral, F., Salgia, R., Elias, A.D., Skarin, A., Sheffner, J., Wen, P., Arinello, P., Bramen, G., Esseltine, D. and Ritz, J. (1995) Immunotoxin therapy of small cell lung cancer (SCLC): Clinical trials of N901-blocked ricin (N901-bR). *Fourth International Symposium on Immunotoxins*, Myrtle Beach, SC.

Matzku, S., Brüggen, J., Bröcker, E.B. and Sorg, C. (1987) Criteria for selecting monoclonal antibodies with respect to accumulation in melanoma tissue. *Cancer Immunol. Immunother.*, **24**, 151–157.

Matzku, S., Moldenhauer, G., Kalkhoff, K., Canavari, S., Colnaghi, M., Schuhmacher, J. and Bihl, H. (1990) Antibody transport and internalization into tumors. *Br. J. Cancer*, **62**, 1–5.

Matzku, S., Tilgen, W., Kalthoff, H., Schmiegel, W.H. and Bröcker, E.B. (1988) Dynamics of antibody transport and internalization. *Int. J. Cancer* (*Suppl.*), **2**, 11–14.

Matzku, S., Brocker, E.B., Bruggen, J., Dippold, W.G. and Tilgen, W. (1986) Modes of binding and internalization of monoclonal antibodies to human melanoma cell lines. *Cancer Res.*, **46**, 3848–3854.

May, R.D., Wheeler, H.T., Finkelman, F.D., Uhr, J.W. and Vitetta, E.S. (1991) Intracellular routing rather than cross-linking or rate of internalization determined the potency of immunotoxins directed against different epitopes of sIgD on murine B cells. *Cell. Immunol.*, **135**, 490–500.

Melino, G., Hobbs, J.R., Radford, M., Cooke, K.B., Evans, A.M., Castello, M.A and Forrest, D.M. (1984) Drug targeting for 7 neuroblastoma patients using human polyclonal antibodies. *Protides. Biol. Fluids*, **32**, 413–416.

Milenic, D.E., Yokota, T., Filpula, D.R., Finkelman. M.A.J., Dodd, S.W., Wood, J.F., Whitlow, M., Snoy, P. and Schlom, J. (1991) Construction, binding properties, metabolism, and tumor targeting of a single chain Fv derived from the pancarcinoma monoclonal antibody CC49. *Cancer Res.*, **51**, 6363–6371.

Murphy, J.R., Bishai, W., Borowski, M., Miyanohara, A., Boyd, J. and Nagle, S. (1986) Genetic construction, expression, and melanoma-selective cytotoxicity of a diphtheria toxin-related alpha-melanocyte-stimulating hormone fusion protein. *Proc. Natl. Acad. Sci. USA*, **83**, 8258–8262.

Myklebust, A.T., Godal, A., Pharo, A., Juell, S. and Fodstad, O. (1993) Eradication of small cell lung cancer cells from human bone marrow with immunotoxins. *Cancer Res.*, **53**, 3784–3788.

Naglich, J.G., Metherall, J.E., Russell, D.W. and Eidels, L. (1992) Expression cloning of a diphtheria toxin receptor: identity with a heparin-binding EGF-like growth factor precursor. *Cell*, **69**, 1051–1061.

Ogata, M., Fryling, C.M., Pastan, I. and FitzGerald, D.J. (1992) Cell-mediated cleavage of Pseudomonas exotoxin between Arg^{279} and Gly^{280} generates the enzymatically active fragment which translocates to the cytosol. *J. Biol. Chem.*, **267**, 25396–25401.

Ohkuma, S. and Poole, B. (1978) Flourescence probe measurement of the intralysosomal pH in living cells and the perturbation of pH by various agents. *Proc. Natl. Acad. Sci. USA*, **75**, 3327–3331.

Oldham, R.K., Lewis, M., Orr, D.W., Avner, B., Liao, S.K., Ogden, J.R., Avner, B. and Birch, R. (1988) Adriamycin custom-tailored immunoconjugates in the treatment of human malignancies. *Mol. Biother.*, **1**, 103–113.

Oldham, R.K., Lewis, M., Orr, D.W., Liao, S.K., Ogden, J.R., Hubbard, W.H. and Birch, R. (1989) Individually specified drug immunoconjugates in cancer treatment. *Int. J. Biol. Markers*, **4**, 65–77.

Oratz, R., Speyer, J.L., Wernz, C.J., Hochster, H., Meyers, M., Mischak, R. and Spitler, L.E. (1990) Antimelanoma monoclonal antibody-ricin-A chain immunoconjugate (XMMME-001-RTA) plus cyclophosphamidein the treatment of metastatic malignant melanoma: results of a phase II trial. *J. Biol. Response Mod.*, **9**, 345–354.

Orlandi, R., Guessow, D.H., Jones, P.T. and Winter, G. (1989) Cloning immunoglobulin variable domains for expression by the polymerase chain reaction. *Proc. Natl. Acad. Sci. USA*, **86**, 3833–3837.

Orr, D., Oldham, R.K., Lewis, M., Ogden, J.R., Liao, S.K., Leung, K., Dupere, S., Birch, R. and Avner, B. (1989) Phase I trial of mitomycin C immunoconjugates cocktails in human malignancies. *Mol. Biother.*, **1**, 229–240.

Pack, P. and Plückthun, A. (1992) Miniantibodies: Use of amphipathic helices to produce functional, flexibly linked dimeric F_v fragments with high affinity in *Escherichia coli*. *Biochemistry*, **31**, 1579–1584.

Pai, L.H., Bookman, M.A., Ozols, R.F. Young, R.C., Smith, J.W., Longo, D.L., Gould, B., Frankel, A., McClay, E.F., Howell, S., Reed, E., Willingham, M.C., Fitzgerald, D.J. and Pastan, I. (1991) Clinical evaluation of intraperitoneal Pseudomonas exotoxin immunoconjugate OVB3-PE in patients with ovarian cancer. *J. Clin. Oncol.*, **9**, 2095–2103.

Pai, L.H., Wittes, R., Setser, A., Willingham, M.C. and Pastan, I. (1996) Treatment of advanced solid tumors with immunotoxin LMB-1: an antibody linked to Pseudomonas exotoxin. *Nat.-Med.*, **2**, 350–353.

Pastan, I., Chaudhary, V. and FitzGerald, D.J. (1992) Recombinant toxins as novel therapeutic agents. *Annu. Rev. Biochem.*, **61**, 331–354.

Pastan, I., Lovelace, E.T., Gallo, M.G., Rutherford, V., Magnani, J.L. and Willingham, M.C. (1991) Characterization of monoclonal antibodies B1 and B3 that react with mucinous adenocarcinomas. *Cancer Res.*, **51**, 3781–3787.

Pastan, I.H., Pai, L.H., Brinkmann, U. and FitzGerald, D.J. (1995) Recombinant toxins: new therapeutic agents for cancer. *Annals New York Acad. Sci. USA*, **758**, 345–354.

Perentesis, J.P, Miller, S.P. and Bodley, J.W. (1992) Protein toxin inhibitors of protein synhesis. *BioFactors*, **3**, 173–184.

Pimm, M.V. (1988) Drug monoclonal antibody conjugates for cancer therapy: potentials and limitations. *CRC Crit. Rev. Therap. Drug. Carrier Syst.*, **5**, 189–227.

Plückthun, A. (1992) Mono- and bivalent antibody fragments produced in *Escherichia coli*: Engineering, folding and antigen binding. *Immunol. Rev.*, **130**, 151–188.

Poznansky, M.J. and Juliano, R.L. (1984) Biological approaches to the controlled delivery of drugs. *Pharmacol. Rev.*, **36**, 277–336.

Preijers, F.W.M.B., Tax, W.J.M., de Witte, T., Janssen, A., v.d. Heijden, H., Vidal, H., Wessels, J.M.C. and Capel, P.J.A. (1988) Relationship between internalization and cytotoxicity of ricin A-chain immunotoxins. *Br. J. Haematol.*, **70**, 289–294.

Press, O.W., Martin, P.J., Thorpe, P.E. and Vitetta, E.S. (1988) Ricin A-chain containing immunotoxins directed against different epitopes on the CD2 molecule differ in their ability to kill normal and malignant cells. *J. Immunol.*, **141**, 4410–4417.

Press, O.W., Farr, A.G., Borroz, K.I., Anderson, S.K. and Martin, P.J. (1989) Endocytosis and degradation of monoclonal antibodies targeting human B-cell malignancies. *Cancer Res.*, **49**, 4906–4912.

Press, O.W., Vitetta, E.S., Farr, A.G., Hansen, J.A. and Martin, P.J. (1986) Evaluation of ricin A chain immunotoxins directed against human T cells. *Cell. Immunol.*, **102**, 10–20.

Ramakrishnan, S. and Houston, L.L. (1984) Comparison of the selective cytotoxic effects of immunotoxins containing ricin A chain or pokeweed antiviral protein and anti-Thy-1.1 monoclonal antibodies. *Cancer Res.*, **44**, 201–208.

Reiter, Y., Brinkmann, U., Jung, S.H., Lee, B., Kasprzyk, P.G., King, C.R. and Pastan, I. (1994) Improved binding and antitumor activity of a recombinant anti-erbB2 immunotoxin by disulfide stabilization of the Fv fragment. *J. Biol. Chem.*, **269**, 18327–18331.

Riechmann, L., Clark, M., Waldmann, H. and Winter, G. (1988) Reshaping human antibodies for therapy. *Nature*, **332**, 323–327.

Roth, J.A., Ames, R.S., Fry, K., Lee, H.M. and Scannon, P.J. (1988) Mediation of reduction of spontaneous and experimental pulmonary metastases by ricin A-chain immunotoxin 45-2D9-RTA with potentiation by systemic monensin in mice. *Cancer Res.*, **48**, 3496–3501.

Sausville, E.A., Headlee, D., Stetler-Stevenson, M., Jaffe, E.S., Solomon, D., Figg, W.D., Herdt, J., Kopp, W.C., Rager, H., Steinberg, S.M., Ghetie, V., Schindler, J., Uhr, J., Wittes, E. and Vitetta, E.S. (1995) Continuous infusion of the anti-CD22 immunotoxin IgG-RFB4-SMPT-dgA in patients with B-cell lymphoma: A phase I study. *Blood*, **85**, 3457–3465.

Scheper, R.J., Broxterman, H.J., Scheffer, G.L., Kaaijk, P., Dalton, W.S., van Heijningen, T.H.M., Kalken, C.K., Sovak, M.L., de Vries, E.G.E., van der Valk, P., Meijer, C.J.L.M. and Pinedo, H.M. (1993) Overexpression of a M_r 110,000 vesicular protein in non-P-glycoprotein-mediated multidrug resistance. *Cancer Res.*, **53**, 1475–1479.

Schmidt, M., Hynes, N.E., Groner, B. and Wels, W. (1996) A bivalent single-chain antibody-toxin specific for ErbB-2 and the EGF receptor. *Int. J. Cancer*, **65**, 538–546.

Schneck, D., Butler, F., Dugan,W., Littrel, D., Dorbecker, S., Peterson, B., Bowsher, R., Delong, A. and Zimmermann, J. (1989) Phase I studies with a murine monoclonal antibody vinca conjugate (KS1/4-DAVLB) in patients with adenocarcinoma. *Antibody, Immunoconj., Radiopharm.*, **2**, 93–100.

Schnell, R., Hatwig, M.T., Radszuhn, A., Cebe, F., Drillich, S., Schön, G., Bohlen, H., Tesch, H., Hansmann, M.L., Schindler, J., Ghetie, V., Uhr, J., Diehl, V., Vitetta, E.S. and Engert, A. (1995) A clinical phase I study of an anti-CD25-deglycosylated ricin A-chain immunotoxin (RFT5-SMPT-dgA) in patients with refractory Hodgkin's disease. *Fourth International Symposium on Immunotoxins*, Myrtle Beach, SC.

Schroff, R.W., Foon, K.A., Beatty, S.M., Oldham, R.K. and Morgan, A.C. (1985) Human anti-murine immunoglobulin responses in patients receiving monoclonal antibody therapy. *Cancer Res.*, **45**, 879–885.

Shawler, D.L., Bartholomew, R.M., Smith, L.M. and Dillman, R.O. (1985) Human immune response to multiple injections of murine monoclonal IgG. *J. Immunol.*, **135**, 1530–1535.

Sheldon, K., Marks, A. and Baumal, R. (1989) Sensitivity of multidrug resistant KB-C1 cells to an antibody-dextran-adriamycin conjugate. *Anticancer Res.*, **9**, 637–642.

Shih, L.B., Goldenberg, D., Xuan, H., Lu, H.W.Z., Mattes, M.J. and Hall, T.C. (1994) Internalisation of an intact doxorubicin immunoconjugate. *Cancer Immunol. Immunother.*, **38**, 92–98.

Shockley, T.R., Lin, K., Nagy, J.A., Tompkins, R.G., Yarmush, M.L. and Dvorak, H.F. (1992) Spatial distribution of tumor-specific monoclonal antibodies in human melanoma xenografts. *Cancer Res.*, **52**, 367–376.

Siegall, C.B., Chace, D., Mixan, B., Garrigues, U., Wan, H., Paul, L., Wolff, E., Hellström, I. and Hellström, K.E. (1994) *In vitro* and *in vivo* characterization of BR96 sFv-PE40. *J. Immunol.*, **152**, 2377–2384.

Simpson, J.C., Dascher, C., Roberts, L.M., Lord, J.M. and Balch, W.E. (1995) Ricin cytotoxicity is sensitive to recycling between the endoplasmic reticulum and the golgi complex. *J. Biol. Chem.*, **270**, 20078–20083.

Sivam, G.P., Martin, P.J., Reisfeld, R.A. and Mueller, B.M. (1995) Therapeutic efficacy of a doxorubicin immunoconjugate in a preclinical model of spontaneous metastatic human melanoma. Cancer Res., **55**, 2352–2356.

Skilleter, D.N., Price, R.J., Parnell, G.D. and Cumber, A.J. (1989) The low uptake of an abrin A chain immunotoxin by rat hepatic cells *in vivo* and *in vitro*. *Cancer Lett.*, **46**, 161–166.

Smith, A., Waibel, R., Westera, G., Martin, A., Zimmerman, A.T. and Stahel, R.A. (1989) Immunolocalisation and imaging of small cell lung cancer xenografts by the IgG2a monoclonal antibody SWA11. *Br. J. Cancer*, **59**, 174–178.

Smyth, M.J., Pietersz, G.A. and McKenzie, I.F.C. (1987) Use of vasoactive agents to increase tumor pefusion and the anti-tumor efficacy of drug-monoclonal antibody conjugates. *J. Natl. Cancer Inst.*, **79**, 1367–1373.

Smyth, M.J., Pietersz, G.A., Classon, B.J. and McKenzie, I.F.C. (1986) Specific targeting of chlorambucil to tumors with the use of monoclonal antibodies. *J. Natl. Cancer Inst.*, **76**, 503–510.

Spitler, L.E. (1988) Clinical studies: solid tumors. *Cancer Treat. Res.*, **37**, 493–514.

Spitler, L.E., del Rio, M., Khentigan, A. Wedel, N.I., Brophy, N.A., Miller, L.L., Harkonen, W.S., Rosendorf, L.L., Lee, H.M., Mischak, R.P., Kawahata, R.T., Stoudemire, J.B., Fradkin, L.B., Bautista, E.E. and Scannon, P.J. (1987) Therapy of patients with malignant melanoma using a monoclonal antimelanoma antibody-ricin A chain immunotoxin. *Cancer Res.*, **47**, 1717–1723.

Starling, J.J., Hinson, N.A., Marder, P., Maciak, R.S. and Laguzzo, B.C. (1988) Rapid internalization of antigen immunoconjugate complexes is not required for anti-tumor activity of monoclonal antibody-drug conjugates. *Antibody Immunoconj. Radiopharm*, **1**, 311–324.

Stirpe, F. and Barbieri, L. (1986) Ribosome-inactivation proteins up to date. *FEBS Lett.*, **195**, 1–8.

Sung, C., van Osdol, W.W., Saga, T., Neumann, R.D., Dedrick, R.L. and Weinstein, J.N. (1994) Streptavidin distribution in metastatic tumors pretargeted with a biotinylated monoclonal antibody: Theoretical and experimental pharmacokinetics. *Cancer Res.*, **54**, 2166–2175.

Takahashi, T., Yamaguchi, T., Kitamura, K., Suzuyama, H., Homda, M., Yokota, T., Kotanagi, H., Takahashi, M. and Hashimoto, Y. (1988) Clinical application of monoclonal antibody-drug conjugates for immunotargeting chemotherapy of colorectal carcinoma. *Cancer*, **61**, 881–888.

Tartakoff, A. and Vassalli, P. (1978) Comparative studies of intracellular transport of secretory proteins. *J. Cell. Biol.*, **79**, 694–707.

Tepler, I., Schwartz, G., Parker, K., Charette, J., Kadin, M.E., Woodworth, T.G. and Schnipper, L.E. (1994) Phase I trial of an interleukin-2 fusion toxin (DAB486IL-2) in hematologic malignancies: complete response in a patient with Hodgkin's disease refractory to chemotherapy. *Cancer*, **73**, 1276–1285.

Theuer, C.P., Kreitman, R.J., FitzGerald, D.J. and Pastan, I. (1993) Immunotoxins made with a recombinant form of Pseudomonas exotoxin A that do not require proteolysis for activity. *Cancer Res.*, **53**, 340–347.

Thielemans, K., Maloney, D.G., Meeker, T., Fujimoto, J., Doss, C., Warnke, R.A., Bind, J., Gralow, J., Miller, R.A. and Levy, R. (1984) Strategies for production of monoclonal anti-idiotype antibodies against human B-cell lymphomas. *J. Immunol.*, **133**, 495–501.

Thomas, G.D., Chappel, M.J., Dykes, P.W., Ramsden, D.B., Godfrey, K.R., Ellis, J.R.M. and Bradwell, A.R. (1989) Effect of dose, molecular size, affinity, and protein binding on tumor uptake of antibody or ligand: A Biomathematical Model. *Cancer Res.*, **49**, 3290–3296.

Thorpe, E.J., Wallace, P.M.-, Knowles, P.P., Relf, M.G., Brown, A.N.F., Watson, G.J., Blakey, D.C. and Newell, D.R. (1988) Improved antitumor effects of immunotoxins prepared with deglycosylated ricin A-chain and hindered disulfide linkages. *Cancer Res.*, **48**, 6396–6403.

Thorpe, P.E. and Ross, W.C.J. (1982) The preparation and cytotoxic properties of antibody-toxin conjugates. *Immunol. Rev.*, **62**, 119–158.

Thorpe, P.E., Wallace, P.M., Knowles, P.P., Relf, M.G., Brown, A.N., Watson, G.J., Knyba, R.E., Wawrzynczak, E.J. and Blakey, D.C. (1987) New coupling agents for the synthesis of immunotoxins containing a hindered disulfide bond with improved stability *in vivo*. *Cancer Res.*, **47**, 5924–5931.

Till, M., May, R.D., Uhr, J.W., Thorpe, P.E. and Vitetta, E.S. (1988) An assay that predicts the ability of monoclonal antibodies to form potent ricin A chain-containing immunotoxins. *Cancer Res.*, **48**, 1119–1123.

Tjandra, J.J., Pietersz, G.A., Teh, T.J., Cuthbertson, A.M., Sullivan, J.R., Penfold, C. and McKenzie, I.F.C. (1989) Phase I clinical trial of drug monoclonal antibody conjugates in patients with advanced colorectal carcinoma: A preliminary report. *Surgery*, **106**, 533–545.

Trail, P.A., Willner, D., Lasch, S.J., Henderson, A.J., Hofstead, S., Casazza, A.M., Firestone, R.A., Hellström, I. and Hellström, K.E. (1993) Cure of xenografted human carcinomas by BR96-doxorubicin immunoconjugates. *Science*, **261**, 212–214.

Upeslacis, J. and Hinman, L. (1988) Chemical modification of antibodies for cancer chemotherapy. In Salzmann, (ed), *Annual Reports in Medicinal Chemistry*, Vol. 23, Academic Press, pp. 151–160.

Wales, R., Roberts, L.M. and Lord, J.M. (1993) Addition of an endoplasmic reticulum retrieval sequence to ricin A chain significantly increases its cytotoxicity to mammalian cells. *J. Biol. Chem.*, **268**, 23986–23990.

Wargalla, U.C. and Reisfeld, R.A. (1989) Rate of internalization of an immunotoxin correlates with cytotoxic activity against human tumor cells. *Proc. Natl. Acad. Sci. USA*, **86**, 5146–5150.

Wawrzynczak, E.J. and Thorpe, P.E. (1987) Methods for preparing immunotoxins: effect of the linkage on activity and stability. In C.-W. Vogel, (ed), *Immunoconjugates. Antibody Conjugates in Radioimmaging and Therapy of Cancer*, Oxford University Press , N.Y., pp. 28–55.

Wawrzynczak, E.J., Zangemeister-Wittke, U., Waibel, R., Henry, R.V., Parnell, G.D., Jones, M. and Stahel, R.A. (1992) Molecular and biological properties of an abrin A chain immunotoxin designed for therapy of human small cell lung cancer. *Br. J. Cancer*, **66**, 361–366.

Weil-Hillman, G., Runge, W., Jansen, F.K. and Vallera, D.A. (1985) Cytotoxic effect of anti–M_r 67,000 protein immunotoxins on human tumors in a nude mouse model. *Cancer Res.*, **45**, 1328–1336.

Weiner, L.M., O'Dwyer, J., Kitson, J., Comis, R.L., Frankel, A.E., Bauer, R.J., Konrad, M.S. and Groves, E.S. (1989) Evaluation of an anti-breast carcinoma monoclonal antibody 260F9-recombinant ricin A chain immunoconjugate. *Cancer Res.*, **49**, 4062–4067.

Wels, W., Beerli, R., Hellmann, P., Schmidt, M., Marte, B.M., Kornilova, E.S., Hekele, A., Mendelsohn, J., Groner, B. and Hynes, N.E. (1995) EGF receptor and $p185^{erbB-2}$-specific single-chain antibody toxins differ in their cell-killing activity on tumor cells expressing both receptor proteins. *Int. J. Cancer*, **60**, 137–144.

Wels, W., Harwerth, I.M., Mueller, M., Groner, B. and Hynes, N.E. (1992) Selective inhibition of tumor cell growth by a recombinant single-chain antibody-toxin specific for the erbB-2 receptor. *Cancer Res*, **52**, 6310–6317.

Weltman, J.K., Pedroso, P., Johnson, S.A., Davignon, D., Fast, L.D. and Leone, L.A. (1987) Rapid screening with indirect immunotoxin for monoclonal antibodies against human small cell lung cancer. *Cancer Res.*, **47**, 5552–5556.

Werlen, R.C., Lankinen, M., Smith, A., Chernushevich, I., Standing, K.G., Blakey, D.C., Shuttleworth, H., Melton, R.G., Offord, R.E. and Rose, K. (1995) Site-specific immunoconjugates. *Tumor Targeting*, **1**, 251–258.

Williams, D.P., Parker, K., Bacha, P., Bishai, W., Borowski, M., Genbauffe, F., Strom, T.B. and Murphy, J.R. (1987) Diphtheria toxin receptor binding domain substitution with interleukin-2: genetic construction and properties of a diphtheria toxin-related interleukin-2 fusion protein. *Protein Engineering*, **1**, 493–498.

Willner, D. Trail, P.A., Hofstead, S.J., King, H.D., Lasch, S.J., Braslawsky, G.R., Greenfield, R.S., Kaneko, T. and Firestone, R.A. (1993) (6-Maleimidocaproyl)hydrazone of doxorubicine- A new derivative for the preparation of immunoconjugates of doxorubicin. *Bioconjugate Chem.*, **4**, 521–527.

Winkler, U., Gottstein, C., Schön, G., Kaoo, U., Wolf, J., Hansmann, M.-L., Bohlen, P., Thorpe, P., Diehl, V. and Engert, A. (1994) Successful treatment of disseminated human Hodgkin's disease in SCID mice with deglycosylated ricin A-chain immunotoxins. *Blood*, **83**, 466–475.

Yokota, T., Milenic, D.E., Whitlow, M. and Schlom, J. (1992) Rapid tumor penetration of a single-chain Fv and comparison with other immunoglobulin forms. *Cancer Res.*, **52**, 3402-3408.

Zangemeister-Wittke, U., Lehmann, H.P., Waibel, R., Wawrzynczak, E.J. and Stahel, R.A. (1993) Action of a CD24-specific deglycosylated ricin A chain immunotoxin in conventional and novel models of small cell lung cancer xenograft. *Int. J. Cancer*, **53**, 521–528.

Zangemeister-Wittke, U., Frösch, B., Collinson, A.R., Waibel, R., Schenker, T. and Stahel, R.A. (1994) Immunotoxins recognizing a new epitope on the 140-kDa isoform of the neural cell adhesion molecule have potent cytotoxic effects against small cell lung cancer cells. *Br. J. Cancer*, **69**, 32–39.

Zhou, S.Q., Wang, N.Q., Liu, T. and Dong, Z.W. (1992) Experimental study of tumor directed therapy with gastric cancer monoclonal antibody-mitomycin conjugate combined with propanolol or angiotensin II. *Yao Hsueh Hsueh Pao*, **27**, 891–894.

Zhu, Z., Kralovec, J., Ghose, T. and Mammen, M. (1995) Inhibition of Epstein-Barr-virus-transformed human chronic lymphocytic leukemic B cells with mAb.

Zimmermann, S., Wels, W., Froesch, B.A., Gerstmayer, B., Stahel, R.A. and Zangemeister-Wittke, U. (1997) A novel immunotoxin recognising the epithelial glycoprotein-2 has potent anti-tumoural activity on chemotherapy resistant lung cancer. *Cancer Immunol. Immunother.*, **44**, 1–9.

4. SELECTIVE DRUG DELIVERY USING TARGETED ENZYMES FOR PRODRUG ACTIVATION

NATHAN O. SIEMERS and PETER D. SENTER

Bristol-Myers Squibb Pharmaceutical Research Institute, 3005 First Avenue, Seattle, Washington 98121, USA

INTRODUCTION

A major limitation in the chemotherapeutic treatment of cancer results from the lack of tumor specificity displayed by the drugs currently in use (DeVita, 1993). Because of this, a great deal of research has focused on the development of new chemotherapeutic agents that are able to more effectively exploit differences between neoplastic and normal tissues. One approach has been to prepare inactive drug precursors, known as prodrugs, that can be activated by enzymes or physiological conditions associated with cancer cells and tumor masses (Sinhababu *et al.*, 1996). Unfortunately, progress in this area has been hampered by difficulties in finding suitably selective prodrug activation pathways. Alternatively, several drug targeting strategies are based on the preferential expression of a variety of antigens on tumor cell surfaces. Monoclonal antibodies (mAbs) against these antigens have been used to deliver chemotherapeutic drugs (Hellström *et al.*, 1995; Canevari *et al.*, 1994), potent plant and bacterial toxins (Brinkmann *et al.*, 1994; Siegall *et al.*, 1995), and radionuclides (Schubiger *et al.*, 1995) to tumors. However, with few exceptions (Press *et al.*, 1993), clinical success with such mAb-based targeting approaches has proven to be elusive, since mAbs often do not penetrate well into solid tumors and can localize in very heterogeneous manners (Dvorak *et al.*, 1991; Shockley *et al.*, 1991).

These limitations have prompted considerable research into the potential of using mAbs for the delivery of enzymes to tumor cell surfaces. The enzymes are selected for their abilities to convert anticancer prodrugs into active antitumor agents. Thus, the necessary conditions for selective prodrug activation are created when the conjugate binds to tumor-associated antigens. This approach has been called antibody directed enzyme-prodrug therapy (ADEPT), or antibody directed catalysis (ADC). In the past several years, many enzyme/prodrug combinations have been utilized in this targeting strategy (reviewed in Senter *et al.*, 1993; Jungheim *et al.*, 1994; Bagshawe *et al.*, 1994). In this chapter, we will provide an overview of this rapidly developing field.

MAB–ENZYME CONJUGATES FOR PRODRUG ACTIVATION

The basic scheme of prodrug activation by targeted enzymes is illustrated in Figure 1. The first step involves systemic administration of a mAb–enzyme conjugate. The immunoconjugate can be prepared by chemically linking whole antibody, Fab, Fab′ or $F(ab')_2$ fragments to an enzyme of interest. Alternatively, the conjugate can be a fusion protein, produced by fusing antibody variable region genes to the gene encoding the enzyme and expressing the recombinant protein. Depending on the pharmacokinetics of the particular conjugate being used, it may take anywhere from several hours to several days for the conjugate to localize in tumors and clear from non-target tissue. The second step of this therapy is systemic prodrug administration. Ideally, the prodrug should be non-toxic, resistant to the action of endogenous enzymes, and be converted into active drug only by the targeted enzyme.

Therapy with mAb–enzyme conjugates for prodrug activation may offer numerous advantages over other antibody-based approaches. Since the enzyme behaves as a catalyst, a single conjugate molecule at a tumor site should be able to generate a large amount of active drug. In contrast, the amount of active drug

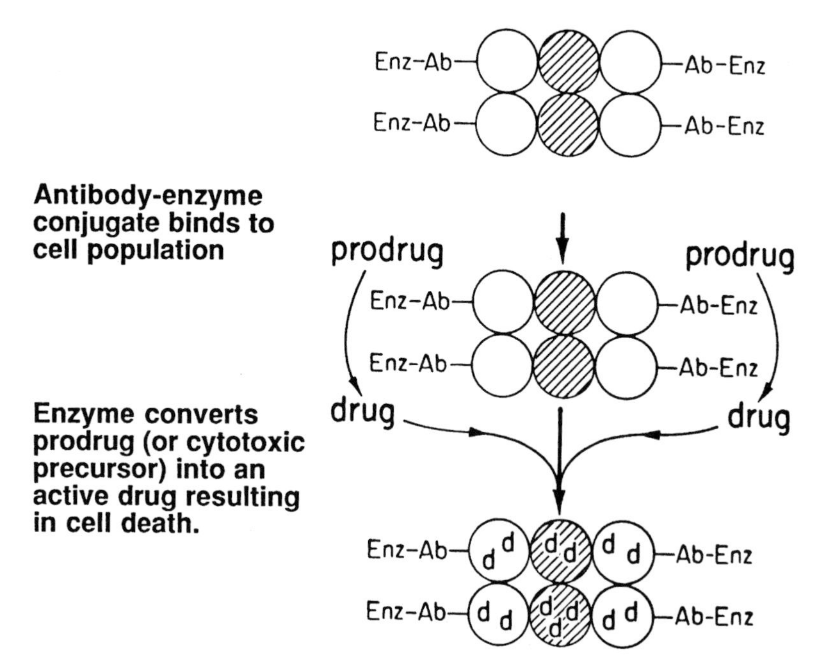

Figure 1 Generation of cytotoxic agents by targeted enyzmes. Hatched circles represent cells that have not bound antibody–enzyme conjugate, and internalized drug is abbreviated with the letter “d”.

that can be delivered to tumors through a direct mAb–drug conjugate is necessarily limited by the small amount of conjugate that localizes within human tumor masses. Another possible advantage is that the targeted drug is not covalently bound to the immunoconjugate. As a small molecule, it will be free to diffuse throughout a tumor, even if the enzyme conjugate is primarily confined to the tumor periphery, or is not bound to all of the cells within a tumor mass.

GENERATION OF CHEMOTHERAPEUTIC AGENTS BY TARGETED ENZYMES

The first demonstration that a cytotoxic agent could be generated by a targeted enzyme was reported before the advent of mAb technology (Philpott *et al.*, 1973). The enzyme glucose oxidase was conjugated to polyclonal antisera. Glucose oxidase produces hydrogen peroxide as a byproduct of the oxidation of glucose. When antigen-positive cells were treated with the immunoconjugate in the presence of glucose, little cytotoxicity was observed. However, the generation of hydrogen peroxide could be coupled to iodine production via the enzyme lactoperoxidase and iodide. In the presence of immunoconjugate, glucose, iodide salts, and lactoperoxidase, significant cytotoxic activity was observed on antigen-positive cells. In addition, it was shown that in mixed cell populations, preferential killing of antigen positive cells took place.

Since these studies, many applications of targeted oxidase enzymes have been reported (Ito *et al.*, 1989). Glucose oxidase and lactoperoxidase were linked to a pan-leukocyte specific mAb and it was shown that the conjugates in combination with glucose and iodide led to cytotoxic activities on T-cells (Ito *et al.*, 1990). Immunologically specific cytotoxic activities have also been obtained with conjugates of xanthine oxidase in combination with xanthine or hypoxanthine (Battelli *et al.*, 1988; Dinota *et al.*, 1990). Evidence that the activities were due to the formation of reactive oxygen species was based on the abilities of catalase and superoxide dismutase to neutralize the cytotoxic effects. Consistent results were obtained in recently reported *in vivo* studies using a subcutaneous tumor model in rabbits (Yoshikawa *et al.*, 1995). It was found that antitumor activity could be obtained by localized administration of xanthine oxidase followed by systemic hypoxanthine treatment.

Collectively, these studies indicate that significant cytotoxic activities can be obtained using targeted oxidases in combination with glucose, iodide, xanthine and hypoxanthine as substrates. A substantial body of evidence suggest that these combinations lead to the formation of cytotoxic oxidative species. This approach shows promise for the selective elimination of T-cells from bone marrow. The *in vivo* activities appear to be promising, in light of the abundance of serum proteins capable of neutralizing the active oxygen species formed.

The strategy of using targeted enzymes for the generation of cytotoxic agents has been extended to include more conventional anticancer drugs (reviewed in Senter *et al.*, 1993; Jungheim *et al.*, 1994; Bagshawe *et al.*, 1994). Table 1

Table 1 Targeted enzymes for prodrug activation

Enzyme	*Released drug*	*Reference*
Alkaline phosphatase	Etoposide	Senter *et al.*, 1988, 1989; Haisma *et al.*, 1992b
	Mitomycin	Senter *et al.*,1989; Sahin *et al.*, 1990
	Doxorubicin	Senter, 1990
	Phenol mustard	Wallace *et al.*, 1991
Arylsulfatase	Etoposide	Senter *et al.*, 1995b
	Phenol mustard	Senter *et al.*, 1995b
Carboxypeptidase A	Methotrexate	Haenseler *et al.*, 1992; Keufner *et al.*, 1989; Esswein *et al.*, 1991; Vitols *et al.*, 1995; Perron *et al.*, 1996
Carboxypeptidase G2	Nitrogen mustards	Antoniw *et al.*, 1990; Bagshawe *et al.*, 1988; Springer *et al.*, 1990, 1991, 1995; Sharma *et al.*, 1990, 1991; Eccles *et al.*, 1994; Blakey *et al.*, 1995
Cytosine deaminase	5-Fluorouracil	Senter *et al.*, 1991; Wallace *et al.*, 1994
β-Galactosidase	Daunorubicin	Gesson *et al.*, 1994
	Doxorubicin	Azoulay *et al.*, 1995
β-Glucosidase	Cyanide	Rowlinson-Busza *et al.*, 1992
	Daunorubicin	Leenders *et al.*, 1995
β-Glucuronidase	Phenol mustard	Roffler *et al.*, 1991, 1992
	Doxorubicin	Bosslet *et al.*, 1994
	Epirubicin	Haisma *et al.*, 1992a
β-Lactamase	Vinca alkaloid	Meyer *et al.*, 1992, 1993; Shepherd *et al.*, 1991; Jungheim *et al.*, 1992
	Nitrogen mustards	Kerr *et al.*, 1995; Vrudhula *et al.*, 1993b; Alexander *et al.*, 1991; Svensson *et al.*, 1992
	Doxorubicin	Svensson *et al.*, 1995; Kerr *et al.*, 1995; Vrudhula *et al.*, 1995; Rodrigues *et al.*, 1995b; Senter *et al.*, 1995a
	Carboplatinum analogue	Hanessian *et al.*, 1993
	Paclitaxel	Rodrigues *et al.*, 1995a
Nitroreductase	Actinomycin	Mauger *et al.*, 1994
	Mitomycin	Mauger *et al.*, 1994
	Nitrogen mustards	Mauger *et al.*, 1994
	Enediyne	Hay *et al.*, 1995
Penicillin amidase	Doxorubicin	Kerr *et al.*, 1990; Vrudhula *et al.*, 1993
	Melphalan	Kerr *et al.*, 1990; Vrudhula *et al.*, 1993
	Palytoxin	Bignami *et al.*, 1992

summarizes the enzyme/prodrug combinations that have been reported. Prodrug forms of important anticancer agents such as doxorubicin, etoposide, methotrexate, 5-fluorouracil, melphalan, and paclitaxel have been described. In addition, the feasibility of using prodrugs of agents that have either failed or would likely fail in the clinic due to systemic toxicity is evident from studies utilizing prodrugs of agents such as palytoxin, potent alkylating agents (phenol and phenylenediamine mustards), and cyanide. In principle, this delivery strategy should broaden the scope of drugs that can be used clinically, since drug release takes place in a site-selective manner.

The *in vitro* activities of some mAb–enzyme/prodrug combinations are illustrated in Figure 2. Etoposide phosphate (Figure 2A) provided the first demonstration that an anticancer prodrug could be activated in a site specific manner by a mAb–enzyme conjugate (Senter *et al.*, 1988). It was found that etoposide phosphate was approximately 100-fold less cytotoxic than etoposide on H3347 colorectal carcinoma cells. Significant levels of cytotoxic activity were obtained by exposing the cells to L6-alkaline phosphatase, a conjugate that bound to cell-surface antigens, prior to etoposide phosphate treatment. This activation was immunologically specific, since a non-binding control conjugate (1F5-alkaline phosphatase) did not effect prodrug activation. Similar results were obtained with prodrugs of 5-fluorouracil on H2981 cells (Figure 2B) and doxorubicin on H2987 cells (Figure 2C) which were activated by mAb conjugates of cytosine deaminase (Senter *et al.*, 1991) and β-lactamase (Svensson *et al.*, 1995), respectively. These results are indicative of many other such *in vitro* studies, and illustrate the wide range of anticancer drugs that can be generated by mAb–enzyme conjugates.

IN VIVO THERAPEUTIC EFFICACY

The potential utility of mAb–enzyme/prodrug therapy for the treatment of human cancers has now been clearly demonstrated in a large number of *in vivo* therapeutic models. Several therapy regimens, using mAb conjugates of β-lactamase and carboxypeptidase enzymes, are presented below in detail. In nearly every case reported, the antitumor activities are significantly greater than what can be achieved at the highest possible systemic dose of prodrug or the parent drug. Experimentally, it is necessary in each mAb–enzyme/prodrug system and therapy model to demonstrate that any anti-tumor effects observed are immunologically specific and not due simply to an altered method of administration of the drug. In general, immunological specificity is demonstrated using non-binding mAb conjugates prepared in the same manner as the binding counterpart, since macromolecules localize non-specifically within solid tumor masses (Dvorak, 1990).

β-Lactamases have proven to be valuable prodrug converting agents due to their high activities as well as their broad substrate specificities, allowing for the activation of multiple prodrug substrates. A wide range of anticancer drugs have been appended to cephalosporins, such that β-lactamase-catalyzed hydrolysis leads to drug elimination through a fragmentation reaction. The first report of *in vivo* activity in a mAb–enzyme/prodrug system utilized the β-lactamase enzyme from *Enterobacter cloacae* and a cephalosporin–vinca alkaloid prodrug (Meyer *et al.*, 1993). The enzyme was linked to antibody Fab′ fragments recognizing the CEA, TAG-72, and KS1/4 antigens on tumor tissues. The therapeutic effects of

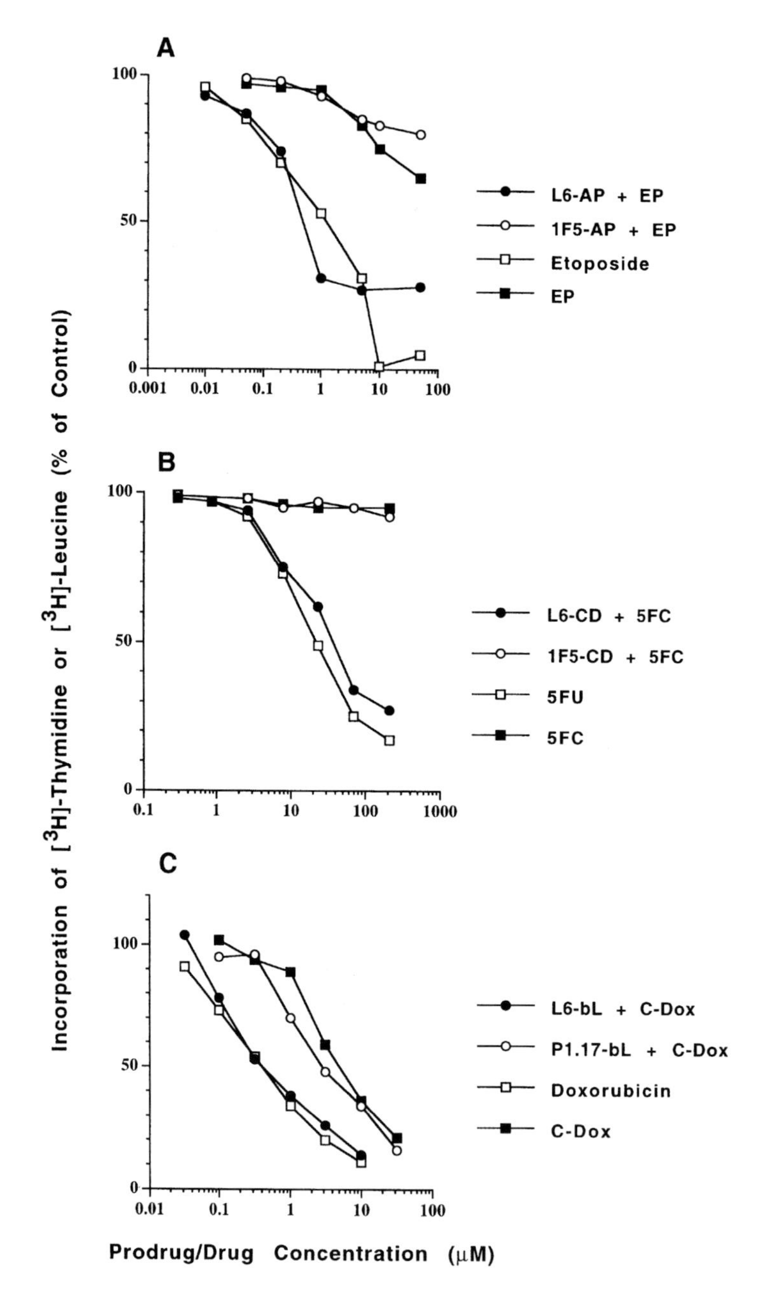

Figure 2 *In vitro* cytotoxicity of mAb–enzyme/prodrug combinations on carcinoma cell lines (L6 antigen positive, 1F5 and P1.17 antigen negative). (A) Effects of etoposide phosphate (EP), etoposide, and combinations of mAb–alkaline phosphatase (mAb–AP) conjugates with EP on H3347 colorectal carcinoma cells (Senter *et al.*, 1988). (B) Effects of 5-fluorocytosine (5FC), 5-fluorouracil (5FU), and combinations of mAb–cytosine deaminase (mAb–CD) conjugates with 5FC on H2981 lung adenocarcinoma cells (Senter *et al.*, 1991). (C) Effects of doxorubicin, C-Dox, and combinations of mAb–β-lactamase (mAb–bL) conjugates with C-Dox on H2987 lung adenocarcinoma cells (Svensson *et al.*, 1995).

each of these mAb–enzyme conjugates in combination with the vinca prodrug were studied in nude mouse models of human colorectal carcinoma. In all cases, the therapeutic effects of the antitumor mAb–β-lactamase conjugate in combination with the vinca prodrug were superior to drug therapy, prodrug alone, and to non-binding IgG enzyme conjugates with prodrug (Figure 3A). The effects were also superior to those obtained when the vinca drug was attached directly to the mAb. Long term regressions were obtained in several of the dosing regimens, even in animals having tumors as large as 700 mg in size at the initiation of therapy. It is impressive that such results were obtained, in light of the fact that the maximum tolerated dose of the prodrug was only slightly higher than that of the drug itself. A potential explanation for the therapeutic efficacy of the mAb–β-lactamase conjugate/vinca alkaloid combination is that the prodrug is cleared much more slowly than the active drug, resulting in intratumoral prodrug concentrations that are much higher than that of the drug, when administered systemically. Detailed drug and prodrug pharmacokinetic studies are needed to verify that this is indeed the case.

In related studies, prodrugs of doxorubicin and phenylenediamine mustard were evaluated in combination with an anti-melanoma mAb–β-lactamase conjugate (Kerr *et al.*, 1995). The Fab′ fragment was derived from 96.5, a high affinity mAb that recognizes the p97 melanotransferrin antigen present on many melanomas (Brown *et al.*, 1981; Rose *et al.*, 1986). *In vitro* cytotoxicity experiments showed that the doxorubicin prodrug C-Dox was approximately 9-fold less toxic than doxorubicin on 3677 melanoma cells, while the nitrogen mustard prodrug (CCM) was 26 times less toxic than phenylenediamine mustard (PDM). Therapy studies in nude mice bearing s.c. 3677 tumor xenografts showed that the 96.5-β-lactamase/C-Dox combination was much more effective than doxorubicin or a non-binding control conjugate/C-Dox combination. In fact, doxorubicin by itself had negligible effects on tumor growth.

The effects of 96.5-β-lactamase with the mustard prodrug CCM were even more pronounced. Regressions were observed in 100% of the treated mice at doses that caused no apparent toxicity (Figure 3B). Four out of the five mice in this treatment arm remained tumor free at day 120, the end of the observation period. Significant antitumor effects were even seen in mice that had large (800 mm^3) tumors before the first prodrug treatment (Figure 3C).

A number of *in vivo* experiments have been reported using bacterial carboxypeptidase G2 (CPG2) for prodrug activation. ICR12, a mAb that recognizes the HER-2 antigen on some breast carcinomas was covalently linked to CPG2, and tested for *in vivo* antitumor activity (Eccles *et al.*, 1994). In this case, conjugate cleared very slowly from the blood, requiring a 12–14 day delay before a nitrogen mustard prodrug could be administered. Even after such a delay, the conjugate tumor : blood ratio was only 2 : 1. Despite this poor localization, a single dose of a glutamic acid mustard prodrug resulted in long term regressions in the animals treated with the anti HER-2–CPG2 conjugate (Figure 3D).

The *in vivo* results described in this section demonstrate that treatment with mAb–enzyme/prodrug combinations can result in therapeutic effects that are far greater than that obtained by systemic drug administration. The effects have been

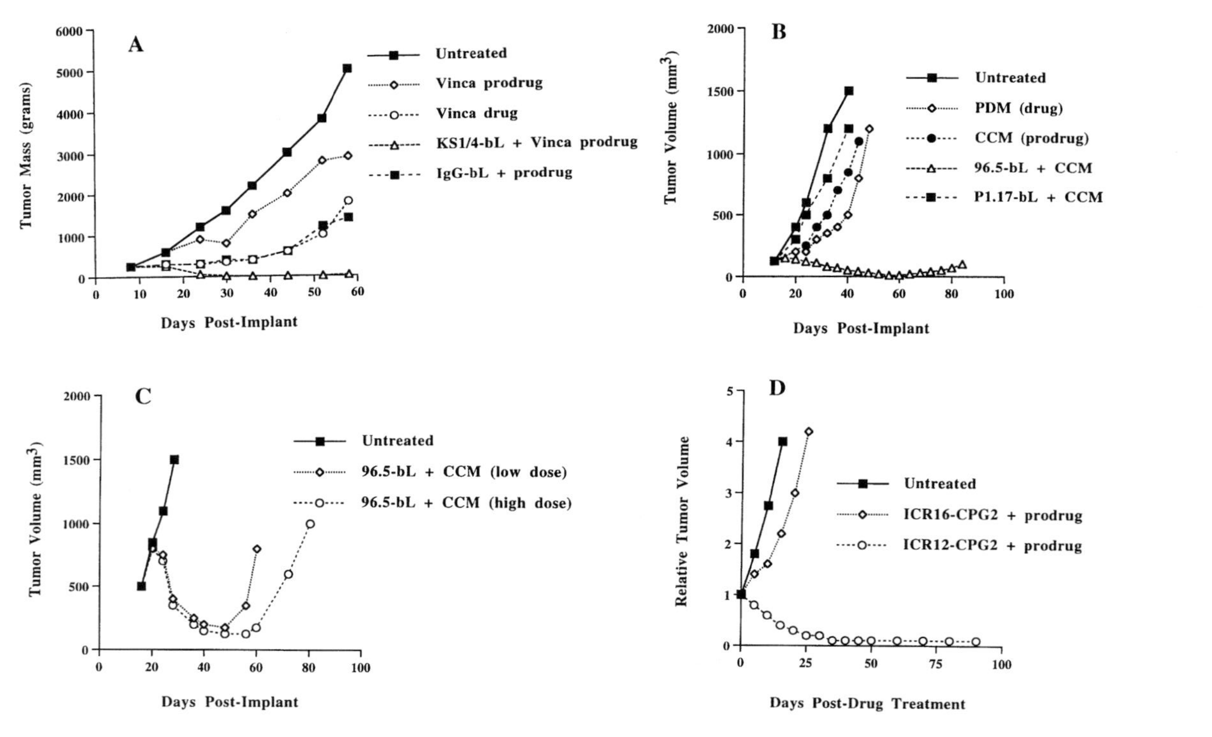

Figure 3 *In vivo* activities of mAb–enzyme/prodrug therapy. (A) Mice with s.c. LS174T (KS1/4 antigen positive) colorectal carcinoma xenografts were treated with vinca drug, vinca prodrug, KS1/4-β-lactamase (bL)+vinca prodrug, or a non-binding antibody–β-lactamase conjugate+vinca prodrug (Meyer *et al.*, 1993). (B) Effects of phenylenediamine mustard (PDM) and a PDM prodrug in combination with antibody–β-lactamase conjugates (96.5, binding; P1.17, nonbinding) in animals with s.c. 3677 melanoma xenografts (Kerr *et al.*, 1995). (C) Therapy of large (800 mm^3) 3677 xenografts with the 96.5–β-lactamase conjugate and PDM prodrug (Kerr *et al.*, 1995). (D) Treatment of mice bearing s.c. MDA MB 361 breast carcinoma xenografts with ICR16 (non-binding) and ICR12 (binding)-carboxypeptidase G2 (CPG2) conjugates in combination with a glutamic acid mustard prodrug (Eccles *et al.*, 1994).

shown to be immunologically specific, and systemic toxicity is manageable. The data presented also show that a single mAb–enzyme conjugate is capable of generating a panel of anticancer drugs that differ in their mechanisms of activity. This may prove to be of importance, since chemotherapeutic protocols usually involve the use of drug combinations.

PHARMACOKINETIC STUDIES

One of the most challenging aspects in optimizing the therapeutic effects of mAb–enzyme/prodrug combinations stems from the fact that in its simplest form, there are three variables: the mAb–enzyme dose, the length of time allowed for conjugate localization and clearance, and the prodrug dose. Some approaches have even applied a separate agent to clear blood-borne mAb–enzyme before prodrug administration. Thus, the multi-component nature of the approach adds a degree of complexity not normally found in single-agent chemotherapy. Several pharmacokinetic studies have been performed to address these issues, and to provide insight towards the design of optimal therapeutic protocols.

Biodistribution studies were performed with a prodrug of benzoic acid mustard in mice treated with a mAb–CPG2 conjugate that bound to choriocarcinoma tumor xenografts (Antoniw *et al.*, 1990). The concentration of mAb–CPG2 was higher in tumors than in other tissues 56 h post conjugate administration. Consistent with this was the finding that the active drug to prodrug ratio was highest in tumors of mice that received conjugate followed by prodrug. In fact, the tumor was the only tissue where the level of prodrug was found to be below the limit of detection. The studies are consistent with active drug being generated intratumorally by a mAb–enzyme conjugate/prodrug combination. However, it was also noted that active drug concentrations in a number of normal tissues were actually greater than that found in the tumor mass. This suggests that substantial levels of prodrug activation took place outside of the tumor.

Experiments were undertaken to improve tumor : normal tissue conjugate ratios by inducing conjugate clearance (Sharma *et al.*, 1994; Rogers *et al.*, 1995). This was achieved by modifying an A5B7–F(ab$'$)$_2$ CPG2 conjugate with galactose in order to accelerate hepatic clearance of the conjugate from the blood. As expected, the galactosylated conjugate cleared from the circulation very rapidly, but did not exhibit an improved tumor : blood ratio compared to ungalactosylated conjugate. In addition, the absolute amount delivered to the tumor xenografts was significantly less. Uptake of the galactosylated conjugate in the liver could be blocked by pretreating the animals with large amounts of a bovine submaxillary mucin. Substantial levels of the galactosylated conjugate then were maintained for several hours in the blood. Once the hepatic clearance functions were restored, the galactosylated conjugate was rapidly cleared. Under these conditions, tumor : blood ratios of 45 : 1 (24 h) and 110 : 1 (72 h) were achieved in mice that had subcutaneous tumor xenografts.

An alternative method to improve conjugate tumor : normal tissue ratios is to form rapidly cleared immune complexes once the mAb–enzyme conjugate has localized within the tumor mass. An A5B7–CPG2 mAb–enzyme conjugate (ungalactosylated) could be efficiently cleared from the bloodstream with a galactosylated anti-CPG2 antibody that was administered 19 h after conjugate injection. There was little affect on conjugate levels in the tumor, resulting in a three-fold improvement in the tumor : blood ratios.

Another study of the utility of clearing agents in mAb–enzyme/prodrug therapy involved a mAb–cytosine deaminase conjugate that was capable of converting 5-fluorocytosine into 5-fluorouracil (Wallace *et al.*, 1994). The mAb–cytosine deaminase conjugate contained the whole L6 mAb, and displayed very slow plasma clearance in mice with concomitant low tumor : blood ratios. Clearance of L6–cytosine deaminase was greatly accelerated by injecting an antibody against the L6 idiotype 24 h after conjugate administration. The blood levels of L6–cytosine deaminase dropped by a factor of 40–70, while the tumor levels were mostly maintained. As a result, very high doses of 5-fluorocytosine could be administered without any apparent toxicity. This led to vastly greater intratumoral 5-fluorouracil concentrations than those obtained in animals receiving systemic 5-fluorouracil therapy (Figure 4A). The prodrug : drug ratio was very high in the blood, kidneys and livers of mice that received the L6–cytosine deaminase/anti-idiotype/5-fluorocytosine combination, while the reverse was true for the tumor. Importantly, the intratumoral 5-fluorouracil concentration in the conjugate treated mice was 17–25 fold higher than the normal tissues tested. Finally, much lower amounts of 5-fluorouracil were generated in a tumor model that was apparently L6 antigen negative compared to the antigen positive line.

Pharmacokinetic studies have also been performed in mice that received a recombinant mAb–β-glucuronidase fusion protein and a doxorubicin prodrug (Bosslet *et al.*, 1994). In this study, the levels of fusion protein containing both a functional binding domain and enzymatic reactivity could be measured using a direct enzyme immunoassay with a chromogenic glucuronidase substrate. It was also possible to measure crude enzyme activity from tissue samples. After injection of the fusion protein in mice bearing either colorectal or stomach carcinoma xenografts, the enzyme immunoassay indicated favorable tumor/tissue conjugate ratios at 72 h and a greater than 200 : 1 tumor : blood distribution at 168 h. There was, however, significant conjugate uptake in the liver, which gradually decreased to basal levels after 168 h.

In this study it was also possible to measure the levels of doxorubicin and a doxorubicin prodrug in various tissues after prodrug administration. Although high doxorubicin levels in the livers, lungs, and kidneys were measured in the animals treated with the fusion protein/prodrug regimen, the levels were lower than that seen with a maximum tolerated dose of doxorubicin alone. At the same time, intratumoral levels of doxorubicin could be increased to roughly five times the level attainable with systemic doxorubicin treatment. While the therapeutic effects of this treatment protocol were impressive, the results are complicated by the fact that β-glucuronidase is endogenously expressed in necrotic human tumor xenografts (Bosslet *et al.*, 1995).

A study of the levels of doxorubicin and its cephalosporin prodrug C-Dox in two lung carcinoma models was undertaken in animals that received mAb–β-lactamase

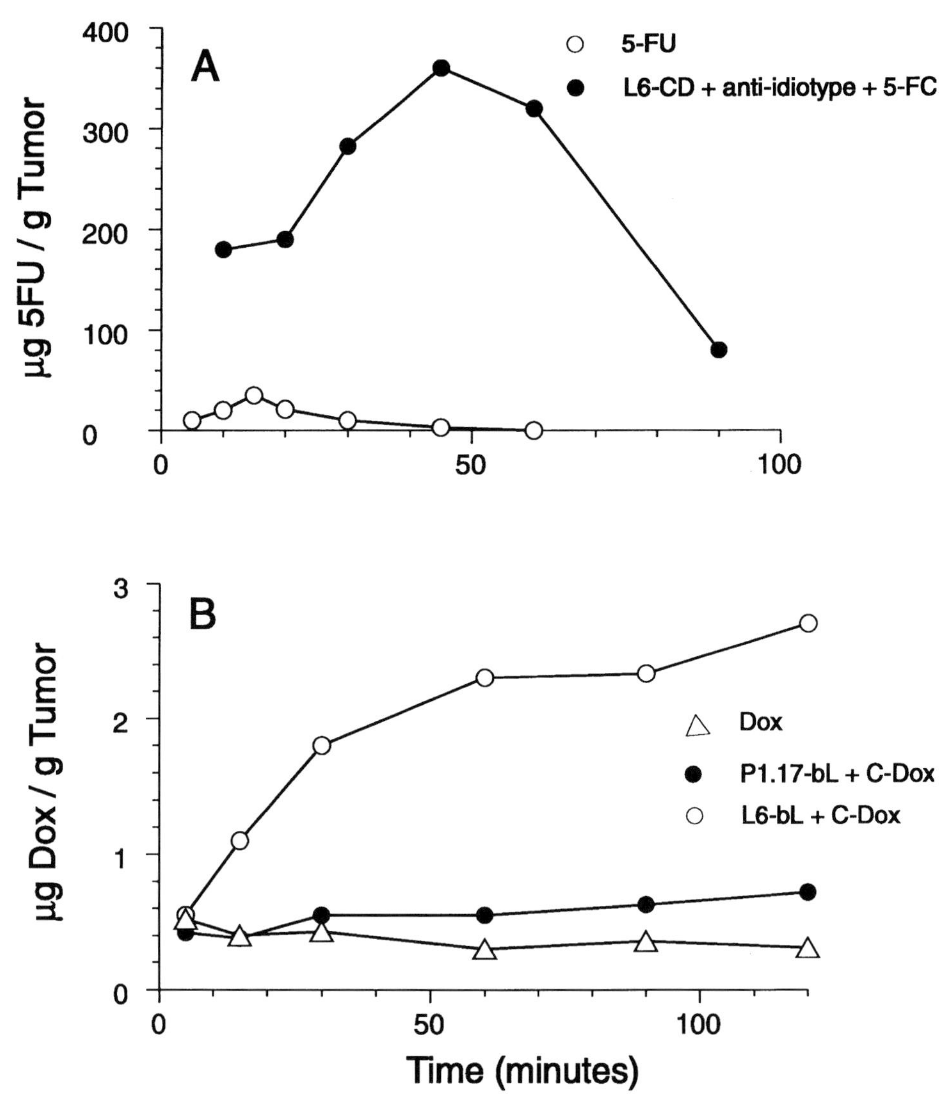

Figure 4 Intratumor drug levels generated by mAb–enzyme/prodrug treatment compared to systemic drug administration. (A) Tumor levels of 5-FU in mice treated with 5-FU (90 mg/kg) or with 300 µg of L6-cytosine deaminase (CD), anti-idiotypic antibody 13B (200 µg, 24 h later), and 5-FC (800 mg/kg, 48 h after L6 conjugate administration) (Wallace *et al.*, 1994). (B) Levels of doxorubicin (Dox) present in H2981 lung carcinoma xenografts after treatment with Dox, P1.17-β-lactamase (bL) conjugate (non-binding) and the Dox prodrug C-Dox, or L6–β-lactamase conjugate (binding)+C-Dox (Svensson *et al.*, 1995).

conjugates (Svensson *et al.*, 1995). In this study, HPLC was used to measure both prodrug and drug levels in tissue extracts. At doses of conjugate that were demonstrated not to lead to significant depletion of blood-borne prodrug, intratumoral doxorubicin concentrations in animals receiving the mAb–β-lactamase/C-Dox combination were increased five-fold compared to simple doxorubicin treatment (Figure 4B).

EFFECTS OF ENZYME ACTIVITY ON INTRATUMORAL PRODRUG ACTIVATION

Theoretical studies concerning pharmacokinetic models of mAb–enzyme/prodrug approaches and the effects that enzyme kinetic parameters have on prodrug activation have been reported (Yuan *et al.*, 1991; Baxter *et al.*, 1996). One of the interesting conclusions of this work was that under certain conditions less efficient enzymes should produce a greater ratio of drug released at the tumor vs. the bloodstream. With highly efficient enzymes (high k_{cat} and/or low K_m values) all of the available prodrug is converted to drug in both the tumor and bloodstream. The resulting blood and tumor drug levels are more dependent on tumor : blood prodrug distribution, which is not expected to be favorable, rather than the tumor : blood distribution of the mAb–enzyme conjugate. Low activity enzymes, on the other hand, will result in only a fraction of the prodrug being converted to drug, and yield drug tumor : blood ratios more dependent on the mAb–enzyme tissue distribution.

While these conclusions are potentially very important, they have not been confirmed *in vivo*, and are difficult to test experimentally. Several reports describe the production of a series of prodrug derivatives having a range of kinetic parameters with a given enzyme (Springer *et al.*, 1994, 1995; Vrudhula *et al.*, 1993a, 1995). However, even closely related prodrug derivatives will likely have different chemical and pharmacological properties, making any conclusions based on therapeutic efficacy tenuous. One method has now been described that should provide a means to measure the effect that enzyme parameters have on prodrug activation *in vivo* (Siemers *et al.*, 1996). The authors performed random mutagenesis on a five amino acid region of *Enterobacter cloacae* P99 β-lactamase that lies proximal to the enzyme active site. The resulting library was screened for activity, and a panel of mutant enzymes with a 20-fold range of kinetic parameters was obtained. As the mAb conjugates of these mutants would be expected to possess the same pharmacokinetic properties as the parent mAb–β-lactamase, a controlled test of enzyme kinetics vs. *in vivo* efficacy may be possible with conjugates of these enzymes.

CONSIDERATIONS FOR OPTIMIZED THERAPY

It is likely that standard chemical protein cross-linking techniques will lead to considerable heterogeneity in conjugate composition and to variations from one preparation to the next. To circumvent this problem, site specific chemical coupling methods and several techniques for recombinant fusion protein preparation have now been developed. For example, a chemical conjugation method has been developed to link *Enterobacter cloacae* β-lactamase to an antibody Fab′ fragment by taking advantage of the reactivity of the amino-terminal threonine present on the enzyme (Mikolajczyk *et al.*, 1994). Selective oxidation of the threonine was achieved using low concentrations of periodate, leading to the formation of an aldehyde group that was further derivatized. The activated aldehyde was then coupled to a mAb Fab' fragment containing a single sulfhydryl group in the hinge region. This led to the formation of a highly homogeneous chemical conjugate in yields that were much higher than those obtained using nonspecific coupling procedures. A similar process has been described for the preparation of mAb–CPG2 conjugates (Werlen *et al.*, 1994). In this case, the N-terminal alanine of the enzyme was mutated to threonine to allow the site-specific oxidation. The conjugate thus formed had superior binding characteristics compared to a non-specifically produced conjugate.

Recombinant DNA technology has been used for the preparation of fusion proteins having well defined characteristics. A single-chain sFv antibody–β-lactamase fusion protein was constructed and expressed using the L6 antitumor antibody and *Bacillus cereus* β-lactamase (Goshorn *et al.*, 1993). The fusion protein was expressed in soluble form in *Escherichia coli* and purified by affinity chromatography with an anti-idiotypic mAb column. The L6–β-lactamase fusion protein retained enzymatic activity, and was able to effect prodrug activation *in vitro*.

In a related study, a disulfide-stabilized Fv (dsFv) β-lactamase fusion protein was produced in *Escherichia coli* (Rodrigues *et al.*, 1995). The antibody fragments used in the construction were from a humanized mAb of murine origin, and were fused to the RTEM-1 class A β-lactamase. The fusion protein recognized the p185 (HER2) antigen, which is present on many breast and ovarian carcinomas. One of the several disulfide-linked variants investigated in the study bound as well as the wild type Fv. *In vitro* cytotoxicity assays with the combination of the fusion protein and a cephalosporin-doxorubicin prodrug demonstrated a specific and potent cytotoxic effect toward cells expressing the HER2 antigen. Although no animal therapy studies have been reported, enzyme immunoassay-based pharmacokinetic studies were performed with the fusion protein in nude mice. The conjugate cleared very quickly (α-phase half life 0.23 h; β-phase half life 1.27 h), consistent with its small size. These data suggest that elimination of blood-borne fusion protein should take place in a relatively short time span.

It is quite likely that enzymes of bacterial origin will be immunogenic in humans. This has provided the basis for preparing a recombinant mAb–human β-glucuronidase fusion protein in which the carboxy termini of a mAb heavy chain fragment was linked to a β-glucuronidase subunit (Bosslet *et al.*, 1992). The material was purified by affinity chromatography and had a molecular weight of 250 kDa, consistent with spontaneous dimerization. Interchain disulfide bond formation was apparently incomplete, since gel electrophoresis under denaturing conditions indicated the presence of two protein bands with molecular weights of 125 and 250 kDa. The recombinant protein retained the high avidity binding and the immunohistochemical characteristics of the parent antibody. The enzyme portion of the protein was also shown to be active. As mentioned earlier, high tumor : blood ratios were obtained with the fusion protein, but blood clearance was slow (Bosslet *et al.*, 1994).

One of the more exciting prospects for mAb–enzyme/prodrug therapy is the production of catalytic antibodies (known as abzymes) that are capable of effecting prodrug activation. Typically these agents are created by immunization of mice with a transition state analog of the reaction that is desired, followed by isolation of active antibodies by hybridoma techniques. The first report described a catalytic antibody capable of releasing chloramphenicol from an ester prodrug (Miyashita *et al.*, 1993). Other workers have reported catalytic antibodies capable of activating prodrugs of 5-fluorodeoxyuridine and nitrogen mustards (Campbell *et al.*, 1994; Wentworth *et al.*, 1996).

Unfortunately, the efficiencies of the catalytic antibodies described so far are probably orders of magnitude lower than what would be necessary for clinical efficacy. For example, a catalytic antibody has been described that is capable of hydrolyzing an ester prodrug of the cytotoxic drug 5-fluorodeoxyuridine (Campbell *et al.*, 1994). The prodrug consisted of a D-valine ester, which was intended to be stable towards endogenous esterases. It was demonstrated that the prodrug/abzyme combination was capable of exerting an antibacterial effect under conditions where either prodrug or antibody alone had no activity. However, the conditions described to achieve the results employed a 5% mole ratio of antibody, which is equal to a 20 : 1 ratio of mAb to prodrug by weight. Also, the study does not provide evidence for true catalysis in the therapy model, as less than one turnover of each antibody variable region would have been required to produce the observed effects. It therefore seems apparent that further progress in the field of catalytic antibodies is required before the technology will be applicable for prodrug activation.

A CLINICAL TRIAL OF MAB–ENZYME/PRODRUG THERAPY

A pilot scale clinical trial was performed using a combination of A5B7–CPG2 and a glutamic acid mustard prodrug (Bagshawe *et al.*, 1995). The trial was besieged by many problems; there was batch to batch variations in conjugate characteristics, severe complications resulted from attempts at immunosupressive therapy, and the formulation of prodrug in DMSO limited its dosage. Aside from the formulation problems, the prodrug alone was well tolerated in the seven patients that received it. The dose limiting toxicity was myelosupression (WHO grade 4 in several patients). Neutralizing antibody responses were seen in all patients within 10 days, except those treated with cyclosporin. Some partial and mixed responses were reported.

CONCLUSION

A variety of enzyme/prodrug combinations have now been described for site-selective drug generation. This targeting strategy may have some advantages over strategies involving direct attachment of cytotoxic agents to mAbs, since the targeted enzyme can greatly amplify the number of drug molecules delivered to each cell. In addition, it may be possible to eliminate tumor cells that are either antigen negative or that have bound insufficient quantities of conjugate, since the prodrug is converted to active drug extracellularly. The model studies described in this chapter indicate that mAb–enzyme conjugates in combination with anticancer prodrugs result in higher intratumoral drug concentrations than that achieved by systemic drug administration, and that correspondingly better antitumor activities can be obtained. While a number of challenging problems remain to be resolved, it seems apparent that mAb–enzyme conjugates will have clinical utility for selective drug targeting.

REFERENCES

Alexander, R.P., Beeley, N.R.A., O'Driscoll, M., O'Neill, F.P., Millican, T.A., Pratt, A.J. and Willenbrock, F.W. (1991) Cephalosporin nitrogen mustard carbamate prodrugs for "ADEPT". *Tetrahedron Lett.* **32**, 3269–3272.

Antoniw, P., Springer, C.J., Bagshawe, K.D., Searle, F., Melton, R.G., Rogers, G.T., Burke, P.J. and Sherwood, R.F. (1990) Disposition of the prodrug 4-(bis(2-chloroethyl)amino)benzoyl-L-glutamic acid and its active parent drug in mice. *Br. J. Cancer* **62**, 909–914.

Azoulay, M., Florent, J.C., Monneret, C., Gesson, J.P., Jacquesy, J.C., Tillequin, F., Koch, M., Bosslet, K., Czech, J. and Hoffman, D. (1995) Prodrugs of anthracycline antibiotics suited for tumor specific activation. *Anti-Cancer Drug Des.* **10**, 441–450.

Bagshawe, K.D., Springer, C.J., Searle, F., Antoniw, P., Sharma, S.K., Melton, R.G. and Sherwood, R.F. (1988) A cytotoxic agent can be generated selectively at cancer sites. *Br. J. Cancer.* **58**, 700–703.

Bagshawe, K.D., Sharma, S.K., Springer, C.J. and Rogers, G.T. (1994) Antibody directed enzyme prodrug therapy (ADEPT). A review of some theoretical, experimental and clinical aspects. *Ann. Oncol.* **5**, 879–891.

Bagshawe, K.D., Sharma, S.K., Springer, C.J. and Antoniw, P. (1995) Antibody directed enzyme prodrug therapy: a pilot-scale clinical trial. *Tumor Targeting* **1**, 17–29.

Battelli, M G., Abbondanza, A., Tazzari, P.L., Donota, A., Rizzi, S., Grassi, M. and Stirpe, F. (1988) Selective cytotoxicity of an oxygen-radical-generating enzyme conjugated to a monoclonal antibody. *Clin. Exp. Imm.* **73**, 128–133.

Baxter, L.T. and Jain, R.K. (1996) Pharmacokinetic analysis of the microscopic distribution of enzyme-conjugated antibodies and prodrugs: comparison with experimental data. *Br. J. Cancer* **73**, 447–456.

Bignami, G.S., Senter, P.D., Grothaus, P.G., Fischer, K.J., Humphreys, T. and Wallace, P.M. (1992) N-(4'-Hydroxyphenyl)palytoxin: A palytoxin prodrug that can be activated by a monoclonal antibody-penicillin G amidase conjugate. *Cancer Res.* **52**, 5759–5764.

Blakey, D.C., Davies, D.H., Dowell, R.I., East, S.J., Burke, P.J., Sharma, S.K., Springer, C.J., Mauger, A.B. and Melton, R.G. (1995) Anti-tumour effects of an antibody–carboxypeptidase G2 conjugate in combination with phenol mustard prodrugs. *Br. J. Cancer* **72**, 1083–1088.

Bosslet, K., Czech, J., Lorenz, P., Sedlacek, H.H., Shuermann, M. and Seemann, G. (1992) Molecular and functional characterization of a fusion protein suited for tumor specific prodrug activation. *Br. J. Cancer* **65**, 234–238.

Bosslet, K., Czech, J. and Hoffmann, D. (1994) Tumor-selective prodrug activation by fusion protein-mediated catalysis. *Cancer Res.* **54**, 2151–2159.

Bosslet, K., Czech, J. and Hoffmann, D. (1995) A novel one-step tumor-selective prodrug activation system. *Tumor Targeting* **1**, 45–50.

Brinkmann, U. and Pastan, I. (1994) Immunotoxins against cancer. *Biochim. Biophys. Acta* **1198**, 27–45.

Brown, J.P., Nishiyama, K., Hellström, I. and Hellström, K.E. (1981) Structural characterization of human melanoma-associated antigen p97 with monoclonal antibodies. *J. Immun.* **127**, 539–545.

Campbell, D.A., Gong, B., Kochersperger, L.M., Yonkovich, S., Gallop, M.A. and Schultz, P.G. (1994) Antibody-catalyzed prodrug activation. *J. Am. Chem. Soc.* **116**, 2165–2166.

Canevari, S., Colombatti, M. and Colnaghi, M.I. (1994) Immunoconjugates: lessons from animal models. *Ann. Oncol.* **5**, 698–701.

Deonarain, M.P. and Epenetos, A.A. (1994) Targeting enzymes for cancer therapy: old enzymes in new roles. *Br. J. Cancer* **70**, 786–794.

DeVita, V.T. (1993) Principles of chemotherapy. In V.T. DeVita, S. Hellman, and S.A. Rosenberg, (eds), *Cancer: Principles and Practice of Oncology*, J.B. Lippencott, Philadelphia, pp. 276–292.

Dinota, A., Tazzari, P.L., Abbondanza, A., Battelli, M.G., Gobbi, M. and Stirpe, F. (1990) Bone marrow pruging by a xanthine oxidase-antibody conjugate. *Bone Marrow Transplant.* **6**, 31–36.

Dvorak, H.F. (1990) Leaky tumor vessels: consequences for tumor stroma generation and for solid tumor therapy. *Prog. Clin. Biol. Res.* **354A**, 317–330.

Dvorak, H.F., Nagy, J.A. and Dvorak, A.M. (1991) Structure of solid tumors and their vasculature: implications for therapy with monoclonal antibodies. *Cancer Cells* **3**, 77–85.

Eccles, S.A., Court, W.J., Box, G.A., Dean, C.J., Melton, R.G. and Springer, C.J. (1994) Regression of established breast carcinoma xenografts with antibody-directed enzyme prodrug therapy against c-erbB2 p185. *Cancer Res.* **54**, 5171–5177.

Esswein, A., Hansler, E., Montejano, Y., Vitols, K.S. and Huennekens, F.M. (1991) Construction and chemotherapeutic potential of carboxypeptidase-A/monoclonal antibody conjugate. *Adv. Enzyme Regul.* **31**, 3–12.

Gesson, J.-P., Jacquesy, J.-C., Mondon, M., Petit, P., Renoux, B., Andrianomenjanahary, S., Dufat-Trinh Van, H., Koch, M., Michel, S., Tillequin, F., Florent, J.-C., Monneret, C., Bosslet, K., Czech, J. and Hoffman, D. (1994) Prodrugs of anthracyclines for chemotherapy via enzyme-monoclonal antibody conjugates. *Anticancer Drug Des.* **9**, 409–423.

Goshorn, S.C., Svensson, H.P., Kerr, D.E., Somerville, J.E., Senter, P.D. and Fell, H.P. (1993) Genetic construction, expression, and characterization of a single chain anti-carcinoma antibody fused to β-lactamase. *Cancer Res.* **53**, 2123–2127.

Haenseler, E., Esswein, A., Vitols, K.S., Montejano, Y., Mueller, B.M., Reisfeld, R.A. and Huennekens, F.M. (1992) Activation of methotrexate-α-alanine by carboxypeptidase A-monoclonal antibody conjugate. *Biochemistry* **31**, 891–897.

Haisma, H.J., Boven, E., van Muijen, M., de Jong, J., van der Vijgh, W.J.F. and Pinedo, H.M. (1992a) A monoclonal antibody-β-glucuronididase conjugate as activator of the prodrug epirubicin-glucuronide for specific treatment of cancer. *Br. J. Cancer* **66**, 474–478.

Haisma, H.J., Boven, E., van Muijen, M., De Vries, R. and Pinedo, H.M. (1992b) Analysis of a conjugate between anti-carcinoembrionic antigen monoclonal antibody and alkaline phosphatase for specific activation of the prodrug etoposide phosphate. *Cancer Immunol. Immunother.* **34**, 343–348.

Hanessian, S. and Wang, J. (1993) Design and synthesis of cephalosporin-carboplatinum prodrug activatable by a β-lactamase. *Can. J. Chem.* **71**, 896–906.

Hay, M.P., Willson, W.R. and Denny, W.A. (1995) A novel enediyne prodrug for antibody-directed enzyme prodrug therapy (ADEPT) using E. coli B nitroreductase. *Bioorganic and Medicinal Chem. Letters* **5**, 2829–2834.

Hellström, I., Hellström, K.E., Siegall, C.B. and Trail, P.A. (1995) Immunoconjugates and immunotoxins for therapy of cancer. *Adv. Pharm.* **33**, 349–388.

Ito, H., Morizet, J., Coulombel, L., Goavec, M., Rousseau, V., Bernard, A. and Stanislawski, M. (1989) An immunotoxin system intended for bone marrow purging composed of glucose oxidase and lactoperoxidase coupled to monoclonal antibody 097. *Bone Marrow Transplant.* **4**, 519–527.

Ito, H., Morizet, J., Coulombel, L. and Stanislawski, M. (1990) T cell depletion of human bone marrow using an oxidase-peroxidase enzyme immunotoxin. *Bone Marrow Transplant.* **6**, 395–398.

Jungheim, L.N., Shepherd, T.A. and Meyer, D.L. (1992) Synthesis of acylhydrazido-substituted cephems. Design of cephalosporin-vinca alkaloid prodrugs: substrates for an antibody-targeted enzyme. *J. Org. Chem.* **57**, 2334–2340.

Jungheim, L.N. and Shepherd, T.A. (1994) Design of anticancer prodrugs: substrates for antibody targeted enzymes. *Chem. Rev.* **94**, 1553–1566.

Kerr, D.E., Senter, P.D., Burnett, W.V., Hirschberg, D.L., Hellström, I. and Hellström, K.E. (1990) Antibody-penicillin-V-amidase conjugates kill antigen-positive tumor cells when combined with doxorubicin phenoxyacetamide. *Cancer Immunol. Immunother.* **31**, 202–206.

Kerr, D.E., Schreiber, G.L., Vrudhula, V.M., Svensson, H.P., Hellström, I., Hellström, K.E. and Senter, P.D. (1995) Regressions and cures of melanoma xenografts following treatment with monoclonal antibody β-lactamase conjugates in combination with anticancer prodrugs. *Cancer Res.* **55**, 3558–3563.

Kuefner, U., Lohrmann, U., Montejano, Y.D., Vitols, K.S. and Huennekens, F.M. (1989) Carboxypeptidase-mediated release of methotrexate from methotrexate-α-peptides. *Biochemistry* **28**, 2288–2297.

Leenders, R.G.G., Scheeren, H.W., Houba, P.H.J., Boven, E. and Haisma, H.J. (1995) Synthesis and evaluation of novel daunomycin-phosphate-sulfate-β-glucuronide and -β-glucoside prodrugs for application in ADEPT. *Bioorganic and Medicinal Chem. Letters* **5**, 2975–2980.

Mauger, A.B., Burke, P.J., Somani, H.H., Friedlos, F. and Knox, R.J. (1994) Self-immolative prodrugs: candidates for antibody-directed enzyme prodrug therapy in conjuction with a nitroreductase enzyme. *J. Med. Chem.* **37**, 3452–3458.

Meyer, D.L., Jungheim, L.N., Mikolajczyk, S.D., Shepherd, T.A., Starling, J.J. and Ahlem, C. (1992) Preparation and characterization of β-lactamase-Fab′ conjugates for the site specific activation of oncolytic agents. *Bioconjugate Chem.* **3**, 42–48.

Meyer, D.L., Jungheim, L.N., Law, K.L., Mikolajczyk, S.D., Shepherd, T.A., Mackensen, D.G., Briggs, S.L. and Starling, J.J. (1993) Site-specific prodrug activation by antibody-β-lactamase conjugates: regression and long-term growth inhibition of human colon carcinoma xenograft models. *Cancer Res.* **53**, 3956–3963.

Mikolajczyk, S.D., Meyer, D.L., Starling, J.J., Law, K.L., Rose, K., Dufour, B. and Offord, R.E. (1994) High yield, site-specific coupling of N-terminally modified β-lactamase to a proteolytically derived single-sulfhydryl murine Fab′. *Bioconjugate Chem.* **5**, 636–646.

Miyashita, H., Karaki, Y., Kikuchi M. and Fujii, I. (1993) Prodrug activation via catalytic antibodies. *Proc. Natl. Acad. Sci. USA* **90**, 5337–5340.

Perron, M.J. and Page, M. (1996) Activation of methotrexate-phenylalanine by monoclonal antibody-carboxypeptidase. A conjugate for the specific treatment of ovarian carcinoma *in vitro*. *Br. J. Cancer* **73**, 281–287.

Philpott, G.W., Shearer, W.T., Bower, R.W. and Parker C.W. (1973) Selective cytotoxicity of hapten-substituted cells with an antibody-enzyme conjugate. *J. Immunol.* **111**, 921–929.

Press, O.W., Eary, J.F., Appelbaum, F.R., Martin, P.J., Badger, C.C., Nelp, W.B., Glenn, S., Butchko, G., Fisher, D., Porter, B., Matthews, D.C., Fisher, L.D. and Bernstein, I.D. (1993) Radiolabeled-antibody therapy of B-Cell Lymphoma with autologous bone marrow support. *New Eng. J. Med.* **329**, 1219–1224.

Rodrigues, M.L., Carter, P., Wirth, C., Mullins, S., Lee, A. and Blackburn, B.K. (1995a) Synthesis and β-lactamase-mediated activation of a cephalosporin-taxol prodrug. *Chemistry and Biology* 2, 223–227.

Rodrigues, M.L., Presta, L.G., Kotts, C.F., Wirth, C., Mordenti, J., Osaka, G., Wong, W.L.T., Nuijens, A., Blackburn, B. and Carter P. (1995b) Development of a humanized disulfide-stabilized anti-p185HER2Fv-β-lactamase fusion protein protein for activation of a cephalosporin doxorubicin prodrug. *Cancer Res.*, **55**, 63–70.

Roffler, S.R., Wang, S.-M., Chern, J.-W., Yeh, M.-Y. and Tung, E. (1991) Anti-neoplastic glucuronide prodrug treatment of human tumor cells targeted with a monoclonal antibody-enzyme conjugate. *Biochem. Pharmacol.* **42**, 2062–2065.

Roffler, S.R., Chern, J.-W., Yeh, M.-Y., Ng, J.C. and Tung, E. (1992) Specific activation of glucuronide prodrugs by antibody-targeted enzyme conjugates for cancer therapy. *Cancer Res.* **52**, 4484–4491.

Rogers, G.T., Burke, P.J., Sharma, S.K., Koodie, R. and Boden, J.A. (1995) Plasma clearance of an antibody-enzyme conjugate in ADEPT by monoclonal anti-enzyme: its effect on prodrug activation *in vivo*. *British J. Cancer* **72**, 1357–1363.

Rose, T.M., Plowman, G.D., Teplow, D.B., Dreyer, W.J., Hellström, K.E. and Brown, J.P. (1986) Primary structure of the human melanoma-associated antigen p97 (melanotransferrin) deduced from the mRNA sequence. *Proc. Natl. Acad. Sci. USA* **83**, 1261–1265.

Rowlinson-Busza, G., Bamias, A., Krausz, T. and Epenetos, A.A. (1992) Antibody-guided enzyme nitrile therapy (Agent): *in vitro* cytotoxicity and *in vivo* tumour localisation. In: A.A. Epenetos (Ed.), *Monoclonal Antibodies 2. Applications in Clinical Oncology*, Chapman and Hall, London, pp. 111–118.

Sahin, U., Hartmann, F., Senter, P., Pohl, C. Engert, A., Diehl, V. and Pfreundschuh, M. (1990) Specific activation of the prodrug mitomycin phosphate by a bispecific anti-CD30/anti-alkaline phosphatase monoclonal antibody. *Cancer Res.* **50**, 6944–6948.

Schubiger, P.A. and Smith, A. (1995) Optimising the radioimmunotherapy of malignant disease: the broadening choice of carrier and effector moieties. *Pharm. Acta. Helv.* **70**, 203–217.

Senter, P.D., Saulnier, M.G., Schreiber, G.J., Hirschberg, D.L., Brown, J.P., Hellström, I. and Hellström, K.E. (1988) Anti-tumor effects of antibody–alkaline phosphatase conjugates in combination with etoposide phosphate. *Proc. Natl. Acad. Sci. USA.* **85**, 4842–4846.

Senter, P.D., Schreiber, G.J., Hirschberg, D.L., Ashe, S.A., Hellström, K.E. and Hellström, I. (1989) Enhancement of the *in vitro* and *in vivo* antitumor activities of phosphorylated mitomycin c and etoposide derivatives by monoclonal antibody-alkaline phosphatase conjugates. *Cancer Res.* **49**, 5789–5792.

Senter, P.D. (1990) Activation of prodrugs by antibody-enzyme conjugates: a new approach to cancer therapy. *FASEB* , 188–193.

Senter, P.D., Su, P.C.D., Katsuragi, T., Sakai, T., Cosand, W.L., Hellström, I. and Hellström, K.E. (1991) Generation of 5-fluorouracil from 5-fluorocytosine by monoclonal antibody-cytosine deaminase conjugates. *Bioconjugate Chem.* **2**, 447–451.

Senter, P.D., Wallace, P.M., Svensson, H.P., Vrudhula, V.M., Kerr, D.E., Hellström, I. and Hellström, K.E. (1993) Generation of cytotoxic agents by targeted enzymes. *Bioconjugate Chem.* **4**, 3–9.

Senter, P.D., Svensson, H.P., Schreiber, G.J., Rodriguez, J.L. and Vrudhula, V.M. (1995a) Poly(ethylene glycol)-doxorubicin conjugates containing *β*-lactamase-sensitive linkers. *Bioconjugate Chem.* **6**, 389–394.

Senter, P.D., Vrudhula, V.M., Wallace, P.M., Sommerville, J.E., Wang, I.K. and Lowe, D.A. (1995b) Sulfated etoposide and nitrogen mustard prodrugs and their activation by streptomyces arylsulfatase. *Drug Delivery* **2**, 110–116.

Sharma, S.K., Bagshawe, K.D., Burke, P.J., Boden, R.W. and Rogers, G.T. (1990) Inactivation and clearance of an anti-CEA carboxypeptidase G2 conjugate in blood after localisation in a xenograft model. *British J. Cancer* **61**, 659–662.

Sharma, S.K., Bagshawe, K.D., Springer, C.J., Burke, P.J., Rogers, G.T., Boden, J.A., Antoniw, P., Melton, R.G. and Sherwood, R.F. (1991) Antibody directed enzyme prodrug therapy (ADEPT): a three phase system. *Dis. Markers* **9**, 225–231.

Sharma, S.K., Bagshawe, K.D., Burke, P.J., Boden, J.A., Rogers, G.T., Springer, C.J., Melton, R.G. and Sherwood, R.F. (1994) Galactosylated antibodies and antibody-enzyme conjugates in antibody-directed enzyme prodrug therapy. *Cancer Supp.* **73**, 1114–1120.

Shepherd, T.A., Jungheim, L.N., Meyer, D.L. and Starling, J.J. (1991) A novel targeted delivery system utilizing a cephalosporin oncolytic prodrug activated by an antibody *β*-lactamase conjugate for the treatment of cancer. *Bioorg. Med. Chem. Letters* **1**, 21–26.

Shockley, T.R., Lin, K., Nagy, J.A., Tompkins, R.G., Dvorak, H.F. and Yarmush, M.L. (1991) Penetration of tumor tissue by antibodies and other immunoproteins. *Ann. N.Y. Acad. Sci.* **618**, 367–382.

Siegall, C.B., Wolff, E.A., Gawlak, S.L., Paul, L., Chace, D. and Mixan, B. (1995) Immunotoxins as cancer chemotherapeutic agents. *Drug Devel. Res.* **34**, 210–219.

Siemers, N.O., Yelton, D.E., Bajorath, J. and Senter, P.D. (1996) Modifying the specificity and activity of the *Enterobacter cloacae* P99 *β*-lactamase by mutagenesis within an M13 phage vector. *Biochemistry* **35**, 2104–2111.

Sinhababu, A.K. and Thakker, D.R. (1996) Prodrugs of anticancer agents. *Adv. Drug. Del. Rev.* in press.

Springer, C.J., Antoniw, P., Bagshawe, K.D., Searle, F., Bisset, G.M.F. and Jarman M. (1990) Novel prodrugs which are activated to cytotoxic alkylating agents by carboxypeptidase G2. *J. Med. Chem.* **33**, 677–681.

Springer, C.J., Bagshawe, K.D., Sharma, S.K., Searle, F., Boden, J.A. Antoniw, P., Burke, P.J., Rogers, G.T., Sherwood, R.F. and Melton, R.G. (1991) Ablation of human choriocarcinoma xenografts in nude mice by antibody-directed enzyme prodrug therapy (ADEPT) with three novel compounds. *Eur. J. Cancer* **27**, 1361–1366.

Springer, C.J., Niculescu-Duvaz, I. and Pedley, R.B. (1994) Novel prodrugs of alkylating agents derived from 2-fluoro- and 3-fluorobenzoic acids for antibody-directed enzyme prodrug therapy. *J. Med. Chem.* **37**, 2361–2370.

Springer, C.J., Dowell, R., Burke, P.J., Hadley, E., Davies, D.H., Blakey, D.C., Melton, R.G. and Niculescu-Duvaz, I. (1995) Optimization of alkylating agent prodrugs derived from phenol and aniline mustards: a new clinical candidate prodrug (ZD2767) for antibody-directed enzyme prodrug therapy (ADEPT). *J. Med. Chem.* **38**, 5051–5065.

Svensson, H.P., Kadow, J.F., Vrudhula, V.M., Wallace, P.M. and Senter, P.D. (1992) Monoclonal antibody-β-lactamase conjugates for the activation of a cephalosporin mustard prodrug. *Bioconjugate Chem.* **3**, 176–181.

Svensson, H.P., Vrudhula, V.M., Emsweiler, J.E., MacMaster, J.F., Cosand, W.L., Senter. P.D. and Wallace, P.M. (1995) *In vitro* and *in vivo* activities of a doxorubicin prodrug in combination with monoclonal antibody-β-lactamase conjugates. *Cancer Res.* **55**, 2357–2365.

Vitols, K.S., Haag-Zeino, B., Baer, T., Montejano, Y.D. and Huennekens, F.M. (1995) Methotrexate-α-phenylalanine: optimization of methotrexate prodrug for activation by carboxypeptidase A-monoclonal antibody conjugate. *Cancer Res.* **55**, 478–481.

Vrudhula, V.M., Senter, P.D., Fischer, K.J. and Wallace, P.M. (1993a) Prodrugs of doxorubicin and melphalan and their activation by a monoclonal antibody-penicillin-G amidase conjugate. *J. Med. Chem.* **36**, 919–923.

Vrudhula, V.M., Svensson, H.P., Kennedy, K.A., Senter, P.D. and Wallace, P.M. (1993b) Antitumor activities of a cephalosporin prodrug in combination with monoclonal antibody-β-lactamase conjugates. *Bioconjugate Chem.* **4**, 334–340.

Vrudhula, V.M., Svensson, H.P. and Senter, P.D. (1995) Cephalosporin derivatives of doxorubicin as prodrugs for activation by monoclonal antibody β-lactamase conjugates. *J. Med. Chem.* **38**, 1380–1385.

Wallace, P.M. and Senter, P.D. (1991) *In vitro* and *in vivo* activities of monoclonal antibody-alkaline phosphatase conjugates in combination with phenol mustard phosphate. *Bioconjugate Chem.* **2**, 349–352.

Wallace, P.M., MacMaster, J.F., Smith, V.F., Kerr, D.E., Senter, P.D. and Cosand, W.L. (1994) Intratumoral generation of 5-fluorouracil mediated by an antibody-cytosine deaminase conjugate in combination with 5-fluorocytosine. *Cancer Res.* **54**, 2719–2723.

Wentworth, P., Datta, A., Blakey, D., Boyle, T., Partridge, L.J. and Blackburn, G.M. (1996) Toward antibody-directed "abzyme" prodrug therapy, ADAPT: Carbamate prodrug activation by a catalytic antibody and its *in vitro* application to human tumor cell killing. *Proc. Natl. Acad. Sci. USA* **93**, 799–803.

Werlen, R.C., Lankinen, M., Rose, K., Blakey, D., Shuttleworth, H., Melton, R. and Offord, R.E. (1994) Site-specific conjugation of an enzyme and an antibody fragment. *Bioconjug. Chem.* **5**, 411–417.

Yoshikawa, T., Kokura, S., Tainaka, K., Naito, Y. and Kondo, M. (1995) A novel cancer therapy based on oxygen radicals. *Cancer Res.* **55**, 1617–1620.

Yuan, F., Baxter, L.T. and Jain, R.K. (1991) Pharmacokinetic analysis of two-step approaches using bifunctional and enzyme-conjugated antibodies. *Cancer Res.* **51**, 3119–3130.

5. IMMUNOSCINTIGRAPHY AND RADIOIMMUNOTHERAPY

ANGELIKA BISCHOF DELALOYE

Service de Médecine Nucléaire, Centre Hospitalier Universitaire Vaudois, CH-1011 Lausanne, Switzerland

INTRODUCTION

Since the first attempts in the early seventies at tumor visualization with I-131 labeled polyclonal antibodies directed against carcino-embryonic antigen (CEA) in tumor bearing animals (Goldenberg *et al.*, 1974; Mach *et al.*, 1974), followed some years later by using the same principle for tumor scintigraphy in humans (Goldenberg *et al.*, 1978; Mach *et al.*, 1980; Begent *et al.*, 1980) immunoscintigraphic techniques have tremendously improved but the method is still not part of routine patient management in diagnosis, follow-up or therapy of neoplastic disease. More successful, at least in European countries, was the use of monoclonal antibodies directed against the non-specific cross-reacting antigen (NCA) 95 present on neutrophils in the detection of infectious foci (Becker *et al.*, 1994; Bläuenstein *et al.*, 1995; Boubaker *et al.*, 1995) and for the early diagnosis of tumor involvement of bone marrow by medullary scintigraphy (Reske *et al.*, 1993). Visualization of neutrophils and their precursors is not the purpose of the present overview which is dedicated to the specific interaction of antibodies with tumor associated antigens. It is evident that tumor cells must express the antigen for which the antibody is specific and that these tumor cells must be accessible to the antibody. The larger the tumor the more difficult is tissue penetration by macromolecules such as proteins (Jain, 1990). In the following, the stepwise improvement of the methods used for imaging and also treatment of tumors with radiolabeled antibodies raised against tumor associated antigens will be described first, followed by presentation of clinical data in the most extensively studied tumors and considerations on the potential contribution to patient management, be it in clinical work up or treatment of tumors.

The elaboration of the hybridoma technique by Köhler and Milstein (1975) made it possible to prepare monoclonal antibodies (mAb) specific for a given epitope of the antigen. Promising results obtained in selected patients with colorectal cancer, melanoma and ovarian tumors, made the medical community believe that Paul Ehrlichs theory of the "magic bullet" had been realized. But as soon as non selected patients with occult tumors were studied, it became obvious that the technique had to be improved in order to reliably distinguish tumor from non tumor uptake. In fact, even with monoclonal antibodies the absolute tumor uptake remained very low, in the order of 10^{-2}% or less of the injected activity, and it was often difficult to distinguish tumor from circulating activity.

The presence of circulating antigen was thought to hinder antibodies from reaching the tumor by being complexed and subsequently removed from blood by the reticulo-endothelial system. We did not find any difference in plasma clearance (9.5 h) of an I-123 labeled anti-CEA Fab fragment in patients with low ($\leqslant$50 ng/mL) and high ($\geqslant$500 ng/mL) circulating CEA (Delaloye *et al.*, 1987). Bosslet *et al.* (1988) showed in an *in vitro* experiment that binding of anti-CEA mAbs on human carcinoma cell lines was not inhibited by pre-incubation with a 20 molar excess of serum CEA. Inversely, binding to tumor associated CEA of mAbs directed against CEA–non-specific cross-reacting antigens could be inhibited under identical conditions. This suggests that mAbs directed against CEA–NCA present higher affinity to circulating CEA than the specific anti-CEA mAbs tested in this experiment. In patients with increased circulating antigen Davidson *et al.* (1991) found that tumor uptake of the labeled antibody used was preserved despite high levels of circulating immune complexes. Behr *et al.* (1996) confirmed that targeting sensitivity of monoclonal anti-CEA antibodies was not affected by complexation. They also showed that complexation was dependent from the affinity of the antibodies. Clearance of the anti-CEA antibodies was also related to the tumor type, patients with colorectal cancer clearing the antibody significantly faster from blood and whole body (17.6 ± 12.6 and 53.2 ± 30.1 h, respectively) than all other tumor types (44.2 ± 23.7 and 114.6 ± 59.7 h, respectively). This phenomenon has only been observed with anti-CEA antibodies. The authors postulate that different CEA-expressing tumors might produce heterogeneous CEA molecules and that the variability in mAb clearance could be due to varying clearance rates of these CEA subspecies.

IMPROVEMENT OF TUMOR TO BACKGROUND RATIO

Background Subtraction

Background subtraction by using a second label representing circulating activity (Tc-99m human serum albumin) was proposed to overcome poor differentiation of tumor uptake from circulating activity (Goldenberg *et al.*, 1978). This technique, however, was soon recognized as a possible source of errors being able to cause false positive as well as false negative results (Mach *et al.*, 1980; Jones *et al.*, 1982; Ott *et al.*, 1983). Instead of subtracting background by the means of a second label other methods were suggested to decrease circulating activity such as using fragments or injecting a second antibody. Later-on two- or three-step labeling techniques were proposed.

Antibody Fragments

Mach *et al.* (1980) were able to show that with either $F(ab')_2$ or Fab fragments, tumors could be equally well detected as with intact purified polyclonal goat anti-CEA antibodies. Tumor uptake expressed as percent of injected dose per gram (%ID/g) was comparable between intact (0.002–0.004%ID/g) and F(ab')$_2$ fragments (0.0015–0.0016%ID/g), whereas for Fab fragments absolute tumor uptake was sensibly lower (0.0006%ID/g). With this latter fragment, however, the tumor to serum ratio was higher than that obtained with intact IgG. These first results with polyclonal antibodies were later-on confirmed with monoclonal antibodies in the nude mouse model (Buchegger *et al.*, 1983; Andrew *et al.*, 1986) as well as in patients with colorectal carcinoma (Delaloye *et al.*, 1986; Buraggi *et al.*, 1987). Buchegger *et al.* (1983) showed in a nude mouse experiment that tumor to normal organ ratios were inversely proportional to the size of the antibody molecule, they increased from 7 (intact IgG) to 25 for the F(ab')$_2$ fragment and 85 for the Fab fragment, all labeled with I-131. The specificity indices calculated by dividing these ratios by the corresponding ratios obtained for simultaneously injected irrelevant IgG and its fragments, all labeled with I-125, were 3.4 for intact mAb, 8.2 for F(ab')$_2$ and 19 for Fab. Behr *et al.* (1995) confirmed that higher sensitivity in detecting liver metastases with Tc-99m labeled anti-CEA mAb fragments, F(ab')$_2$ and F(ab') in comparison with intact IgG was not due to higher absolute uptake but to faster uptake and blood clearance. One of the problems with fragments is the high kidney uptake (Figure 1). This can be decreased to a certain extent by administration of cationic acids, which allowed to reduce nephrotoxicity in colon cancer bearing nude mice treated with Y-90 labeled anti-CEA mAb as IgG, F(ab')$_2$ and Fab (Behr *et al.*, 1996).

Recently a trivalent antibody construct has been generated by a novel approach using polyoxime chemistry (Werlen *et al.*, 1996). This construct is formed of three Fab' fragments of a monoclonal antibody directed against CEA. In the nude mouse model initial tumor uptake of this construct is comparable to that of the intact mAb from which the Fab' were produced, residence time is shorter, but tumor/normal tissue ratios are up to 10 fold higher than those seen for the intact parent antibody. The potential advantage of this trivalent construct over monovalent Fab' fragments lies in the higher absolute tumor uptake and the longer residence time, but this favorable biokinetic data need still to be confirmed in human studies.

Second Antibody Injection

Begent *et al.* (1980) used a liposomally entrapped second antibody directed against the labeled anti-tumor antibody to accelerate clearance of non-tumor-bound first antibody without affecting its clearance from the tumor itself. After a transitory increase of liver activity during the first four hours after the second antibody administration, clearance was accelerated and tumor imaging improved. Despite confirmation of these results by others (Senekowitsch *et al.*, 1987; Stewart *et al.*, 1990; Tromholt *et al.*, 1991), injection of a second antibody never gained widespread interest in immunoscintigraphy with labeled anti-tumor antibodies, but it introduced the principle of chasing circulating antibody which was used later-on in multi-step labeling techniques (Goodwin *et al.*, 1988).

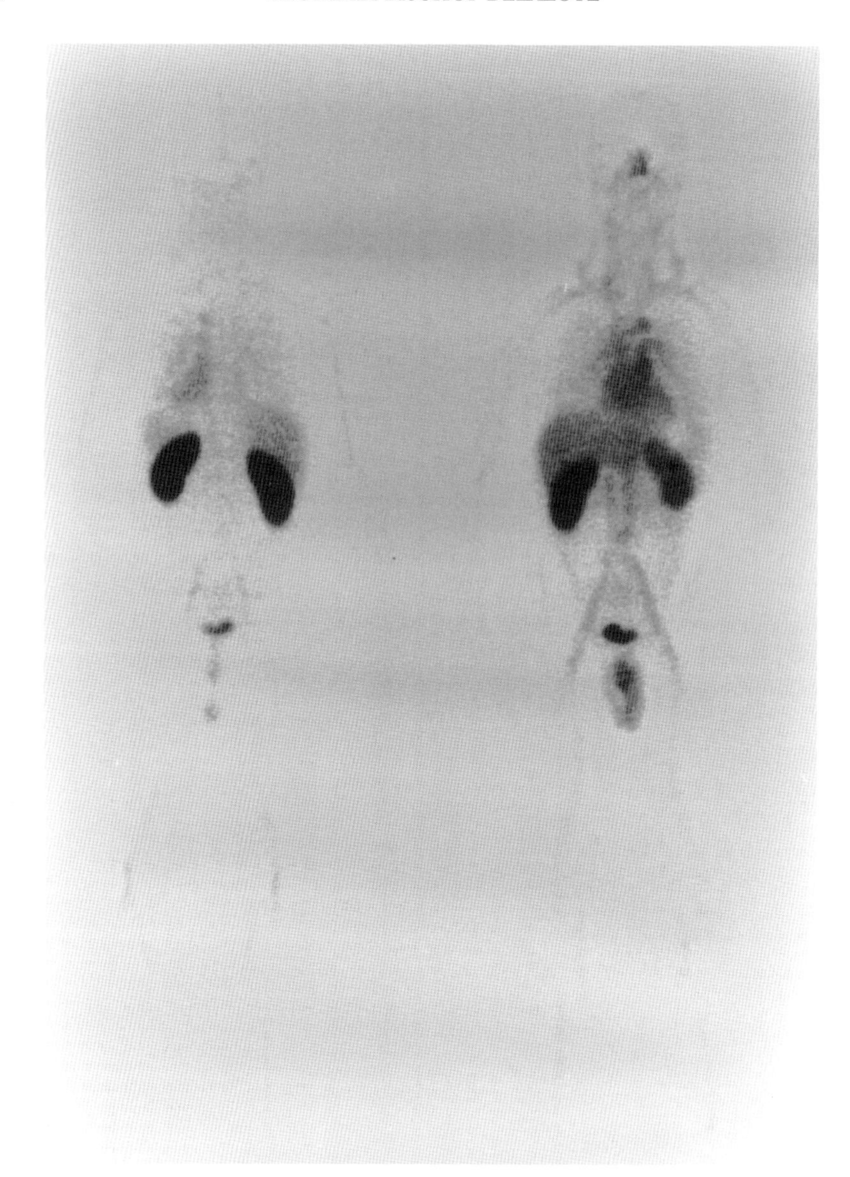

Figure 1 Anterior and posterior whole body scintigraphy of a patient 5 h after injection of a Tc-99m labeled anti-CEA Fab′ fragment. Besides circulating activity and some uptake in organs such as liver, spleen and lungs this figure shows the prominent retention of Fab in renal parenchyma.

Multi-Step Labeling Techniques

Maximal tumor concentrations of mAb are achieved within the first 24 h in humans, but blood-born background remains generally high so that tumor to background ratios are often insufficient for high quality imaging. This is even more important when considering mAb labeled with beta-, or even worse with alpha-emitters for therapeutic purposes. In fact, irradiation of normal tissues, especially bone marrow is one of the major limiting factors besides low tumor uptake of mAb in radioimmunotherapy, particularly of solid tumors. In multi-step labeling techniques tumors are pretargeted with unlabeled mAbs having a high affinity noncovalent binding site for a small, rapidly excreted effector molecule, which is radiolabeled and administered after having given enough time to the cold mAb to concentrate in the tumor. Removal of the macromolecule-binder conjugate from the circulation with a polyvalent "chase" macromolecule before giving the labeled effector molecule greatly improves the target-to-blood ratio. In an excellent editorial, Goodwin (1995) demonstrates the principle of these techniques. He also lists the most important presently studied mAb based pretargetting molecules as well as their corresponding effector molecules and discusses the complexity of the system. He indicates the following prerequisites for effector molecules: they must be small, hydrophilic, rapidly diffusable, quickly excreted solely by the kidneys and have little or no concentration in any normal tissues. The uptake of the pretargeted mAb by the tumor must be rapid (hours) and its subsequent release from the tumor must be slow (days). Figures 2–4 (Goodwin, 1995, with permission) show the time-activity curves in tumor and blood obtained in mice after injecting directly labeled mAb (Figure 2) and pre-targeted mAb in a two-step approach, without chase (Figure 3) and a three-step approach, with chase (Figure 4). The AUC (area under the curve) ratios of tumor

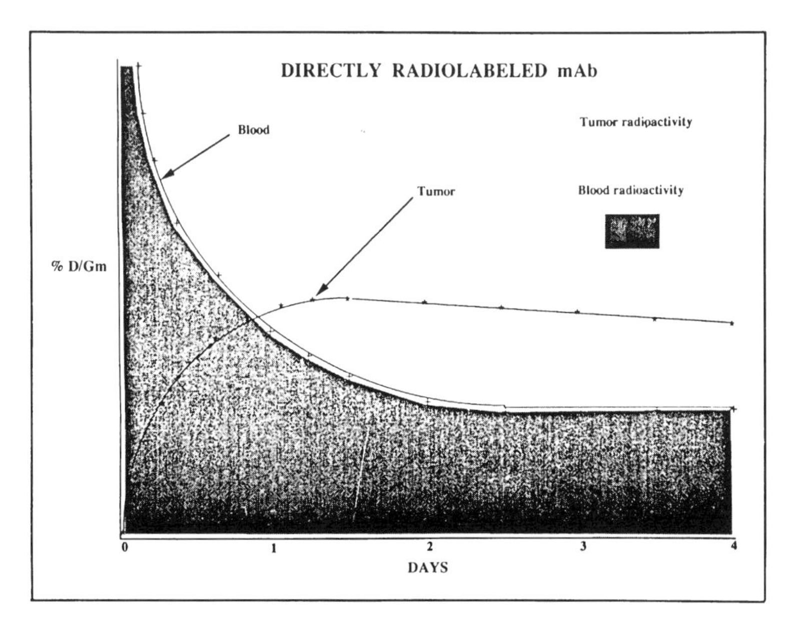

Figure 2 Time-concentration curves in tumor mice of covalent conjugates of directly labeled mAb in blood and tumor. The high tumor concentration is offset by high blood levels, giving a TR ~3/1 compared to ~24/1 for pretargeting. (with permission, Goodwin, 1995)

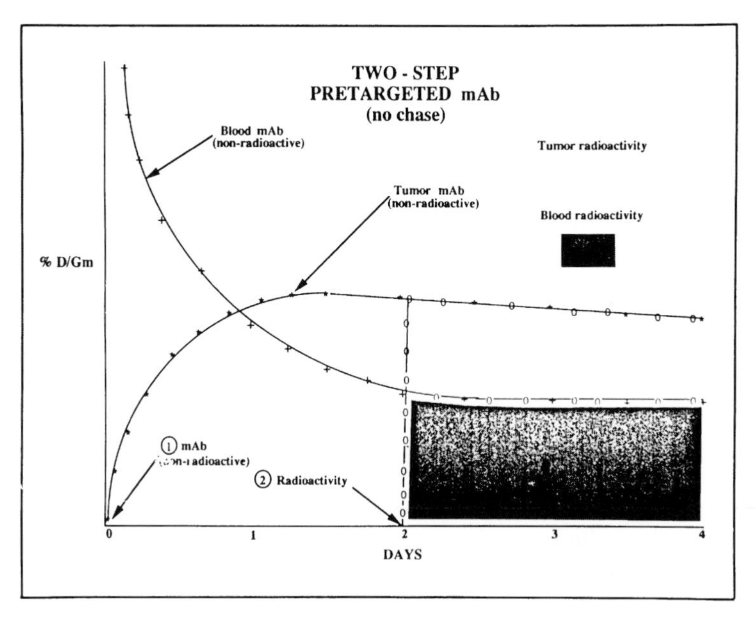

Figure 3 Pharmacokinetics of two-step pretargeting. A long waiting period is needed for blood mAb levels to fall. There is no radioactivity present during the mAb localization phase. (with permission, Goodwin, 1995)

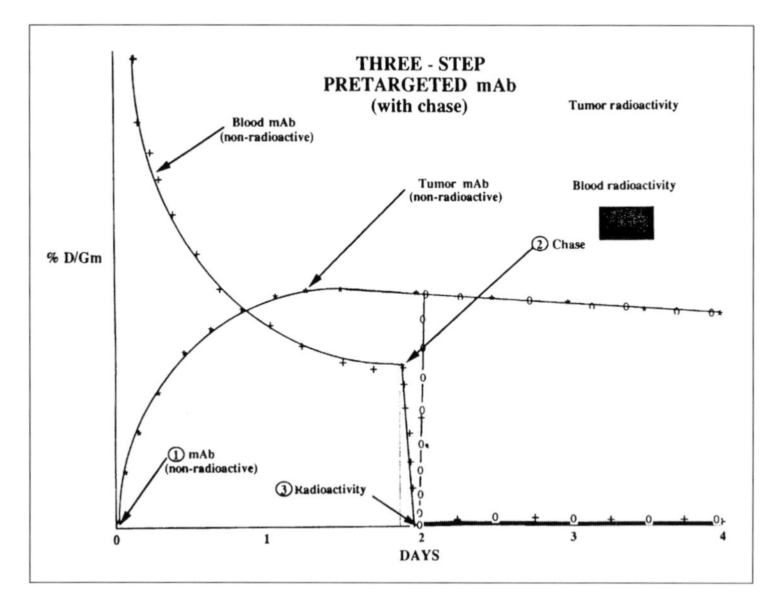

Figure 4 Pharmacokinetics of three-step pretargeting. Rapid uptake at 3 h and slow release of hapten from the tumor is shown over 4 days with pretargeted mAb. Note the large difference between the rates of diffusion into and out of the tumor: very rapid uptake (hours) compared to very slow loss (days) from the tumor. Blood levels are low at all times. (with permission, Goodwin, 1995)

to blood time activity curves are approximately 3 : 1 for directly labeled and 24 : 1 for pretargeted antibodies. Examples of pretargetting macromolecule conjugates and corresponding effector molecules include bivalent mAb/hapten (Goodwin *et al.*, 1988; Le Doussal *et al.*, 1989; Stickney *et al.*, 1991; Schuhmacher *et al.*, 1995) mAb-avidin/biotin (Hnatowitch *et al.*, 1987) or mAb-biotin/avidin (Paganelli *et al.*, 1988, 1991, 1992; Kalofonos *et al.*, 1990; Koch *et al.*, 1992) and mAb-oligonucleotide/antisense oligonucleotide (Bos *et al.*, 1994). Bispecific antibodies may be obtained by coupling, through covalent linkage, Fab′ fragments of the tumor antibody to fragments of another antibody raised against a hapten (Brennan *et al.*, 1985) or by the production of hybrid hybridomas (Milstein and Cuello, 1983). Monovalent as well as bivalent haptens have been proposed. Le Doussal *et al.* (1989) used a bivalent In-111-DTPA hapten which interacts with two rather than one single bispecific antibody conjugate already concentrated in the tumor. By this dual interaction they form a sort of bridge that "blocks" the conjugate within the tumor. These authors also observed that the radiolabeled haptens bound with higher affinity to cell-bound rather than to circulating bivalent antibodies and they called this phenomenon "affinity enhancement". It is essential to choose the right amount of the bi-specific mAb, and even more the optimal moment for the injection of the labeled hapten, especially when it is not preceded by a chase step. In the nude mouse model a delay of 24 h gave the best results with this compound, whereas in patients injections of di-DTPA-In-111 followed administration of the unlabeled bispecific anti-CEA immunoconjugate by 4–5 days. A striking contrast enhancement was obtained with this method in comparison with the directly labeled antibody directed against the same epitope (Peltier *et al.*, 1993) which allowed scintigraphic detection of even smaller lesions of medullary thyroid carcinoma expressing CEA. Results were particularly improved by substantially lowering non-specific liver uptake which favored the visualization of liver metastases. Despite the fact that absolute tumor uptake was not increased, the investigators expect the method to be promising for radioimmunotherapy by the substantial decrease of normal tissue exposure.

Other investigators (Somasundaram *et al.*, 1993; Schuhmacher *et al.*, 1995) have developed bispecific antibodies and a gallium chelate to be used with the single photon emitter Ga-67 as well as with the positron emitter Ga-68. With a specific antibody directed against the glycoprotein CD44v that is associated with a rat pancreas carcinoma cell line (1.1ASML) these authors found in the nude mouse model an increase of tumor to normal tissue ratios (blood, liver) of about 5 in comparison with the directly labeled antibody. Again, they observed contrast enhancement only, but not an increase of absolute tumor uptake which remained unchanged.

The avidin–biotin multi-step labeling techniques take advantage of the extreme affinity of the avidin–biotin system ($k = 10^{15}$ L/mol). Several strategies have been described using either biotinylated antibodies and labeled avidin/streptavidin (Paganelli *et al.*, 1988, 1991, 1992) or streptavidin-conjugated mAb followed by injection of radiolabeled biotin (Kalofonos *et al.*, 1990). In the three-step approach proposed by Paganelli *et al.* (1991) the cold, biotinylated mAb is injected first, followed one to five days later by the intravenous administration of unlabeled avidin in two steps: A small dose of avidin (1 mg) is injected first, in order to precipitate any biotinylated mAb still in circulation. Ten minutes later a larger amount of avidin (4 mg) is given to target the biotinylated mAb on tumor cells. One to three days later labeled biotin is injected and allowed to interact with the avidin that is now bound on the biotinylated mAb on tumor cells. This three-step protocol favors clearance of the unbound antibody from the blood at a moment when tumor uptake is supposed to be the highest (usually 24–48 h after injection). Two- as well as three-step labeling techniques with the avidin-biotin system allow to clear circulating radioactivity very rapidly through the kidneys. About 50% of the activity is eliminated in the urine during the first hour after injection. Liver activity is even lower than with the above mentioned use of bispecific anti-CEA antibodies administered without a chase step. It is in the order of 1–2% of injected activity, and tumors can be visualized as early as 20 min after injection of the labeled biotin. Repeating the avidin chase diminished even more non-target radioactivity in the nude mouse model, which allows to anticipate further improvement of tumor to normal tissue radiation dose ratio in radioimmunotherapy (Kobayashi *et al.*, 1995). The immunogenicity of these compounds may limit their repeated use. This may be a major drawback in radioimmunotherapy, where fractionated antibody administration might increase efficiency, as shown in animal experiments (Buchegger *et al.*, 1990; Buchsbaum *et al.*, 1995). Marshall *et al.* (1996) have tried to reduce the immune response of mice by using a polyethylene glycol modified galactosylated (gal) streptavidin. The number of mice that elicited an anti-gal-streptavidin response was reduced, but unfortunately this modified compound no longer allowed to clear the biotinylated antibody from circulation and antibody uptake by the tumor was decreased.

Besides the higher contrast achieved with these methods, which is important for efficient immunoscintigraphy, but even more for radioimmunotherapy, one might also expect multi-step labeling techniques to be useful in the presence of human anti-mouse antibodies (HAMA). At least as long as titers are not too elevated, only a part of the unlabeled antibody will be complexed by HAMA after injection and then be trapped by the reticulo-endothelial system, mostly in the liver (Kupffer cells) where it will no longer be available for subsequent labeling with the radiolabeled hapten. There might still be a significant amount of mAb left that is able to interact with the tumor bound antigen where it can be subsequently labeled by the radioactive hapten.

CLINICAL RESULTS

Immunoscintigraphy

Immunoscintigraphy (IS) is of little value in the diagnosis of primary cancers, but may be used for staging in well defined situations because of the possibility of whole body imaging. The major indications are the detection of recurrences and/or metastases in patients with new symptoms and/or rising tumor markers and the pre-operative evaluation of patients scheduled for surgical treatment of isolated local recurrence or liver metastases.

Colorectal carcinoma

This is the most extensively studied field in immunoscintigraphy. Various antibodies have been used, the most often directed against epitopes of CEA or the tumor associated glycoprotein-72 (TAG-72), generally labeled with Indium-111 or Technetium-99m. I-131 was rapidly abandoned because of its low photon yield and unnecessary irradiation due to its beta emission, I-123 was used in only a few studies because of its restricted availability and relatively high costs. In primary tumors the sensitivity and specificity of immunoscintigraphy with either antibody and radiolabel is usually high, in the order of more than 80%, but the diagnosis of primary tumors was never based on immunoscintigraphy. This method was exceptionally able to detect previously unknown distant disease, but these rare cases probably do not justify its routine use for staging. Adding peroperative probe techniques (radioimmunoguided surgery, RIGS) to immunoscintigraphy allowed to increase diagnostic efficiency (Prati *et al.*, 1995).

The main interest of immunoscintigraphy lies in the early detection of recurrence in patients with new symptoms and/or increasing serum tumor markers (Bischof Delaloye *et al.*, 1989; Lind *et al.*, 1991; Hasemann *et al.*, 1992; Doerr *et al.*, 1990; Steinsträsser *et al.*, 1995). The sensitivity of immunoscintigraphy for the detection of local recurrence in the extra-hepatic abdomen is high, up to 90%, occult cancers are detected with a sensitivity of 55–70%. This compares favorably with the results of methods based on morphological changes such as ultrasonography (US) and computerized tomography (CT). Visualisation of liver metastases is more difficult because of uptake in normal parenchyma. It is nevertheless possible to detect metastatic involvement (Figure 5) well before it is evident on CT (Bischof Delaloye *et al.*, 1989).

In a study comparing the performance of imaging with conventional diagnostic modalities and immunoscintigraphy with a Tc-99m-labeled anti-CEA Fab′ fragment, Moffat *et al.* (1996) could show that sensitivity of immunoscintigraphy was superior to that of conventional imaging in the extra-hepatic abdomen (55% vs 32%) and pelvis (69% vs 48%) and that it was complementary in the liver. Among 122 patients with known disease, the positive predictive value was significantly higher when both modalities, CT and IS, were positive (98%) compared with each one alone (68% to 70%). In 88 patients with occult disease, imaging accuracy was significantly enhanced by adding immunoscintigraphy to the conventional approach (61% vs 33%). Potential clinical benefit was demonstrated in 89 of 210 patients (42%). In a European multicenter study on immunoscintigraphy with the Tc-99m labeled anti-CEA intact mAb BW 431/26 (Steinsträsser and Oberhausen, 1995) including 730 patients investigated in 51 centers, results of immunoscintigraphy were compared with those of the conventional work up. In 17% of the patients, immunoscintigraphy was the only technique that allowed to visualize the tumor whereas it offered additional information in 24–52% of the cases. In patients scheduled for curative surgery of local recurrence or

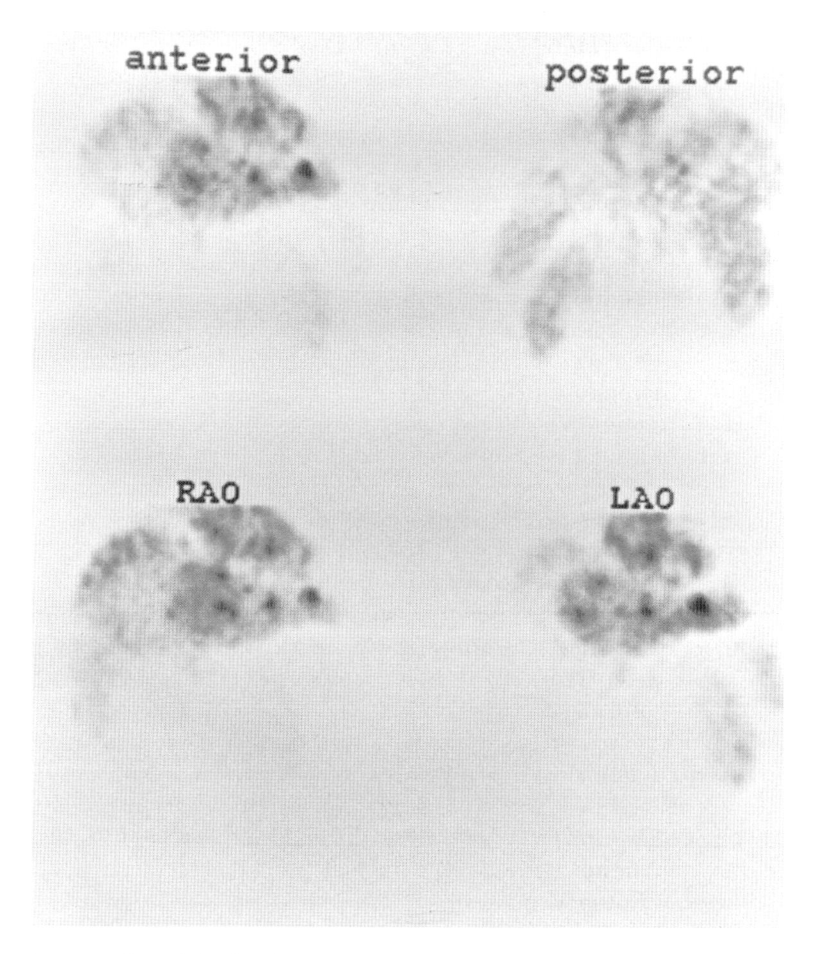

Figure 5 Reprojected single photon emission tomography images of the liver of a patient having previously been submitted to partial right hepatectomy for metastases of colon cancer which is the cause of reduced tracer uptake in this area. Multiple hot spots are shown in the left liver on the anterior, right and left anterior oblique (RAO, LAO) projections 23 h after injection of a Tc-99m labeled anti-CEA Fab′ fragment. At this moment CT was entirely negative, metastases were confirmed at follow-up.

metastases, immunoscintigraphy can confirm the presence of disease limited to one site or detect multiple sites and thus induce a change of the therapeutic approach. Development of human anti-mouse antibodies (HAMA) with repeated use of immunoscintigraphy may prevent from using this method for regular follow-up studies. HAMA appear more frequently with intact mAb than with fragments. In the study of Steinsträsser and Oberhausen (1995), 15% of the patients injected once with the Tc-99m-labeled intact anti-CEA antibody developed HAMA. This was the case in only 2 of 210 patients studied with the Tc-99m-labeled anti-CEA Fab' fragment in the study of Moffat (1996), whereas none of the 19 assessable patients having undergone two injections of the labeled mAb fragment showed increasing

HAMA titers. This low HAMA rate with small fragments raises the question on the opportunity of using chimeric or even further "humanized" antibodies. In our experience chimeric and murine mAb that are directed against the same epitope of CEA had comparable biokinetics in patients with CEA producing tumors (Buchegger *et al.*, 1995). We could not show any advantage in radioimmunoimaging as well as activity measurements in tissue specimen by using the chimeric (human IgG_4 subclass) in comparison with the murine mAb CE25/B7 which is directed against the epitope Gold 4 of CEA and from which the chimera had been derived. Nevertheless, the development of HAMA which certainly has to be faced with multiple administrations can change the biokinetics of the radiolabeled mAb in a way that tumor targeting becomes insufficient. This is particularly important when using these compounds in a therapeutic setting. For this reason the future development of immunoscintigraphy, but even more of radioimmunotherapy, will depend on the availability of reshaped mAb (Slavin-Chiorini *et al.*, 1995).

Ovarian carcinoma

Average sensitivity, specificity and accuracy of immunoscintigraphy reported in ovarian carcinoma are $\geqslant 75\%$ (Bischof Delaloye *et al.*, 1995a). Several studies have compared the performance of immunoscintigraphy with that of other diagnostic modalities in the detection of recurrent ovarian cancer. Massuger *et al.* (1990) found a sensitivity of 67% for immunoscintigraphy and of 53% and 23% for CT and US, respectively. In a study on correlative imaging in the diagnosis of ovarian cancer recurrences, Peltier *et al.* (1992) showed respective sensitivity, specificity and accuracy of 89%, 73% and 85% for immunoscintigraphy; 43%, 75% and 43% for US; and 57%, 67% and 59% for CT. Low *et al.* (1995) compared the accuracy of breath-hold gadolinium- and perflubron-enhanced magnetic resonance (MR) imaging, immunoscintigraphy with indium-111-anti-TAG-72 intact mAb and contrast material-enhanced CT for the detection of ovarian cancer prior to laparotomy. They found a per patient accuracy of 75%, 86% and 81% for MR, IS and CT, respectively. For the detection of individual tumor sites MR was the most sensitive method (81%) compared with CT (51%) and IS (50%). The combination of MR and IS detected 89% sites of tumor involvement confirmed at laparotomy. All these studies converge in revealing the poor sensitivity of CT which nevertheless remains one of the first-line investigations in routine medical practice. De Rosa *et al.* (1995) raise the question on the actual role of the CT scan versus second look surgery and conclude that with respect to the high false negative rate (53% in their study), especially for omental, mesenteric and peritoneal implants as well as for bowel infiltration, the CT scan does not provide sufficiently accurate diagnostic information to replace second-look laparotomy. The exact role of immunoscintigraphy in this context, however, remains to be further elucidated.

Lymphoma

The stage of B-cell non-Hodgkin's lymphoma at diagnosis is one of the important factors to determine treatment as well as prognosis (The International

Non-Hodgkin's Lymphoma Prognostic Factor Project, 1993). Screening for relapse with presently available methods has been shown to lack sensitivity (Weeks *et al.*, 1991). Most of relapses occur in new sites, methods that are targeted to specific sites are thus insufficient. Immunoscintigraphy is one of the methods, besides the Gallium scan and positron emission tomography (PET), that allow total body imaging. Sensitivities of up to 90% have been reported in small series (Blend *et al.*, 1995). These results need to be confirmed in larger trials. Becker *et al.* (1995) have compared the results of Gallium scans and immunoscintigraphy with a Tc-99m labeled anti-CD-22 antigen monoclonal antibody Fab′ fragment in patients with B-cell lymphoma. They found an overall sensitivity of 80% for both tracers, with an advantage for immunoscintigraphy in low grade and for Ga-67-citrate in high grade lymphoma.

Malignant melanoma

In a survey on immunoscintigraphy of melanoma we found that melanoma was detected with a sensitivity and specificity of about 70% and 90%, respectively (Bischof Delaloye *et al.*, 1995a). Few publications indicate a much lower sensitivity in cutaneous (Boni *et al.*, 1995) as well as in ocular melanoma (Schaling *et al.*, 1990). When comparing their results in 102 patients with histologically proven cutaneous and 35 with ocular melanoma, Löffler *et al.* (1996) found a much lower sensitivity for ocular (50%) than for cutaneous melanoma (89%) with false positive images in two patients, one with subretinal hemorrhage, the other with Wegener's granulomatosis. No correlation of immunoscintigraphic data with histological features or immunoreactivity patterns could be established, but antigenic differences between cutaneous and ocular melanoma were evident and probably account for the observed results.

In recent years the interest in immunoscintigraphy of malignant melanoma has diminished. The increased availability of PET using F-18-fluoro-deoxy-glucose (FDG) may have contributed to this decline of interest. In fact, higher accuracy of tumor detection has been reported with this technique in comparison with immunoscintigraphy (Scheidhauer *et al.*, 1996). PET with FDG or other radiopharmaceuticals labeled with positron emitters will certainly change the management of cancer patients, but it seems unlikely that these radiotracers will completely replace monoclonal antibodies. Tumor imaging with radionuclide antibody conjugates is based on an antibody–antigen interaction, it has thus the potential of specific tissue characterization and prepares the grounds for systemic radiotherapy.

Radioimmunoguided Surgery

Due to the low absolute tumor accumulation of radiolabeled mAb and the still relatively poor spatial resolution of the presently available gamma-cameras, small tumor foci cannot reliably be detected even with single photon emission tomography (SPECT). The absence of anatomic landmarks does not allow to precisely indicate the location of a given hot spot, which can thus often not be found during surgery, especially in the presence of post-operative changes in the case of re-intervention for suspected recurrence. Intra-operative detection of tumor foci with a hand-held gamma-probe opens a new dimension to radioimmunodiagnosis. Some conditions have to be fulfilled in order to obtain reliable data: Tumor-to-background ratio has to be high enough to allow a clear distinction between tumor and surrounding structures, uptake of the antibody has to be tumor specific to avoid the resection of tumor free-tissues, the physical and technical conditions of the probe must be adapted to surface counting without contamination by deeper structures, it must be able to undergo sterilization, should be easy to handle in the operation theater, etc.

Arnold *et al.* (1996) report on the comparison of traditional surgical exploration and RIGS in patients with either primary or recurrent colorectal carcinoma following injection of I-125 labeled CC49 mAb, a second generation antibody raised against TAG-72. In 41 patients with primary tumors, traditional surgical exploration detected 45 sites of disease (1.1 sites/patient) compared with 153 RIGS-positive sites (3.7 sites/patient). In 45 patients with recurrent disease, traditional exploration found 116 sites (2.6 sites/patient) whereas RIGS detected 184 sites (4.1 sites/patient). Except for liver metastases RIGS detected more tissue involved, the highest proportion of RIGS-positive tissue was found in the gastrohepatic ligament and celiac nodes, areas which are usually not routinely resected. These authors (Bertsch *et al.*, 1996) also found that RIGS might challenge the traditional staging of colorectal cancer by identifying different disease patterns. In a group of 16 patients, traditional en bloc resection was performed in only 5 whereas in 11 patients additional extraregional tumor tissue was removed because of positive gamma-probe results. After a median follow-up of 37 months all 16 patients were alive with no evidence of disease in 14. In comparison, 14 of 16 other patients where the probe measurements indicated unresectable disease were dead at follow-up, two alive with disease. Tumor extent in these two groups was obviously very different at the time of laparotomy, but one may speculate that the prognosis of the first group would have been worse if the additional sites revealed by RIGS only would not have been resected during routine laparotomy. Probe-positivity of lymph nodes with mAb directed against TAG-72, however, has also been reported without metastatic involvement. Cote *et al.* (1996) reported on the analysis of 57 resected regional lymph nodes from patients with primary or recurrent colorectal carcinoma. Fourteen of 17 lymph nodes showing metastatic involvement on routine histologic examination (hematoxylin–eosin staining, HE) were RIGS positive. Ten of 25 RIGS positive/HE negative nodes demonstrated occult metastases after serial section immunohistochemistry analysis. A positive RIGS reading in lymph nodes that showed no evidence of tumor by routine histopathologic examination was significantly associated with the presence of occult metastases. Intra-operative radioimmunoguided detection of occult metastases with anti TAG-72 antibodies, however, may be hampered by the fact that B72.3 immunoreactivity has been observed in the germinal centers of benign abdominal lymph nodes associated with gastrointestinal disease, 49% of the cases of colonic adenocarcinoma and 12% of benign gastrointestinal disease (Loy and Haege, 1995).

RADIOIMMUNOTHERAPY

In recent years the interest of the medical community has shifted from immunoscintigraphy to radioimmunotherapy (RIT) (Bischof Delaloye *et al.*, 1995b). This has several reasons: mortality rates of various solid tumors such as colorectal, ovarian and lung cancers have not changed significantly despite new therapeutic regimen, the success of RIT in lymphomas raises the hope to transfer efficient treatment with radiolabeled antibodies to solid tumors and finally the support to this expectancy given by successful RIT of xenografted human tumors in athymic mice (Knox *et al.*, 1995). Due to the relatively poor radiosensitivity of these tumors and the low uptake of the radiopharmaceutical no therapeutic efficiency is to be expected from RIT alone, at least in bulky tumors. In micrometastases the situation is different due to the relative increase of uptake per gram tumor tissue in small metastases in comparison with larger tumors. The higher relative uptake is more favorable in smaller tumors because of the decreased interstitial pressure and the higher relative number of cells expressing the antigen, with less necrosis and fibrosis. Several approaches have been made to increase tumor uptake, but so far the only way to increase efficacy is by increasing the tumor to normal tissue activity ratio. Fractionated therapy may also help to deliver a higher total amount of radioactivity to the tumor (DeNardo *et al.*, 1990; Buchegger *et al.*, 1995; Buchsbaum *et al.*, 1995). Furthermore attempts have been made to increase the expression of the tumor associated antigen by immunomodulation. Interferon gamma as well as alpha have been shown to upregulate antigen expression in tumors which might be used for RIT (Oettgen, 1991; Galmiche *et al.*, 1996; Murray *et al.*, 1995). These techniques have essentially been tested in xenografted nude mice but have not yet been used in large trials in humans. Again two- or three-step labeling of the antibodies is a very promising technique for RIT. The decrease of non-target irradiation contributes to the better tolerance of the therapy and allows multiple treatment courses. It becomes thus mandatory to develop techniques that use recombinant antibodies which are less immunogenic than intact murine IgG. When using antibody fragments, absolute tumor uptake decreases proportionally to the size of the fragment. Due to the lower immunogenicity of fragments, radioactivity can be administered in multiple fractions that allow to increase the area under the activity-time curve, which corresponds to the total tumor residence time of radioactivity.

The most often proposed radionuclides for RIT are beta emitters. Alpha emitters would be preferable on theoretical grounds essentially because of their high linear energy transfer (LET). In order to be authorized to use alpha emitters one must be certain that non-target activity is minimal. Further efforts need to be taken in order to ascertain that non-tumor activity can be reduced to a level which allows the use of alpha emitters. The same is true for radionuclides decaying by electron capture and emitting Auger electrons. Besides low background activity, these radionuclides, in order to be efficient, need to be conjugated to internalizing antibodies (Matzku *et al.*, 1990; Novak-Hofer *et al.*, 1994). Daghigian *et al.* (1996) performed cytotoxicity experiments *in vitro* with I-125 labeled antibody A33, an internalizing antibody which detects a restricted determinant on colon cancer cells. A33 accumulates in cytoplasmic vehicles which transport the activity close to the nucleus. The increase of radiation dose to the nucleus was calculated as by a factor 3 compared to the average dose calculated based on the assumption of a uniform distribution on the cell membrane. The cytoplasm of antigen-negative normal cells shielded the nucleus from the electrons emitted from extracellular I-125. This shielding effect was 30 times less for I-131. These findings suggest an improvement of the tumor-to-normal tissue (especially bone marrow) dose with I-125-labeled mAb A33.

Probably because of the extensive experience gained with I-131 in treating benign and malignant thyroid disorders as well as its availability at low costs this radionuclide has been one of the most often used in RIT, even if it is far from being the ideal isotope with regard to its physical characteristics. The beta emission is of low energy with an optimal range of 2.6–5.0 mm. This makes this radionuclide more favorable for the treatment of micrometastases than bulky tumors. The high energy gamma emission (364 keV, 82%) accounts for two thirds of the absorbed dose equivalent (Humm, 1986) and contributes significantly to irradiation of normal tissues, especially bone marrow, and to the radiation exposure of medical, technical and nursing staff. In order to limit the radiation exposure of the environment, patients treated with therapeutic activities of I-131 need to be hospitalized in specially equipped rooms for several days, which increases the costs of the procedure.

Potential radiopharmaceuticals for RIT including their optimal therapy range in uniformly targeted tumor spheres (O'Donoghue *et al.*, 1995) are listed on Table 1.

Table 1 Beta-emitters for radioimmunotherapy

Radionuclide	*Half-Life*	*Optimal Range* (mm)
P-32	14 d	18–30
Cu-67	62 h	1.6–2.8
Y-90	64 h	28–42
Ag-111	7.45 d	7–13
I-131	8 d	2.6–5.0
Sm-153	46.8 h	2.8–5.0
Ho-166	26.8 h	18–25
Lu-177	6.7 d	1.2–3.0
Re-186	90 h	7–12
Re-188	17 h	23–32

So far, clinical trials often failed to show convincingly positive results. A number of factors may contribute to the apparent lack of efficacy in many RIT trials. The antibody concentration in the tumor is due to the antibody bound on antigen-positive tumor cells and to the antibody in the blood circulating through the tumor and diffusing/convecting into the interstitial spaces (Muthuswamy *et al.*, 1996). In larger solid tumors access of radiolabeled antibodies to antigen-positive cells may be difficult due to high interstitial pressure, antibody distribution may be heterogeneous. After injection the radionuclide–antibody conjugate preferentially accumulates in the periphery of the tumor and energy of beta particles will in part be deposited outside the tumor. This energy loss is more important with higher energy beta emitters such as Yttrium-90 than with lower mean beta energy emitters such as Copper-67. Another question is the sensitivity of the tumor cells to low-dose-rate irradiation. In mice bearing three different kinds of tumors, Knox *et al.* (1993) have correlated the relative efficacy of low-dose-rate irradiation versus equivalent doses of high-dose-rate irradiation with the extent of arrest of cells in G2 phase of the cell cycle. They found that murine 38C13 B-cell lymphoma had the shortest tumor volume doubling and cell cycle times, the most G2 arrest after low-dose-rate irradiation, more G2 arrests with low-dose-rate than high-dose-rate irradiation, and the highest alpha/beta ratio. Neither the HT29 human colorectal xenograft nor the SNB75 human glioblastoma xenograft was significantly more sensitive to low-dose-rate than fractionated high-dose-rate irradiation. The combination of radioimmunotherapy with external beam radiation (Buchegger *et al.*, 1995) and/or chemotherapy (Order *et al.*, 1987), hyperthermia (Wilder *et al.*, 1993), with administration of vasoactive substances to modify tumor to normal tissues perfusion ratio (Gilgien *et al.*, 1988; Burton *et al.*, 1988) or cytokines to increase vascular permeability or antigen expression (Oettgen *et al.*, 1991; Galmiche *et al.*, 1996) have been proposed to enhance the efficacy of RIT.

Clinical Results

Objective response rates between 20% and 85% of patients with B-cell lymphoma failing multiple chemotherapy regimens have been observed (DeNardo *et al.*, 1990; Czuczman *et al.*, 1993; Kaminski *et al.*, 1993, 1996; Press *et al.*, 1993, 1995) whereas in patients with solid tumors results were much less convincing. This difference is, among others, due to the higher radiosensitivity of these tumors, but the radiobiologic mechanisms that are responsible for radioresponsiveness are still unclear. Macklis *et al.* (1994) have shown that some malignant lymphoma lines are highly sensitive to low-dose-rate irradiation and that a portion of the cytotoxicity appears to be mediated by the induction of radiation-induced apoptosis. Press *et al.* (1995) determined the administered therapeutic activity of I-131 in a way to deliver 25–31 Gy of irradiation to the normal organ receiving the greatest dose of radiation. They infused 2.5 mg/kg anti-B1 labeled with 12.765-29.045 GBq I-131, followed by autologous stem cell reinfusion. Eighteen of 21 treated patients had objective responses, including 16 complete remissions.

Progression-free survival was 62% and overall survival 93% with a median follow-up of 2 years. All patients had advanced stage lymphomas that had failed multiple chemotherapy regimens. Inversely, Kaminski *et al.* (1993, 1996) chose to treat their patients with non marrow ablative doses. They calculated the dose to deliver a specified whole body irradiation predicted by a tracer dose. 75 cGy was established as the maximally tolerated whole-body dose. The administration of the therapeutic dose of I-131 labeled anti-B1 (1.258–5.957 GBq) was preceded by infusion of 685 mg unlabeled mAb. 22/28 patients responded including 14 complete remissions. Median duration of complete remission exceeded 16.5 months with six patients in complete remission 16–31 months after treatment. Non-myeloablative treatment was well tolerated except for dose limiting hematological toxicity. About 40% of patients with multiple infusions developed HAMA. This confirms the necessity of developing chimeric or humanized mAb (Juweid *et al.*, 1995).

Several authors have published encouraging results of radioimmunotherapy of human colon tumor xenografts in nude mice, but the results in human studies remain very modest. Higher radiation doses need to be delivered to tumors that are less sensitive than lymphomas, but dose escalation is limited by myelotoxicity. The various strategies developed to enhance tumor uptake or at least tumor to normal tissue ratio have already been discussed. Only few objective responses have been induced in carcinomas, which were most often categorized as stabilization of tumor progression (Delaloye *et al.*, 1987; Breitz *et al.*, 1992; Murray *et al.*, 1994; Meredith *et al.*, 1995; Juweid *et al.*, 1996). Induction of complete remission was essentially observed in ovarian carcinomas with microscopic or occult residual disease with intraperitoneal injection of I-131, Y-90 or Lu-177 labeled mAb (Hird *et al.*, 1993; Crippa *et al.*, 1995; Meredith *et al.*, 1996). DeNardo *et al.* (1994) reported on I-131 labeled chimeric mAb L6, an antibody that binds to adenocarcinoma, including 50% of breast cancers, and demonstrates high efficiency in mediating antibody dependent toxicity. They were able to obtain objective tumor regression in the majority of patients with metastatic breast cancer, probably through the combined effect of RIT and direct stimulation of immunoeffector cell function. Local administration of the therapeutic compound has obviously an advantage over systemic infusion. After direct intratumoral injection of I-131 labeled anti-tenascin antibodies in patients with malignant glioma, Riva *et al.* (1995) observed an overall response rate of 40% in recurrent as well as newly diagnosed tumors.

Radioimmunotherapy is a promising tool in treatment of various cancers, especially B-cell lymphomas and leukemias, but probably also as an adjuvant treatment modality in minimal residual disease of solid tumors. Continuing research should compare different radionuclides and antibodies in the same system (Bischof Delaloye *et al.*, 1997) as well as the relative efficacy and toxicity of equivalent doses of RIT and external beam irradiation. It should be determined how heterogeneity of radiolabeled antibody deposition affects the absorbed dose distribution and relevant biological end points. Strategies for increasing and optimizing the therapeutic index of RIT have to be developed as well as combined treatment modalities (Knox, 1995).

CONCLUSION

The use of monoclonal antibodies for diagnostic imaging and radioimmunotherapy is the subject of ongoing multidisciplinary research. Despite the numerous advances in our knowledge, the handling of these compounds by the human body is far from being completely elucidated. To overcome the mechanisms that hinder the mAb from reaching its target, the cell bound antigen, is the subject of many efforts produced by the various groups that devote their research to this field. Despite these difficulties, the specificity of the antibody-antigen reaction together with the relative ease of labeling the antibody with a large variety of radionuclides without changing its immunologic properties offer a unique potential for the characterization of tumor versus non-tumor tissue by nuclear medicine imaging techniques and for the efficient treatment of tumors by systemic radiotherapy.

REFERENCES

Andrew, S.M., Pimm, M.V., Perkins, J.P. *et al.* (1986) Comparative imaging and biodistribution studies with an anti-CEA monoclonal antibody and its $F(ab')_2$ and Fab fragments in mice with colon carcinoma xenografts. *Eur. J. Nucl. Med.*, **12**, 168–175.

Arnold, M.W., Hitchcock, C.L., Young, D.C. *et al.* (1996) Intra-abdominal patterns of disease dissemination in colorectal cancer identified using radioimmunoguided surgery. *Dis. Colon Rectum*, **39**, 509–513.

Becker, W., Goldenberg, D.M. and Wolf, F. (1994) The use of monoclonal antibodies and antibody fragments in the imaging of infectious lesions. *Semin. Nucl. Med.*, **24**, 142–153.

Becker, W.S., Behr, T.M., Cumme, F. *et al.* (1995) ^{67}Ga Citrate *versus* ^{99m}Tc-labeled LL2-Fab′ (Anti-CD22) Fragments in the Staging of B-Cell Non-Hodgkin's Lymphoma[1]. *Cancer Res. (suppl.)*, **55**, 5771s–5773s.

Begent, R.H., Searle, F., Stanway, G. *et al.* (1980) Radioimmunolocalization of tumours by external scintigraphy after administration of ^{131}I antibody to human chorionic gonadotrophin: Preliminary communication. *J. R. Soc. Med.*, **73**, 624–630.

Begent, R.H., Keep, P.A., Green, A.J. *et al.* (1982) Liposomally entrapped second antibody improves tumour imaging with radiolabelled (first) antitumour antibody. *Lancet*, **2**(8301), 739–742.

Behr, T., Becker, W., Hannapel, E., Goldenberg, D.M. and Wolf, F. (1995) Targeting of liver metastases of colorectal cancer with IgG, $F(ab')_2$, and F(ab′) anti-carcinoembryonic antigen antibodies labeled with ^{99m}Tc: the role of metabolism and kinetics. *Cancer Res. (suppl.)*, **55**, 5777s–5785s.

Behr, T.M., Sharkey, R.M., Blumenthal, R.D. *et al.* (1996) Improvement of the therapeutic efficacy of radiometal-conjugated antibody fragments and peptides by overcoming their nephrotoxic potential. *Eur. J. Nucl. Med.*, **23**, 1118 (abstract).

Behr, T.M., Sharkey, R.M., Juweid, M.I. *et al.* (1996) Factors influencing the pharmacokinetics, dosimetry, and diagnostic accuracy of radioimmunodetection and radioimmunotherapy of carcinoembryonic antigen-expressing tumors. *Cancer Res.*, **56**, 1805–1816.

Bertsch, D.J., Burak, W.E. Jr., Young, D.C., Arnold, M.W. and Martin, E.W. Jr. (1996) Radioimmunoguided surgery for colorectal cancer. *Ann. Surg. Oncol.*, **3**, 310–316.

Bischof Delaloye, A., Delaloye, B., Buchegger, F. *et al.* (1989) Clinical value of immunoscintigraphy in colorectal carcinoma patients: a prospective study. *J. Nucl. Med.*, **30**, 1646–1656.

Bischof Delaloye, A. and Delaloye, B. (1995)a Tumor imaging with monoclonal antibodies. *Semin. Nucl. Med.*, **25**, 144–164.

Bischof Delaloye, A. and Delaloye, B. (1995)b Radiolabelled monoclonal antibodies in tumour imaging and therapy: out of fashion? (Review). *Eur. J. Nucl. Med.*, **22**, 571–580.

Bischof Delaloye, A., Delaloye, B., Buchegger, F. *et al.* (1997) Comparison of 67Cu- and 125I-labeled anti-CEA monoclonal antibody biodistribution in patients with colorectal tumors. *J. Nucl. Med.*, **38**, 847–853.

Bläuenstein, P., Locher, J.T., Seybold, K. *et al.* (1995) Experience with the iodine-123 and technetium-99m labelled anti-granulocyte antibody MAb47: a comparison of labelling methods. *Eur. J. Nucl. Med.*, **22**, 690–698.

Blend, M.J., Hyun, J., Kozloff, M., Levi, H. *et al.* (1995) Improved staging of B-cell non-Hodgkin's lymphoma patients with 99mTc-labeled LL2 monoclonal antibody fragment. *Cancer Res.* (*suppl.*), **55**, 5764s–5770s.

Boni, R., Huch Boni, R.A., Steinert, H. *et al.* (1995) Anti-melanoma monoclonal antibody 225.28S immunoscintigraphy in metastatic melanoma. *Dermatology*, **191**, 119–123.

Bos, E.S., Kuijpers, W.H.A., Meesters-Winters, M. *et al.* (1994) *In vitro* evaluation of DNA–DNA hybridization as a two-step approach in radioimmunotherapy of cancer. *Cancer Res.*, **54**, 3479–3486.

Bosslet, K., Steinsträsser, A., Schwarz, A. *et al.* (1988) Quantitative considerations supporting the irrelevance of circulating serum CEA for the immunscintigraphic visualisation of CEA expressing carcinomas. *Eur. J. Nucl. Med.*, **14**, 523–528.

Boubaker, A., Bischof Delaloye, A., Blanc, C.H. *et al.* (1995) Immunoscintigraphy with antigranulocyte monoclonal antibodies for the diagnosis of septic loosening of hip prosthesis. *Eur. J. Nucl. Med.*, **22**, 139–147.

Breitz, H.B., Weiden, P.L., Vanderheyden, J.L. *et al.* (1992) Clinical experience with rhenium-186-labeled monoclonal antibodies for radioimmunotherapy: results of phase I trials. *J. Nucl. Med.*, **33**, 1099–1112.

Brennan, M., Davison, P.F. and Paulus, H. (1985) Preparation of bispecific antibodies by chemical recombination of monoclonal immunoglobulin G1 fragments. *Science*, **229**, 81–83.

Buchegger, F., Haskell, C.M., Schreyer, M. *et al.* (1983) Radiolabeled fragments of monoclonal antibodies against carcinoembryonic antigen for localization of human colon carcinoma grafted into nude mice. *J. Exp. Med.*, **158**, 413–427.

Buchegger, F., Pèlegrin, A., Delaloye, B., Bischof Delaloye, A. and Mach, J.P. (1990) 131I labeled F(ab′)2 fragments are more efficient and less toxic than intact anti-CEA antibodies in radioimmunotherapy of large human colon carcinoma grafted in nude mice. *J. Nucl. Med.*, **31**,1035–1044.

Buchegger, F., Rojas, A., Bischof Delaloye, A. *et al.* (1995) Combined radioimmunotherapy and radiotherapy of human colon carcinoma grafted in nude mice. *Cancer Res.*, **55**, 83–89.

Buchegger, F., Mach, J.P., Pèlegrin, A. *et al.* (1995) Radiolabeled chimeric anti-CEA monoclonal antibody compared with the original mouse monoclonal antibody for surgically treated colorectal carcinoma. *J. Nucl. Med.*, **36**, 420–429.

Buchsbaum, D., Khazaeli, M.B., Liu, T. *et al.* (1995) Fractionated radioimmunotherapy of human colon carcinoma xenografts with ^{131}I-labeled monoclonal antibody CC49. *Cancer Res.* (*suppl.*), **55**, 5881s–5887s.

Buraggi, G.L., Callegaro, L., Turrin, A. *et al.* (1987) Immunoscintigraphy of colorectal carcinoma with F(ab′)$_2$ fragments of anti-CEA monoclonal antibody. *Cancer Detect. Prev.*, **10**, 335–345.

Burton, M.A., Gray, B.N. and Coletti, A. (1988) Effect of angiotensin II on blood flow in the transplanted sheep squamous cell carcinoma. *Eur. J. Cancer Clin. Oncol.*, **24**, 1373–1376.

Cote, R.J., Houchens, D.P., Hitchcock, C.L., Saad, A.D., Nines, R.G., Greenson, J.K., Schneebaum, S., Arnold, M.W. and Martin, E.W. Jr. (1996) Intraoperative detection of occult colon cancer micrometastases using 125I-radiolabeled monoclonal antibody CC49. *Cancer,* **77**, 613–620.

Crippa, F., Bolis, G., Seregni, E. *et al.* (1995) Single-dose intraperitoneal radioimmunotherapy with the murine monoclonal antibody I-131 Mov 18: clinical results in patients with minimal residual disease of ovarian cancer. *Eur. J. Cancer,* **31A**, 686–690.

Czuczman, M.S., Straus, D.J., Divgi, C.R. *et al.* (1993) Phase I dose escalation trial of iodine 131-labeled monoclonal antibody OKB7 in patients with non-Hodgkin's lymphoma. *J. Clin. Oncol.*, **11**, 2021–2029.

Daghighian, F., Barendswaard, E., Welt, S. *et al.* (1996) Enhancement of radiation dose to the nucleus by vesicular internalization of Iodine-125-labeled A33 monoclonal antibody. *J. Nucl. Med.*, **37**, 1052–1057.

Davidson, B.R., Babich, J., Young, H. *et al.* (1991) The effect of circulating antigen and radiolabel stability on the biodistribution of an indium labelled antibody. *Br. J. Cancer*, **65**, 850–856.

De Rosa, V., Mangoni di Stefano, M.L., Brunetti, A. *et al.* (1995) Computed tomography and second-look surgery in ovarian cancer patients. Correlation, actual role and limitations of CT scan. *Eur. J. Gynaecol. Oncol.*, **16**, 123–129.

Delaloye, B., Bischof Delaloye, A., Buchegger, F. *et al.* (1986) Detection of colorectal carcinoma by emission-computerized tomography after injection of ^{123}I-labeled Fab or F(ab')$_2$ fragments from monoclonal anti-carcinoembryonic antigen antibodies. *J. Clin. Invest.*, **77**, 301–311.

Delaloye, B., Bischof Delaloye, A., Buchegger, F. *et al.* (1987) Biokinetics of iodine labeled anti-CEA monoclonal antibody fragments in patients with CEA producing tumors. In, R. Höfer and H. Bergmann (eds), *Radioiaktive Isotope in Klinik und Forschung* (*17. Band, 2. Teil*), Egermann, Austria, pp. 749–755.

Delaloye, B., Bischof Delaloye, A., Pettavel, J. *et al.* (1987) Intra-arterial administration of 131-I-anti-CEA-MAb as a therapeutic approach to liver metastases. *Nucl-Med.*, **26**, 52 (abstract).

DeNardo, G.L., DeNardo, S.J., O'Grady, L.F., Levy, N.B., Adams, G.P. and Mills, S.L. (1990) Fractionated radioimmunotherapy of B-cell malignancies with I-131-Lym-1. *Cancer Res.*, **50**, 1014–1016.

DeNardo, S.J., Mirick, G.R., Kroger, L.A. *et al.* (1994) The biologic window for ChL6 radio-immunotherapy. *Cancer* (*Phila.*), **73**, 1023–1032.

Doerr, R.J., Abdel-Nabi, H. and Merchant, B. (1990) Indium ZCE-025 immunoscintigraphy in occult recurrent colorectal cancer with elevated carcinoembryonic antigen level. *Arch. Surg.*, **125**, 226–229.

Galmiche, M.C., Vogel, C.A., Bischof Delaloye, A. *et al.* (1996) Combined effects of interleukin-3 and interleukin-11 on hematopoiesis in irradiated mice. *Exp. Hematology*, **24**, 1298–1306.

Gilgien, W., Bischof Delaloye, A., Pettavel, J. *et al.* (1988) Enhancement of tumor/parenchyma perfusion ratio by betablockade in liver metastases. *Eur. J. Nucl. Med.*, **14**, 245 (abstract).

Goldenberg, D.M., Preston, D.F., Primus, F.J. *et al.* (1974) Photoscan localization of GW-39 tumors in hamsters using radiolabeled anticarcinoembryonic antigen immunoglobulin G. *Cancer Res.*, **34**, 1–9.

Goldenberg, D.M., DeLand, F.H., Kim, E.E. *et al.* (1978) Use of radiolabeled antibodies to carcino-embryonic antigen for the detection and localization of diverse cancers by external photoscanning. *N. Engl. J. Med.*, **298**, 1384–1386.

Goodwin, D.A., Meares, C.F., McCall, M.J. *et al.* (1988) Pretargeted immunoscintigraphy of murine tumors with Indium-111-labeled bifunctional haptens. *J. Nucl. Med.*, **29**, 226–234.

Goodwin, D.A. (1995) Tumor pretargeting: Almost the bottom line. *J. Nucl. Med.*, **36**, 876–879.

Haseman, M.K., Brown, D.W., Keeling, C.A. and Reed, N.L. (1992) Radioimmunodetection of occult carcinoembryonic antigen producing cancer. *J. Nucl. Med.*, **33**, 1750–1757.

Hird, V., Maraveyas, A., Snook, D. *et al.* (1993) Adjuvant therapy of ovarian cancer with radioactive monoclonal antibody. *Br. J. Cancer*, **68**, 403–406.

Hnatowitch, D.J., Virzi, F. and Ruschkowski, M. (1987) Investigations of avidin and biotin for imaging applications. *J. Nucl. Med.*, **28**, 1294–1302.

Humm, J.L. (1986) Dosimetric aspects of radiolabeled antibodies for tumor therapy. *J. Nucl. Med.*, **27**, 1490–1497.

Jain, R.K. (1990) Physiological barriers to delivery of monoclonal antibodies and other macro-molecules in tumors. *Cancer Res.* (*suppl.*), **50**, 814s–819s.

Jones, B.E., Green, A., Vernon, P. *et al.* (1982) Technical problems encountered in the use of a dual isotope background subtraction technique in antibody screening (radioimmunodetection). *Nucl. Med. Commun.*, **3**, 24 (abstract).

Juweid, M., Sharkey, R.M., Markowitz, A. *et al.* (1995) Treatment of non-Hodgkins lymphoma with radiolabeled murine, chimeric, or humanized LL2, an anti-CD22 monoclonal antibody. *Cancer Res.* (*suppl.*), **55**, 5899s–5907s.

Juweid, M.E., Sharkey, R.M., Behr, T. *et al.* (1996) Radioimmunotherapy of patients with small-volume tumors using iodine-131-labeled anti-CEA monoclonal antibody NP-4 F(ab')$_2$. *J. Nucl. Med.*, **37**, 1504–1510.

Kalofonos, H.P., Ruschkowski, M., Siebecker, D.A. *et al.* (1990) Imaging of tumor in patients with Indium-111-labeled biotin and streptavidin-conjugated antibodies: Preliminary communication. *J. Nucl. Med.*, **31**, 1791–1996.

Kaminsky, M.S., Zasadny, K.R., Francis, I.R. *et al.* (1993) Radioimmunotherapy of B-cell lymphoma with [^{131}I]anti-B1 (anti-CD20) antibody. *N. Engl. J. Med.*, **329**, 459–565.

Kaminsky, M.S., Zasadny, K.R., Francis, I.R. *et al.* (1996) Iodine-131-anti-B1 radioimmunotherapy for B-cell lymphoma. *J. Clin. Oncol.*, **14**, 1974–1981.

Knox, S.J., Sutherland, W. and Goris, M.L. (1993) Correlation of tumor sensitivity to low-dose-rate irradiation with G2/M-phase block and other radiobiological parameters. *Radiat. Res.*, **135**, 24–31.

Knox, S.J. (1995) Overview of studies on experimental radioimmunotherapy. *Cancer Res.* (*suppl.*), **55**, 5832s–5836s.

Kobayashi, H., Sakahara, H., Endo, K., Yao, Z.S. and Konishi, J. (1995) Repeating the avidin "chase" markedly improved the biodistribution of radiolabeled biotinylated antibodies and promoted the excretion of additional background radioactivitiy. *Eur. J. Cancer*, **31A**(10), 1689–1696.

Koch, P. and Maecke, H.R. (1992) ^{99m}Tc labeled biotin conjugate in a tumor pretargeting approach with monoclonal antibodies. *Angewandte Chemie*, **31**, 1507–1509.

Köhler, G. and Milstein, C. (1975) Continuous cultures of fused cells secreting antibodies of predefined specificity. *Nature*, **256**, 495–497.

Le Doussal, J.M., Martin, M., Gautherot, E., Delaage, M. and Barbet, J. (1989) *In vitro* and *in vivo* targeting of radiolabeled monovalent and divalent haptens with dual specificity monoclonal antibody conjugates: Enhanced divalent hapten affinity for cell-bound antibody conjugate. *J. Nucl. Med.*, **30**, 1358–1366.

Lind, P., Lechner, P., Arian-Schad, K. *et al.* (1991) Anti-carcinoembryonic antigen immunoscintigraphy (Technetium-99m-monoclonal antibody BW 431/26) and serum CEA levels in patients with suspected primary and recurrent colorectal carcioma. *J. Nucl. Med.*, **32**, 1319–1325.

Löffler, K.U., Brautigam, P., Simon, J., Althauser, S.R., Moser, E. and Witschel, H. (1994) Immunszintigraphische Ergebnisse beim okularen im Vergleich zum kutanen Melanom. *Ophthalmologe*, **91**, 529–532.

Low, R.N., Carter, W.D., Saleh, F. and Sigeti, J.S. (1995) Ovarian cancer: comparison of findings with perfluorocarbon-enhanced MR imaging, In-111-CYT-103 immunoscintigraphy, and CT. *Radiology*, **195**, 391–400.

Loy, T.S. and Haege, D.D. (1995) B72.3 immunoreactivity in benign abdominal lymph nodes associated with gastrointestinal disease. *Dis. Colon Rectum*, **38**, 983–987.

Mach, J.P., Carrel, S., Merenda, C. *et al.* (1974) *In vivo* localisation of radiolabelled antibodies to carcinoembryonic antigen in human colon carcinoma grafted into nude mice. *Nature*, **248**, 704–706.

Mach, J.P., Carrel, S., Forni, M. *et al.* (1980) Tumor localization of radiolabeled antibodies against carcinoembryonic antigen in patients with carcinoma. *N. Engl. J. Med.*, **303**, 5–10.

Mach, J.P., Forni, M., Ritschard, J. *et al.* (1980) Use and limitations of radiolabeled anti-CEA antibodies and their fragments for photoscanning detection of human colorectal carcinomas. *Oncodevelopmental Biol. Med.*, **1**, 49–69.

Macklis, R.M., Beresford, B.A. and Humm, J.L. (1994) Radiobiologic studies of low-dose-rate 90Y-lymphoma therapy. *Cancer* (*suppl.*), **73**, 966–973.

Marshall, D., Pedley, R.B., Boden, J.A., Boden, R., Melton, R.G. and Begent, R.H. (1996) Polyethylene glycol modification of a galactosylated streptavidin clearing agent: effects on immunogenicity and clearance of a biotinylated anti-tumour antibody. *Br. J. Cancer*, **73**, 565–572.

Massugger, L.F.A.G., Kenemans, P., Claessens, R.A.M.J. *et al.* (1990) Immunoscintigraphy of ovarian cancer with Indium-111-labeled OV-TK 3 $F(ab')_2$ monoclonal antibody. *J. Nucl. Med.*, **31**, 1802–1810.

Matzku, S., Moldenhauer, G., Kalthoff, H. *et al.* (1990) Antibody transport and internalization into tumours. *Br. J. Cancer*, **62**, suppl. X, 1–5.

Meredith, R.F., Khazaeli, M.B., Plott, W.E. *et al.* (1995) Initial clinical evaluation of iodine-125-labeled chimeric 17-1A for metastatic colon cancer. *J. Nucl. Med.*, **36**, 2229–2233.

Meredith, R.F., Partridge, E.E., Alvarez, R.D. *et al.* (1996) Intraperitoneal radioimmunotherapy of ovarian cancer with Lutetium-177-CC49. *J. Nucl. Med.*, **37**, 1491–1496.

Milstein, C. and Cuello, A.C. (1983) Hybrid hybridomas and their use in immunohistochemistry. *Nature*, **305**, 537–540.

Moffat, F.L. Jr., Pinsky, C.M., Hammershaimb, L. *et al.* (1996) Clinical utility of external immunoscintigraphy with the IMMU-4 Technetium-99m Fab′ antibody fragment in patients undergoing surgery for carcinoma of the colon and rectum: results of a pivotal, phase III trial. The Immunomedics Study Group. *J. Clin. Oncol.*, **14**, 2295–2305.

Murray, J.L., Macey, D.J., Grant, E.J. *et al.* (1995) Enhanced TAG-72 expression and tumor uptake of radiolabeled monoclonal antibody CC49 in metastatic breast cancer patients following α-interferon treatment. *Cancer* (*suppl.*), **55**, 5925s–5928s.

Murray, J.L., Macey, D.J., Kasi, L.P. *et al.* (1994) Phase II radioimmunotherapy trial with ^{131}I-CC49 in colorectal cancer. *Cancer* (*suppl.*), **73**, 1057–1066.

Muthuswamy, M.S., Robertson, P.L., Ten Haken, R.K. and Buchsbaum, D.J. (1996) A quantitative study of radionuclide characteristics for radioimmunotherapy from 3D reconstructions using serial autoradiography. *Int. J. Radiation Oncology Biol. Phys.*, **35**, 165–172.

Novak-Hofer, I., Amstutz, H.P., Morgenthaler, J.J. and Schubiger, P.A. (1994) Internalization and degradation of monoclonal antibody chCE7 by human neuroblastoma cells. *Int. J. Cancer*, **57**, 427–432.

O'Donoghue, J.A., Bardiès, M. and Wheldon, T.E. (1995) Relationships between tumor size and curability for uniformely targeted therapy with beta-emitting radionuclides. *J. Nucl. Med.*, **36**, 1902–1909.

Oettgen, H.F. (1991) Cytokines in clinical cancer therapy. Current opinion (1991). *Immunology*, **3**, 699–705.

Order, S.E., Klein, J.L. and Leichner, P.K. (1987) Hepatoma: model for radiolabeled antibody in cancer treatment. *Natl. Cancer Inst. Monogr.*, **G3**, 37–41.

Ott, R.J., Grey, L.J., Zivanovic, M.A. *et al.* (1983) The limitations of the dual radionuclide subtraction technique for the external detection of tumours by radioiodine-labelled antibodies. *Br. J. Radiol.*, **56**, 101–108.

Paganelli, G., Riva, P., Deleide, G. *et al.* (1988) *In vivo* labelling of biotinylated monoclonal antibodies by radioactive avidin: A strategy to increase tumor radiolocalization. *Int. J. Cancer* (*suppl.*), **2**, 121–125.

Paganelli, G., Magnani, P., Zito, F. *et al.* (1991) Three-step monoclonal antibody tumor targeting in carcinoembryonic antigen-positive patients. *Cancer Res.*, **51**, 5960–5966.

Paganelli, G., Belloni, C., Magnani, P. *et al.* (1992) Two-step tumour targeting in ovarian cancer patients using biotinylated monoclonal antibodies and radioactive streptavidin. *Eur. J. Nucl. Med.*, **19**, 322–329.

Peltier, P., Wiharto, K., Dutin, J.P. *et al.* (1992) Correlative imaging study in the diagnosis of ovarian cancer recurrences. *Eur. J. Nucl. Med.*, **19**, 1006–1010.

Peltier, P., Curtet, C., Chatal, J.F. *et al.* (1993) Radioimmunodetection of medullary thyroid cancer using a bispecific anti-CEA/anti-indium-DTPA antibody and an Indium-111-labeled DTPA dimer. *J. Nucl. Med.*, **34**, 1267–1273.

Prati, U., Roveda, L., Cantoni, A. *et al.* (1995) Radioimmunoassisted follow-up and surgery vs. traditional examinations and surgery after radical excision of colorectal cancer. *Anticancer Res.*, **15**, 1081–1085.

Press, O.W., Eary, J.F., Appelbaum, F.R. *et al.* (1993) Radiolabeled-antibody therapy of B-cell lymphoma with autologous bone marrow support. *N. Engl. J. Med.*, **329**, 1219–1224.

Press, O.W., Eary, J.F., Appelbaum, F.R. *et al.* (1995) Phase II trial of ^{131}I-B1 (anti-CD20) antibody therapy with autologous stem cell transplantation for relapsed B cell lymphomas. *Lancet*, **G346**, 336–340.

Reske, S.N., Karstens, J.H., Henrich, M.M. *et al.* (1993) Nachweis des Skelettbefalls maligner Erkrankungen durch Immunszintigraphie des Knochenmarks. *Nuklearmedizin*, **32**,111–119.

Riva, P., Arista, A., Franceschi, G. *et al.* (1995) Local treatment of malignant gliomas by direct infusion of specific monoclonal antibodies labeled with ^{131}I: comparison of the results obtained in recurrent and newly diagnosed tumors. *Cancer Res.* (*suppl.*), **55**, 5952s–5956s.

Schaling, D.F., van der Pol, J.P., Jager, M.J. *et al.* (1990) Radioimmunoscintigraphy and immunohistochemistry with melanoma-associated monoclonal antibodies in choroidal melanoma: A comparison of the clinical and immunohistochemical results. *Br. J. Ophthalmol.*, **74**, 538–541.

Scheidhauer, K., Groth, W., Pietrzyk, U. *et al.* (1996) Diagnosis of metastatic malignant melanoma: FDG-PET in comparison to immunoscintigraphy. *Eur. J. Nucl. Med.*, **23**, 1106 (abstract).

Schuhmacher, J., Kilévnyi, G., Matys, R. *et al.* (1995) Multistep tumor targeting in nude mice using bispecific antibodies and a gallium chelate suitable for immunoscintigraphy with positron emission tomography. *Cancer Res.*, **55**, 115–123.

Senekowitsch, R., Bode, W., Reidel, G. *et al.* (1987) Improved radioimmunoscintigraphy of human mammary carcinoma xenografts after injection of an anti-antibody. *Nucl-Med.*, **26**, 13–19.

Slavin-Chiorini, D.C., Kashmiri, S.V., Schlom, J. *et al.* (1995) Biological properties of chimeric domain-deleted anticarcinoma immunoglobulins. *Cancer Res.* (*suppl.*), **55**, 5957s–5967s.

Somasundaram, C., Matzku, S., Schuhmacher, J. *et al.* (1993) Development of a bispecific monoclonal antibody against a gallium-67 chelate and the human melanoma associated antigen p97 for potential use in pretargeted immunoscintigraphy. *Cancer Immunol. Immunother.*, **36**, 337–345.

Steinsträsser, A. and Oberhausen, E. (1995) Anti-CEA labelling kit BW 431/26. Results of the European multicenter trial. *Nucl-Med.*, **34**, 232–242.

Stewart, S.J.W., Sivolapenko, G.B., Hird, V. *et al.* (1990) Clearance of 131I-labeled murine monoclonal antibody from patient's blood by intravenous human anti-murine immunoglobulin antibody. *Cancer Res.*, **50**, 563–567.

Stickney, D.R., Anderson, L.D., Slater, J.B. *et al.* (1991) Bifunctional antibody: A binary radiopharmaceutical delivery system for imaging colorectal carcinoma. *Cancer Res.*, **51**, 6650–6655.

The International Non-Hodgkin's Lymphoma Prognostic Factor Project. (1993) A predictive model for aggressive non-Hodgkin's lymphoma. *N. Engl. J. Med.*, **329**, 987–994.

Tromholt, N. and Selmer, J. (1991) Biological background subtraction improves immunoscintigraphy by subsequent injection of antigen. *J. Nucl. Med.*, **32**, 2318–2321.

Weeks, J.C., Yeap, B.Y., Canellos, G.P. and Shipp, M.A. (1991) Value of follow-up procedures in patients with large-cell lymphoma who achieve a complete remission. *J. Clin. Oncol.*, **9**, 1196–1203.

Werlen, R.C., Lankinen, M., Offord, R.E., Schubiger, P.A., Smith, A. and Rose, K. (1996) Preparation of a trivalent antigen-binding construct using polyoxime chemistry: improved biodistribution and potential for therapeutic appplications. *Cancer Res.*, **56**, 809–815.

Wilder, R.B., Langmuir, V.K., Mendonca, H.L., Goris, M.L. and Knox, S.J. (1993) Local hyperthermia and SR 4233 enhance the antitumor effets of radioimmunotherapy in nude mice with human colonic adenocarcinoma xenografts. *Cancer Res.*, **53**, 3022–3027.

6. CLINICAL STUDIES IN ONCOLOGY

JAN SCHMIELAU and WOLFF SCHMIEGEL

Medizinische Klinik, Ruhruniversität Bochum, Knappschaftskrankenhaus, In der Schornau 23-25, 44892 Bochum, Germany

INTRODUCTION

The growth behavior and metastatic competence of tumor cells in contrast to their normal counterparts has been thought to be reflected by expression of specific surface molecules. It has been assumed that malignant cells could be distinguished from their benign counterparts by tumor associated antigens (TAA). When evaluating mAb against these TAA for crossreactivity it became obvious that the majority of TAA are not tumor-specific. Since proteins have been identified which are more common on tumor cells, while still detectable on normal cells, an operational window of specificity has been an objective for therapy. Progress has been made in characterizing tumor-specific mutations which result in peptide motifs recognized by antibodies only on tumor cells (Urban and Schreiber, 1992). Preclinical studies revealed a strong antitumor potential of monoclonal antibodies (mAb) directed towards TAA but their therapeutic use has been far less successful. Antibody therapy was first initiated in patients suffering from hematopoietic diseases and occasional remissions raised expectations which were not fulfilled especially in solid tumors. After about 15 years of clinical trials, until recently the clinical results have been largely disappointing with only a few sustained clinical responses observed. Particular limitations are held responsible for this disadvantage of mAb in malignant diseases, some of them might be overcome in the near future. As with other antitumoral agents mAb trials were carried out as phase I/II clinical trials in progressive disease, characterized by large tumor burden with wide-spread metastatic or infiltrative disease. This restrains mAb delivery to the target cells since the molecular weight of about 150 kDa prevents diffusion comparable to most chemotherapeutic agents. Since a high antibody density is a prerequisite for cytotoxic effects (Capone *et al.*, 1984; Herlyn *et al.*, 1985b) a weak antigen expression favors a tumor escape. The mAb have to ensure a strong affinity that otherwise hinders the mAb from deep invasion into the tumor (Jain, 1990). Circulating free antigen as well as human antimouse antibodies (HAMA) also might prevent antitumor activity by inactivating mAb by complex formation. When mAb are administered, antigenic modulation such as internalization or shedding might prevent cytotoxic events as well as further mAb binding. Due to an advanced stage of the disease or previous chemotherapy patients often live in an immunodeficient state which possibly results in an inadequacy of host effector cell mechanisms. Genetic instability with phenotypic heterogeneity is a common finding in tumors which has accompanied the evaluation of mAb in tumor therapy since the early beginning (Heppner, 1984). Phenotypic heterogeneity might lower immunogenicity, when target antigen expression is decreased, which accounts for escape from cytotoxic effects of antigen-negative tumor cell variants (Fleuren *et al.*, 1995).

Effector Mechanisms

Several aspects have to be considered regarding antibody-dependent tumor suppression. Most notable, antibodies mediate cytotoxic effects which are linked to the Fc-region and are depending on the Ig-class and subclass. Effector cells for IgG are infiltrating as well as circulating macrophages/monocytes, neutrophils and NK-cells. Their variable milieu-dependent behavior partially results in a different expression pattern of characterized Fc-receptors (Fanger *et al.*, 1989). The heterogenic IgG-receptors (FcγR) are subdivided in three groups, which are all members of the Ig supergene family (Lin *et al.*, 1994). All human FcγR bind human IgG1 and IgG3, which are most effective in mediating antibody-dependent cell-mediated cytotoxicity (ADCC), with comparable affinity (Shaw *et al.*, 1988). The FcγRI (CD64) is predominantly found on monocytes and macrophages and binds monomeric human IgG1 and IgG3 and murine IgG2a and IgG3 with high avidity. As a common receptor on hematopoietic cells, including monocytes, neutrophils, platelets and B cells, FcγRII (CD32) have a low affinity to monovalent human IgG1 and IgG3 and murine IgG2b, including subgroups binding human IgG2 and murine IgG1 with similar affinity. They seem to especially bind immune complexes and interact in multiple receptor-ligand formations. For FcγRI and FcγRII IFN-γ acts as a competent enhancer of cytotoxicity, while introducing FcγRII-mediated cytotoxicity to neutrophils. The FcγRIII (CD16) appears as a transmembrane and as a phosphatidylinositol glycan (PIG) membrane anchored receptor which is released upon specific phospholipase C action. Its transmembrane form is present on NK-cells and tissue macrophages but not monocytes, neutrophils and eosinophils and express a low affinity to monomeric human IgG1 and IgG3 and murine IgG3 more than to IgG2a. Neutrophils express the PIG-anchored FcγRIII which ensures tight contact of neutrophils to IgG-covered antigen and thereby supports effects mediated through the FcγRII. NK-cells seem to exclusively express transmembrane FcγRIII which links its function especially to ADCC that is otherwise also established by FcγRI and FcγRII. Subsequent signal pathways are related to receptor aggregation and receptor-associated proteins, which initiate phosphorylation of protein tyrosine kinases (Daeron, 1997). Concerning ADCC with murine mAb, human effector cells were generally less effective than mouse effector cells where only a few mAb of murine IgG2a, IgG3 and IgG1 isotype generated ADCC (Herlyn *et al.*, 1985a).

The classic complement cascade could be initiated by binding of the C1q-subunit to the CH2-domain of aggregated IgG molecules or to the CH3-domain of a single IgM-molecule. Cell lysis in complement-mediated cytotoxicity (CDC) occurs as a consequence of formation of the membrane attack complex which consists of a tetramolecular C5b, C6, C7, C8-complex and up to twelve C9 molecules that assemble a membrane perforating tube (Müller-Eberhard, 1988). Cellular cytotoxicity utilizes a C9-related protein, perforin, which is cytotoxic by itself, and is supplemented by secreted enzymes and lymphokines like the tumor necrosis factor α (TNF-α) and interferon γ (IFN-γ). Among murine antibodies the mAb of the IgM and IgG3-class are most effective in mediating cytotoxicity in conjugation with human complement followed by IgG2a and IgG2b (Herlyn *et al.*, 1985).

Apart from being a potential target for cytotoxic mechanisms, growth factor receptors are essential for cell proliferation and could be blocked by mAb (Mendelsohn, 1990). This particularly applies for tumors where increased growth results from overexpression of normally functioning growth factor receptors. Additionally, growth can be assisted by autocrine loops, as known for the EGF-receptor and its ligand TGF-α (Rodeck *et al.*, 1990) or neuroendocrine factors (Mulshine *et al.*, 1989) and decreased by anti-receptor antibodies. Other targets for blocking antibodies are the transferrin receptor (Saavage *et al.*, 1987; Taetle and Honeyset, 1987) or the IL-2-receptor in T cell leukemia (Waldmann *et al.*, 1988). Withdrawal of required growth factors or their displacement by competitive mAb lead to sustained growth. Cell cycle arrest, e.g. in B cell lymphomas when surface IgM is cross-linked (Goodnow, 1992), or apoptosis, induced in lymphoid tumor cells (Trauth *et al.*, 1989), have also been established as a specific function of receptor binding mAb.

Additional strategies may evolve from activation of cytotoxic T cells with antibodies against the TCR/CD3-complex (Ellenhorn *et al.*, 1988). Furthermore, cytokines could be linked to tumor-specific antibodies which locally activate effector cells such as NK-cells (Becker *et al.*, 1996).

Although HAMA has been mentioned to be of inhibiting nature, antibodies which arise against the antibody idiotype are of therapeutic use and not restricted to the xenogenetic origin. According to the network hypothesis anti-idiotypic antibodies play an essential role in the regulation of the immune response (Jerne, 1974). Anti-idiotypic antibodies (Ab2) against an antitumor antibody (Ab1) represent an internal image of the antigen and themselves are targets of human anti-anti-idiotypic antibodies (Ab3). These Ab3 antibodies arise in anti-idiotypic vaccination with tumor antigen-defined mAb and might have one of the aforementioned direct antitumor effects (Herlyn *et al.*, 1989). The anti-idiotype network is not restricted to humoral immune response and CTL activity may also be induced by such a vaccination.

Studies in Hematopoietic Diseases

Acute and chronic as well as lymphatic and leukemic diseases have been treated with intravenous administration of unmodified mAb. Overall, no consistent response could be observed, but therapy has been well-tolerated and an occasional partial (PR) or complete (CR) remission has occurred. In almost all cases with leukemic disease a transient depletion of malignant cells from the peripheral blood has been detected. Bone marrow, lymphatic organs and infiltrated tissue were accessible, as indicated by studies with labeled antibodies, but density on target cells was mostly low. Following high doses of radionuclid-labeled mAb stem cell transplantation occasionally has been necessary. Furthermore, antigen modulation has to be considered, since several antibodies are targeted to receptor molecules and subsequently exposed to internalization.

In 1980 the first treatment of a patient with mAb was reported (Nadler *et al.*, 1980). The mAb was generated against cells obtained from that patient with *diffuse poorly differentiated lymphocytic lymphoma* (D-PDL), and was found to react with about 10% of D-PDL and CLL but not with normal hematopoietic cells. This murine IgG2a mAb exerts CDC on tumor cells but failed to mediate ADCC. Single doses of up to 150 mg in two courses revealed a considerable reduction of circulating tumor cells which lasted only for several hours. A final administration of 1.5 g in advanced disease yielded a comparable quantity of tumor cell bound mAb. Severe side effects as well as sustained therapeutic effects were not observed.

Further clinical experience was obtained in patients with *acute lymphocytic leukemia* (ALL) who were treated with an antibody against the common acute lymphoblastic leukemia antigen (CALLA) CD10 (Ritz *et al.*, 1981). The murine IgG1 J-5 applied in a single dose of 1–170 mg per patient successfully decreased $CALLA^+$ blasts but $CALLA^-$ lymphoblasts persisted in large amounts. However, lymphoblasts re-expressed CALLA when J-5 was undetectable in the serum, which has been attributed to internalization of the antigen. Even though binding of J-5 to bone marrow cells occurred, regression was not achieved.

Ball *et al.* (1983) treated three patients with *acute myelogenous leukemia* (AML) with murine IgG2a and introduced murine IgM into treatment schedules. Up to three different IgM mAb were applied in a single patient and one patient received IgG2a and IgM mAb. The mAb were specific for divergent differentiation antigens on normal and malignant myeloid cells. Apart from a transient decrease of blasts in the peripheral blood no further success could be achieved. Although the IgM mAb mediated no significant CDC with human complement *in vitro*, blood sampling suggested cell lysis and as the IgM mAb bound normal and malignant cells, bridging of target and effector cells might have occurred. This results compare to a trial with two murine IgM mAb which caused transient but no sustained effects and revealed mAb-mediated reduction of serum complement (Greenaway *et al.*, 1994).

Scheinberg *et al.* (1991) enrolled nine patients with AML and one patient with *chronic myelomonocytic leukemia* (CMML) for the use of the murine IgG2a M195 against the CD33 glycoprotein expressed on myelogenous leukemias, monocytes and myeloid progenitor cells. Multiple doses of up to 10 mg/m^2 with a maximum total amount of 72 mg were given, with a first dose of radiolabeled antibody to study organ delivery. A marked uptake into the bone marrow could be demonstrated, but apart from a transient peripheral blast depletion no substantial response has been achieved. *Ex vivo* quantification of the administered labeled antibody revealed most of it to be already internalized, which compares to *in vitro* assays where CDC and ADCC are circumvented by rapid internalization of mAb. Moreover, this recommends the anti-CD33 antibody for a therapy with toxin-conjugates or radionuclide conjugates. Efforts to improve the cytotoxicity of the original antibody lead to the humanized M195, which has an eight-fold higher affinity for CD33 than the murine M195 and has the capability to mediate ADCC *in vitro* (Caron *et al.*, 1994). Treatment of 12 patients with AML and one patient with CML consisted in six repetitive infusions of up to 10 mg/m^2 M195. As measured in previous studies, the initially applied labeled antibody showed a marked uptake in bone marrow and an internalization into CD33-positive blasts of up to 56% of total bind activity 24 h after infusion. One patient experienced a reduction of bone marrow blasts and reverted from AML FAB-M6 to the preexisting myelodysplastic syndrome. Another patient showed stable disease during therapy.

Several studies used mAb binding to CD5 expressed on T cells, lymphocytes of B CLL and well differentiated B cell lymphoma, in the therapy of lymphatic diseases. Patients with T cell leukemia experienced transient reduction of leukemia cells but no sustained effect was observed when they were treated with the anti-Leu-1 mAb against CD5 (Miller *et al.*, 1981; Levy and Miller, 1983). No advantage was seen when anti-Leu-1 was combined with one or two other mAb which do not induce antigenic modulation.

Since cutaneous T cell lymphomas (CTCL) express CD5 at high levels, the corresponding mAb anti-Leu-1 has been used in a trial with seven patients (Miller and Levy, 1981; Miller *et al.*, 1983). Four to 17 doses with a total amount of 13–761 mg IgG2a mAb induced a PR of short duration in four patients where development of HAMA was made responsible for tumor escape from therapy. Cyclophosphamide was administered but was unable to suppress HAMA production in the one evaluable patient. Moreover, restrained clearance of mAb-coated target cells from the blood was observed and suspected to be due to myelosuppression with decreased numbers of effector cells. HAMA were differentiated in two patients with PR and anti-idiotypic antibodies were measured with an amount of about 5% of total HAMA. The mechanism of antitumor response remained unclear since no inflammatory reactions except in one case were seen in regressing skin lesions.

Dillmann *et al.* extended first trials with mAb T101 in two patients with CLL which revealed only transient decrease of lymphocytes (Dillman *et al.*, 1982) when they also included patients with CTCL (Dillman *et al.*, 1986). Several single doses of 10–500 mg mAb were given in up to 4 courses. Among 10 patients with CLL and 13 patients with CTCL only two and four, respectively, minimal responses were observed apart from intermittent changes of lymphocyte count. MAb T101 does not use human effector cells or complement to mediate ADCC or CDC, is rapidly internalized, and does not inhibit cell growth. The antibody immune response was only seen in patients with CTCL and largely consisted of an anti-idiotype response (Shawler *et al.*, 1985). This anti-idiotype response even occurred in patients with minimal response and it can be speculated about a possible involvement of the idiotypic network.

Foon *et al.* (1984) obtained similar results when they noticed a minor response of the skin lesions and HAMA in all 4 patients with CTCL. But one has to state that at least the humoral response has not been identified as anti-idiotypic (Schroff *et al.*, 1985). No clinical or immune response evolved in 10 patients with CLL.

In another study with CTCL and different T cell malignancies one CR and one PR occurred among 13 patients while low levels of HAMA developed in the patient with CR and increasing amounts of HAMA rendered further therapy ineffective (Bertram *et al.*, 1986). Summarizing the results of the above studies, the mechanisms which account for a moderate therapeutic success with the mAb T101 have remained unclear.

Knox *et al.* (1991) treated seven patients with CTCL with a chimeric anti-CD4 IgG1. While all patients showed some clinical improvement, two patients had a PR, one lasting only nine days, the other at least 12 weeks. This was observed with the maximal single dose of 80 mg six times during three weeks. Immune suppression induced by therapy was associated with opportunistic infections in two patients. Even though peripheral CD4 positive T cells were covered with antibody, no significant depletion of T cells occurred. Mixed lymphocyte reaction was reduced, if it had not already been low before treatment, but response to keyhole limpet hemocyanin was unchanged during antibody application, which in two patients repeatedly failed to induce T cell proliferation. Thus, anti-CD4 does not seem to suppress the ability to generate an immune response to a co-administered antigen.

The CAMPATH-1 rat antibody family against CD52 i.e. IgM, IgG2a and IgG2b with a rather weak specificity have served for clinical trials with various lymphoid malignancies, including ALL, CLL, and non-Hodgkin lymphoma (NHL) (Hale *et al.*, 1983; Dyer *et al.*, 1989). CD52 is a phosphatidylinositolglycan (PIG)-anchored glycoprotein (PIG-AP) expressed on normal T and B lymphocytes, monocytes, and the majority of B cell NHL. All mAb have been proven to implement CDC with human complement but only the CAMPATH-1 mAb IgG2b mediated ADCC where it binds to the human FcγRIII. All mAb induced a transient depletion of blood leukemic cells but only the mAb of IgG2b isotype generated a sustained lymphopenia as well as reduction of bone marrow infiltration and splenomegaly (Dyer *et al.*, 1989). Improvement in bone marrow function with cleared infiltration was observed in four of nine and reduced malignant cells in one additional patient. Splenomegaly, when initially found, also ameliorated, but only moderate effects on lymph node or extranodal involvement were noted. Antibody response was detected in two patients compared to no antibody response in a trial with the chimeric IgG1 against CAMPATH-1 in two patients with NHL (Hale *et al.*, 1988). The latter treatment cleared lymphoma cells from the blood and bone marrow with relatively low amounts of total mAb (85 and 126 mg). Splenomegaly, and in one patient lymphadenopathy resolved. Escape from immune response might at least in part have been due to PIG-AP-deficient T cell variants without the PIG-anchored CD52 (Hertenstein *et al.*, 1995).

The murine mAb LYM-1 recognizes a HLA-DR antigen on normal and malignant B cells. This IgG2a mAb mediates ADCC and CDC with human sources as well as direct antiproliferative effects. When 10 patients with B cell lymphoma were treated weekly for 4 weeks with single doses of 58.1–464.8 mg mAb, penetration of extravascular tumor tissue was poor and clinical outcome with three minor responses appeared to correlate with the capability of infiltrating cells to mediate ADCC (Hu *et al.*, 1989). As observed in other studies even at high doses antigenic saturation was not achieved in extravascular tissue, but a transient decrement of circulating lymphoma cells which did not lead to infections during therapy. Since complement consumption was not observed, it was concluded that ADCC and the mononuclear phagocyte system account for therapeutic effects.

The surface molecule CD20, expressed on normal and malignant B cells, avoids some of the above-mentioned obstacles. It has multiple trans-membrane domains and is neither shed from the cell surface nor internalized, so that prolonged continued therapy is feasible. Consequently, Press *et al.* (1987) administered 1F5, a murine IgG2a mAb against CD20, by continuous intravenous infusions to four patients with different refractory *B cell lymphoma* in a dose escalating manner over five to ten days. Those patients with circulating tumor cells experienced an approximately 90% elimination from the circulation within hours after beginning of therapy, which relapsed two to three days after the antibody infusion. The patient with the highest achieved dose revealed also a 90% reduction of lymph node disease but bone marrow remained almost unchanged. While yielding an efficacious treatment of lymph node cells with normal antigen expression, compared to peripheral blood B cells, the antibody lacks any effect on the grossly involved bone marrow, where only minor CD20 expression was detected. This demonstrates heterogeneity of tumor cell antigen expression as a considerable problem in immunotherapy. The duration of remission of a patient with a maximal total dose of 2,380 mg persisted only six weeks. Noteworthy, none of the patients developed tumor regression within the bone marrow, even when binding site saturation on cells with a mean CD20 expression 3,5 times above normal was elevated from 4% to 61% with an increased antibody dose from 10 to 100 mg/m^2/d. The number of normal B cells, when not already at low levels, neutrophils and platelets decreased in all patients. Complement showed a considerable consumption in patients with elimination of leukemic cells. Several subsequent clinical studies using radio-immunoconjugates of anti-CD20 mAb revealed CR in the range of three quarters of the patients after bone marrow ablation and stem cell transplantation (Press *et al.*, 1995).

A chimeric anti-CD20 antibody has been used for a phase I study conducted by Maloney *et al.* (1994) in relapsed B cell lymphoma. The antibody which, consists of the variable region of a murine anti-CD20 and a human IgG1-k constant region, exhibits a considerable increase in CDC and an ADCC activity that is about 1,000-fold more effective than that achieved with murine antibody (Reff *et al.*, 1994). Six out of 15 patients treated with a single dose of mAb revealed a tumor regression including two PR. Immunohistochemical evaluation of lymph nodes from nine patients, which received 100 mg/m^2 or more, was performed two weeks after treatment. Except for a biopsy with no detectable tumor and another with tumor necrosis, the anti-CD20 antibody was identified in all cases but one. B cells were rapidly removed from the peripheral blood lasting for two to three months, while no effect was observed on the amount of T cells, complement or immunoglobulins.

A different strategy has been designed for the treatment of *adult T cell leukemia* (ATL), which is characterized by an overexpression of IL-2-receptors. Activated

T cells present an increased high affinity, multichain IL-2-receptor expression, composed of a 55 kD and 75 kD protein, which is not present on most normal resting cells. Waldmann *et al.* (1988) used the murine IgG2a anti-Tac against the 55 kD protein, which sufficiently blocks IL-2-induced proliferation but does not mediate CDC or ADCC, in the therapy of nine patients with HTLV-I induced ATL. Three responses with one CR and one PR evolved during sequential infusions with a maximal total dose of 490 mg and various treatment protocols of up to 445 days. An additional three patients displayed a transient decline of malignant T cells in the blood stream. Growth inhibition by antagonizing IL-2 has been presumed.

Since IL-6 is known to be a potent growth factor for tumor cells in multiple myeloma Klein *et al.* (1991) treated one patient with *plasma cell leukemia* for two months with daily infusions of two murine anti-IL-6 mAb, IgG2b and IgG1. *Ex vivo* proliferation assays with bone marrow tumor cells harvested pre-treatment demonstrated a completely IL-6-dependent proliferation. During therapy with increasing amounts of monoclonal antibody a loss of tumor cells in the S-phase, which was 4.5% on day 0, was observed on day 10, followed by a recurrence of 2% cells in the S-phase on day 60. The application of anti-IL-6-antibody was paralleled by a considerable increase of biologically inactive IL-6 serum levels, due to blocking antibodies. This has been reflected by reduction in serum calcium, serum monoclonal IgG, and C-reactive protein levels. The progression of disease has been suggested in part to be due to the development of HAMA against both antibodies. In a second trial 10 patients with multiple myeloma were treated with multiple injections with a total dose of up to 1,510 mg (Bataille *et al.*, 1995). Three patients had an antiproliferative effect with significant reduction of the myeloma cell labelling index within the bone marrow, with one 30% regression of tumor mass. Antiproliferative effects were associated with complete inhibition of the C-reactive protein.

Anti-idiotypic antibodies

Since most human B lymphocytic malignancies are thought to be derived from a single transformed B cell, it is assumed that they express unique immunoglobulins in a uniform pattern. Conceivably, antibodies against those immunoglobulin idiotopes would be almost tumor specific and suitable for passive immunotherapy. This therapy remains cumbersome, since antibodies have to be obtained individually for each patient (Miller *et al.*, 1982). Whole tumor cells as well as idiotype proteins were used for immunization mostly in mice, but also in sheep. Studies mainly focused on direct antibody effects, while observations of the arising idiotype network were not performed. Additionally, direct antiproliferative effects of antibodies against surface-bound immunoglobulin of B cells have been described (Vitetta and Uhr, 1994).

Rankin *et al.* (1985) used spleen cells from two patients with advanced B cell NHL for immunizing in mice. The derived mAb of isotype IgG2a were able to transiently depress circulating malignant cells, but only a minor reduction of tumor masses was observed when they were given in multiple doses up to a total amount of 3.8 and 5.8 g. In a study comprising various B cell lymphomas 16 patients were evaluated upon multiple infusions with a total dose of 400–15,500 mg of murine monoclonal IgG1, IgG2a or IgG2b anti-idiotypic antibody (Miller *et al.*, 1982; Meeker *et al.*, 1985; Brown *et al.*, 1989). Tissue penetration was demonstrated in six of eight patients by biopsy and modulation of Ig expression could be seen in several samples. As has been recognized before, the antibodies were capable of clearing malignant cells from the blood. However, as soon as HAMA emerged, neither functional activity nor tissue penetration could be observed any longer. Several lymphoma secreted a considerable amount of idiotypic immunoglobulin into the serum, i.e. with a concentration of up to 243 μg/ml. Tumor response to anti-idiotypic antibody was not obtained until eliminating of idiotypic immunoglobulin was achieved with 50–400 mg anti-idiotypic antibody. One CR persisted for six years following a total dose of 400 mg IgG2b. Five patients received more than one anti-idiotypic mAb, among those three had PR lasting more than six months. Four additional patients had a PR for up to 25 months.

Regarding intra-individual responses to the same therapy schedule, thirteen patients with B cell lymphoma were treated with one or two murine monoclonal anti-idiotypic antibodies of isotype IgG1, IgG2a or IgG2b in two subsequent courses with a total amount of antibody of $2 \times 1{,}680$–$2 \times 5{,}757$ mg. Since mutations in the immunoglobulin's variable region gene is a property of normal as well as certain malignant B cells, emergence of idiotype-negative lymphoma cells was assumed to be a major disadvantage of the therapy. Therefore the chemotherapeutic agent chlorambucil was added in the beginning of the second course to especially address those idiotype-negative B cells. An idiotype selection indeed occurred in several cases, but no recruitment of idiotype-negative cells into the cell cycle took place, and chlorambucil did not seem to have any major effect on either tumor response or effector function. Overall the clinical benefit with one CR, eight PR, and two minor responses compares to previous studies.

Following the strategy of anti-idiotypic mAb production against B cells a mAb against the variable domain of the TCR from a patient with TCLL has been raised (Janson *et al.*, 1989). This murine IgG2a mAb has been administered three times with a total dose of 12 mg. The patient revealed a transient 50% reduction of circulating T cells, and showed a steady decrease of up to 80% following the second infusion. Reaction to the third application was not evaluable.

The characterization of the surface immunoglobulins on B cells from NHL as a target for antibodies has been extended in an attempt with idiotype vaccination (Kwak *et al.*, 1992). Nine patients which had minimal residual disease or a CR after chemotherapy received multiple subcutaneous injections of their tumor-derived mAb, conjugated to a carrier protein (keyhole limpet hemocyanin) and mixed with an immunologic adjuvant. In seven of nine patients a sustained idiotype-specific antibody and/or a cell-mediated response was demonstrated. In two patients with measurable disease the tumors regressed completely.

Recently established anti-idiotypic mAb reactive with B cells from multiple patients might overcome the practical obstacles and are included in ongoing studies. Major pitfalls are an anti-idiotype selection as described above and a relapse of the disease, which eventually originate from idiotype-negative pre-B cells. An idiotypic variations due to somatic mutations in the variable region might also cause an escape from therapy (Meeker *et al.*, 1985).

Studies in Solid Tumors

Melanoma

Initial experiences were obtained with mAb against p240, a glycoprotein found on most melanoma cells. Oldham *et al.* (1984) treated eight patients with the murine IgG2a mAb 9.2.27 where doses of more than 50 mg revealed *in vivo* localisation in subcutaneous tumors in six patients but tumor response was not observed. Goodman *et al.* (1985) combined two murine anti-melanoma mAb against p240 (IgG1) and p97 (IgG2a). Although occasionally anti-idiotypic reaction was measured the lack of any tumor response might have been due to the inability of the infused mAb to mediate CDC and ADCC.

The murine IgG3 mAb R24 against the disialoganglioside GD3 surpasses a variety of other murine antibodies tested regarding strong activation of human complement and ADCC. Gangliosides are glycolipids particularly expressed as membrane constituents of neuroectodermal tissue, including melanocytes and astrocytes. Since they have also been demonstrated in melanomas and astrocytomas with occasional high expression, clinical studies used these as targets for mAb. The anti-GD3 mAb has been applied to 21 patients with metastatic melanoma as reported by Houghton *et al.* (1985) and updated by Vadhan-Raj (Vadhan-Raj *et al.*, 1988). It has been administered at three dose levels (8, 80, 240, or 400 mg/m^2) over a period of two weeks, provoking inflammatory responses with urticaria, pruritus, erythema, subcutaneous ecchymoses, or local pain, probably due to nerve binding. Responses were detected first after 4 weeks and as late as 10 weeks after beginning of treatment, including four PR, lasting from 6 to 46 weeks. HAMA developed in 20 out of 21 patients after an average 14 days. Interestingly, three of four patients with PR established HAMA later than 20 days afterwards. Target antigen expression in pretreated biopsies revealed 100% GD3 positive melanoma cells in 13 out of 21 patients, and only one patient had less than 50% positive cells. In general, progression or recurrence of disease did not present with a GD3-negative population of melanoma cells, but in one case GD3 was absent in a biopsy of relapsed tissue and recovered during long-term cell culture. Biopsies demonstrated the penetration of the antibody to be dose-dependent, being not detectable at 1 mg/m^2/d applied mAb, and showed complement deposition, mast cell activation, and increased infiltration of T lymphocytes in tumor sites.

The mAb R24 was also used in a study with malignant melanoma and one apudoma (Dippold *et al.*, 1985, 1988). It induced a local inflammatory response in cutaneous tumor sites and pain in bulky tumor masses when given at a single dose of up to 200 mg and a total dose of 440 mg. Following a treatment of usually three weeks, two patients had tumor regression and the patient with apudoma a shrinkage of one liver metastasis. MG-21 is another murine anti-GD3 mAb of

IgG3 isotype which lyses cells via CDC and ADCC and has been infused in eight patients with malignant melanoma (Goodman *et al.*, 1987). No *in vivo* localization of the mAb and consecutively no anti-tumor effect has been detected. The murine IgG2a mAb ME-36.1 recognizes GD3 and GD2 and induces CDC and ADCC as well as growth inhibition. While 6 out of 13 patients with malignant melanoma were proven to have mAb delivery to tumor sites one CR has been observed (Lichtin *et al.*, 1988). This patient revealed post-infusion B cells which could be stimulated by goat anti-idiotypic ME-36.1 antibodies and then produced antibodies specific for GD2. As this suggests that tumor response might have been related to anti-GD2-antibody induction, an anti-idiotypic response could have been involved.

Since the gangliosides GD2 and GM2 are immunogenic in man Irie *et al.* (Irie and Morton, 1986; Irie *et al.*, 1989) were able to develop B-lymphoblastoid cells which produce human IgM mAb and have used them for local therapy. Three PR and one CR among eight patients (Irie and Morton, 1986) and one CR in two patients with malignant melanoma (Irie *et al.*, 1989) were achieved with multiple intralesional injections. Interestingly, the patient with CR (Irie *et al.*, 1989) developed anti-idiotypic IgG antibodies in response to human IgM injections which suggests a T cell-dependent B cell activation. As seen in other studies, IgM antibodies are less effective than IgG in systemic applications (Dyer *et al.*, 1989) but might induce antibody and T cell responses in response to an induced internal image of the original TAA.

The murine IgG3 mAb 3F8 which mediates CDC and ADCC was also raised against GD2 and applied in 17 patients with neuroblastoma or malignant melanoma (Cheung *et al.*, 1987). The study was closed at a mAb dose of 100 mg/m^2 because all patients receiving this dose level developed arterial hypertension. All patients complained about pain which usually involved tumor sites and especially the abdomen, back, and extremities, independent of mAb dose. Patients with less than 1,000 U/ml HAMA, measured by ELISA, had the expected side effects as well as a therapeutic benefit from repeated infusions. Reassessment of the clinical responses (Cheung *et al.*, 1992) revealed two PR. In a phase II trial Cheung *et al.* (1991) treated 16 patients with neuroblastoma. Among 13 evaluable patients one PR occurred. When patients were evaluated according to disease sites, 2 out of 7 with bone lesions and 3 out of 8 with bone marrow involvement showed tumor shrinkage (Cheung *et al.*, 1994).

The GD2 serves as an antigen for the murine IgG2a mAb 14G2a with CDC- and ADCC-activity *in vitro*. Twelve patients with malignant melanoma were treated with a total dose of 10–120 mg (Saleh *et al.*, 1992b). Apart from neurologic symptoms, including abdominal/pelvic pain and two cases of reversible motor neuropathy, one PR was observed. In patients with neuroblastoma this mAb was applied at a total dose of 100–400 mg/m^2 (Handgretinger *et al.*, 1992). Three out of nine patients were treated in an adjuvant setting with a relapse after one year or later. Among the remaining six patients two patients each revealed either CR or PR. Furthermore, all patients in both regimens where the mAb was applied in multiple infusions developed HAMA. Murray *et al.* (1994) treated 18 patients suffering from malignant melanoma, neuroblastoma, or osteosarcoma with continuous infusions over 5 days with a total dose of 50–200 mg/m^2 mAb 14G2a. Although all but two patients had detectable tumor sites in immunoscintigraphy only two neuroblastoma patients had PR. It is noteworthy that 16 out of 18 patients developed anti-idiotypic antibodies where high levels correlated with clinical response. Side effects usually were severe abdominal pain and allergic reactions in all studies using anti-GD2 mAb.

Gastrointestinal cancer

The monoclonal antibody 17-1A, which is unable to distinguish malignant from normal cells, has been one of the most extensively tested antibodies in solid tumors within clinical trials. Nevertheless, promising results have been recently obtained with this murine IgG2a monoclonal antibody, which binds to an epithelial glycoprotein (EGP40) involved in intercellular adhesion (Litvinov *et al.*, 1994). Frequently observed abdominal symptoms such as diarrhea have been linked to its low specificity. MAb 17-1A otherwise has been shown to predominantly bind to a wide range of colorectal cancer cells and to exhibit effective ADCC *in vitro*. First clinical studies with relatively low doses of 15–200 mg confirmed a histologically proven tumor lysis of liver metastases in one eligible patient, while having no severe side effects (Sears *et al.*, 1982). Extending the total dose up to 1,000 mg, given in a single dose except in two patients, three CR were observed in a subsequent trial with 20 patients suffering from gastrointestinal cancer (Sears *et al.*, 1984). A following clinical trial with 20 patients with colorectal cancer, given a single 200–850 mg infusion, has shown a minor success with one PR, one mixed response and one stable disease (Sears *et al.*, 1985).

When the mAb 17-1A was administered to patients with metastatic adenocarcinoma of the pancreas at a dose of 400 mg with or without PBL, four PR were obtained (Sindelar *et al.*, 1986). In contrast to subsequent studies, development of anti-idiotypic antibodies did not improve the clinical outcome where three of the four responders had no anti-idiotypic antibodies. In a trial with mAb-preincubated PBL at an amount of 400 mg, tumor response was restricted to a decrease in serum tumor marker levels in two patients (Tempero *et al.*, 1986). The mAb 17-1A has also been tested in combination with chemotherapy (5-fluorouracil, adriamycin and mitomycin) at a single dose of 400 mg (Paul *et al.*, 1986). As there was no clinical response with the mAb alone an indirect indication for a possible mAb effect occurred when one of two responding patients with progressive disease under chemotherapy converted to PR after a single mAb injection. To ensure sufficient mAb delivery to the tumor Weiner *et al.* (1993) treated 28 patients with metastatic pancreatic cancer with an excessive amount of up to 12 g with multiple single injections of 500 mg. As seen before, only minimal effects were observed with one PR.

Verrill *et al.* (1986) treated 22 patients suffering from metastatic gastrointestinal cancer with 17-1A-armed effector cells following leukopheresis with a single dose of 200 mg mAb. Having observed no CR or PR they recognized a positive correlation of anti-idiotypic response and stable or only slowly progressive disease in contrast to rapid progression.

Douillard *et al.* (1986) reviewed five clinical trials with the mAb 17-1A in 95 patients with gastrointestinal cancer. In an initial trial they infused a single dose of 500 mg mAb associated with autologous PBL in 25 patients, obtaining one CR and four PR. In three subsequent studies 49 patients received either 500 mg mAb with or without PBL followed by a second infusion with 100 mg mAb without PBL on day seven, or 150 mg on day one with 50 mg per day thereafter for four days. Only one PR was detected in the trial with 30 patients receiving 500 mg mAb associated with PBL and additional 100 mg one week later. In a fifth trial up to four different murine IgG2a mAb were mixed with 150–200 mg each and infused with preincubated leukopheresed cells. These mAb, 17-1A, 73-3, 19-9, and 55-2, all mediate ADCC. Two CR, one in colon carcinoma and one in pancreatic cancer were obtained.

Another series of trials with the mAb 17-1A has been introduced in patients with colorectal cancer (Frödin *et al.*, 1988) and reviewed by Mellstedt *et al.* (1991). Frödin *et al.* (1988) applied the mAb preincubated with PBL to ten patients in repeated courses up to four times with a maximal total dose of 1,000 mg. One CR, two PR, and one stable disease were detected which led to a prolonged median survival time of 19 months for these patients compared to 7 months for nonresponding patients. An additional 20 patients were treated with multiple infusions at a total dose of less than 2 g, with one minor response observed (Mellstedt *et al.*, 1991). In order to amplify antibody delivery to the tumor, which depends in part on integrated serum concentration over time, an escalating total dose of up to 12 g was applied in another 18 patients. Including stable disease into the evaluation of response there was a tendency towards a lower response rate with only one stable disease.

Similar observations were made when LoBuglio *et al.* (1988) administered a single dose of 400 mg up to four times during 8 days in 25 patients with gastrointestinal cancer. They attained one CR and four stable diseases, evaluated six weeks after treatment. All responding patients belonged to the group that had received two infusions, i.e. 800 mg, while all patients with a total dose of 1,200 or 1,600 mg showed a progressive disease. An interpretation assuming that the optimal total dose is below 1,000 mg would be contradictory to the believed importance of sufficient antibody binding to the tumor. More insights into tumor response were obtained when the patients' immune response was considered. As reviewed below, mAb infusions induce reactions of the idiotypic network with anti-idiotypic (Ab_2) and anti-anti-idiotypic (Ab_3) immune response, the latter being able to bind to the target tumor antigen. When tumor patients treated with mAb 17-1A (Mellstedt *et al.*, 1991) were evaluated for the presence of Ab_3, a strong correlation between Ab_3 production and tumor response was found (Frödin *et al.*, 1991). Fourtyone out of 43 patients developed Ab_2, but Ab_3 was detected in only 20 out of 43 patients, suggesting that reduced tumor response with high doses of therapeutic mAb might be due to suppressed Ab_3 response.

The murine IgG2a mAb GA733 binds to the same glycoprotein as 17-1A does, also mediates ADCC, but has a higher affinity. With this advantage, gastrointestinal symptoms were more evident but no tumor response could be documented in five patients (Herlyn *et al.*, 1991). The TAA CA19-9 detects a ganglioside which is preferentially found on adenocarcinomas of the pancreas and colon. Its elevated serum levels have been used as a tumor marker (Kalthoff *et al.*, 1986), but no satisfactory tumor delivery of the antibody nor response has been found (Herlyn *et al.*, 1991). Saleh *et al.* (1993) treated 21 patients with metastatic gastrointestinal cancer with a murine IgG2a mAb D612 raised against another membrane glycoprotein expressed on normal and malignant gastrointestinal cells, but only six stable diseases were observed after six weeks.

Büchler *et al.* (1990) conducted four trials in patients suffering from adenocarcinomas of the pancreas with the murine IgG1 mAb 494/32 which binds to a carbohydrate epitope in about 90% of human pancreatic cancers and mediates ADCC. Among three trials with 87 patients and a maximal total dose of 1,000 mg one PR was observed. A transient stable disease in about one third of patients has been noted which indicated that mAb effects in pancreatic cancer might be restricted to less advanced stages. Therefore and with respect to the known ADCC activity of the antibody *in vitro*, a randomized controlled study in an adjuvant situation with resected pancreatic cancer has been performed (Büchler *et al.*, 1991). When 29 patients were treated with a total dose of 370 mg mAb and compared to 32 patients of a control group, no benefit of the mAb 494/32 therapy was observed.

Lung and other cancer

Several autocrine growth factors have been identified in lung cancer (Mulshine *et al.*, 1989) among those the gastrin releasing peptide which has been used as a target for a murine mAb in a clinical trial (Mulshine *et al.*, 1988). Although radiolabeled mAb imaging revealed increased radioactivity in tumor sites while gastrin serum levels were decreased no anti-tumor effects have been recognized. In a second trial when multiple infusions were administered with a total dose of 3,000 mg, one CR was obtained among 12 patients (Kelley *et al.*, 1993).

The murine IgG2a mAb KS1/4 has been applied in a clinical trial with non-small cell lung carcinoma where patients were also treated with its immunoconjugate with methotrexate (Elias *et al.*, 1990). MAb KS1/4 binds to an antigen found on normal epithelial cells and a variety of malignant cells. It is especially expressed by adenocarcinomas and some squamous cell carcinomas of the lung and exhibits a homology greater than 50% to the antigen recognized by GA733. Biopsies revealed a dose dependent tissue delivery into tumor and colonic mucosa which could be demonstrated in nearly all patients at a single dose level of 500 and 1,000 mg 24 h after infusion and was accompanied by complement deposition. Only one possible response was detected in a patient treated with the immunoconjugate.

Endocrine gastrointestinal tumors often grow slowly, cause symptoms without wide-spread disease and frequently exhibit rather large liver metastases at the time of diagnosis. This limited character of the disease stimulated a study with a murine IgG2a mAb derived from immunization with tumor cells from a patient with Verner–Morrison's syndrome (Juhlin *et al.*, 1994). MAb infusions temporarily increased but afterwards decreased the most evident symptom of diarrhea.

The autopsy six weeks after therapy confirmed no changes of the liver metastases. However, peritoneal metastasis previously seen during multiple laparatomies had vanished. This might indicate the therapeutic potential of mAb in early metastatic or even minimal residual disease in a compartment accessible to mAb and largely equipped with immunocompetent cells.

Studies with Chimeric and Humanized Antibodies

A frequently appearing obstacle in antibody therapy is HAMA response which might render further antibody therapy ineffective. Removing xenogeneic domains of the constant regions of mAb and replacing them by human domains leads to a decreased immune response and clearance from circulation. This could be achieved by chemical or recombinant DNA-techniques as described in Chapter 1. Since antitumor effects have been predominantly ascribed to the Fc-domain of the antibody cytotoxic mechanisms of human effectors are intensified with chimeric mAb. Furthermore, advances in genetic engineering techniques permit grafting of complementarity-determining regions (CDR) of only small, epitope determining, xenogeneic fragments into a human framework.

The results of the studies in hematopoietic diseases have been reviewed above. Some aspects of the chimeric mAb in those studies will be outlined in the following. When chimeric mAb were introduced to human therapy in initial attempts with rat/human CAMPATH-1H smaller amounts of mAb were needed for sufficient clearance of malignant cells from blood circulation, and one CR and one PR were achieved in two patients (Hale *et al.*, 1988). Similarly, in a trial with chimeric anti-CD20 mAb (Maloney *et al.*, 1994) no immune response was detectable. Comparable results were obtained with less mAb and only a single application as opposed to repeated infusions (Press *et al.*, 1987). The effect may supposedly be due to a marked increase of ADCC capability. In a trial with chimeric anti-CD4 mAb (Knox *et al.*, 1991) an immune response against three different moieties was detected. Two out of seven patients developed anti-idiotype, antimouse and antihuman (allotypic) constant region antibodies. A third patient only expressed antimouse antibodies. While PR was obtained with the maximum dose of mAb those patients had no immune response. The humanized anti-CD33 mAb was not immunogenic (Caron *et al.*, 1994) whereas the murine counterpart showed HAMA in four of six cases tested (Scheinberg *et al.*, 1991). Furthermore, an enhanced ADCC activity might be claimed for one minor response but rapid internalization otherwise would have prevented cytotoxic effects.

A somewhat different approach was used in a patient with B cell NHL where the Fab-fragment of a mouse anti-idiotypic mAb was coupled with a complete human IgG molecule (Hamblin *et al.*, 1987). This univalent Fab/IgG construct was given four times at a total dose of 1,890 mg and a PR was observed after six weeks. As seen in other studies infusion of the antibody was accompanied by a transient fall in the number of circulating malignant cells. A rise of C3c suggested CDC as one important mechanism. No immune response against the IgG was detected.

LoBuglio *et al.* (1989) performed a trial with mouse/human chimeric mAb composed of the murine 17-1A variable domains and human constant domains that did not improve clinical response but showed some pharmacological benefits. Since the chimeric mAb caused only a modest anti-idiotypic response it had a ca. 6-fold circulation time as compared to the murine mAb.

Goodman *et al.* (1993) used a chimeric derivate of the murine mAb L6 which has been used before in non-small-cell lung, colon and breast cancer (Goodman *et al.*, 1990). This chimeric mAb possessed an increased cytotoxic potential with human effectors but did not change clinical outcome. Pharmacokinetics appeared to be similar to the murine analogue and immune response was markedly reduced but HAMA against the variable region still appeared in 4 out of 18 patients.

The anti-GD2 mAb 14G2a which has been used in patients with neuroblastoma and malignant melanoma gave some promising results but also generated HAMA in all patients (Handgretinger *et al.*, 1992; Saleh *et al.*, 1992b). The chimeric analogue of the mAb yielded comparable CDC but improved ADCC. While a single dose of 5–100 mg of the chimeric mAb ch14.18 was not able to induce tumor response in 13 patients with metastatic melanoma, eight patients developed a rather weak antibody response directed at the variable region (Saleh *et al.*, 1992a). In a second trial two CR and PR, respectively, and one minor response were observed in nine patients with neuroblastoma representing similar clinical results documented in the previous study with the murine mAb (Handgretinger *et al.*, 1995). When administered repeatedly no HAMA was detected. This illustrated the advantage of chimeric mAb, since patients were effectively treated with up to 4 courses with 8–12 weeks between cycles of 5 days with up to 50 mg/m^2 per day.

Since the overexpression of the *c-erbB*-2 (HER2) gene has been linked with an unfavorable outcome in patients with breast cancer its protein product has been chosen as a target for mAb therapy (Baselga *et al.*, 1996). Humanization of the anti-p185HER2 mAb prevented immune response in all 43 patients. While antitumor mechanisms were probably related to its known ADCC and antiproliferative effects *in vitro*, one CR, four PR and two minor responses were obtained.

Studies with Anti-idiotypic Antibodies

Different strategies have been employed to intervene in the idiotype network for therapeutic use. First of all, anti-idiotypic antibodies were applied in clinical studies in terms of a passive treatment of antibody- covered malignant B cells as targets. Those anti-idiotypic antibodies were intended to attract the components of Fc-mediated effectors to the B cell, but further participation in active immunization is not excluded. Investigations of anti-idiotype reactions were intensified during clinical studies with antibodies targeting cellular antigens, since evidence was obtained that this immunization corresponds with clinical benefit. In a second step anti-idiotypic antibodies (Ab2) were administered, mirroring an autologous antibody response with an internal image of the antigen. Such a procedure has the advantage of a higher specificity and the potential to break the immunologic tolerance to tumor-associated antigens, which is frequently found in tumor patients.

Idiotype vaccination

Targeting tumor associated antigens with mAb was considered to be a passive approach in immunotherapy. However, a growing number of reports suggest the importance of immunizing effects. Anti-idiotypic antibody (Ab2) arising against the therapeutic antibody (Ab1) mimics an epitope of the tumor antigen, which in turn might induce an anti-anti-idiotypic response. This consists of a humoral component of the idiotypic antibody network (Ab3), but moreover, challenges T cells binding to Ab2 (T_3) and the original TAA.

This has been confirmed in a clinical study with the monoclonal antibody 17-1A and 24 patients with metastatic colorectal carcinoma (Fagerberg *et al.*, 1995). The murine monoclonal antibody 17-1A and its chimeric form with a human IgG1 constant region represented the Ab1 in six different therapy schedules. The cytokines IL-2 or GM-CSF were added in some patients, the latter coincided with a benefit concerning tumor response. The most promising results were obtained comparing the tumor response with the arising idiotypic network. Taking together all groups, two CR, two PR and one minor response have been evaluated. Exclusively, these patients developed or significantly increased preexisting anti-idiotypic antibody secretion. The same phenomenon accounts for T cells reactive with Ab2 (T_3), where patients who lacked any detectable T cell response progressed in disease, except for one patient treated with supplementary GM-CSF. *In vitro* T cell proliferation in response to incubation with purified antigen occurred in four of the five T cell cultures isolated from patients with a clinical response. Classification of T cells was not reported.

The high molecular weight melanoma associated antigen (HMW-MAA) is expressed in a high degree on melanoma lesions with restricted distribution in normal tissue. This tumor associated antigen served as a target for antibodies, which have been used in clinical studies for idiotype vaccination. Mittelman *et al.* (1990) applied the murine anti-HMW-MAA monoclonal antibody MF11-30 in two consecutive studies to patients with metastatic melanoma. The antibody was injected subcutaneously at an amount of 0.5, 1, and 3 mg at day 0, 7, and 28 and up to 10 times to achieve a detectable immunization. Registrating a dose dependent anti-idiotypic antibody level with escalating single doses of up to 4 mg, and three minor responses, the authors conducted a second trial with 2 mg subcutaneously administered MF11-3 with the same time schedule and a maximum of 12 injections in 19 evaluable patients (Mittelman *et al.*, 1990). Separating the 16 patients, which developed detectable Ab3, into two groups according to their Ab3 level, a significantly prolonged average survival time of 55 weeks compared to 19 weeks was noted with Ab3 levels of 1 : 8 and more. One CR was obtained.

In a subsequent study 23 evaluable patients received 2 mg of the murine anti-HMW-MAA monoclonal antibody mAb 763.74 conjugated with keyhole limpet hemocyanin and additional bacillus Calmette-Guérin subcutaneously on days 0, 7, and 28 and up to 12 times, when immunization appeared to be low (Mittelman *et al.*, 1992). Although anti-idiotype response against the Ab2 has been obtained in 18 patients, sera from four of those patients did not react with melanoma cells. Among the remaining 14 patients three patients revealed a PR, and a significantly prolonged survival was seen compared to the nine patients with no anti-idiotype or no anti-tumor response. T_H-cells may play a role since IgG and IgM have been detected, which indicates an isotype switch.

Anti-idiotype vaccination

The reported relevance of anti-idiotype reactions for clinical results suggests even more effects involved in sufficient tumor suppression than merely Fc-regulated mechanisms and receptor-depending effects. Clinical studies with antibodies against tumor associated antigens provide interesting data on this issue, but at the same time they show that in some studies only a minor quantity of individuals react against the idiotype region. Different mechanisms are thought to play a role in tumor escape from the immune surveillance, among those tolerance against major tumor antigen epitopes may be essential. Ab2 vaccines, which partially mimic tumor antigens, are highly specific compared to tumor vaccines, and potentially counteract preexisting immunologic tolerance.

Following initial evidence of a beneficial effect of anti-idiotype vaccination with antibodies against the murine anti-17-1A antibody in patients with gastrointestinal cancer a clinical study of patients with advanced colorectal carcinoma has been conducted (Herlyn *et al.*, 1987). After four subcutaneous injections in weeks 0, 1, 2, and 5 with aluminum hydroxide precipitated polyclonal goat anti-idiotypic antibody (Ab2) to mAb 17-1A (Ab1) all patients independent of the applied dose of Ab2 developed anti-anti-idiotypic antibodies (Ab3). Some patients already had low serum levels of antibodies reactive with surface antigen of cultured tumor cells before therapy, among those one out of four patients tested revealed antibody binding to the 17-1A antigen and the Ab2, respectively. So, Ab2 vaccination in itself in some cases represents a booster injection. Six patients were classified as having a PR, and another seven patients showed an arrest of metastases after vaccination, but a second course of vaccination was accompanied by a chemotherapy with 5-fluorouracil and folic acid, which prevents conclusions about the benefit of the mAb application.

The carcinoembryogenic antigen (CEA) is an antigen commonly associated with colon carcinoma, but since it is a normal oncofetal antigen immune responses are reported only rarely. With the help of anti-idiotypic antibodies specific epitopes of the original antigen could be used for immunization. This has been performed in the context of a clinical study with a murine monoclonal IgG1 antibody (Ab2), which mimics an epitope of the human CEA (Foon *et al.*, 1995). Aluminum hydroxide precipitated anti-idiotypic antibody was injected intracutaneously four times every other week using 1, 2, or 4 mg. Even though the majority of patients had significantly elevated CEA serum levels before treatment, in nine out of twelve patients an anti-CEA/Ab3 antibody response was measured.

A response of T cells, which were characterized as mainly CD4 positive, was observed in seven patients. These T cells exhibited proliferation *in vitro* when coincubated with Ab2, while CEA reactive T cells were induced in only four patients. All patients experienced a progression of their already metastatic colon carcinoma.

A clinical trial with anti-idiotype vaccination has been conducted in an adjuvant concept with colon carcinoma patients after surgery and no evidence of disease (Somasundaram *et al.*, 1995). Thirteen patients were subcutaneously injected with aluminum hydroxide precipitated polyclonal goat Ab2 against murine Ab1 GA733 in weeks 1, 2, 3, and 6 with doses of 0.5–8 mg. Seven patients developed Ab3 that bound specifically to goat Ab2 GA733 and inhibited Ab1 GA733 binding to colon carcinoma cells, thus bound to GA733, whereas only five serum samples bound to colon carcinoma cells. One patient's serum contained antibodies reactive with colon carcinoma before treatment, but only afterwards recognized GA733. As the induction of antigen specific T cells has been described above, here two patients yielded a T helper cell response, whereas another two patients showed suppressor effects. Interestingly, these suppressing, probably T cell mediated effects evolved in patients which demonstrated antigen-reactive T cells before therapy. Concerning the clinical evaluation, four patients showed a recurrence of the disease, two having an Ab3 response with either T helper cell or suppressor cell action, while the other nine patients showed no evidence of disease after 39–86 months of observation.

Monoclonal Antibodies and Biologic Response Modifiers

Cytokines

Cytokines mediate a variety of effects in immune response, which could be utilized for tumor therapy. These mechanisms cover (1) effector cell activation and proliferation, support of inflammation by (2) increased expression of adhesion molecules, (3) elevated blood vessel proliferation or permeability, and (4) chemotaxis, as well as (5) enhanced susceptibility of modulated target cells.

The lymphokine IL-2 is known to stimulate cells with natural killer cell activity (NK cells), then called lymphokine activated K cells (LAK). These LAK cells are much more efficient in cellular cytotoxicity than normal NK cells. This does not only account for MHC unrestricted cytotoxicity, but also for FcR mediated ADCC. Therefore Vlasveld *et al.* (1995) combined an anti-CD19 mAb specific for B cells with continuous intravenous infusion of IL-2 in the therapy of 7 patients with NHL. They noticed a sustained increase in the number of circulating NK-cells and a T cell activation in the majority of patients. Saturated CD19 mAb binding was documented in four of five patients tested by lymph node biopsy, indicating adequate tissue penetration and mAb dosing. However, T cells remained unchanged in lymph nodes and only a slight NK-cell infiltration was detected. This was reflected by one PR and one mixed response and compared to therapy schedules with either mAb or IL-2 alone. As the antitumor effect occurred before

NK-cell activation one has to consider different mechanisms of growth control. The immunoglobulin receptor complex on B cells has been shown to mediate cell cycle arrest or apoptosis depending on the stage of differentiation when immunoglobulins were cross-linked. The CD19 is part of the immunglobulin receptor complex. Studies with various anti-CD19 mAb have been demonstrated to induce cell cycle arrest in B cell lymphomas depending on affinity and epitope specificity (Ghetie *et al.*, 1994). A therapy with mAb which induce cell cycle arrest might be followed by a relapse of the disease from resting cells when they are released from growth inhibiting effects.

Ziegler *et al.* (1992) treated 15 patients with various adenocarcinomas with the murine IgG2a mAb L6 and IL-2. Since earlier studies revealed the mAb L6 to remain in the tumor from two to three weeks after administration they set up a considerable delay for IL-2 treatment of 7 days after a 7 day therapy with mAb. ADCC was significantly elevated and one PR in a patient with colorectal cancer and one mixed response in a patient with breast cancer was noted. Bajorin *et al.* (1990) treated twenty patients suffering from metastatic melanoma with recombinant human IL-2 and anti-GD3 antibody. While inducing a lymphocytosis, ADCC for the applied antibody was only enhanced in two patients, one having stable disease for 8 months, the other a minor response. LAK activity improved in fifteen patients and did not correlate with tumor response. Otherwise HAMA developed in all patients and seemed to be accelerated by IL-2, since it appeared earlier than in previous studies, beginning on the fifth day. Overall, two minor responses and one PR were observed, suggesting that this therapy was not superior compared to either agent alone.

Some tumors are largely infiltrated by macrophages, which do have the capacity of ADCC as well, but only after stimulation. The macrophage-colony stimulating factor (M-CSF) is a potent activator of ADCC and increases the number of circulating monocytes *in vivo*. Minasian *et al.* (1995) administered M-CSF in combination with anti-GD3 antibody to patients with metastatic melanoma. Compared to the results revealed with IL-2, no clinical benefit has been observed apart from a marked monocytosis, obtaining three patients with minor or mixed responses. The potentiation of ADCC of the murine IgG2a against a gastrointestinal glycoprotein by M-CSF observed *in vitro* stimulated a clinical phase II trial with 14 patients with metastatic gastrointestinal cancer (Saleh *et al.*, 1995). Apart from a commonly seen monocytosis with an expansion of the CD16 positive cell population again no antitumor effect was observed. The granulocyte-monocyte-colony stimulating factor (GM-CSF) augments ADCC of lymphocytes and monocytes (Masucci *et al.*, 1989). Ragnhammar *et al.* (1993) evaluated its clinical benefit in mAb 17-1A-treated patients with metastatic colorectal cancer. During the first seven days they were able to detect an increased ADCC of blood monocytes. Thereafter, they observed a return to base level of ADCC during 10 days of subcutaneous GM-CSF injections of 250 mg/m^2. Fiftytwo patients were treated with an escalating dose of mAb 17-1A with up to 12 g, but clinical response did not correlate with the total dose applied. Comparing patients from different therapy schedules who received less than 2 g mAb 17-1A there seemed to be a slight benefit for those with additional administration of GM-CSF, i.e. one complete and three minor responses and six stable diseases with the single therapy and one complete, partial and minor response each and three stable diseases with the combined treatment.

An additive effect has been expected for anti-idiotypic antibodies in B cell lymphoma in combination with IFN-α, since both are known for independent activity in follicular NHL (Brown *et al.*, 1989). Infusions of murine IgG1, IgG2a and IgG2b were given three times a week for usually three to four weeks with a total dose of 1,680–8,400 mg, while IFN-α was administered intramuscularly three times a week for eight weeks at 12×10^6 U/m^2. While a trend to a higher tumor response was observed, the emergence of idiotype-negative clones was not prevented. Two CR, and seven PR were achieved among 12 patients.

IFN-α is known to stimulate IgG Fc receptor expression and ADCC and therefore has been used in combination with the anti-GD3 mAb R24 in patients with malignant melanoma (Caulfield *et al.*, 1990). Among 15 patients no persistent clinical response was observed. Rather weak localization of R24 in the tumor was demonstrated without consistent changes of cellular infiltrates after treatment. An IFN-α dose-dependent neutropenia was accompanied by a decrease in T-suppressor and T_H-cells which might also be related to a CDC-dependent lysis of T cells and no changes of ADCC activity were recognized.

Fc receptor expression and ADCC as well as nonspecific monocyte cytotoxicity are also enhanced by IFN-γ. Four studies used this ability in order to improve mAb 17-1A effects. Low doses of IFN-γ were found to stimulate ADCC, whereas high doses of up to $8 \cdot 10^7$ U/m^2 IFN-γ over four consecutive days did not (Weiner *et al.*, 1986). Apart from a transient decrease of serum tumor markers no tumor response was detected in 27 patients with colorectal or pancreatic cancer. In a second attempt Weiner *et al.* (1988) applied $1 \cdot 10^6$ U/m^2 IFN-γ over four consecutive days with a total of 450 mg mAb to 19 patients with colorectal cancer, but again apart from enhanced ADCC no tumor response was noted. A similar study design was adjusted in a trial with 30 patients suffering from pancreatic cancer (Tempero *et al.*, 1990) where one CR was obtained. Saleh *et al.* (1990) extended treatment of IFN-γ to a 15 day period and combined it with four infusions of 400 mg mAb but failed to detect any significant tumor response. One considerable obstacle among others already mentioned might have been a down-regulating effect of IFN-γ for EGP40 recognized in colorectal cancer cells (Kortner *et al.*, 1995).

TNF-α has attracted considerable attention by its ability to induce a hemorrhagic necrosis in tumors. Moreover, it is an ubiquitous, pre-inflammatory cytokine with pleiotropic effects (Vilcek and Lee, 1991). Apart from inducing thrombocytosis and damage of tumor blood vessels and thereby causing necrosis, it increases the permeability for macromolecules (Clark *et al.*, 1988; Brett *et al.*, 1989) which promotes the accumulation of antibodies in the tumor (Jain, 1990). As known for macrophages and neutrophils exposed to IFN-γ (Graziano and Fanger, 1987), the TNF-α-induced increase in ADCC of macrophages does not correspond to an increased Fc-receptor-expression but an additional production of hydrogen peroxide (Hoffman and Weinberg, 1987). Furthermore, TNF-α mediates an elevated adhesion molecule expression which influences the recruitment of effector cells in tumor tissue (Pober *et al.*, 1986; Bevilacqua *et al.*, 1987). Chemotaxis could be promoted by either endothelial cells or fibroblasts secreting monocyte chemoattractant protein-1 (MCP-1) upon TNF-α exposure (Strieter *et al.*, 1989).

In order to augment tumor infiltration Minasian *et al.* (1994) added TNF-α to the anti-GD3 antibody therapy of malignant melanoma in seven patients. One patient developed a tumor lysis syndrome within hours after treatment, but the remaining patients had tumor progression. *Ex vivo* experiments demonstrated an increased respiratory burst of neutrophils after treatment, although antibody-independent cytotoxicity did not change. ADCC of neutrophils from treated patients were not tested.

A high target antigen density has been shown to be a prerequisite for antibody-dependent antitumor effects (Capone *et al.*, 1984; Herlyn *et al.*, 1985b). As overexpression of growth factor receptors is frequently seen in tumors, the epidermal growth factor receptor (EGFR) protein expression has been found to be TNF-α-inducible in pancreatic cancer cells (Schmiegel *et al.*, 1993) and facilitates advanced ADCC with an anti-EGFR antibody (Schmiegel *et al.*, 1997). Administration of a constant amount of TNF-α in addition to increasing doses of the murine IgG2a anti-EGFR antibody mAb 425 with a maximal single dose of 320 mg has been tested with a total dose of 400, 800 or 1,600 mg (Schmiegel *et al.*, 1997). One CR and one stable disease among 26 patients with advanced pancreatic cancer have been observed. Furthermore, a significant reduction in tumor growth correlated with the mAb dose and led to a prolonged median survival in the groups with 800 and 1,600 mg total dose.

Antibodies as biologic response modifiers

Anti-ganglioside antibodies are of great interest in clinical studies, because they bind preferentially to cells of neuroectodermal origin, including melanomas, and astrocytomas as well. Evidence has been raised that these antibodies also stimulate T cell proliferation by a mechanism so far unknown (Welte *et al.*, 1987). MAb against GD3 but not GD2 induced a T cell proliferation which could be augmented by exogenous IL-2 and phytohemagglutinin. This effect has been ascribed to CD8 positive T cells (Hersey *et al.*, 1986).

Antibodies against the components of the TCR complex are capable of either inhibiting or stimulating the T cell function. While high doses of the anti-CD3 mAb OKT3 are used for inhibition of acute allograft rejection, i.e. 5–10 mg for 10 or more days, low doses are immunostimulatory including activation of tumor specific T cells *in vivo* (Ellenhorn *et al.*, 1990). Therefore OKT3 was used in the therapy of several solid tumors with a maximum total dose of 0.4 mg. Apart from early activation markers detected on lymphocytes no functional T cell activation and subsequent tumor response was seen (Urba *et al.*, 1992). Richards *et al.* (1990) applied an anti-CD3 mAb to patients with solid tumors with repeated doses of 50 or 100 mg every 14 days and obtained one PR.

Minimal Residual Disease

Even in cases where the tumor masses could be completely removed the implicated cure often fails. This has been assigned to micrometastases, which are difficult to detect, and are the origin of tumor relapse. Great efforts have been made to evaluate adjuvant therapies in such minimal disease. Chemo- and radiotherapy are potent inhibitors of proliferating cells and prolonged therapy schedules intend to consider different growth characteristics. Although resting cells are almost omitted from those cytotoxic effects, they are distinguished from normal cells by unique or tumor specific antigens and might be susceptible to immunotherapy. Concerning this, promising approaches have been made with immunocytological methods by detecting tumor cells with mAb against tumor associated antigens. An extreme sensitivity of detecting a single tumor cell in about 200,000 normal cells has been reached by immunocytological procedures in breast and colon cancer when investigating the bone marrow (Schlimok *et al.*, 1987). These findings have been extended to gastrointestinal tumors, where disseminated disease could be observed as well in the peritoneal cavity at time of surgery (Juhl *et al.*, 1994). Both studies revealed a correlation of occurrence of stained cells in the bone marrow or peritoneal cavity with conventional risk factors, such as distant metastases or lymph node involvement. Concerning the growth characteristics of micrometastases, proliferative fractions of tumor cells obtained from bone marrow were very small (Pantel *et al.*, 1993), which means a disadvantage for conventional adjuvant chemo- or radiotherapy. Nevertheless, the determination of high risk patients may be of clinical importance in the decision of admitting them to adjuvant therapy. Furthermore, it may help to establish better criteria for success in adjuvant therapeutic trials.

Adjuvant Therapy

The studies reported so far have been mostly phase I or II clinical trials, and therapeutic efforts have been challenged by advanced tumor stages. This disadvantageous situation especially comprises insufficient accessibility of cells in a tumor burden where antibody delivery through convection is restricted by elevated interstitial pressure and predominantly achieved by comparably slow diffusion (Jain, 1990). In early stages metastases are small, supplied with a normal distribution of blood vessels and, while lymphatic vessels are mostly functioning, interstitial pressure remains normal. Such micrometastases are scattered in mesenchymal compartments or the mononuclear phagocytic system where effector cells and complement are abundantly present. Regarding such a less compact tissue like the bone marrow as an indicator for prognosis of a disseminated tumor disease, clinical studies in hematopoietic malignancies have shown a dose-dependent, sufficient targeting of antibody. Furthermore, micrometastatic cells appear to be non-cycling or dormant, and resist antiproliferative treatment (Pantel *et al.*, 1993). Antibody-dependent cytotoxic events are generated independent of the cell cycle.

The murine monoclonal IgG2a antibody against the EGP40 antigen has been applied safely without severe side effects and is the first antibody evaluated in an adjuvant situation. Patients with completely resected Dukes' stage C colorectal cancer ($T_XN_XM_0$, UICC stage III) were enrolled for a prospective randomized trial (Riethmüller *et al.*, 1994). Among 166 eligible patients 76 were treated with 500 mg 17-1A mAb followed by four boosters of 100 mg each in 4 week intervals, beginning two weeks after surgery. During a median follow-up of 5 years the death rate was 36%, compared to 51% in the control group, which means a 30% reduction of mortality. This matches the magnitude obtained by standard chemotherapy (Moertel *et al.*, 1995). Interestingly, the 17-1A mAb was unable to protect against a local recurrence, whereas showing a clear benefit in the prevention of distant metastases. This might be due to a poor blood supply of the connective scar tissue, which means minor antibody targeting, probably also suppressed by increased interstitial tissue pressure with a defective lymphatic system, and a reduced extent of effector cell invasion. Local metastatic satellites might otherwise have a relatively large volume, which makes them less sensitive in contrast to dispersed cells of distant micrometastases.

HAMA

The human immune system recognizes murine mAb as xenoantigens and therefore HAMA has to be expected in immunocompetent patients. However, detectable levels of HAMA have been observed frequently but several studies documented response only in a minority of cases. There was no consistent pattern of therapeutic effects when HAMA occurred. This ranged from probably no interaction up to situations where specific response was abrogated once HAMA had been detected. Within several clinical trials the development of HAMA coincided with tumor response. Since immune response is directed against the constant and the variable region, including the antigen-binding region, an anti-idiotypic reaction has also to be considered. Moreover, anti-idiotypic immune response may constitute an internal image of the original TAA and therefore booster antitumor and humoral as well as T cell activities (Herlyn *et al.*, 1989). For therapeutic purposes, humanized or human mAb show the least immune complex formation and clearance from circulation, with a benefit for direct antitumor strategies but also immunization. In order to prevent HAMA, several studies used chimeric mAb as reviewed before. Furthermore, vaccinations with xenogeneic Ab2 are able to evoke an immune response, but this might be dominated by reactions to the constant region and thereby repress anti-idiotype response. Therefore a human anti-idiotypic mAb (Ab2) against a murine mAb, specific for a TAA was applied to patients with metastatic colorectal cancer (Robins *et al.*, 1991). While this Ab2 prevented antibody response against the constant, but also variable region, evidence for a T cell response has been collected.

Drug-induced immune suppression is another strategy to prevent HAMA. While high doses of cyclophosphamide are known to reduce HAMA (Thistlethwaite Jr *et al.*, 1984) they did not prevent antibody immune response in a trial performed by Miller *et al.* (1983) but were able to delay antibody clearance from circulation. Cyclosporin A significantly reduced HAMA, including anti-idiotypic response, in patients treated with a mAb against CEA (Ledermann *et al.*, 1988). There was no difference in the pattern of HAMA specificity, i.e. whether this was directed against the constant or the variable region. Deoxyspergualin, a fermentation product which has been isolated from Bacillus laterosporus and which blocks antibody immune response, suppressed or delayed HAMA in response to treatment with the murine IgG2a mAb L6 (Dhingra *et al.*, 1995). While $CD4^+/CD8^+$ T-cell subsets remained unchanged, this reduction, including anti-idiotypic response, was linked to a block of B cell differentiation into immunoglobulin-producing cells. The effect appeared to be dose-dependent with no clinical difference in antitumor activity of L6.

Clinical Aspects

Administration

Murine mAb have a plasma half-life of about 24 hours, sometimes up to 48 h whereas chimeric mAb show half-lifes of 5–6 days (Trang *et al.*, 1990). Assuming direct antitumor effects, the mAb should be applied repeatedly over an appropriate period of time in order to achieve a large area under the curve concerning plasma level over time. In addition to plasma concentration and other factors, tissue concentration is built up by diffusion, which is slow for macromolecules like IgG, and convection where tumors usually have elevated interstitial pressure (Jain, 1990). Therefore, quantities of grams are needed as a total dose for sufficient mAb saturation of the tumor since Fc-dependent cytotoxicity correlates to surface bound mAb (Herlyn *et al.*, 1985). Bolus infusions are well tolerated by most patients if circulating cells do not express the antigen which might trigger immediate cytokine release by effector cells and could be rather toxic. This leads to a widely used infusion time of 2–6 hours. Nevertheless, several studies imply that tumor response does not depend on large amounts of mAb but show superior results when given less than 2 g cumulative mAb dose (Mellstedt *et al.*, 1991). This indicates that immune response is rather complex and further mechanisms have to be concerned in antibody-dependent tumor defense. Some evidence outlines anti-idiotype response as an important factor. Therefore a protocol that meets the characteristics of a vaccination with a single injection followed by several booster injections rather than a continuing high dose application might be the optimal therapy schedule.

Toxicity and side effects

Several mechanisms could be responsible for initiating side effects in mAb administration. This includes hypersensitive reactions, immune complex related effects, and tumor lysis syndrome. Anaphylactic reactions account for about 1% of side-effects but are easily managed by discontinuing the infusion and by administering corticosteroids and antihistamins. Allergic effects may also appear as hypotension, tachycardia, bronchospasm, fever, urticaria or pruritus where prophylactic use of corticosteroids or antihistamins can prevent symptoms during following applications.

Additional side effects may relate to the specific action of the mAb. For instance mAb 17-1A, which is used in colorectal cancer patients, causes diarrhea dependent on infusion speed in about one third of the patients. Fever, chills, sweats, rigors, prostration, and occasional hypotension and dyspnoe occur in patients treated with mAb against circulating leukemia cells and begin hours after infusion. This dose-related reaction is probably caused by the release of various cytokines. In studies with anti-GD2 mAb pain was thought to be mediated by mAb binding to peripheral nerve fibers and required administration of morphine (Handgretinger *et al.*, 1995).

CONCLUSION

The most appropriate challenge for antibody therapy was revealed to be their administration in minimal residual tumor disease. After about 15 years of clinical trials with only sporadic success one of the first mAb raised against colorectal carcinoma cells has achieved a reputation in the immunotherapy of malignant disease (Riethmüller *et al.*, 1994). Other mAb expressing at least comparable antitumor effects in preclinical studies will corroborate the intended approach. Furthermore, development of chimeric mAb with a constant human domain will reduce HAMA while anti-idiotypic response remains unaffected and strenghten Fc-dependent effector mechanisms of unconjugated mAb.

Future trials will be focused on treatment of minimal residual disease following surgical, chemotherapeutic and radiotherapeutic approaches where the initially mentioned obstacles are of minor relevance. The studies performed so far reveal that antibodies can be administered safely without serious adverse effects. A combination with chemotherapy could impair cellular repair mechanisms and increase the sensitivity to cytotoxic effects mediated by mAb. MAb may also attack dormant cells which are spared by antiproliferative drugs. Hence, mAb which represent an option as an adjuvant therapeutic agent in colorectal cancer will be brought to clinical use in combination with chemotherapy (Schöber *et al.*, 1996) and radiation therapy.

REFERENCES

Bajorin, D.F., Chapman, P.B., Wong, G., Coit, D.G., Kunicka, J., Dimaggio, J., Cordon, C.C. *et al.* (1990) Phase I evaluation of a combination of monoclonal antibody R24 and interleukin 2 in patients with metastatic melanoma. *Cancer Res.* **50**: 7490–7495.

Ball, E.D., Bernier, G.M., Cornwell, G.d., McIntyre, O.R., O'Donnell, J.F. and Fanger, M.W. (1983) Monoclonal antibodies to myeloid differentiation antigens: *in vivo* studies of three patients with acute myelogenous leukemia. *Blood* **62**: 1203–1210.

Baselga, J., Tripathy, D., Mendelsohn, J., Baughman, S., Benz, C.C., Dantis, L., Sklarin, N.T. *et al.* (1996) Phase II study of weekly intravenous recombinant humanized anti-p185^{HER2} monoclonal antibody in patients with HER2/neu-overexpressing metastatic breast cancer. *J. Clin. Oncol.* **14**: 737–744.

Bataille, R., Barlogie, B., Lu, Z.Y., Rossi, J.F., Lavabre, B.T., Beck, T., Wijdenes, J., Brochier, J. and Klein, B. (1995) Biologic effects of anti-interleukin-6 murine monoclonal antibody in advanced multiple myeloma. *Blood* **86**: 685–691.

Becker, J., Varki, N., Gillies, S., Furukawa, K. and Reisfeld, R. (1996) An antibody-interleukin 2 fusion protein overcomes tumor heterogeneity by induction of a cellular immune response. *Proc. Natl. Acad. Sci. USA* **93**: 7826–7831.

Bertram, J.H., Gill, P.S., Levine, A.M., Boquiren, D., Hoffman, F.M., Meyer, P. and Mitchell, M.S. (1986) Monoclonal antibody T101 in T cell malignancies: a clinical, pharmacokinetic, and immunologic correlation. *Blood* **68**: 752–761.

Bevilacqua, M.P., Pober, J.S., Mendrick, D.L. and Cotran, R.S. (1987) Identification of an inducible endothelial-leukocyte adhesion molecule. *Proc. Natl. Acad. Sci. USA* **84**: 9238–9242.

Brett, J., Gerlach, H., Nawroth, P., Steinberg, S., Godman, G. and Stern, D. (1989) Tumor necrosis factor/cachectin increases permeability of endothelial cell monolayers by a mechanism involving regulatory proteins. *J. Exp. Med.* **169**: 1977–1991.

Brown, S.L., Miller, R.A., Horning, S.J., Czerwinski, D., Hart, S.M., McElderry, R., Basham, T. *et al.* (1989) Treatment of B cell lymphomas with anti-idiotype antibodies alone and in combination with alpha interferon. *Blood* **73**: 651–661.

Büchler, M., Friess, H., Malfertheiner, P., Schultheiss, K.-H., Muhrer, K.-H., Kraemer, H.-P. and Beger, H.-G. (1990) Studies of pancreatic cancer utilizing monoclonal antibodies. *Int. J. Pancreatol.* **7**: 151–157.

Büchler, M., Friess, H., Schultheiss, K.H., Gebhardt, C., Kuebel, R., Muhrer, K.H., Winkelmann, M. *et al.* (1991) A randomized controlled trial of adjuvant immunotherapy (murine monoclonal antibody 494/32) in resectable pancreatic cancer. *Cancer* **68**: 1507–1512.

Büchler, M., Kübel, R., Klapdor, R. *et al.* (1989) Immunotherapy of pancreatic cancer with monoclonal antibody BW 494: results from a multicentric phase I-II trial. In H.G. Beger, M. Büchler, R.A. Reisfeld and G. Schulz, (eds.), *Cancer Therapy*, Heidelberg, Springer: 32–41.

Büchler, M., Kübel, R., Malfertheiner, P. *et al.* (1988) Immuntherapie des fortgeschrittenen Pankreaskarzinoms mit dem monoklonalen Antikörper BW 494. *Dtsch. Med. Wochenschr.* **113**: 374–380.

Capone, P., Papsidero, L. and Chu, T. (1984) Relationship between antigen density and immunotherapeutic response elicited by monoclonal antibodies against solid tumors. *J. Natl. Cancer Inst.* **72**: 673–677.

Caron, P.C., Jurcic, J.G., Scott, A.M., Finn, R.D., Divgi, C.R., Graham, M.C., Jureidini, I.M. *et al.* (1994) A phase 1B trial of humanized monoclonal antibody M195 (anti-CD33) in myeloid leukemia: specific targeting without immunogenicity. *Blood* **83**: 1760–1768.

Caulfield, M.J., Barna, B., Murthy, S., Tubbs, R., Sergi, J., Medendorp, S. and Bukowski, R.M. (1990) Phase Ia-Ib trial of an anti-GD3 monoclonal antibody in combination with interferon-alpha in patients with malignant melanoma. *J. Biol. Response Modif.* **9**: 319–328.

Cheung, N.K., Burch, L., Kushner, B.H. and Munn, D.H. (1991) Monoclonal antibody 3F8 can effect durable remissions in neuroblastoma patients refractory to chemotherapy: a phase II trial. *Prog. Clin. Biol. Res.* **366**: 395–400.

Cheung, N.K., Kushner, B.H., Yeh, S.J. and Larson, S.M. (1994) 3F8 monoclonal antibody treatment of patients with stage IV neuroblastoma: a phase II study. *Prog. Clin. Biol. Res.* **385**: 319–328.

Cheung, N.K., Lazarus, H., Miraldi, F.D., Abramowsky, C.R., Kallick, S., Saarinen, U.M., Spitzer, T. *et al.* (1987) Ganglioside GD2 specific monoclonal antibody 3F8: a phase I study in patients with neuroblastoma and malignant melanoma (see comments). *J. Clin. Oncol.* **5**: 1430–1440.

Cheung, N.K.V., Lazarus, H., Miraldi, F.D., Berger, N.A., Abramowski, C.R., Saarinen, U.M., Spitzer, T., Strandjord, S.E. and Coccia, P.F. (1992) Reassessment of patient response to monoclonal antibody 3F8. *J. Clin. Oncol.* **10**: 371.

Clark, M.A., Chen, M.-J., Crooke, S.T. and Bomalaski, J.S. (1988) Tumour necrosis factor (cachectin) induces phospholipase A_2 activity and synthesis of a phospholipase A_2-activating protein in endothelial cells. *Biochem. J.* **250**: 123–132.

Daeron, M. (1997) Fc receptor biology. *Annu. Rev. immunol.* **15**: 203–234.

Dhingra, K., Fritsche, H., Murray, J.L., LoBuglio, A.F., Khazaeli, M.B., Kelley, S., Tepper, M.A. *et al.* (1995) Phase I clinical pharmacological study of suppression of human antimouse antibody response to monoclonal antibody L6 by deoxyspergualin. *Cancer Res.* **55**: 3060–3067.

Dillman, R.O., Beauregard, J., Shawler, D.L., Halpern, S.E., Markman, M., Ryan, K.P., Baird, S.M. and Clutter, M. (1986) Continuous infusion of T101 monoclonal antibody in chronic lymphocytic leukemia and cutaneous T cell lymphoma. *J. Biol. Response Modif.* **5**: 394–410.

Dillman, R.O., Shawler, D.L., Sobol, R.E., Collins, H.A., Beauregard, J.C., Wormsley, S.B. and Royston, I. (1982) Murine monoclonal antibody therapy in two patients with chronic lymphocytic leukemia. *Blood* **59**: 1036–1045.

Dippold, W.G., Bernhard, H., Dienes, H.P. and Meyer zum Büschenfelde, K.-H. (1988) Treatment of patient with malignant melanoma by monoclonal ganglioside antibodies. *Eur. J. Cancer Clin. Oncol.* **21**: S65–S67.

Dippold, W.G., Knuth, K.R.A. and Meyer zum Büschenfelde, K.-H. (1985) Inflammatory tumor response to monoclonal antibody infusion. *Eur. J. Cancer Clin Oncol.* **21**: 907–912.

Douillard, J.Y., Lehur, P.A., Vignoud, J., Blottiere, H., Maurel, C., Thedrez, P., Kremer, M. and Le Mevel, B. (1986) Monoclonal antibodies specific immunotherapy of gastrointestinal tumors. *Hybridoma* **5** (Suppl. 1): S139–149.

Dyer, M.J., Hale, G., Hayhoe, F.G. and Waldmann, H. (1989) Effects of CAMPATH-1 antibodies *in vivo* in patients with lymphoid malignancies: influence of antibody isotype. *Blood* **3**: 1431–1439.

Elias, D.J., Hirschowitz, L., Kline, L.E., Kroener, J.F., Dillman, R.O., Walker, L.E., Robb, J.A. and Timms, R.M. (1990) Phase I clinical comparative study of monoclonal antibody KS1/4 and KS1/4-methotrexate immunconjugate in patients with non-small cell lung carcinoma. *Cancer Res.* **50**: 4154–4159.

Ellenhorn, J., Schreiber, H. and Bluestone, J. (1990) Mechanism of tumor rejection in anti-CD_3 monoclonal antibody-treated mice. *J. Immunol.* **144**: 2840–2846.

Ellenhorn, J.D., Hirsch, R., Schreiber, H. and Bluestone, J.A. (1988) *In vivo* administration of anti-CD3 prevents malignant progressor tumor growth. *Science* **242**: 569–571.

Fagerberg, J., Hjelm, A.L., Ragnhammar, P., Froedin, J.E., Wigzell, H. and Mellstedt, H. (1995) Tumor regression in monoclonal antibody-treated patients correlates with the presence of anti-idiotype-reactive T lymphocytes. *Cancer Res.* **55**: 1824–1827.

Fanger, M.W., Shen, L., Graziano, R.F. and Guyre, P.M. (1989) Cytotoxicity mediated by human Fc receptors for IgG. *Immunol. Today* **10**: 92–99.

Ferrone, S., Chen, Z.J., Liu, C.C., Hirai, S., Kageshita, T. and Mittelman, A. (1993) Human high molecular weight – melanoma associated antigen mimicry by mouse anti-idiotypic monoclonal antibodies MK2-23. Experimental studies and clinical trials in patients with malignant melanoma. *Pharmacol. Ther.* **57**: 259–290.

Fleuren, G., Gorter, A., Kuppen, P., Litvinov, S. and Warnaar, S. (1995) Tumor heterogeneity and immunotherapy of cancer. *Immunol. Rev.* **145**: 91–122.

Foon, K.A., Chakraborty, M., John, W.J., Sherratt, A., Kohler, H. and Bhattacharya, C.M. (1995) Immune response to the carcinoembryonic antigen in patients treated with an anti-idiotype antibody vaccine. *J. Clin. Invest.* **96**: 334–342.

Foon, K.A., Schroff, R.W., Bunn, P.A., Mayer, D., Abrams, P.G., Fer, M., Ochs, J. *et al.* (1984) Effects of monoclonal antibody therapy in patients with chronic lymphocytic leukemia. *Blood* **64**: 1085–1093.

Frödin, J.-E., Faxas, M.-E., Hagström, B., Lefvert, A.-K., Masucci, G., Nilsson, B., Steinitz, M., Unger, P. and Mellstedt, H. (1991) Induction of anti-idiotypic (ab_2) and anti-anti-idiotypic (ab_3) antibodies in patients treated with the mouse monoclonal antibody 17-1A (ab_1). Relation to the clinical outcome – an important antitumoral effector function? *Hybridoma* **10**: 421–431.

Frödin, J.-E., Harmenberg, U., Biberfeld, P., Christensson, B., Lefvert, A.K., Rieger, A., Shetye, J., Wahren, B. and Mellstedt, H. (1988) Clinical effects of monoclonal antibodies (MAb 17-1A) in patients with metastatic colorectal carcinomas. *Hybridoma* **7**: 309–321.

Ghetie, M.-A., Picker, L.J., Richardson, J.A., Tucker, K., Uhr, J.W. and Vitetta, E.S. (1994) Anti-CD19 inhibits the growth of human B cell tumor lines *in vitro* and of Daudi cells in SCID mice by inducing cell cycle arrest. *Blood* **83**: 1329–1336.

Goodman, G.E., Beaumier, P., Hellstroem, I., Fernyhough, B. and Hellstroem, K.E. (1985) Pilot trial of murine monoclonal antibodies in patients with advanced melanoma. *J. Clin. Oncol.* **3**: 340–352.

Goodman, G.E., Hellstroem, I., Brodzinsky, L., Nicaise, C., Kulander, B., Hummel, D. and Hellstroem, K.E. (1990) Phase I trial of murine monoclonal antibody L6 in breast, colon, ovarian, and lung cancer. *J. Clin. Oncol.* **8**: 1083–1092.

Goodman, G.E., Hellström, I., Hummel, D., Brodzinski, L., Yeh, M.Y. and Hellström, K.E. (1987) Phase-I trial of monoclonal antibody MG-21 directed against a melanoma associated GD3 ganglioside antigen. *Proc. Annu. Meet. Am. Soc. Clin. Oncol.* **6**: 209.

Goodman, G.E., Hellstrom, I., Yelton, D.E., Murray, J.L., O'Hara, S., Meaker, E., Zeigler, L. *et al.* (1993) Phase I trial of chimeric (human–mouse) monoclonal antibody L6 in patients with non-small-cell lung, colon, and breast cancer. *Cancer Immunol. Immunother.* **36**: 267–273.

Goodnow, C.C. (1992) Transgenic mice and analysis of B cell tolerance. *Ann. Rev. Immunol.* **10**: 489–518.

Graziano, R.F. and Fanger, M.W. (1987) FcγRI and FcγRII on monocytes and granulocytes are cytotoxic trigger molecules for tumor cells. *J. Immunol.* **139**: 3536–3541.

Greenaway, S., Henniker, A.J., Walsh, M. and Bradstock, K.F. (1994) A pilot clinical trial of two murine monoclonal antibodies fixing human complement in patients with chronic lymphatic leukaemia. *Leuk. Lymphoma* **13**: 323–331.

Hale, G., Dyer, M.J., Clark, M.R., Phillips, J.M., Marcus, R., Riechmann, L., Winter, G. and Waldmann, H. (1988) Remission induction in non-Hodgkin lymphoma with reshaped human monoclonal antibody CAMPATH-1H. *Lancet* **2**: 1394–1399.

Hale, G., Swirsky, D.M., Hayhoe, F.G.J. and Waldmann, H. (1983) Effects of monoclonal anti-lymphocyte antibodies *in vivo* in monkeys and humans. *Mol. Biol. Med.* **1**: 321–324.

Hamblin, T.J., Cattan, A.R., Glennie, M.J., MacKenzie, M.R., Stevenson, F.K., Watts, H.F. and Stevenson, G.T. (1987) Initial experience in treating human lymphoma with a chimeric univalent derivative of monoclonal anti-idiotype antibody. *Blood* **69**: 790–797.

Handgretinger, R., Anderson, K., Lang, P., Dopfer, R., Klingebiel, T., Schrappe, M., Reuland, P. *et al.* (1995) A phase I study of human/mouse chimeric antiganglioside GD2 antibody ch14.18 in patients with neuroblastoma. *Eur. J. Cancer* **31A**: 261–267.

Handgretinger, R., Baader, P., Dopfer, R., Klingebiel, T., Reuland, P., Treuner, J., Reisfeld, R.A. and Niethammer, D. (1992) A phase I study of neuroblastoma with the anti-ganglioside GD2 antibody 14.G2a. *Cancer Immunol. Immunother.* **35**: 199–204.

Hekman, A., Honselaar, A., Vuist, W.M., Sein, J.J., Rodenhuis, S., ten Bokkel Huinink, W.W., Somers, R., Ruemke, P. and Melief, C.J. (1991) Initial experience with treatment of human B cell lymphoma with anti-CD19 monoclonal antibody. *Cancer Immunol. Immunother.* **32**: 364–372.

Heppner, G.H. (1984) Tumor heterogeneity. *Cancer Res.* **44**: 2259–2265.

Herlyn, D., Herlyn, M., Steplewski, Z. and Koprowski, H. (1985a) Monoclonal anti-human tumor antibodies of six isotypes in cytotoxic reactions with human and murine effector cells. *Cell Immunol.* **92**: 105–114.

Herlyn, D., Powe, J., Ross, A.H., Herlyn, M. and Koprowski, H. (1985b) Inhibition of tumor growth by IgG2a monoclonal antibodies correlates with antibody density on tumor cells. *J. Immunol.* **134**: 1300–1304.

Herlyn, D., Sears, H.F., Ernst, C.S., Iliopoulos, D., Steplewski, Z. and Koprowski, H. (1991) Initial clinical evaluation of two murine IgG2a monoclonal antibodies for immunotherapy of gastrointestinal carcinoma. *Am. J. Clin. Oncol.* **14**: 371–378.

Herlyn, D., Wettendorff, M. and Koprowski, H. (1989) Modulation of cancer patients' immune responses by anti-idiotypic antibodies. *Int. Rev. Immunol.* **4**: 347–357.

Herlyn, D., Wettendorff, M., Schmoll, E., Iliopoulos, D., Schedel, I., Dreikhausen, U., Raab, R. *et al.* (1987) Anti-idiotype immunization of cancer patients: modulation of the immune response. *Proc. Natl. Acad. Sci. USA* **84**: 8055–8059.

Hersey, P., Schibeci, S.D., Thownsend, P., Burns, C. and Cheresh, D.A. (1986) Potentiation of lymphocyte responses by monoclonal antibodies to the ganglioside GD3. *Cancer Res.* **46**: 6083–6090.

Hertenstein, B., Wagner, B., Bunjes, D., Duncker, C., Raghavachar, A., Arnold, R., Heimpel, H. and Schrezenmeier, H. (1995) Emergence of CD52-, phosphatidylinositolglycan-anchor-deficient T lymphocytes after *in vivo* application of Campath-1H for refractory B cell non-Hodgkin lymphoma. *Blood* **86**: 1487–1492.

Hoffman, M. and Weinberg, J.B. (1987) Tumor necrosis factor α induces increased hydrogen peroxid production and Fc receptor expression, but not increased Ia antigen expression by peritoneal macrophages. *J. Leukoc. Biol.* **42**: 704–707.

Houghton, A.N., Mintzer, D., Cordon, C.C., Welt, S., Fliegel, B., Vadhan, S., Carswell, E. *et al.* (1985) Mouse monoclonal IgG3 antibody detecting GD3 ganglioside: a phase I trial in patients with malignant melanoma. *Proc. Natl. Acad. Sci. USA* **82**: 1242–1246.

Hu, E., Epstein, A.L., Naeve, G.S., Gill, I., Martin, S., Sherrod, A., Nichols, P. *et al.* (1989) A phase 1a clinical trial of LYM-1 monoclonal antibody serotherapy in patients with refractory B cell malignancies. *Hematological Oncology* **7**: 155–166.

Irie, R.F., Matsuki, T. and Morton, D.L. (1989) Human monoclonal antibody to ganglioside GM2 for melanoma treatment (letter). *Lancet* **1**: 786–787.

Irie, R.F. and Morton, D.L. (1986) Regression of cutaneous metastatic melanoma by intralesional injection with human monoclonal antibody to ganglioside GD2. *Proc. Natl. Acad. Sci. USA* **83**: 8694–8698.

Jain, R.K. (1990) Physiological barriers to delivery of monoclonal antibodies and other macromolecules. *Cancer Res.* **50** (suppl.): 814s–819s.

Janson, C.H., Tehrani, M.J., Mellstedt, H. and Wigzell, H. (1989) Anti-idiotypic monoclonal antibody to a T cell chronic lymphatic leukemia. Characterization of the antibody, *in vitro* effector functions and results of therapy. *Cancer Immunol. Immunother.* **28**: 225–232.

Jerne, N.K. (1974) Towards the network theory of the immune system. *Ann. Immunol.* **124**: 373–389.

Juhl, H., Stritzel, M., Wroblewski, A., Henne, B.D., Kremer, B., Schmiegel, W., Neumaier, M. *et al.* (1994) Immunocytological detection of micrometastatic cells: comparative evaluation of findings in the peritoneal cavity and the bone marrow of gastric, colorectal and pancreatic cancer patients. *Int. J. Cancer* **57**: 330–335.

Juhlin, C., Papanicolaou, V., Arnberg, H., Klareskog, L., Loerelius, L.E., Rastad, J., Oberg, K. and Akerstroem, G. (1994) Clinical and biochemical effects *in vivo* of monoclonal antitumor antibody in Verner-Morrison's syndrome. *Cancer* **73**: 1346–1352.

Kalthoff, H., Kreiker, C., Schmiegel, W.H., Greten, H. and Thiele, H.G. (1986) Characterization of CA 19-9 bearing mucins as physiological exocrine pancreatic secretion products. *Cancer Res.* **46**: 3605–3607.

Kelley, M., Avis, I., Linnoila, R., Richardson, G., Snider, R., Phares, J., Ashburn, R. *et al.* (1993) Complete response in a patient with small cell lung cancer (SCLC) treated on a Phase II trial using a murine monoclonal antibody (2A11) directed against gastrin-releasing peptide (GRP). *Proc. Annu. Meet. Am. Soc. Clin. Oncol.* **12**: A1133.

Klein, B., Wijdenes, J., Zhang, X.G., Jourdan, M., Boiron, J.M., Brochier, J., Liautard, J. *et al.* (1991) Murine anti-interleukin-6 monoclonal antibody therapy for a patient with plasma cell leukemia. *Blood* **78**: 1198–1204.

Knox, S.J., Levy, R., Hodgkinson, S., Bell, R., Brown, S., Wood, G.S., Hoppe, R. *et al.* (1991) Observations on the effect of chimeric anti-CD4 monoclonal antibody in patients with mycosis fungoides. *Blood* **77**: 20–30.

Kortner, G., Arnold, R. and Simon, B. (1995) Regulation of the adhesion molecule EGP40 promoter by gamma-interferon in colon-carcinoma cells. *Gastroenterology* **108**: A492.

Kwak, L.W., Campbell, M.J., Czerwinski, D.K., Hart, S., Miller, R.A. and Levy, R. (1992) Induction of immune responses in patients with B cell lymphoma against the surface-immunoglobulin idiotype expressed by their tumors. *N. Engl. J. Med.* **327**: 1209–1215.

Ledermann, J.A., Begent, R.H.J., Bagshawe, K.D., Riggs, S.J., Searle, F., Glaser, M.G., Green, A.J. and Dale, R.G. (1988) Repeated antitumor antibody therapy in man with suppression of the host response by Cyclosporin A. *Br. J. Cancer.* **58**: 654–657.

Levy, R. and Miller, R.A. (1983) Biological and clinical implications of lymphocyte hybridomas: tumor therapy with monoclonal antibodies. *Annu. Rev. Med.* **34**: 107–116.

Lichtin, A., Iliopoulos, D., Guerry, D., Elder, D., Herlyn, D. and Steplewski, Z. (1988) Therapy of melanoma with an anti-melanoma ganglioside monoclonal antibody: a possible mechanism of a complete response. *Proc. Annu. Meet. Am. Soc. Clin. Oncol.* **7**: 247.

Lin, C.-T., Shen, Z. and Unkeless, J.C. (1994) Fc receptor-mediated signal transduction. *J. Clin. Immunol.* **14**: 1–13.

Litvinov, S.V., Velders, M.P., Bakker, H.A.M., Fleuren, G.J. and Warnaar, S.O. (1994) Ep-CAM: a human epithelial antigen is a homophilic cell-cell adhesion molecule. *J. Cell. Biol.* **125**: 437–446.

LoBuglio, A.F., Saleh, M.N., Lee, J., Khazaeli, M.B., Carrano, R., Holden, H. and Wheeler, R.H. (1988) Phase I trial of multiple large doses of murine monoclonal antibody CO17-1A. I. Clinical aspects. *J. Natl. Cancer. Inst.* **80**: 932–936.

LoBuglio, A.F., Wheeler, R.H., Trang, J., Haynes, A., Rogers, K., Harvey, E.B., Sun, L., Ghrayeb, J. and Khazaeli, M.B. (1989) Mouse/human chimeric monoclonal antibody in man: kinatics and immune response. *Proc. Natl. Acad. Sci. USA* **86**: 4220–4224.

Maloney, D.G., Brown, S., Czerwinski, D.K., Liles, T.M., Hart, S.M., Miller, R.A. and Levy, R. (1992) Monoclonal anti-idiotype antibody therapy of B cell lymphoma: the addition of a short course of chemotherapy does not interfere with the antitumor effect nor prevent the emergence of idiotype-negative variant cells. *Blood* **80**: 1502–1510.

Maloney, D.G., Liles, T.M., Czerwinski, D.K., Waldichuk, C., Rosenberg, J., Grillo, L.A. and Levy, R. (1994) Phase I clinical trial using escalating single-dose infusion of chimeric anti-CD20 monoclonal antibody (IDEC-C2B8) in patients with recurrent B cell lymphoma. *Blood* **84**: 2457–2466.

Masucci, G., Wersäll, P., Ragnhammer, P. and Mellstedt, H. (1989) Granulocyte-macrophage-colony-stimulating factor augments cytotoxic capacity of lymphocytes and monocytes in antibody-dependent cellular cytotoxicity. *Cancer Immunology. Immunotherapy*. **29**: 288–292.

Meeker, T., Lowder, J., Cleary, M.L., Steward, S., Warnke, R., Sklar, J. and Levy, R. (1985) Emergence of idiotype variation during treatment of B cell lymphoma with anti-idiotype antibodies. *N. Engl. J. Med.* **312**: 1658–1665.

Meeker, T.C., Lowder, J., Maloney, D.G., Miller, R.A., Thielemans, K., Warnke, R. and Levy, R. (1985) A clinical trial of anti-idiotype therapy for B cell malignancy. *Blood* **65**: 1349–1363.

Mellstedt, H., Frödin, J.-E., Masucci, G., Ragnhammar, P., Fagerberg, J., Hjelm, A.L., Shetyc, J., Wersäll, P. and Osterborg, A. (1991) The therapeutic use of monoclonal antibodies in colorectal carcinoma. *Semin. Oncol.* **18**: 462–477.

Mendelsohn, J. (1990) The epidermal growth factor receptor as a target for therapy with antireceptor monoclonal antibodies. *Semin. Cancer. Biol.* **1**: 339–344.

Miller, R. A. and Levy, R. (1981) Response of cutaneous T cell lymphoma to therapy with hybridoma monoclonal antibody. *Lancet* **2**: 226–230.

Miller, R.A., Maloney, D.G., McKillop, J. and Levy, R. (1981) *In vivo* effects of murine hybridoma monoclonal antibody in a patient with T cell leukemia. *Blood* **8**: 78–86.

Miller, R.A., Maloney, D.G., Warnke, R. and Levy, R. (1982) Treatment of B cell lymphoma with monoclonal anti-idiotype antibody. *N. Engl. J. Med.* **306**: 517-522.

Miller, R.A., Oseroff, A.R., Stratte, P.T. and Levy, R. (1983) Monoclonal antibody therapeutic trials in seven patients with T cell lymphoma. *Blood* **2**: 988–995.

Minasian, L.M., Szatrowski, T.P., Rosenblum, M., Steffens, T., Morrison, M.E., Chapman, P.B., Williams, L., Nathan, C.F. and Houghton, A.N. (1994) Hemorrhagic tumor necrosis during a pilot trial of tumor necrosis factor-alpha and anti-GD3 ganglioside monoclonal antibody in patients with metastatic melanoma. *Blood* **83**: 56–64.

Minasian, L.M., Yao, T.J., Steffens, T.A., Scheinberg, D.A., Williams, L., Riedel, E., Houghton, A.N. and Chapman, P.B. (1995) A phase I study of anti-GD3 ganglioside monoclonal antibody R24 and recombinant human macrophage-colony stimulating factor in patients with metastatic melanoma. *Cancer* **75**: 2251–2257.

Mittelman, A., Chen, Z.J., Kageshita, T., Yang, H., Yamada, M., Baskind, P., Goldberg, N. *et al.* (1990) Active specific immunotherapy in patients with melanoma. A clinical trial with mouse antiidiotypic monoclonal antibodies elicited with syngeneic anti-high-molecular-weight-melanoma-associated antigen monoclonal antibodies (published erratum appears in *J. Clin. Invest.*, 1991 Feb;87(2):757). *J. Clin. Invest.* **86**: 2136–2144.

Mittelman, A., Chen, Z.J., Yang, H., Wong, G.Y. and Ferrone, S. (1992) Human high molecular weight melanoma-associated antigen (HMW-MAA) mimicry by mouse anti-idiotypic monoclonal antibody MK2-23: induction of humoral anti-HMW-MAA immunity and prolongation of survival in patients with stage IV melanoma. *Proc. Natl. Acad. Sci. USA* **89**: 466–470.

Moertel, C.G., Fleming, T.R., Macdonald, J.S., Haller, D.G., Laurie, J.A., Tangen, C.M., Ungerleider, J.S. *et al.* (1995) Fluorouracil plus levamisole as effective adjuvant therapy after resection of stage III colon carcinoma: a final report. *Ann. Intern. Med.* **122**: 321–326.

Müller-Eberhard, H.J. (1988) Molecular organization and function of the complement system. *Ann. Rev. Biochem.* **57**: 321–347.

Mulshine, J.L., Cuttitta, F., Avis, I., Treston, A.M., Kasprzyk, P., Carrasquillo, J.A. *et al.* (1988) Phase I evaluation of an anti-gastrin releasing peptide (GRP) monoclonal antibody (MoAb) in patients with advanced lung cancer. *Proc. Annu. Meet. Am. Soc. Clin. Oncol.* **7**: 213.

Mulshine, J.L., Treston, A.M., Natale, R.B., Kasprzyk, P.G., Avis, I., Nakanishi, Y. and Cuttitta, F. (1989) Autocrine growth factors as therapeutic targets in lung cancer. *Chest* **96**: 31S–34S.

Murray, J.L., Cunningham, J.E., Brewer, H., Mujoo, K., Zukiwski, A.A., Podoloff, D.A., Kasi, L.P. *et al.* (1994) Phase I trial of murine monoclonal antibody 14G2a administered by prolonged intravenous infusion in patients with neuroectodermal tumors. *J. Clin. Oncol.* **12**: 184–193.

Nadler, L.M., Stashenko, P., Hardy, R., Kaplan, W.D., Button, L.W., Kufe, D.W., Antman, K.H. and Schlossman, S.F. (1980) Serotherapy of a patient with monoclonal antibody directed against a human lymphoma-associated antigen. *Cancer Res.* **40**: 3147–3154.

Oldham, R.K., Foon, K.A., Morgan, A.C., Woodhouse, C.S., Schroff, R.W., Abrams, P.G., Fer, M. *et al.* (1984) Monoclonal antibody therapy of malignant melanoma: *in vivo* localization in cutaneous metastasis after intravenous administration. *J. Clin. Oncol.* **2**: 1235–1244.

Pantel, K., Schlimok, G., Braun, S. *et al.*, (1993) Differential expression of proliferation-associated molecules in individual micrometastatic carcinoma cells. *J. Natl. Cancer Inst.* **17**: 1419–1424.

Paul, A.R., Engstrom, P.F., Weiner, L.M., Steplewski, Z. and Koprowski, H. (1986) Treatment of advanced measurable or evaluable pancreatic carcinoma with 17-1A murine monoclonal antibody alone or in combination with 5-fluorouracil, adriamycin and mitomycin (FAM). *Hybridoma* **5**: S171–S174.

Pober, J.S., Gimbrone, M.A., Lapierre, L.A., Mendrick, D.L., Fiers, W., Rothlein, R. and Springer, T.A. (1986) Overlapping patterns of activation of human endothalial cells by interleukin 1, tumor necrosis factor, and immune interferon. *J. Immunol.* **137**: 1893–1896.

Press, O.W., Appelbaum, F., Ledbetter, J.A., Martin, P.J., Zarling, J., Kidd, P. and Thomas, E.D. (1987) Monoclonal antibody 1F5 (anti-CD20) serotherapy of human B cell lymphomas. *Blood* **69**: 584–591.

Press, O.W., Eary, J.F., Appelbaum, F.R., Martin, P.J., Nelp, W.P., Glenn, S., Fisher, D.R. *et al.* (1995) Phase II trial of ^{131}I-B1 (anti-CD20) antibody therapy with autologous stem cell transplantation for relapsed B cell lymphomas. *Lancet* **346**: 336–340.

Ragnhammar, P., Fagerberg, J., Froedin, J.E., Hjelm, A.L., Lindemalm, C., Magnusson, I., Masucci, G. and Mellstedt, H. (1993) Effect of monoclonal antibody 17-1A and GM-CSF in patients with advanced colorectal carcinoma – long-lasting, complete remissions can be induced. *Int. J. Cancer* **53**: 751–758.

Rankin, E.M., Hekman, A., Somers, R. and W.W., t. B.H. (1985) Treatment of two patients with B cell lymphoma with monoclonal anti-idiotype antibodies. *Blood* **65**: 1373–1381.

Reff, M.E., Carner, K., Chambers, K.S., Chinn, P.C., Leonard, J.E., Raab, R., Newman, R.A., Hanna, N. and Anderson, D.R. (1994) Depletion of B cells *in vivo* by chimeric mouse human monoclonal antibody to CD20. *Blood* **83**: 435.

Richards, J.M., Vogelzang, N.J. and Bluestone, J.A. (1990) Neurotoxicity after treatment with muromonab-CD3. *N. Engl. J. Med.* **323**: 487–489.

Riethmüller, G., Schneider, G.E., Schlimok, G., Schmiegel, W., Raab, R., Hoeffken, K., Gruber, R. *et al.* (1994) Randomised trial of monoclonal antibody for adjuvant therapy of resected Dukes' C colorectal carcinoma. German Cancer Aid 17-1A Study Group (see comments). *Lancet* **343**: 1177–1183.

Ritz, J., Pesando, J.M., Sallan, S.E., Clavell, L.A., Notis-McConarty, J., Rosenthal, P. and Schlossman, S.F. (1981) Serotherapy of acute lymphoblastic leukemia with monoclonal antibody. *Blood* **58**: 141–152.

Robins, R.A., Denton, G.W., Hardcastle, J.D., Austin, E.B., Baldwin, R.W. and Durrant, L.G. (1991) Antitumor immune response and interleukin 2 production induced in colorectal cancer patients by immunization with human monoclonal anti-idiotypic antibody. *Cancer Res.* **51**: 5425–5429.

Rodeck, U., Williams, N., Murthy, U. and Herlyn, M. (1990) Monoclonal antibody 425 inhibits growth stimulation of carcinoma cells by exogenous EGF and tumor-derived EGF/TGF-a. *J. Cell. Biochem.* **44**: 69–79.

Saavage, C.A., Mendelsohn, J.C., Leslie, J.F. and Trowbridge, I.S. (1987) Effects of monoclonal antibodies that block transferrin receptor function on the *in vivo* growth of a syngeneic murine leukemia. *Cancer Res.* **47**: 747–753.

Saleh, M.N., Khazaeli, M.B., Grizzle, W.E., Wheeler, R.H., Lawson, S., Liu, T., Russel, C. *et al.* (1993) A phase I clinical trial of murine monoclonal antibody D612 in patients with metastatic gastrointestinal cancer. *Cancer Res.* **53**: 4555–4562.

Saleh, M.N., Khazaeli, M.B., Wheeler, R.H., Allen, L., Tilden, A.B., Grizzle, W., Reisfeld, R.A. *et al.* (1992a) Phase I trial of the chimeric anti-GD2 monoclonal antibody ch14.18 in patients with malignant melanoma. *Hum. Antibodies Hybridomas* **3**:, 19–24.

Saleh, M.N., Khazaeli, M.B., Wheeler, R.H., Bucy, R.P., Liu, T., Everson, M.P., Munn, D.H., Schlom, J. and LoBuglio, A.F. (1995) Phase II trial of murine monoclonal antibody D612 combined with recombinant human monocyte colony-stimulating factor (rhM-CSF) in patients with metastatic gastrointestinal cancer. *Cancer Res.* **55**: 4339–4346.

Saleh, M.N., Khazaeli, M.B., Wheeler, R.H., Dropcho, E., Liu, T., Urist, M., Miller, D.M. *et al.* (1992b) Phase I trial of the murine monoclonal anti-GD2 antibody 14G2a in metastatic melanoma. *Cancer Res.* **52**: 4342–4347.

Saleh, M.N., LoBuglio, A.F., Wheeler, R.H., Rogers, K.J., Haynes, A., Lee, J.Y. and Khazaeli, M.B. (1990) A phase II trial of murine monoclonal antibody 17-1A and interferon-gamma: clinical and immunological data. *Cancer Immunol. Immunother.* **32**: 185–190.

Samonigg, H., Wilders-Trusching, M., Loibner, H., Plot, R., Rot, A., Kuss, I., Werner, G. *et al.* (1992) Immune response to tumor antigens in a patient with colorectal cancer after immunization with anti-idiotype antibody. *Clin. Immunol. Immunopathol.* **65**: 271–277.

Scheinberg, D.A., Lovett, D., Divgi, C.R., Graham, M.C., Berman, E., Pentlow, K., Feirt, N. *et al.* (1991) A phase I trial of monoclonal antibody M195 in acute myelogenous leukemia: specific bone marrow targeting and internalization of radionuclide. *J. Clin. Oncol.* **9**: 478–490.

Schlimok, G., Funke, I., Holzmann, B., Goettlinger, G., Schmidt, G., Haeuser, H., Swierkot, S. *et al.* (1987) Micrometastatic cancer cells in bone marrow: *in vitro* detection with anti-cytokeratin and *in vivo* labeling with anti-17-1A monoclonal antibodies. *Proc. Natl. Acad. Sci. USA* **84**: 8672–8676.

Schmiegel, W., Roeder, C., Schmielau, J., Rodeck, U. and Kalthoff, H. (1993) Tumor necrosis factor α induces the expression of transforming growth factor α and the epidermal growth factor receptor in human pancreatic cancer cells. *Proc. Natl. Acad. Sci. USA* **90**: 863–867.

Schmiegel, W., Schmielau, J., Henne-Bruns, D., Juhl, H., Roeder, C. Buggisch, P., Onur, A., Kremer, B., Kalthoff, H. and Jensen, E.V. (1997) Cytokine-mediated enhancement of epidermol growth factor receptor expression provides an immunological approach to the therapy of pancreatic cancer. *Proc. Natl. Acad. Sci. USA* **94**: 12622–12626.

Schöber, C., Schulze, M., Schlimok, G., Holz, E., Gruber, R., Riethmöller, G., Graubner, M. and Schmoll, H.-J. (1996) Open label pilot study of monoclonal antibody 17-1A in combination with 5-fluorouracil/levamisol or 5-FU/folinic acid in patients with advanced colorectal cancer. *Proc. Am. Soc. Clin. Oncol.* **15**: 229.

Schroff, R.W., Foon, K.A., Beatty, S.M., Oldham, R.K. and Morgan Jr., A.C. (1985) Human anti-murine immunoglobulin responses in patients receiving monoclonal antibody therapy. *Cancer Res.* **45**: 879–885.

Schulz, G., Büchler, M., Muhrer, K.H. *et al.* (1988) Immunotherapy of pancreatic cancer with monoclonal antibody BW 494. *Int. J. Cancer* **Suppl. 2**: 89–94.

Sears, H.F., Atkinson, B., Mattis, J., Ernst, C., Herlyn, D., Steplewski, Z., Haeyry, P. and Koprowski, H. (1982) Phase-I clinical trial of monoclonal antibody in treatment of gastrointestinal tumours. *Lancet* **1**: 762–765.

Sears, H.F., Herlyn, D., Steplewski, Z. and Koprowski, H. (1984) Effects of monoclonal antibody immunotherapy on patients with gastrointestinal adenocarcinoma. *J. Biol. Response. Modif.* **3**: 138–150.

Sears, H.F., Herlyn, D., Steplewski, Z. and Koprowski, H. (1985) Phase II clinical trial of a murine monoclonal antibody cytotoxic for gastrointestinal adenocarcinoma. *Cancer Res.* **45**: 5910–5913.

Shaw, D.R., Khazaeli, M.B. and Lobuglio, A.F. (1988) Mouse/human chimeric antibodies to a tumor-associated antigen: biologic activity of the four human IgG subclasses. *J. Natl. Cancer. Inst* **80**: 1553–1559.

Shawler, D.L., Bartholomew, R.M., Smith, L.M. and Dillman, R.O. (1985) Human immune response to multiple injections of murine monoclonal IgG. *J. Immunol.* **135**: 1530–1535.

Sindelar, W.F., Maher, M.M., Herlyn, D., Sears, H.F., Steplewski, Z. and Koprowski, H. (1986) Trial of therapy with monoclonal antibody 17-1A in pancreatic carcinoma: preliminary results. *Hybridoma* **5**: S125–132.

Somasundaram, R., Zaloudik, J., Jacob, L., Benden, A., Sperlagh, M., Hart, E., Marks, G. and Kane (1995) Induction of antigen-specific T and B cell immunity in colon carcinoma patients by anti-idiotypic antibody. *J. Immunol.* **155**: 3253–3261.

Strieter, R.M., Wiggins, R., Phan, S.H., Wharram, B.L., Showell, H.J., Remick, D.G., Chensue, S.W. and Kunkel, S.L. (1989) Monocyte chemotatic protein gene expression by cytokine-treated human fibroblasts and endothelial cells. *Biochem. Biophys. Res. Commun.* **162**: 694–700.

Taetle, R. and Honeyset, J.M. (1987) Effects of monoclonal antitransferrin antibodies on *in vitro* growth of human solid tumor cells. *Cancer Res.* **47**: 2040–2044.

Tempero, M.A., Pour, P.M., Uchida, E., Herlyn, D. and Steplewski, Z. (1986) Monoclonal antibody CO1717-1A and leukopheresis in immunotherapy of pancreatic cancer. *Hybridoma* **5**: S133–S138.

Tempero, M.A., Sivinski, C., Steplewski, Z., Harvey, E., Klassen, L. and Kay, H.D. (1990) Phase II trial of interferon gamma and monoclonal antibody 17-1A in pancreatic cancer: biologic and clinical effects. *J. Clin. Oncol.* **8**: 2019–2026.

Thistlethwaite Jr, J., Cosimi, A., Delmonico, F., Rubin, R., Talkoff-Rubin, N., Nelson, P., Fang, L. and Russell, P. (1984) Evolving use of OKT3 monoclonal antibody for treatment of renal allograft rejection. *Transplantation* **38**: 695–701.

Trang, J.M., LoBuglio, A.F., Wheeler, R.H. *et al.* (1990) Pharmacokinetics of a mouse/human chimeric monoclonal antibody (C-17-1A) in metastaic adenocarcinoma patients. *Pharm Res.* **7**: 587–592.

Trauth, B.C., Klas, C., Peters, A.M. J., Matzku, S., Möller, P., Falk, W., Debatin, K.-M. and Krammer, P.H. (1989) Monoclonal antibody-mediated tumor regression by induction of apoptosis. *Science* **245**: 301–305.

Urba, W.J., Ewel, C., Kopp, W., Smith II, J.W., Steis, R.G., Ashwell, J.D., Creekmore, S.P. *et al.* (1992) Anti-CD3 monoclonal antibody treatment of patients with CD3-negative tumors: a phase Ia/b study. *Cancer Res.* **52**: 2394–2401.

Urban, J.L. and Schreiber, H. (1992) Tumor antigens. *Ann. Rev. Immunol.* **10**: 617–644.

Vadhan-Raj, S., C., C.-C., Carswell, E., Mintzer, D., Dantis, L., Duteau, C., Templeton, M.A. *et al.* (1988) Phase I trial of a mouse monoclonal antibody against GD3 ganglioside in patients with melanoma: induction of inflammatory responses at tumor sites. *J. Clin. Oncol.* **6**: 1636–1648.

Verrill, H., Goldberg, M., Rosenbaum, R., Abbott, R., Simunovic, L., Steplewski, Z. and Koprowski, H. (1986) Clinical trial of Wistar Institute 17-1A monoclonal antibody in patients with advanced gastrointestinal adenocarcinoma: a preliminary report. *Hybridoma* **5**: S175–S183.

Vilcek, J. and Lee, T. H. (1991) Tumor necrosis factor. *J. Biol. Chem.* **26**: 7313–7316.

Vitetta, E.S. and Uhr, J.W. (1994) Monoclonal antibodies as agonists: an expanding role for their use in cancer Therapy. *Cancer Res.* **54**: 5301–5309.

Vlasveld, L.T., Hekman, A., Vyth, D.F., Melief, C.J., Sein, J.J., Voordouw, A.C., Dellemijn, T.A. and Rankin, E.M. (1995) Treatment of low-grade non-Hodgkin's lymphoma with continuous infusion of low-dose recombinant interleukin-2 in combination with the B cell-specific monoclonal antibody CLB-CD19. *Cancer Immunol. Immunother.* **40**: 37–47.

Waldmann, T.A., Goldman, C.K., Bongiovanni, K.F., Sharrow, S.O., Davey, M.P., Cease, K.B., Greenberg, S.J. and Longo, D.L. (1988) Therapy of patients with human T cell lymphotrophic virus I-induced adult T cell leukemia with anti-Tac, a monoclonal antibody to the receptor for interleukin-2. *Blood* **72**: 1805–1816.

Weiner, L.M., Harvey, E., Padavic-Shaller, K., Willson, J.K., Walsh, C., LaCreta, F., Khazaeli, M.B., Kirkwood, J.M. and Haller, D.G. (1993) Phase II multicenter evaluation of prolonged murine monoclonal antibody 17-1A therapy in pancreatic carcinoma. *J. Immunother.* **13**: 110–116.

Weiner, L.M., Moldofsky, P.J., Gatenby, R.A., O'Dwyer, J., O'Brien, J., Litwin, S. and Comis, R.L. (1988) Antibody delivery and effector cell activation in a phase II trial of recombinant γ-interferon and murine monoclonal antibody CO17-1A in advanced colorectal carcinoma. *Cancer Res.* **48**: 2568–2573.

Weiner, L.M., Steplewski, Z., Koprowski, H., Sears, H.F., Litwin, S. and Comis, R.L. (1986) Biologic effects of γ-interferon pre-treatment followed by monoclonal antibody 17-1A administration in patients with gastrointestinal carcinoma. *Hybridoma* **5 (Suppl. 1)**: S65–S77.

Welte, K., Miller, G., Chapman, P., Yuasa, H., Natoli, E., Kunicka, J., Cordon-Cardo, C. *et al.* (1987) Stimulation of T lymphocyte proliferation by monoclonal antibodies against GD3 ganglioside. *J. Immunol.* **139**: 1763–1771.

Ziegler, L.D., Palazzolo, P., Cunningham, J., Janus, M., Itoh, K., Hayakawa, K., Hellstrom, I. *et al.* (1992) Phase I trial of murine monoclonal antibody L6 in combination with subcutaneous interleukin-2 in patients with advanced carcinoma of the breast, colorectum, and lung. *J. Clin. Oncol.* **10**: 1470–1478.

Author	*mAb*	*Ig*	*Ag*	*Total Dose* [mg]	*Disease*	*R/T*	*PR*	*CR*	*HAMA*
Hematopoietic disease									
Miller *et al.*, 1981	anti-Leu-1	mIgG2a	CD5	7	T-ALL	0/1	—	—	1
Levy and Miller, 1983	anti-Leu-1	mIgG2a	CD5	1–92	T-ALL	0/8	—	—	nr
	12E7	mIgG2a	T and B cells						
	4H9	mIgG1	T cells						
Miller and Levy, 1981, Miller *et al.*, 1983	anti-Leu-1	mIgG2a	CD5	13–761	CTCL, DLCL	5/7	4	—	4
Ritz *et al.*, 1981	J-5	mIgG1	CD10	13.5–425	ALL	0/4	—	—	nr
Dyer *et al.*, 1989	CAMPATH-1G CAMPATH-1M	rIgG2b rIgM	CD52	50–875	various lymphoid malignancies	9/20	—	1	2/18
Waldmann *et al.*, 1988	anti-Tac	mIgG2a	IL-2-R (CD25)	100–490	T cell leukemia	3/9	1	1	1
Ball *et al.*, 1983	PMN 6, PMN 29, PM-81, AML-2-23	mIgM mIgG2b	glycolipids and proteins	170–>930	AML	0/3	—	—	1
Scheinberg *et al.*, 1991	M195	mIgG2a	CD33	6–76	AML, CMML	0/10	—	—	4/6
Dillman *et al.*, 1982	T101	mIgG2a	CD5	10–16	CLL	0/2	—	—	—
Dillman *et al.*, 1986	T101	mIgG2a	CD5	40–1,000	CTCL	4/13	—	—	5
				100–500	CLL	2/10	—	—	—
Foon *et al.*, 1984	T101	mIgG2a	CD5	6–400	CLL	0/10	—	—	—
				8–162	CTCL	4/4	—	—	4
Bertram *et al.*, 1986	T101	mIgG2a	CD5	10–>660	T cell malignancies (incl. CTCL), one B-CLL	4/13	1	1	3
Nadler *et al.*, 1980	AB89		B-cell lymphoma	2,000	B-NHL	0/1	—	—	nr

Author	*mAb*	*Ig*	*Ag*	*Total Dose* [mg]	*Disease*	*R/T*	*PR*	*CR*	*HAMA*
Press *et al.*, 1987	1F5	mIgG2a	CD20	52–2,380	NHL	2/4	1	—	1
Hekman *et al.*, 1991		mIgG2a	CD19	225–1,000	B-NHL	2/6	1	—	—
Hu *et al.*, 1989	LYM-1	mIgG2a	HLA-Dr	232.4–1,859.2	B-NHL	3/10	—	—	—
Greenaway *et al.*, 1994	WM-63, WM-66	mIgM	CD48, T/B-cell gp	6–91	B-CLL	0/7	—	—	—
Rankin *et al.*, 1985	anti-idiotypic	mIgG2a	B-cell Ig	3,800–5,800	B-NHL	2/2	—	—	—
Miller *et al.*, 1982, Meeker *et al.*, 1985 Brown *et al.*, 1989	anti-idiotypic	mIgG1, mIgG2a, mIgG2b	B-cell Ig	400–15,500	B-NHL	10/16	7	1	4
Maloney *et al.*, 1992	anti-idiotypic	mIgG1 mIgG2a mIgG2b rIgG1	B-cell Ig	1,680–5,757	B-NHL	10/13	3[1]	—	—
Janson *et al.*, 1989	F1	mIgG2a	TCR (T_υ α/β)	12	T-CLL	0/1	—	—	1
Klein *et al.*, 1991	B-E4/B-E8	mIgG	IL-6	400+400	multiple myeloma	0/1	—	—	1
Bataille *et al.*, 1995	B-E8/mab-8	mIgG	IL-6	40–1,510	multiple myeloma	1/10	—	—	1
Melanoma, Neuroblastoma									
Oldham *et al.*, 1984	9.2.27	mIgG2a	p240	361–860	melanoma	0/9	—	—	3
Goodman *et al.*, 1985	96.5	mIgG2a	p97	212 each	melanoma	0/5	—	—	4/4
	48.7	mIgG1	p240						
Irie and Morton 1986	L72	hIgM	GD2	2–14,66 (0.16–7.4 per nodule)	melanoma	6/8	3	1	—

[1] After first course; second course with additional chemotherapy.

Author	*mAb*	*Ig*	*Ag*	*Total Dose* [mg]	*Disease*	*R/T*	*PR*	*CR*	*HAMA*
Irie *et al.*, 1989	L55	hIgM	GD2	6.5–8.0	melanoma	1/2	—	1	—
Cheung *et al.*, 1987,	3F8	mIgG3	GD2	5–100/m^2 daily over	melanoma, neuroblastoma	2/17	2	—	17
Cheung *et al.*, 1992				2–4 days					
Cheung *et al.*, 1991,	3F8	mIgG3	GD2	50–200/m^2	neuroblastoma	2/16	1	—	6/10
Cheung *et al.*, 1994				50/m^2 per cycle					
Handgretinger *et al.*, 1992	14G2a	mIgG2a	GD2	100–400	melanoma	4/9[2]	2	2	9
Murray *et al.*, 1994	14G2a	mIgG2a	GD2	50–200	melanoma, neuroblastoma, osteosarkoma	4/18	2	—	16
Saleh *et al.*, 1992b	14G2a	mIgG2a	GD2	10–120	melanoma	2/12	1	—	12
Lichtin *et al.*, 1988	ME-36.1	mIgG2a	GD2, GD3	25–500	melanoma	1/13	—	1	13
Houghton *et al.*, 1985, Vadhan-Raj *et al.*, 1988	R24	mIgG3	GD3	12.8–880	melanoma	6/21	4	—	20
Dippold *et al.*, 1985, Dippold *et al.*, 1988	R24	mIgG3	GD3	200–440, 105/m^2	melanoma, one apudoma	3/7	—	—	1
Goodman *et al.*, 1987	MG-21	mIgG3	GD3	35–^{2}700/m^2	melanoma	0/8	—	—	nr
Gastrointestinal Cancer									
Sears *et al.*, 1982	17-1A	mIgG2a	EGP40	15–343	gastrointestinal cancer	1/4	—	—	3
Sears *et al.*, 1984	17-1A	mIgG2a	EGP40	15–1,000	colorectal, gastric, and pancreatic cancer	2/20	2	—	9/19
Sears *et al.*, 1985	17-1A	mIgG2a	EGP40	112–850	colorectal cancer	2/20	1	—	10

[2] 3 patients with adjuvant therapy were not evaluable for response.

Author	*mAb*	*Ig*	*Ag*	*Total Dose* [mg]	*Disease*	*R/T*	*PR*	*CR*	*HAMA*
Verrill *et al.*, 1986	17-1A	mIgG2a	EGP40	200	colorectal, gastric, and pancreatic cancer	0/20	—	—	5/8[3]
LoBuglio *et al.*, 1988	17-1A	mIgG2a	EGP40	400–1,600		1/25	—	1	21
Riethmüller *et al.*, 1994	17-1A	mIgG2a	EGP40	900	colorectal cancer, adjuvant	—/76	—	—	61
Sindelar *et al.*, 1986	17-1A	mIgG2a	EGP40	400 (1,200 in one case)	pancreatic cancer	4/25	4	—	23
Weiner *et al.*, 1993	17-1A	mIgG2a	EGP40	2,000–12,000	pancreatic cancer	1/28	1	—	—
Herlyn *et al.*, 1991	GA733,	mIgG2a	EGP40,	10–300	colorectal,	0/5	—	—	5
	CO19-9		ganglioside	10–600	pancreatic cancer colorectal cancer	0/6	—	—	3
Saleh *et al.*, 1993	D612	mIgG2a	glycoprotein	10–180/m^2	colorectal and gastric cancer	0/21	—	—	18
Paul *et al.*, 1986	17-1A	mIgG2a	EGP40	400	pancreatic cancer	0/8	—	—	nr
Frödin *et al.*, 1988	17-1A	mIgG2a	EGP40	600–1,000	colorectal cancer	3/10	—	1	10
Mellstedt *et al.*, 1991	17-1A	mIgG2a	EGP40	<2,000	colorectal cancer	0/10	—	—	nr
				+ cyclophosphamid		1/10	—	—	nr
				3,600–12,000		0/18	—	—	nr
Douillard *et al.*, 1986	17-1A	mIgG2a	EGP40	350–1,000	colorectal, gastric, biliary tract, and pancreatic cancer	6/74	5	1	44/60
	17-1A, 73-3, 19-9, 55-2	mIgG2a		150–200 each (2, 3, 4 mAb together)	colorectal, gastric, and pancreatic cancer	2/22	—	2	9/11

[3] Anti-idiotypic HAMA.

Author	*mAb*	*Ig*	*Ag*	*Total Dose* [mg]	*Disease*	*R/T*	*PR*	*CR*	*HAMA*
Tempero *et al.*, 1986	17-1A	mIgG2a	EGP40	400	pancreatic cancer	0/18	—	—	13/16
Büchler *et al.*, 1988	BW 494/32	mIgG1	carbohydrate	180–340	pancreatic cancer	0/18	—	—	8/8
Schulz *et al.*, 1988	BW 494/32	mIgG1	carbohydrate	140–490	pancreatic cancer	2/34	—	—	17/18
Büchler *et al.*, 1989	BW 494/32	mIgG1	carbohydrate	370	pancreatic cancer	0/35	—	—	—
Büchler *et al.*, 1991	BW 494/32	mIgG1	carbohydrate	370	pancreatic cancer, adjuvant	—/29	—	—	nr
Lung and others									
Mulshine *et al.*, 1988		mIg	Gastrin releasing peptide (GRP)	12–1,200 /m^2	lung cancer	0/8	—	—	2
Kelley *et al.*, 1993	2A11	mIg	GRP	3,000	lung cancer	1/12	—	1	—
Elias *et al.*, 1990	KS1/4	mIgG2a	40 kD glycoprotein	1,661 except 661 in one case	non-small cell lung cancer	0/6	—	—	5
Juhlin *et al.*, 1994	B7	mIgG2a	autologous VIPoma cells	100 in hepatic artery	malignant VIPoma	1/1	—	—	1
Goodman *et al.*, 1990	L6	mIgG2a	24 kDa surface protein of adenocarcinoma	35–2,800	breast, colon, ovarian, and lung cance	1/18 r	—	1	13
Dhingra *et al.*, 1995	L6	mIgG2a	24 kDa surface protein of adenocarcinoma	1,000–5,000/m^2 + 50/150 mg/m^2 deoxyspergualin	breast, ovarian, gastrointestinal, and lung cancer	0/26	—	—	2(13)/ 24[4]

[4] ELISA (RIA).

Author	*mAb*	*Ig*	*Ag*	*Total Dose* [mg]	*Disease*	*R/T*	*PR*	*CR*	*HAMA*
Chimeric mAb									
Handgretinger *et al.*, 1995	ch14.18	chIgG1	GD2	150–800	neuroblastoma	5/9	2	2	—
Saleh *et al.*, 1992a	ch14.18	chIgG1	GD2	5–100	melanoma	0/13	—	—	8
Goodman *et al.*, 1993	chL6	chIgG1	24 kDa surface protein of adenocarcinoma	35–1,400/m^2	breast, colon, and lung cancer	0/18	—	—	4
LoBuglio *et al.*, 1989	ch17-1A	chIgG1	EGP40	10–120	colorectal cancer	0/10	—	—	1
Baselga *et al.*, 1996	rhuMAb HER2	chIgG1	HER2	1,250[5]	breast cancer	7/46	4	1	—
Hamblin *et al.*, 1987	anti-idiotype	mFab′-hIgG	IgM	1,890	B-NHL	1/1	1	—	—
Maloney *et al.*, 1994	IDEC-C2B8	chIgG1	CD20	16–1,200/m^2	B-NHL	6/15	2	—	—
Knox *et al.*, 1991	Leu 3a	chIgG1	CD4	60–480	Mycosis fungoides	3/7	1	—	3
Caron *et al.*, 1994	HU-M195	huIgG1	CD33	3/m^2–216	AML, CML	1/13	—	—	—
Hale *et al.*, 1988	CAMPATH-1H	chIgG1	CD52	85–126	NHL	2/2	1	1	—
Anti-idiotypic									
Mittelman *et al.*, 1990	MF11-30	mIgG1	murine mAb 225.28	1.5–17.5	melanoma	3/16	—	—	16
Mittelman *et al.*, 1990	MF11-30	mIgG1	murine mAb 225.28	6–24	melanoma	3/21	—	1	16/17
Mittelman *et al.*, 1992, Ferrone *et al.*, 1993	MK2-23	mIgG1	murine mAb 763.74	6–24	melanoma	3/52	3	—	18
Foon *et al.*, 1995	3H1	mIgG1	murine mAb 8019	4–52	colorectal cancer	0/12	—	—	9

[5] Plus maintenance dose of 100 mg weekly for responders until progression.

Author	*mAb*	*Ig*	*Ag*	*Total Dose* [mg]	*Disease*	*R/T*	*PR*	*CR*	*HAMA*
Herlyn *et al.*, 1987	polyclonal goat anti-mouse	goat	murine mAb 17-1A	2–16/cycle	colorectal cancer	6/30	6	—	30
Samonigg *et al.*, 1992	polyclonal goat anti-mouse	goat	murine mAb 17-1A	32	colorectal cancer	0/1	—	—	1
Somasundaram *et al.*, 1995	polyclonal goat anti-mouse	goat	murine mAb GA733	2–16	colon cancer, adjuvant	—/13	—	—	7[6]
Robins *et al.*, 1991	105AD7	hIgG1	murine mAb 791T/36	100–400 i.m.	colorectal cancer	nr/6	nr	nr	—
Frödin *et al.*, 1991	17-1A	mIgG2a	EGP40	1,000–12,000	colorectal cancer	9/43[7]	nr	nr	41/43
Kwak *et al.*, 1992	idiotypic	autologous	B-cell Ig	3,5	B-NHL	2/9[8]	—	2	3
MAb and biologic response modifiers									
Vlasveld *et al.*, 1995	CLB-CD19	mIgG2a	CD19	1,200/m^2+IL-2	NHL	2/7	1	—	—
Ziegler *et al.*, 1992	L6	mIgG2a	adenocarcinoma	1,400/m2+IL-2	breast, lung, colorectal cancer	2/15	1	—	9
Bajorin *et al.*, 1990	R24	mIgG3	GD3	5–60/m^2+IL-2	melanoma	3/20	1	—	20
Minasian *et al.*, 1995	R24	mIgG3	GD3	5–250/m^2+M-CSF	melanoma	3/19	—	—	nr
Saleh *et al.*, 1995	D612	mIgG2a	glycoprotein	120/m^2+M-CSF	colorectal, gastric cancer	0/14	—	—	11
Ragnhammar *et al.*, 1993	17-1A	mIgG2a	EGP40	<2,000+GM-CSF	colorectal cancer	3/15	1	1	—
Weiner *et al.*, 1986	17-1A	mIgG2a	EGP40	400+IFN-γ	colorectal, pancreatic cancer	0/27	—	—	7/11

[6] Human anti-goat not against epitope.
[7] Response included CR, PR, MR and SD.
[8] Minimal residual disease in 2 patients, CR in the remaining patients before therapy.

Author	*mAb*	*Ig*	*Ag*	*Total Dose* [mg]	*Disease*	*R/T*	*PR*	*CR*	*HAMA*
Weiner *et al.*, 1988	17-1A	mIgG2a	EGP40	450+IFN-γ	colorectal cancer	0/19	—	—	—
Saleh *et al.*, 1990	17-1A	mIgG2a	EGP40	1,600+IFN-γ	colorectal cancer	0/15	—	—	13/14
Tempero *et al.*, 1990	17-1A	mIgG2a	EGP40	450+IFN-γ	pancreatic cancer	1/30	—	1	nr
Brown *et al.*, 1989	anti-idiotypic	m/rIgG1 m/rIgG2a	B-cell Ig	1,680–8,400 +IFN-α	B-NHL	11/12	7	2	2
Caulfield *et al.*, 1990	R24	mIgG3	GD3	$80/m^2$+IFN-α	melanoma	0/15	—	—	15
Minasian *et al.*, 1994	R24	mIgG3	GD3	$20/m^2$+TNF-α	melanoma	1/8	—	—	nr
Schmiegel *et al.*, 1996	425	mIgG2a	EGF-R	400–1600+TNF-α	pancreatic cancer	1/25	—	1	9
Richards *et al.*, 1990	muromonab-CD3		CD3	50–100 every 14 d		1/13	1	—	nr
Urba *et al.*, 1992	OKT3	mIgG2a	CD3	0.004–0.4	melanoma, colorectal, renal cell, ovarian pancreatic, non-small cell lung and other cancer	0/36	—	—	8/17

R/T=responders/total.
nr=not reported.

7. CLINICAL STUDIES IN ACUTE AND CHRONIC INFLAMMATION

JANICE C. WHERRY

Schering-Plough Research Institute, Clinical Immunology and Infectious Disease, 2015 Galloping Hill Road, Kenilworth NJ 07033, USA

INTRODUCTION

Monoclonal antibodies (mAbs) and their derivatives are useful in therapeutic applications and as probes of the pathogenesis of acute and chronic inflammatory disorders. Newer therapeutic approaches in treatment of acute and chronic inflammatory disorders stem from technological advances and from advances in understanding the pathophysiology of disorders such as the systemic inflammatory response syndrome (SIRS) and Rheumatoid arthritis (RA). Biotechnologic advances have resulted in newer therapies having a high degree of specificity including: mAb antibodies, recombinant cytokine antagonists, receptors fused to Fc pieces (fusion proteins) and other small peptides.

mAbs have been evaluated for treatment of a number of inflammatory disorders including sepsis/SIRS, RA, systemic lupus erythematosus (SLE), multiple sclerosis (MS), diabetes, Crohn's disease, acute transplant rejection, graft versus host disease (GVHD), vascular ischemia and other disorders. Targets of mAbs therapy include cell surface activation antigens (CD3, CD4, CD5, CD7, CD25, CD52) proinflammatory cytokines (IL-1, TNF-alpha [TNF], IL-6), adhesion molecules (ICAM-1, LFA-1, VLA-4, alpha 4-integrin, L-selectin and others), platelet aggregation (via GPIIb/IIIa), complement C5 activation and endotoxin activity. Although mAbs are specific for their target, other factors such as timing of mAb application relative to inflammatory insult, tissue penetration, host antibody response against the mAb, ability of the mAb to immunoregulate the receptor or cytokine of interest without inhibiting normal homeostatic immune functions and other factors play a key role in the mAb's success in improving acute or chronic inflammatory diseases. Newer strategies include construction of chimeric antibodies (containing murine variable regions and human constant regions, may be less immunogenic), humanization of murine antibodies (only the antigen binding site is of murine origin, therefore less risk of cross-species sensitization), preparation of non-cell depleting mAbs (affords less toxicity due to preservation of the majority of circulating T lymphocytes), alterations in Fc isotype (human g4 and murine g1 are less likely targets of complement activation), construction of primatized mAbs (genetically engineered molecules from monkey and human antibody components), construction of mAb–toxin fusion proteins (potent toxicity is specifically directed to the target of interest), antibody fragments (may be less immunogenic) and combination mAb therapy (synergistic efficacy).

The goal of acute mAb therapy in sepsis, septic shock and SIRS is a one time or early application of a specific mAb. Important issues in the strategy of using mAbs for acute therapy is the speed with which the appropriate subpopulation of patients can be identified, the speed of application of the mAb and the dose/route of application of the mAb. Since patients are not chronically treated with repeat mAb doses, formation of host anti-mAb antibodies is less of an issue. Results of several large multicenter trials of mAbs have been relatively disappointing possibly due to the futility of treatment with mAbs once the cytokine cascade has been activated, the overlapping functions of individual cytokines and the need for the mAb to inhibit abnormal cytokine overactivity without adverse effects on normal immunity.

The overall goal in mAb therapy for chronic diseases is long term disease modification, not short-term improvements in disease activity. Repeat mAb dosing is required which may be difficult if host anti-mAb antibodies are formed. Even with chimeric or humanized mAbs where smaller proportions of the mAb are of foreign origin, anti-isotopic antibodies develop in the host and can limit circulating half-life, interfere with antibody-target interactions or produce allergic reactions.

This chapter includes mAbs in clinical use or in late preclinical testing that are, have been or may be considered for the treatment of chronic inflammatory disorders such as RA and acute inflammatory disorders such as sepsis/SIRS. The first section of this chapter is a brief overview of each of the mAbs of interest and the latter section is divided into separate inflammatory disorders with more detailed descriptions of mAb therapies.

SPECIFIC MONOCLONAL ANTIBODIES

Anti-CD3

CD3 is a 17–29 kD molecule that is part of a complex found on mature T cells and medullary thymocytes. Antibodies reacting with CD3 (muromonal-CD3, Orthoclone OKT3) directed against the epsilon chain of the CD3 molecule block T cell receptor function resulting in potent immunosuppression and a massive cytokine release syndrome (Chatenoud *et al.*, 1983; Norman, 1995; Sgro, 1995; Parlevliet *et al.*, 1995). The immunosuppression is characterized by a profound decrease of peripheral T cells including CD2, CD4 and CD8+ subsets followed by increasing proportions of CD3 negative lymphocytes (Chatenoud *et al.*, 1992). The cytokine release syndrome is characterized by a dose dependent release of TNF, TNF soluble receptors and IL-1 (IL-1 beta) and activation of complement and neutrophils (Chatenoud *et al.*, 1993; Herbelin *et al.*, 1995). Later side effects such as an increased incidence of infections and malignancies may be due to long lasting immunosuppression. Since OKT3 is a murine mAb, treatment results in the formation of anti-murine antibodies (Carey *et al.*, 1995).

Anti-CD3 therapy has been extensively used as an immunosuppressive agent in transplantation (reviewed in Parlevliet and Schellekens, 1992; and in Sgro, 1995). A complete review of clinical studies with anti-CD3 is beyond the scope of this chapter. Recently a chimeric anti-human IgM CD3 single chain antibody (scUCHT1) consisting of the light and heavy variable chain binding domains of anti-human CD3 plus a human IgM Fc region has been developed for clinical use (Ma *et al.*, 1996). This construct retains the specificity of the parent antibody but does not induce cytokine release.

Anti-CD4

CD4+ T cell have been implicated as pivotal in the pathogenesis of RA and other inflammatory disorders (Horneff *et al.*, 1993). The CD4 antigen is the class II MHC receptor and acts as a coreceptor for the T cell antigen receptor complex. The binding of anti-CD4 to the CD4 receptor results in downregulation of T cell activity. Anti-CD4 mAbs have been the prototype mAb studied in RA and in a number of other chronic inflammatory disorders including MS, SLE, psoriasis and diabetes.

The majority of patients treated with anti-CD4 develop an antibody response to anti-CD4, which may limit repeated dose therapy (Kalden and Manger, 1995). Studies of anti-CD4 therapy have in common a profound and long-lasting dose-dependent peripheral lymphopenia (Moreland *et al.*, 1993). The lymphopenia partially recovers with time except for monocytes and CD4+ T cells (Horneff *et al.*, 1993). Peripheral CD4 T cell depletion has been documented at 30 months post therapy (Moreland *et al.*, 1994) and begins to trend towards normal by 5 years after therapy (Moreland *et al.*, 1996). Mitogen and antigen proliferative responses are decreased commensurate with peripheral lymphopenia. Despite the profound long-lasting peripheral lymphopenia, the incidence of serious infectious complications has been relatively rare (Moreland *et al.*, 1995; 1996).

The mechanism of anti-CD4 action is unclear since some studies show that leukopenia does not correlate with magnitude of induced peripheral lymphopenia, with changes in serum cytokine levels, with efficacy (Horneff *et al.*, 1993; 1993a) or with levels of cytokines in inflamed peripheral tissue (Tak *et al.*, 1995).

Small uncontrolled trials with anti-CD4 showed initial promising results in RA, MS, SLE, diabetes and in psoriasis. However, large, randomized, placebo-controlled clinical trials of a murine (B-F5) and a chimeric mAb anti-CD4 (cM-T412) in patients with RA failed to show effectiveness of anti-CD4. A newer strategy involves the use of a non-cell depleting anti-CD4 mAb which binds to CD4 and temporarily reduces excess T cell activation without seriously inhibiting other immune functions. Clinical studies are underway with a non-cell depleting anti-CD4 mAb (OKT4A) for prophylaxis against transplant rejection. A primatized non-cell depleting anti-CD4 mAb (IDEC-CD9.1/SB) has shown promising results in RA patients and a large, randomized, placebo-controlled multicenter trial is in progress. A humanized non-cell depleting anti-CD4 mAb (4162W94) has also shown promising results in RA patients.

Anti-CD5

CD5 is found on all T cells and on a subset of B cells. Features common to clinical studies with anti-CD5 are the profound CD5+ T and B cell depletion with reversible decrease in number of peripheral blood cells, decrease in CD2, CD4, and CD8 lymphocytes and decrease in peripheral cell proliferative responses to antigenic, allogeneic and mitogenic stimuli (Fishwild and Strand, 1994; Cannon *et al.*, 1995). There is no effect upon monocytes or upon spontaneous and pokeweed mitogen-induced immunoglobulin secretion. The effects are reversible within one month as determined by number, phenotype, cytokine and immunoglobulin production and immunoglobin levels.

Animal studies of anti-CD5 treatment demonstrate improvement in RA. Small clinical studies have been performed with immunoconjugates of anti-CD5 (CD5 Plus: mAb anti-CD4+ Ricin A chain) in RA and in GVHD showing favorable results. However, a large randomized trial of the immunoconjugate in RA failed to show efficacy and studies in GVHD patients demonstrates short term (4 week) but not long term (1 year) survival advantage. In addition treatment with immunoconjugate anti-CD5 results in significant toxicities and serious viral infections, especially in GVHD patients.

Anti-IL-2 receptor

The IL-2 receptor (CD25) is expressed on activated T cells. IL-2 receptor bearing cells and soluble-IL-2 receptors are elevated in patients at risk for transplant rejection (Deng *et al.*, 1995; Chang *et al.*, 1996). Anti-IL-2 receptor therapy has been evaluated as prophylaxis in cardiac, hepatic and renal transplant patients and in the prophylaxis and treatment of GVHD. Results are encouraging, particularly with prophylaxis against transplant rejection, where anti-IL-2 therapy was found to be at least as effective as anti-thymocyte globulin (ATG) and anti-CD3 therapy.

Anti-TNF

TNF alpha (TNF) is a 17-kDa protein produced primarily by cells of the macrophage/monocyte lineage and binds to the p55 and p75 receptors found on many types of cells. Activation of the TNF receptor results in cellular activation, expression of adhesion molecules, activation of phagocytosis, degranulation and generation of oxygen free radicals and prostaglandin E2, and proinflammatory cytokine production and release (IL-1, IL-6 and IL-8) (Tracey and Cerami, 1993; Bone, 1991; Nathan and Spron, 1991; Elliott and Maini, 1995).

TNF has been implicated as a primary proinflammatory mediator involved in the pathogenesis of a variety of acute and chronic inflammatory disorders (Strieter *et al.*, 1993; Elliott and Maini, 1994). Injection of TNF accelerates RA and simulates the signs and symptoms of sepsis/SIRS. Treatment of animals with anti-TNF results in prevention and/or improvement of RA, of endotoxin or bacterial or noninfectious challenges in sepsis/SIRS models, of reperfusion injury of transplant rejection and of MS, although several animal studies demonstrate that localized inflammatory challenges may not always be improved by anti-TNF. Elevated serum levels of TNF correlate with disease severity in RA, sepsis, allograft rejection, myocardial infarction, liver and renal transplant rejection, complications of bone marrow transplants (BMT), intestinal ischemia, cardiac surgery and MS. TNF is found in high concentrations in inflamed local tissue in a variety of inflammatory disorders such as acute lung injury, Crohn's disease and is found in high concentrations in infected tissue such as in pneumonia or peritonitis.

Clinical trials have been performed with murine (TNF mAb, CB0006), humanized (CDP571), chimeric (cA2) anti-TNF; anti-TNF antibody fragment (MAK 195F) and recombinant human soluble TNF receptor:Fc constructs (fusion proteins). Studies of murine anti-TNF demonstrate improvement in septic shock, RA and malaria. Studies of chimeric anti-TNF (cA2) and humanized anti-TNF (CDP571) demonstrate dramatic improvements in specific markers of RA disease severity. TNF receptors may be less immunogenic than antibodies or antibody fragments, however, the half-life of the fragments is exceedingly short. Therefore, constructs of TNF receptors fused to Fc pieces have been created to prolong the half life of the receptors. Studies of several different soluble TNF receptor:Fc fusion proteins (p75, p55, or p80) have been performed in RA, in volunteers given endotoxin and in MS. Studies of the anti-TNF antibody fragment have been performed in patients with severe sepsis and GVHD. Large clinical trials are ongoing with many of these anti-TNF moieties.

Anti-adhesion mabs

High levels of adhesion molecules such as ICAM-1 are found in inflamed tissue such as rheumatoid joints, and have been implicated in the pathogenesis of a wide range of acute and chronic inflammatory disorders. Anti-adhesion mAbs have been tested against ICAM-1, LFA-1, alpha 4-integrin, L-selectin and others in acute and chronic inflammatory disorders. Animal models include treatment of RA, MS, SLE, and prophylaxis against graft rejection, ischemia reperfusion, acute colitis and Type I diabetes. Early clinical trials of anti-adhesion mAbs in RA patients show encouraging results.

Anti-IL-6

IL-6 is a 21-kDa glycoprotein produced by lymphocytes, fibroblasts and monocytes and has both inflammatory and anti-inflammatory properties. IL-6 can induce hepatic acute phase protein production (Borden and Chin, 1994), activate the coagulation pathway (van der Poll *et al.*, 1994a), stimulate a pyrogenic response (Dinarello,1989) and inhibit *in vitro* production of proinflammatory cytokines such as TNF and IL-1 (Schindler *et al.*, 1990). IL-6 levels correlate closely with severity and outcome in sepsis/SIRS and with size of brain lesions in stroke patients, thus serving as a surrogate marker of inflammation. Few clinical trials have been performed with anti-IL-6, however a pilot clinical trial of anti-IL-6 in RA demonstrated transient improvement.

Anti-endotoxin

Endotoxin (lipopolysaccharide, LPS) is a part of the cell wall of gram negative bacteria. Endotoxin activates the cascade of proinflammatory mediators in gram negative sepsis leading to fever, hemodynamic instability, end organ failure and death). Conflicting results have been obtained in animal studies of the effects of anti-endotoxin treatment in gram negative sepsis with some studies showing improvement in sepsis with anti-endotoxin treatment and other studies showing anti-endotoxin increases mortality. Results of three large randomized placebo controlled trials of anti-endotoxin mAbs (E5 or HA-1A) in gram negative sepsis have been disappointing with overall no decrease in mortality seen and decreases seen only in small subsets of patients. A large trial of anti-endotoxin (E5) in patients with documented gram negative sepsis is ongoing.

Anti-GPIIb/IIIa

Anti-platelet therapy is effective in preventing acute thromboembolic artery occlusion by inhibiting platelet activation. An anti-platelet Fab mouse/human chimeric mAb (anti-GPIIb/IIIa) has been evaluated in clinical situations predisposing to a high risk of vascular ischemia such as high risk angioplasty, unstable angina and acute myocardial infarction and promising results have been obtained in large clinical trials.

Anti-CD52: CD52 is an activation antigen found on T cells. mAb humanized anti-CD52 therapy (CAMPATH-1H) has been evaluated in RA patients. Results of small uncontrolled studies of anti-CD52 in RA patients demonstrated some efficacy but reports of recurrent synovitis and acute toxicities limit further clinical development.

Similar to anti-CD4 and anti-CD5, anti-CD52 therapy results in profound depletion of peripheral blood mononuclear cells (nearly complete depletion of lymphocytes) followed by gradual repopulation (Burmester and Emmrich, 1993; Matteson *et al.*, 1995). Repopulation of natural killer cells and monocytes return to baseline within 1–2 months. There is a 50% return of B cell within $2^1/_2$ months and complete resolution within 6 months. Profound effects are seen with T cells: CD8 cells recover to 50% by 2 months and CD4 cells return to 20% by 2 months but no further improvements have been noted even up to 3 years (Weinblatt *et al.*, 1996). There has been no correlation between changes in peripheral lymphocyte counts and clinical response.

Acute toxicities may be related to cytokine release and include fever, rigors, nausea and in higher doses, hemodynamic instability, bronchospasm and infections including one fatal opportunistic infection (coccidioidomycosis) following additional immunosuppressive treatment (Isaacs *et al.*, 1996; Watts *et al.*, 1993). At lower doses with a more acceptable safety profile, efficacy is short lived and thus anti-CD52 therapy may be of limited value.

Anti-C5: Anti-C5 blocks the formation of chemotactic and proinflammatory moieties C5a and C5b-9. In animal models, anti-C5 prevents progression of collagen-induced arthritis, onset of SLE and onset of cardiac reperfusion injury.

Anti-CD23: Anti-CD23 binds to CD23, a low-affinity receptor for IgE1 seen in elevated concentrations in RA patients (Plater-Zyberk *et al.*, 1995). Anti-CD23 treatment of mice results in improvement of collagen induced arthritis.

Anti-CD7: CD7 is a pan-T cell antigen appearing on most peripheral blood lymphocytes who have helper, suppressor and cytotoxic activities (Lazarovits *et al.*, 1992). Clinical studies of murine or chimeric mAb anti-CD7 show transient but not lasting efficacy (Kirkham *et al.*, 1991; Kirkham *et al.*, 1992) and infusion related toxicities limit therapy.

Anti-gamma interferon: Anti-gamma interferon or soluble interferon gamma receptors increase survival in a mouse model of SLE.

Combination therapy

Combinations of mAbs have been evaluated in animal and human studies of acute and chronic inflammatory diseases. Anti-CD4 therapy combined with anti-TNF or anti-TNF IgG protein therapy has shown promising results in animal studies of RA. Combined anti-LFA-1 and anti-ICAM-1 treatment prevents graft rejection and diabetes in rat pancreas transplants. Combined anti-CD-2, anti-TNF and anti-IL-2 has been evaluated in treatment of acute GVHD.

RHEUMATOID ARTHRITIS

Rheumatoid arthritis (RA) is a serious debilitating autoimmune disease with an annual incidence of 7 million cases in the US. RA is characterized by chronic inflammation of peripheral synovial joints and systemic symptomatology thought secondary to a T cell autoimmune reaction. Abnormalities of immunity include systemic circulating autoantibodies and proinflammatory cytokines, local accumulation in the synovium of lymphocytes and proinflammatory cytokines (TNF IL-1, IL-6 and IL-8) aided by adhesion molecules such as ICAM-1 resulting in chronic inflammation and connective tissue destruction (Duff, 1994; Ishikawa *et al.*, 1994; Farahat *et al.*, 1993; Arend and Dayer, 1995). IL-1 and TNF stimulate cartilage and bone destroying enzymes and are important in stimulating the acute phase responses by inducing IL-6 to activate hepatic synthesis of acute phase proteins.

Some of the most promising advances in therapy of RA have been with mAbs (rev in Breedveld and van der Lubbe, 1995 and Kalden, 1994). Effectiveness of such therapies have been hampered by toxicity of the mAbs and the chronicity of the inflammation requiring not only acute therapy for "flares" but long term rational and safe disease modifying agents. Accumulation of lymphocytes and proinflammatory cytokines in joints is more difficult to treat than peripheral activated T cells or circulating proinflammatory cytokines. New therapeutic approaches stem from advances in understanding the pathophysiology of the disease (cellular activation, cytokine activity, intracellular signaling, combination therapy), and from technical advances (chimerism, humanization, mAb toxin fusion proteins, soluble receptors, mAb fragments). Encouraging results have been obtained with a non-cell depleting anti-CD4 and from chimeric/humanized or fusion protein anti-TNF. Other mAbs under evaluation include inhibitors of other proinflammatory cytokines (IL-1, IL-6), T cell activation antigens (CD5, CD7, CD52 [CAMPATH-1H]), adhesion molecules and IgE (CD23).

Anti-CD4

T cells, especially CD4+ T cells, initiate and promote inflammation in RA. Improvement in RA has been seen subsequent to T cell depletion by surgical drainage of the thoracic duct (Paulus *et al.*, 1977), by total lymphoid irradiation (Kotzin *et al.*, 1981; Trentham *et al.*, 1981) and by lymphapheresis (Karsh *et al.*, 1981). Evidence of CD4+ T cell involvement in RA include animal studies showing anti-CD4 prevents the onset and progression of RA in a murine collagen-induced arthritis model (Ranges *et al.*, 1985) and histologic studies showing that the RA synovium is infiltrated by predominantly CD4+ T cells (Duke *et al.*, 1982). Injection of anti-CD4 results in specific accumulation in organs with high concentrations of CD4+ T cells (Kinne *et al.*, 1993). Thus, clinical studies of specific immunotherapy for RA have been directed towards modulation of CD4 T cell activity. Putative monoclonal antibody T cell targets include CD4, CD5, CD7, Cdw52 and IL-2R. The prototype mAb, anti-CD4, has undergone extensive clinical testing and results of a multicenter trial using a non-cell depleting anti-CD4 are promising (Levy *et al.*, 1996). Injection of anti-CD4 results in profound and long-lasting depletion of peripheral CD4 cells, however T cell infiltrates and elevated cytokine levels remain in the synovium (Tak *et al.*, 1995; Reece *et al.*, 1996).

Initial small, open-label trials with anti-CD4 therapy demonstrated clinical disease improvement in refractory RA patients (Wendling *et al.*, 1992; Moreland *et al.*, 1993; 1993a; van der Lubbe *et al.*, 1993; van der Lubbe, 1994). In one study, 25 patients with refractory RA treated with low dose methotrexate (<15mg/week) were treated with doses of 10–700 mg anti-CD4 and followed for 6 months (Moreland *et al.*, 1993; 1993a). Adverse events included an immediate profound decrease in CD4+ cells which trended towards normal but remained abnormal in many patients even at the end of the study. Nineteen (19) patients experienced self-limited fever, myalgias, malaise and mild hypotension. Clinical improvement was seen in 43% of patients at 5 weeks and 33% at 6 months (>50% decrease in tender joint counts). Mitogen and antigen proliferative responses were decreased commensurate with peripheral lymphopenia. Twenty-three (23) patients were followed up at 18 and at 30 months post therapy (Moreland *et al.*, 1994). Profound CD4 T cell depletion was still present, particularly in the higher dose groups (300, 700). One death was reported in a 100 mg patient who also received methotrexate and prednisone (18 months post anti-CD4 therapy).

In another open label trial, 32 patients with active RA received 10, 50 or 100 mg anti-CD4 daily for 7 days (van der Lubbe *et al.*, 1993). A significant reduction in clinical markers of disease severity were seen for patients in the higher dose groups (50 and 100 mg). These effects did not correlate with magnitude of induced peripheral lymphocytopenia nor with changes in serum cytokine levels. Clinical improvement was seen at all doses and clinical effect was seen in the presence of unbound circulating mAb. The majority of patients developed an antibody response to anti-CD4.

Although results of the open label trials with anti-CD4 were promising, these results have not been confirmed in large randomized placebo-controlled trials. Three trials have been performed using murine anti-CD4 (Wendling *et al.*, 1996) or chimeric anti-CD4 (Moreland *et al.*, 1995; van der Lubbe *et al.*, 1995) in patients with active RA with or without methotrexate.The study of murine anti-CD4 was performed with 58 active RA patients randomized to receive anti-CD4 (B-F5, 20 mg/day) or placebo for 10 consecutive days (Wendling *et al.*, 1996). Mild flu-like symptoms occurred after anti-CD4 treatment. No significant clinical improvement was seen at one month.

Two multicenter, double-blind, randomized placebo-controlled trials were performed using the chimeric mAb anti-CD4 (cM-T412) but neither demonstrated clinical improvement with anti-CD4 therapy (Moreland *et al.*, 1995; van der Lubbe *et al.*, 1995). In the first trial, patients with refractory RA on low-dose methotrexate therapy were evaluated (Moreland *et al.*, 1995). In this study, 64 patients were randomized to receive anti-CD4 at doses of 5, 10 or 50 mg or to receive placebo given monthly for 3 consecutive months. Adverse events included "flu-like" symptoms (fever, malaise) early post-infusion especially in the 50 mg (29%) and 10 mg (31%) doses. CD4+ T cell depletion was profound (more than a 50% drop in T cells in the 50 mg dose seen even at 6 months). There was no significant difference between treatment groups in clinical response seen at 3 months. Thus, treatment of RA patients with concomitant methotrexate and anti-CD4 did not result in clinical efficacy. In the other study 60 patients were treated with 5 consecutive days of therapy and evaluated over a 6 week interval (van der Lubbe *et al.*, 1995). Thirty (30) patients were selected to participate in a second randomized, double-blind, placebo-controlled study of single-dose anti-CD4 therapy and followed for 9 months. Neither part of the study demonstrated any change in clinical disease activity.

All of these studies have in common profound, long-lasting depletion of CD4 cells (Moreland *et al.*, 1994, Kalden and Manger, 1995), however serious infections rarely occur (Moreland *et al.*, 1995; 1996). Peripheral CD4 T cell depletion has been documented at 30 months post therapy (Moreland *et al.*, 1994) and begins to trend towards normal by 5 years after therapy (Moreland *et al.*, 1996). Lack of efficacy may be due to relative lack of effect of anti-CD4 at the target site resulting in incomplete inhibition of synovial cytokine production (Tak, 1995). Synovial tissue biopsies of patients with active RA who received anti-CD4 revealed significant improvement in synovial inflammatory infiltrates, however, proinflammatory cytokines (IL-1B, TNF) were still found in the synovial tissue and may explain the relative lack of efficacy of anti-CD4.

Newer strategies under evaluation include use of a primatized (genetically engineered monkey/human mAb), non-depleting, anti-CD4 mAb (IDEC-CE9.1/SB 210396) which is being evaluated in clinical trials. The advantage of the non-cell depleting mAb is in the ability to temporarily inhibit excess T cell activation without major inhibition of other immune functions. One hundred twenty-two (122) patients with active RA were enrolled and randomized to receive 40, 80 or 140 mg of the non-cell depleting mAb or placebo twice a week for four weeks (Levy *et al.*, 1996). The mAb was well tolerated, however, treatment was stopped in 3 of 16 patients in the highest-dose group due to rashes. Clinical improvement in RA symptoms (>20% improvement in swollen joints) was seen in 77% of patients in the high dose group, 47% of patients in the 80 mg group and 42% patients in the 40 mg group and 17% in the placebo group. Median time to clinical response was 1 week (140 mg), 2 weeks (80 mg) and $2^1/_2$ weeks (40 mg). A phase III study has been recently initiated using the primatized non-depleting anti-CD4 mAb in RA where patients will receive either 2 times a week doses or weekly doses (or placebo) for one month followed by periodic single doses.

A pilot study of a humanized non-depleting anti-CD4 mAb (4162W94) was evaluated in 24 patients with active RA. Patients were treated with 10 to 300 mg daily for 5 days and disease activity measured (Panayi *et al.*, 1996). The antibody was well tolerated. Clinical responses were seen with the higher doses. Most striking, however, were significant reductions in acute phase reactants. A study of humanized non-depleting anti-CD4 (SB-210396) intraarticular injection was performed (Reece *et al.*, 1996). Twelve (12) patients received the injection which was well tolerated and resulted in improvement in measures of clinical disease activity. Immunohistological evaluations are pending.

Anti-TNF

TNF has been identified as a pivotal proinflammatory cytokine in RA (rev in Elliott *et al.*, 1995). Animal studies demonstrate the importance of TNF in the pathogenesis of RA (Cooper *et al.*, 1992; Elliott *et al.*, 1995; Wooley *et al.*, 1993). Clinical studies of mAb anti-TNF have been encouraging, with the therapy well tolerated and dramatic improvements noted in specific markers of RA severity (Elliott *et al.*, 1994a; Maini *et al.*, 1995; Rankin *et al.*, 1995), however, the long term therapy may be limited by host antibody formation. Clinical studies with anti-TNF include an open-label study of 20 patients (chimeric cA2; Elliott *et al.*,1993), a double blind study of 73 patients (chimeric cA2; Elliott *et al.*, 1994a; Maini *et al.*, 1995) a dose ranging study of patients treated with methotrexate and anti-TNF (chimeric cA2) and a double blind study of 36 patients (humanized CDP571; Rankin *et al.*, 1995). Several TNF receptor-Fc constructs (fusion proteins) have been evaluated in clinical trials with promising results with p55 (Hasler *et al.*, 1996; 1996a; Rau *et al.*, 1996) and p75/p80 (Baumgartner *et al.*, 1996, Moreland *et al.*, 1994a) TNF receptors. A number of clinical trials with either murine, chimeric, humanized or TNF-Fc fusion protein versions of anti-TNF therapy are in progress.

Animal and *in vitro* studies demonstrate the role of TNF in RA. TNF promotes cartilage and bone destruction in RA. The mechanism of action appears to be through stimulation of fibroblast and synovial cell metalloproteinase production and suppression of synthesis of matrix components (Elliott and Maini, 1995). Profound improvement in joint swelling and joint erosion has been demonstrated with mAb anti-TNF therapy in a collagen-induced arthritic mouse model where intra-articular injection of TNF resulted in accelerated disease (Cooper *et al.*, 1992). Systemic administration of anti-TNF leads to both amelioration of established experimental joint disease and prevention of the onset of RA (Elliott and Maini, 1995). Administration of recombinant human TNF receptor:Fc fusion protein in collagen-induced arthritis in mice results in reduced incidence and severity of either established arthritis or onset of new arthritis (Wooley *et al.*, 1993). Mice transgenic for the human TNF gene develop inflammatory arthritis from 4 weeks of age (Keffer *et al.*, 1991).

In vitro analysis of patient samples demonstrate the importance of TNF in the pathogenesis of RA. Elevated serum levels of TNF have been detected in RA patients and correlates with erythrocyte sedimentation rate (ESR), disease severity and degree of anemia (Vreugdenhil *et al.*, 1992). Analysis of patient RA synovial fluid reveals high concentrations of TNF and soluble TNF receptors p55 and p75 and the levels correlate with clinical disease activity (Cope *et al.*, 1992; Saxne *et al.*, 1988). TNF production has been detected in synovial tissue macrophage-like cells (Husby and Williams, 1988; Chu *et al.*, 1991; Cope *et al.*, 1992). Immunohistological staining reveals production of TNF within the RA synovial lining layer, sublining interstitial cells and at the cartilage/pannus junction (Husby and Williams, 1988; Chu *et al.*, 1991). Studies of cultured RA synovium-derived mononuclear cells reveal that treatment with antibodies to TNF results in significant inhibition of pro-inflammatory cytokines (IL-1, GM-CSF, IL-6, IL-8) (Maini *et al.*, 1995).

Clinical studies of anti-TNF therapy (murine, chimeric, humanized, fusion proteins, antibody fragments) are encouraging, with dramatic improvements noted in markers of clinical disease severity.

The initial open label study of chimeric anti-TNF (cA2) was performed in 20 patients with active RA (Elliott *et al.*, 1993). Patients received 20 mg/kg of chimeric anti-TNF and were followed for 8 weeks. Significant improvement was seen in clinical markers of disease activity. Seven (7) of these patients were subsequently enrolled in an open-label trial where patients received 2–4 cycles of anti-TNF where the timing of treatment cycles was determined by disease relapse (Elliott *et al.*, 1994; Maini *et al.*, 1995). Patients were follwed for up to 1 year. Several patients prematurely discontinued (sinusitis, urticaria and development of anti-dsDNA and anti-cardiolipin antibodies). All patient showed improvements in swollen-joint count and serum C-reactive protein greater than 80%. Anti-TNF treatment resulted in restoring peripheral blood mononuclear cell proliferation towards normal (mitogen and recall antigens).

A randomized placebo-controlled double-blind trial of 73 patients treated with one of two doses of chimeric anti-TNF (cA2 at 1 or 10 mg/kg) or placebo resulted in dramatic results (Elliott *et al.*, 1994a; Maini *et al.*, 1995): An intent-to-treat analysis of patients at 4 weeks revealed only 2 of 24 placebo patients (8%) responded in contrast to 11 of 25 patients (44%) treated with 1 mg/kg ($p=0.0083$)

and 19 of 24 patients (79%) treated with 10 mg/kg ($p<0.0001$). With the high dose therapy maximal mean improvements of greater than 60% were seen for tender or swollen joint counts and in serum C-reactive protein. Serum E-selectin and ICAM-1 levels were decreased in patients receiving anti-TNF commensurate with a rise in peripheral lymphocytes (Paleolog *et al.*, 1996). This effect was seen most markedly in patients with ⩾20% improvement in Paulus criteria by week 4. Recent results with chimeric cA2 plus methotrexate in a dose ranging study of 28 RA patients was reported at the American College of Rheumatology (Kavanaugh *et al.*, 1996). Twenty eight (28) patients received anti-TNF in doses of 5–20 mg/kg or placebo along with methotrexate for 12 weeks. Dramatic responses were seen with 81% responders among the anti-TNF groups versus 14% responders in the placebo group.

A randomized, placebo-controlled, double-blind trial of 36 patients with active RA was performed where patients were randomized to receive humanized anti-TNF (CDP571, 1 or 10 mg/kg) or placebo (Rankin *et al.*, 1995). At doses of 10 mg/kg, CDP571 treated patients experienced significant improvements in markers of clinical disease activity at weeks 1 and 2. Minimal adverse events were reported.

To evaluate the mechanism of action of anti-TNF in the synovium, 14 patients with active RA were enrolled in a randomized placebo controlled study of anti-TNF (cA2) effects on synovial tissue (Tak *et al.*, 1996). Patients were randomized to receive anti-TNF or placebo. Synovial biopsies were obtained before and four weeks post therapy. Anti-TNF therapy resulted in a significant improvement in average scores for T cells and for the adhesion molecules, vascular cell adhesion molecules 1 and E-selectin.

Several fusion proteins (TNF receptors fused to Fc) are being evaluated in clinical trials with promising results including fusion of p75 (Moreland *et al.*, 1994a), p55 (Lenercept; Hasler *et al.*, 1996a; Rau *et al.*, 1996) and p80 (Baumgartner *et al.*, 1996) to Fc.

The p75 fusion protein (two p75 soluble TNF receptors fused to IgG1) was evaluated in patients with active RA (Moreland *et al.*, 1994a). Sixteen (16) patients with active RA received the fusion protein at doses ranging from 4 to 32 mg/m^2 followed by twice-weekly subcutaneous dosing for 4 weeks. The fusion protein was well tolerated with the main side effect being a mild injection site reaction. Significant improvements in clinical markers of disease severity were seen but there was no dose response relationship.

The p55 fusion protein (lenercept, two p55 TNF receptors linked to human IgG1-Fc) has been tested in several clinical trials (Hasler *et al.*, 1996; 1996a; Rau *et al.*, 1996). In one open label dose rising study, 64 patients with active RA were treated with a single IV dose of 1–50 mg/kg and improvements in clinical disease activity were seen with the higher doses (Hasler *et al.*, 1996). The fusion protein was well tolerated. An open label long term study of the fusion protein in patients with active RA was performed (Hasler *et al.*, 1996a). Patients received monthly IV infusions of 20 mg for up to one year. Sixty three (63) patients were enrolled but 30 prematurely discontinued primarily for lack of efficacy. A double blind placebo controlled trial of the fusion protein was performed in patients with active RA (Rau *et al.*, 1996). Sixty (60) patients were randomized to receive either the fusion protein or placebo twice a month but only slight differences between the patient groups were seen.

More promising results were obtained with a second fusion protein (soluble p80 TNF receptor: Fc) (Baumgartner *et al.*, 1996). One hundred eighty (180) patients with advanced stage RA were enrolled and treated twice a week for 3 months with subcutaneous doses of 0.25 mg/m^2, 2 mg/m^2 or 16 mg/m^2 of this fusion protein or placebo. Of the 44 patients in the high dose group, 75% had significant improvement in swollen and tender joint counts. Side effects included mild injection site reactions and upper respiratory tract infections, however no antibodies against the fusion protein were detected. Large randomized trials of these fusion proteins in RA patients are expected.

Combined anti-CD4/anti-TNF

Anti-CD4 is being used as a tolerising agent to allow for long term re-application of anti-TNF. Animal trials have been conducted using anti-CD4 plus either anti-TNF (Williams *et al.*, 1994) or anti-TNF-IgG fusion protein (Williams *et al.*, 1995) and results are encouraging. In both combination therapy studies anti-CD4 prevented formation of neutralizing antibodies to anti-TNF or to the anti-TNF fusion protein. Thus, the mechanism of activity of the synergism may be in blockade of antibody response rather than in disease modifying activity targeting CD4+ cells.

Effects of combined anti-CD4 mAb therapy (YTS191.1.2/YTA3.1.2) and anti-TNF mAb (TN3-19.12) were evaluated in an established murine collagen-induced arthritis model (Elliott and Maini, 1994). Anti-TNF alone reduced paw swelling, limb involvement and joint erosion but although anti-CD4 alone did reduce paw-swelling it did not prevent joint erosion. Synergism was seen in combined therapy with significantly greater reduction in paw-swelling and joint erosion. Effects of combined anti-CD-4 mAb therapy and human p55 TNF receptor IgG fusion protein were evaluated in an established collagen-induced arthritis model (Williams *et al.*, 1995). Again, anti-TNF alone reduced paw swelling, limb involvement and joint erosion (reduced from 92% to 50%). Synergism was seen in combined therapy with a decreased incidence of joint erosions (from 100% to 17%).

Anti-adhesion mabs

Cellular adhesion molecules play a major role in the pathogenesis of synovial pannus formation in RA. Selective binding studies show that most infiltrating large cells of synovial tissue and bordering pannus-cartilage junction bind with anti-ICAM-1 (CD54), anti-VLA-4 (Cdw49d) or anti-VLA-5 (Cdw49e) (Ishikawa *et al.*, 1994; Cutolo *et al.*, 1993). In a rabbit study of antigen-induced arthritis, mAb anti-CD18 resulted in improvement of the arthritis and of chronic inflammation (Jasin *et al.*, 1992).

An open-label dose-escalation study of anti-ICAM-1 in RA patients demonstrated improvement in clinical disease markers (Kavanaugh *et al.*, 1994). Thirty two (32) RA patients were enrolled in the open-label dose escalation study. Patients were treated on 5 consecutive days and followed to day 60. Adverse events included a transient cutaneous allergy and peripheral CD3-CD4 lymphocytosis. Clinical improvement was seen in 13 of 23 patients through day 29 and in 9 of 23 patients through day 60. Treatment with anti-ICAM-1 resulted in a transient increase in circulating T cells which normalizes by day 8, and depressed T cell responses (accessory cell-dependent and cell-independent stimuli) probably secondary to depressed IL-2 production (Davis *et al.*, 1995).

Anti-IL-6

IL-6 appears to have both anti-inflammatory and inflammatory activities in the pathogenesis of RA. Due to the pleiotropic effects of IL-6 and the short-lived improvements noted with anti-IL-6 therapy, clinicians have approached anti-IL-6 therapy of RA with caution. However, one small clinical trial has been performed in RA patients which shows transient improvement in RA after IL-6 therapy (Wendling *et al.*, 1993; Wendling *et al.*, 1994).

IL-6 is found in high concentrations in the RA synovial fluid (Farahat *et al.*, 1993; Cutolo *et al.*, 1993) and serum levels of IL-6 are significantly higher than in normal subjects (Cohick *et al.*, 1994). Immunohistochemical staining demonstrates production of IL-6 in RA synovial lining cells and there is also a perivascular distribution in the deeper synovium (Field, 1991). In an animal model of antigen-induced arthritis in rats, administration of IL-6 led to inhibition of arthritis (Mihara *et al.*, 1991). However, IL-6 promotes B cell maturation and production of rheumatoid factor and stimulation of hepatic acute-phase proteins (Wendling *et al.*, 1993; Borden and Chin, 1994). Interestingly, circulating levels of IL-6 show a marked circadian rhythm in RA with high morning levels corresponding to patient symptoms of morning stiffness (Arvidson *et al.*, 1994), possibly accounting for the differences between published studies regarding whether IL-6 levels correlate with disease. Changes in IL-6 and soluble IL-6 receptor levels have been found following methotrexate therapy, commensurate with improvement in clinical disease activity (Crilly *et al.*, 1995) but not following nonsteroidal RA treatment (Cohick *et al.*, 1994).

In a small clinical trial, 5 patients with active RA who had previously received anti-CD4 therapy (B-F5) were enrolled in an open-label study of mAb anti-IL-6 therapy (B-E8, IgG1) (Wendling *et al.*, 1993). Anti-IL-6 was given at 10 mg/day for 10 consecutive days and patients were followed for two months. The anti-IL-6 therapy was well tolerated and there were no significant adverse effects. Transitory improvement in C-reactive protein was seen. An increase in IL-6 levels occurred in four of the six patients commensurate with anti-inflammatory effects. Clinical response was seen in all patients but was short-lived with median response duration of 2 months. The results of an extension of this study to 8 patients was reported with similar results (Wendling *et al.*, 1994).

Anti-CD5

Animal studies demonstrate efficacy of anti-CD5 in an established collagen type-II induced arthritis model where anti-CD5 mAb resulted in a significant decrease in disease severity within 6 days of treatment (Plater-Zyberk *et al.*, 1994). Initial uncontrolled trials of anti-CD5 therapy using an immunoconjugate anti-CD5 (anti-CD5IC: murine mAb anti-CD5+ ricin A chain) were encouraging (Strand *et al.*, 1993), however a large clinical trial failed to show efficacy with the immunoconjugate (Olsen *et al.*, 1994).

The open label clinical study of immunoconjugate anti-CD5 was performed in active RA patients (Strand *et al.*, 1993). Seventy-nine (79) patients received 0.2 or 0.33 mg/kg/day for 5 days. Clinical response was seen at one month (50–68%) and at 6 months (22–25%) even though T cells were restored at one month. Half the patients had significant reduction in either ESR or serum C-reactive protein. There was a high incidence of adverse events including rashes, myalgias and pulmonary edema which may be attributable to the ricin A chain (Kalden, 1994a). Human antibodies were seen in nearly all patients. The multi-center, double-blind, placebo-controlled study of 104 patients with RA was performed using immunoconjugate mAb anti-CD5 (CD5-IC). Patients received doses of 0.2, 2 or 8 mg/m^2/day or placebo for 4 days. Patients treated with placebo fared better than patients treated with the immunoconjugate (Olsen *et al.*, 1994).

Anti-CD52

A total of 111 patients with active and or refractory RA have been evaluated in anti-CD52 (CAMPATH-1H) trials including a trial of 0.3–3.0 mg/day subcutaneously for 10 days (Matteson *et al.*, 1995) and two trails of intravenous therapy in doses ranging from 1–100 mg (Weinblatt *et al.*, 1995; Isaacs *et al.*, 1996), however patients experienced significant toxicities. Immediate toxicities included severe flu-like symptoms, injection site reactions and later adverse events include dose-related infections and a report of one fatal opportunistic infection. Recurrent synovitis after anti-CD52 therapy has been confirmed by histologic evaluation of synovium showing significant T lymphocytic infiltrates even at a time where there is peripheral lymphopenia (Ruderman *et al.*, 1995).

Anti-CD7

Murine anti-CD7 (mAbRFT) or chimeric anti-CD7 (CHH-380) have been evaluated in small clinical trials, however either no clinical response or transient clinical responses have been reported and infusion related toxicities occur (Kirkham *et al.*, 1991; 1992). The mAb was studied in 4 RA patients given doses of 6 mg/day for 14 days and transient improvements were seen in two patients (Kirkham, 1988). Further testing of the mAb revealed clinical responses in 2 of 6 patients given murine anti-CD7 (RFT2) for 15 days (Kirkham *et al.*, 1991). Chimeric anti-CD7 was tested in 10 patients with RA at doses of 4–20 mg/day on days 1 and 7 and although transient improvements were seen in 6 patients these improvements were short-lived (7 days and patients experienced malaise, fever and nausea; Kirkham *et al.*, 1992). CD7 is already present in reduced numbers of cells in peripheral and synovial T cells in RA (Lazarovits and Karsh, 1988; Lazarovits *et al.*, 1992) so that further depletion may not be an effective RA therapy.

Anti-C5

Inhibition of C5 (anti-C5) has been evaluated in a rodent study of arthritis performed using murine mAb anti-C5 (Wang *et al.*, 1995). In the mouse model of collagen-induced arthritis, anti-C5 mAb protected the animals from onset of arthritis and from progression of established arthritis.

Anti-IgE

Inhibition of IgE (anti-CD23) has also been evaluated in animal models of RA. Treatment of arthritic mice (collagen-induced arthritis) results in a dose-related improvement in RA disease severity confirmed by histologic improvement in cellular infiltration of synovium and destruction of bone/cartilage (Plater-Zyberk *et al.*, 1995).

SEPSIS/SYSTEMIC INFLAMMATORY RESPONSE (SIRS)

Sepsis is a major cause of death for critically ill patients despite advances in supportive care, sophisticated intensive care units, advanced surgical and pulmonary/critical care therapies and active inpatient infection control programs (reviewed in Bone, 1991; Blackwell and Christman, 1996). Approximately 400,000 new cases of sepsis/SIRS occur each year in the US and the mortality ranges from 20–50% for patients without shock and from 40–70% for patients with shock (Parillo,1991). Bacterial products and a variety of acute noninfectious insults stimulate the release of a cascade of proinflammatory mediators such as TNF, IL-1, IL-6 and IL-8 (Bone, 1994; Blackwell and Christman, 1996). Noninfectious insults include pancreatitis (Hughes *et al.*, 1996), burns (Yao *et al.*, 1995), radiation (Elliott *et al.*, 1995a), steroids (Gianotti *et al.*, 1996), IL-2 (Reynolds *et al.*, 1995) surgery and trauma (Roumen *et al.*, 1993) and acute hemorrhage and resuscitation (Abraham *et al.*, 1995). The proinflammatory mediators are responsible for the symptomatology of sepsis (SIRS: systemic inflammatory response syndrome, ACCP, 1992), including fever, hemodynamic instability, coagulopathy, end organ dysfunction and failure and death (ACCP, 1992). Strategies in treatment of sepsis/SIRS include neutralization of bacterial products of gram-negative bacteria (endotoxin), inhibition of cytokines (anti-TNF, IL-1 receptor antagonists), and other inhibitors of inflammatory cellular activation. Challenges in the treatment of sepsis/SIRS include the importance of early diagnosis and treatment prior to activation of the proinflammatory cascade, identification of the appropriate subpopulation of patients for a given therapy and the diversity of the underlying conditions (e.g., chronic diseases, immunosuppression, invasive procedures, organisms and site(s) of infection).

Anti-endotoxin

In systemic infections endotoxin and other bacterial products are the key initiators of a cascade of proinflammatory mediators. Thus, therapeutic approaches in treatment of sepsis/SIRS involves strategies to block bacteria and their products from activating the cascade of proinflammatory mediators. Conflicting results have been obtained in studies of anti-endotoxin treatment of septic animals where anti-endotoxin demonstrated improved survival in an animal model of endotoxemia (Romulo *et al.*, 1993: E5) and anti-endotoxin decreased survival in another animal model of endotoxemia (Quenzato *et al.*, 1993: HA-1A). Gram negative organisms "leak" from the gut during sepsis/SIRS, translocate to mesenteric lymph nodes and eventually to the bloodstream, potentially initiating proinflammatory mediators (Swant and Deitch, 1996; Van Leeuwen *et al.*, 1994). mAbs against endotoxin are protective in this model (Gianotti *et al.*, 1996a).Thus, anti-endotoxin might be beneficial either in cases of overwhelming gram negative infection (e.g., meningococcemia) or in situations where gram-negative organisms from the gut stimulate cytokine release. In either case, anti-endotoxin mAb agents would be most effective when given prior to endotoxin activation of the proinflammatory cascade.

Two mAbs directed against bacterial endotoxin have been extensively evaluated in septic patients. Both are IgM mAbs, one is a murine antibody (E5, Greenman *et al.*, 1991; Greenberg *et al.*, 1991; Baumgartner, 1990) and the other is a humanized antibody (HA-1A, Ziegler *et al.*, 1991). Large randomized clinical trials with both mAbs have been performed in patients with gram negative sepsis and although significant improvements in morbidity occur, no statistically significant differences in mortality of the intent to treat populations have been found. However, in subsets of patients with gram negative bacteremia, mortality and end organ failure was improved, although the characteristics of the subsets of patients differed with the two mAbs. For E5 the population of interest was demonstrated to be nonshock bacteremic and nonbacteremic patients, (Baumgartner, 1990), whereas for HA-1A both shock and nonshock patients, but only bacteremic patients, were found to be the subpopulation of interest (Ziegler *et al.*, 1991). The differences between the results of the E5 and HA-1A studies are not readily apparent.

An open label study of mAb anti-endotoxin (E5) was performed in 88 patients with suspected gram-negative sepsis (Kobayashi *et al.*, 1994). Improvement in fever, hemodynamics, TNF and IL-6 circulating levels was seen in most patients. Twenty-three (23) patients with serious gram-negative infections were enrolled in a double-blind placebo-controlled trial of mAb anti-endotoxin (E5) (Greenberg *et al.*, 1991). Patients were randomized to receive 2 doses, 24 h apart of 2.5 mg/kg or 7.5 mg/kg anti-endotoxin or placebo. Mortality was 0/9 for anti-endotoxin treated patients (0/9 high dose, 0/9 low dose) versus 2 deaths among the 9 placebo patients.

Four hundred eighty six (486) patients with gram negative infection were enrolled in a double-blind placebo-controlled multicenter trial of mAb anti-endotoxin (E5) (Greenman *et al.*, 1991). Patients were randomized to receive two doses spaced 24 h apart of either 2 mg/kg anti-endotoxin or placebo. Of the 316 patients with gram negative infection with or without shock there was no difference in mortality. By contrast, among the 137 patients with gram negative infection and no shock there was a 30% reduction in mortality. Organ failure resolution was also significantly improved in this subset of patients (54% versus 30%). Anti-endotoxin was well tolerated and no severe adverse events were related to therapy. In a second double-blind placebo-controlled multicenter trial of mAb anti-endotoxin (E5), 847 patients with known or suspected sepsis without refractory shock were randomized to receive anti-endotoxin (2 mg/kg/day, 2 doses 24 h apart) or placebo and followed for 1 month (Bone *et al.*, 1995). There was no difference between anti-endotoxin treated patients and placebo patients in mortality at one month, however there were significantly greater improvements in end organ failure among patients treated with anti-endotoxin. Decreased mortality was seen in the subset of patients with gram negative bacteremia without hypotension (comprising only 28% of patients in the original population). Acute hypersensitivity reaction occurred in 2.6% of patients and 44% developed anti-murine responses. A meta analysis showed that nonshock patients benefitted from E5 administration by demonstrating earlier organ failure recovery and increased survival (Wedel, 1992).

Five hundred forty three (543) patients with sepsis syndrome were enrolled in a double-blind, placebo-controlled, multicenter trial of anti-endotoxin (HA-1A; Ziegler *et al.*, 1991). Patients were randomized to receive a single does of either mAb (100 mg) or placebo and followed for one month. In the subset of patients with gram negative bacteremia (37%), there was a 39% reduction in mortality versus placebo. In the subset of patients with gram negative bacteremia and shock there was a 42% reduction in mortality and in the subset of patients with gram negative bacteremia, shock and end organ damage a 51% reduction was seen. However, 60% of enrolled patients did not benefit from the anti-endotoxin therapy and may have even had decreased survival. A second trial was prematurely terminated after an interm analysis demonstrated that patients with sepsis due to non-gram negative organisms had increased mortality versus untreated patients (Luce, 1993). Initially, HA-1A was approved for marketing in European countries. A retrospective study of 600 European patients with severe sepsis treated with mAb anti-endotoxin (HA-1A) was published (The French National Registry of HA-1A, 1994). In comparison with historical controls, no decreased mortality in patients with gram negative bacteremia was demonstrated for the HA-1A cohort and increased morality was seen in patients with non-gram negative bacteremic infections treated with HA-1A.

The differences between the results of the E5 and HA-1A studies are not readily apparent. A third multicentered placebo controlled study is nearing completion for mAb anti-endotoxin (E5). In this study meticulous care is being taken to enroll only patients with documented gram negative sepsis.

Anti-TNF

TNF plays a central role in sepsis by mediating the adverse physiologic events that occur during sepsis/SIRS (rev in Wherry *et al.*, 1993; Fisher and Zheng, 1996; Beutler, 1993; Bone, 1991). In human and animal models, injection of purified products of bacteria result in elevated TNF levels occurring commensurate with hemodynamic changes (Suffredini *et al.*, 1989; Michie *et al.*, 1988). Infusion of TNF into cancer patients or into volunteers results in signs and symptoms which simulate sepsis/SIRS including fever, hemodynamic instability, leukopenia, hepatic enzyme abnormality and coagulopathy (Michie *et al.*, 1988; Sherman *et al.*, 1988; van der Poll *et al.*, 1990). Injection of anti-TNF in animals challenged with endotoxin, live *E. coli*, *Pseudomonas aeruginosa*, group A Streptococcus or Staphylococcus results in marked decrease in production of proinflammatory cytokines, reversal of septic neutropenia, reduced neutrophil sequestration, improved mean arterial blood pressure and tissue perfusion, and significant reduction in mortality (van der Poll *et al.*, 1994; Fiedler *et al.*, 1992; Emerson *et al.*, 1992; Hinshaw *et al.*, 1992; Windsor *et al.*, 1994; Stevens *et al.*, 1996). Similar results are obtained with anti-TNF treatment of acute noninfectious (SIRS) challenges such as pancreatitis (Hughes *et al.*, 1996), burns (Yao *et al.*, 1995) or hemorrhage and resuscitation (Abraham *et al.*, 1995). On the other hand, several animal studies show that experimentally induced peritonitis or peritonitis/sepsis is not ameliorated by antibiotics plus anti-TNF (Stack *et al.*, 1995; Remick *et al.*, 1995; Olson *et al.*, 1995), however, the experimental conditions may not correlate with clinical situations. Several studies demonstrate that TNF levels in serum of patients with sepsis correlates with severity of illness and with mortality (Endo *et al.*, 1992; Waage *et al.*,1987; Martin *et al.*, 1994; Pinsky, 1993). Anti-TNF has been evaluated in a number of sepsis trials including murine anti-TNF (TNF mAb, CB0006), humanized mAb (CDP571), mAb fragment (MAK, 195F) and recombinant human soluble TNF receptor:Fc fusion proteins.

A multicenter, open-label, prospective, dose escalating study of murine mAb anti-TNF (CB0006) was performed with patients with severe sepsis or septic shock (Fisher *et al.*, 1993). Eighty (80) patients were treated with 0.1 mg/kg, 1.0 mg/kg or two doses of 1 mg/kg spaced 2 h apart and the antibody was well tolerated. Patients were followed for 1 month and no survival benefit was found. In the subset of patients with high circulating TNF, mortality was lessened.

Two multicenter, randomized placebo-controlled studies of murine anti-TNF (TNFmAb) were performed in sepsis syndrome. In one trial, 994 patients with sepsis syndrome were stratified into shock or nonshock groups then randomized to receive a single infusion of 7.5 or 15 mg/kg of anti-TNF or placebo (Abraham *et al.*, 1993). No decrease in mortality was seen at one month, however, in septic shock patients a significant reduction in mortality was present 3 days after infusion and at 28 days there was a trend towards decreased mortality. Anti-TNF was well tolerated with no immediate hypersensitivity allergic reactions but serum sickness reactions occurred in 2.5% of patients. In the other study, 553 patients were randomized to receive 15 mg/kg or 3.0 mg/kg anti-TNF or placebo as a single intravenous infusion (Sprung *et al.*, 1996). Patients were followed for 28 days and 28 day mortality was reduced in the 3 mg/kg group by 14.5% compared with placebo. A phase III trial is ongoing of anti-TNF in 1900 patients with septic shock.

A randomized placebo controlled dose ranging study was performed with humanized anti-TNF mAb (CDP571) in patients with septic shock (Dhainaut *et al.*, 1995). Forty two (42) patients were randomized to receive anti-TNF (0.1–3.0 mg/kg) or placebo as a single intravenous infusion and patients were followed for a month. The mAb was well tolerated, and although anti-TNF resulted in rapid decrease in circulating TNF levels, no dose response relationship was seen for mortality rates.

An open label study with TNF receptor (TNF receptor p80:Fc) was performed on volunteers given endotoxin (4 ng/kg) (Suffredini *et al.*,1995). Patients received either the TNF receptor (10 mg/m^2, 60 mg/m^2) or placebo. The TNF receptor did not alter endotoxin related symptoms (such as cardiac index, heart rate and sytemic vascular resistance) except for a delay in the febrile response. However, the TNF receptor decreased IL-1, IL-8, IL-1 receptor antagonist, GM-CSF, but not IL-10 circulating levels. A double-blind, placebo-controlled multicenter trial of the TNF receptor was performed in patients with sepsis (Suffredini, 1994). Patients received TNF receptor at 0.15, 0.45 or 1.5 mg/kg or placebo. Dose-related increased mortality was seen (% mortality: placebo: 30%, low dose 30%; medium dose: 48%; high dose: 53%).

Anti-TNF antibody fragment (MAK, 195F) was tested in clinical studies of patients with severe sepsis (Boekstegers *et al.*, 1994; Reinhart *et al.*, 1996). In an uncontrolled trial, 20 patients with severe sepsis were treated with repeated doses of the mAb fragment (11 single doses over 5 days at either 1 mg/kg or 3 mg/kg; Boekstegers *et al.*, 1994). The mAb fragment was well tolerated but there was no dose-response change in cytokine levels or in mortality. One hundred twenty two (122) patients with severe sepsis or septic shock were randomized to receive anti-TNF at 0.1, 0.3 or 1.0 mg/kg or placebo in nine doses every 8 h over 3 days (Reinhart *et al.*, 1996). No differences in mortality were seen at one month, but patients with elevated IL-6 levels ($>$1000 pg/ml) showed improvement with anti-TNF therapy. Anti-TNF antibody was well tolerated and 40% of patients developed anti-murine antibodies. Patients were retrospectively stratified according to baseline TNF, IL-6 and clinical characteristics and only elevated IL-6 levels ($>$1000 pg/ml) were predictive. In this group of patients, 14 day mortality was improved by the anti-TNF fragment. A large multicenter trial of the mAb fragment is ongoing in patients with sepsis and elevated IL-6 levels.

Anti-IL-6

The role of IL-6 in sepsis/SIRS is unclear. Sepsis or endotoxin challenges result in elevated levels of IL-6 (Kuhns *et al.*, 1995) and IL-6 concentrations correlate closely with severity and outcome in sepsis (Hack *et al.*, 1989; Moscovitz *et al.*,

1994). However, injection of IL-6 in experimental animal models does not result in sepsis (Preiser *et al.*, 1991) although partial protection may be afforded by treatment of septic animals or TNF challenged animals with anti-IL-6 or with IL-6 receptors (Libert *et al.*, 1992; van der Poll *et al.*, 1994a).

Jarisch-Herxheimer: The Jarisch-Herxheimer (J-H) reaction is an acute event symptomatically similar to acute sepsis (fever, rigors, hypotension) due to treatment of louse-borne relapsing fever (Borrelia recurrents) with antibiotics. Patients with J-H reactions have increased circulating levels of TNF-alpha, IL-6 and IL-8. Although the reaction is relatively rare, because of the similarity to the symptomatology and cytokine profile of sepsis/SIRS this reaction has been closely studied. Treatment of infected patients with anti-TNF Fab (sheep polyclonal anti-TNF fragments) prior to dosing with antibiotics reduces the incidence of J-H reactions by more than 50%, commensurate with decreases in IL-6, IL-8 and especially TNF (Fekade *et al.*, 1996).

Heat stroke: Patients with heat stroke develop high fevers and significantly high levels of circulating IL-1, TNF and IL-6 (Chang, 1993). mAb anti-cytokine therapy may be beneficial in this clinical setting.

Anti-Adhesion mAbs: Anti-adhesion mAbs may inhibit inflammation caused by endotoxin challenge. For example, in an experimental model of endotoxin-induced uveitis, mAbs against anti-ICAM-1 and anti-LFA-1 dramatically reduced inflammation (Whitcup *et al.*, 1995).

TRANSPLANTATION

Technical advances have resulted in major improvements in the safety of the surgical aspects of transplantation, however, transplant rejection remains a major barrier to transplant success. Transplant rejection is initiated by allogeneic major histocompatibility complex antigens on or within the graft which stimulate CD4+/T helper cells of the host. Subsequent cellular interactions between activated T lymphocytes, antigen presenting cells and proinflammatory mediators (IL-1, IL-2, IL-6, TNF) and anti-inflammatory cytokines (IL-4, IL-5, IL-10 and TNF-beta) results in transplant rejection. Accumulation of host immune cells to the allograft is stimulated by interaction of adhesion molecules on circulating leukocytes and endothelium (Krams *et al.*, 1993). Therapeutic options initially included global immunosuppression therapies (total body irradiation, cyclosporine, FK506 and antilymphocyte globulin) and more specifically included therapies against activated T cells or cytokine targets (Chandler and Passaro, 1993). mAbs against CD3 (OKT3) have been widely used as immunosuppressive agents in the prophylaxis and treatment of transplant rejection. Other mAbs under evaluation include mAbs against activated T cells (anti-IL-2, anti-CD4, including evaluation of a non-depleting anti-CD4 mAb), against proinflammatory cytokines such as TNF and against adhesion molecules.

Anti-CD3

Anti-CD3 (muromonal-CD3, Orthoclone OKT3) therapy has been extensively used as an immunosuppressive agent in transplantation (reviewed in Parlevliet and Schellekens, 1992 and in Sgro, 1995) and is one of the most widely used therapies for transplant rejection. Treatment with anti-CD3 results in a rapid and profound lymphocytopenia possibly due to several mechanisms including complement-dependent cytolysis, cell-mediated antibody-dependent cytolysis, opsonization and phagocytosis by macrophages and inhibition of T cells interacting with antigen presenting cells (Bonnefoy-Berard and Revillard, 1996). Long lasting effects may be due to induction of clonal anergy resulting in long lasting specific unresponsiveness.

OKT3 is very effective as first line therapy for acute renal allograft rejection, for treatment of steroid-resistant rejection as rescue therapy in ATG and ALG (anti-lymphocyte globulin) resistant rejection and as prophylactic therapy. OKT3 has been extensively used in allograft rejection in renal (Haverty *et al.*, 1993; Mochon *et al.*, 1993; Beilman *et al.*, 1993; Uchida *et al.*, 1996; Kumano *et al.*, 1996; Tanabe *et al.*, 1996; Petrie *et al.*,1995; Norman *et al.*, 1993), liver (Wall *et al.*, 1995; Farges *et al.*, 1994), lung (Shennib *et al.*, 1994) and cardiac (Haverty *et al.*, 1993; Shaddy *et al.*, 1993) transplant rejection patients. Side effects include the cytokine release/complement and neutrophil activation phenomenon (Chatenoud, 1993), formation of anti-murine antibodies (Carey *et al.*, 1995) and profound immunosuppression leading to increased infections such as viral infections (Petrie *et al.*, 1995) and increased risk of malignancies.

Anti-CD4

There is interest in evaluating a nondepleting anti-CD4 mAb to avoid the toxicity of prolonged periods of nonspecific immunosuppression. Treatment with a nondepleting anti-rat CD4 mAb (RIB-5/2) was beneficial in an accelerated rat cardiac allograft model (Binder *et al.*, 1996). Clinical studies are underway with a nondepleting anti-CD4 mAb (OKT4A) that does not cause cytokine release or CD4+ cell depletion (Pulito *et al.*, 1996). However, in a mouse model of cardiac allograft rejection, nondepleting anti-CD4 mAbs which resulted in prolonged occupation of CD4 molecules was superior to mAbs (such as OKT4A) which modulate CD4 from the cell surface (Darby *et al.*, 1994).

Anti-IL-2 receptor

Increased circulating IL-2 receptor-bearing cells and increased levels of soluble interleukin 2 receptors are seen in patients with transplant rejection (Deng *et al.*, 1995; Chang *et al.*, 1996). Anti-IL-2 receptor prophylaxis of transplant rejection has been evaluated in heart (Hesse *et al.*, 1995), liver (Nashan *et al.*, 1995; 1996) and renal (van Gelder *et al.*, 1996; Hesse *et al.*, 1995) transplant patients and results are encouraging. Randomized trials comparing mAb anti-IL-2 receptor therapy with OKT3 have been performed in liver (Reding *et al.*, 1993) and heart (van Gelder *et al.*, 1996) transplant patients and anti-IL-2 receptor therapy compares favorably with OKT3 prophylaxis. A humanized anti-IL-2 receptor mAb is also under evaluation (Stock *et al.*, 1996).

Murine anti-IL-2 receptor mAb (BT563) was evaluated as prophylactic therapy in liver transplant patients (Nashan *et al.*, 1995). Thirty-eight (38) patients were enrolled in an open label study of anti-IL-2, 10 mg/day for 12 days in addition to cyclosporine and low dose steroids. No acute rejection episodes occurred and only minimal side effects (minor bacterial infections, no mycotic or viral infections) occurred. Subsequently 32 liver transplant patients were enrolled in a randomized study of murine anti-IL-2 receptor (10 mg/day for 12 days) versus ATG (5 mg/kg/day) for 7 days in addition to cyclosporine and low dose steroids (Nashan *et al.*, 1996). No acute rejection episodes occurred in the group treated with anti-receptor mAb and 5 acute rejections occurred in the ATG group. Survival was 92% at 12 months for both treatment groups. All patients in the mAb receptor group formed an antimurine response and 56% of patients in the ATG group formed anti-rabbit antibodies. Thus anti-IL-2 receptor mAb therapy is at least as effective and possibly preferable to ATG for liver transplant prophylaxis against acute rejection.

The murine anti-IL-2 receptor mAb (BT563) was administered at 10 mg/day prophylactically to 30 heart transplant recipients and 40 renal transplant recipients (Hesse *et al.*, 1995). There was an immediate marked decrease in IL-2 receptors on peripheral blood lymphocytes, returning to normal within 20 days. Anti-murine antibody responses were seen in over half of the patients but did not correlate with clinical response. A double-blind, placebo-controlled randomized study of anti-IL-2 receptor mAb (BT563) was performed as rejection prophylaxis in renal transplant patients (van Gelder *et al.*, 1995). Fifty six (56) patients were randomized to receive either anti-IL-2 mAb or placebo and followed for 3 months. There were significantly less patients in the group treated with anti-receptor mAb with rejection episodes during the 10 day treatment course (0% versus 24%), and the first 4 postoperative weeks (76% versus 96%) and less rejection episodes seen in the mAb receptor group at 3 months (11% versus 28%).

Two randomized trials were performed to compare OKT3 prophylaxis to anti-IL-2 receptor mAb therapy. One hundred twenty nine (129) liver transplant patients were randomized to receive cyclosporine, steroids, azathioprine plus either OKT3, anti-IL-2 mAb (LO-Tact-1) or placebo during the first 10 days post transplantation (Reding *et al.*, 1993). Survival rates at one year were 84% (OKT3), 93% (mAb anti-IL-2 receptor) and 67% (placebo group). Incidence of CMV was highest in the OKT3 group. In a second trial of OKT3 versus anti-IL-2 receptor (murine BT563), 60 heart transplant patients received cyclosporine on day 3 and received either of the mAbs during the first 7 days after transplantation (van Gelder *et al.*, 1996). Patients were followed for an average of 3 years. The two mAbs were similar in overall incidence of rejection and incidence of infectious complications, however the OKT3 treated group experienced a cytokine release syndrome in most patients and the anti-IL-2 receptor treated group experienced earlier rejection than the anti-CD3 treated group.

Anti-TNF

TNF has been implicated in the immunopathology of transplant rejection. Serum TNF levels are significantly elevated during liver allograft rejection (Imagawa *et al.*, 1990), in renal allograft rejection (Maury and Teppo, 1987) and in patients with complications of BMT (Holler *et al.*, 1990). In a rat cardiac transplant model, a single injection of anti-TNF prolonged the time to graft rejection (from 6–12 days) (Wei *et al.*, 1994; Lin *et al.*, 1992). Graft survival in this model was additionally prolonged with combined anti-TNF and cyclosporine therapy (Bolling *et al.*, 1992; Seu *et al.*, 1991). In another similar cardiac transplant model, anti-TNF treatment prolonged graft survival from 7 to 17 days (Imagawa *et al.*, 1991).

Anti-CD7: Anti-CD7 therapy has been evaluated in transplant patients since allogeneic mixed leukocyte reaction are inhibited by anti-CD7 treatment (Lazarovits *et al.*, 1993). An open label randomized study of recipients of first cadaveric renal allografts was performed using human mouse-chimeric mAb anti-CD7 (SDZCHH280). Twenty (20) patients were randomized to receive either anti-CD7 or OKT3. Transplant rejection was delayed to day 35 in anti-CD7 treated patients. Advantages of anti-CD7 compared with anti-CD3 included lack of anti-mouse immunoglobulin reaction and minimal induction of IL-2, IL-6, TNF-alpha and IFN-gamma.

Anti-Adhesion mAbs: Anti-LFA-1 (CD11; 500 μg) protects rejection of transplanted thyroid tissue for 2 months in mice (Talento *et al.*, 1993). At 1.5 years, 50% protection was afforded by the single infusion. Anti-ICAM-1 treatment of rats during the first 5 days post small bowel transplant resulted in a marked decrease in allograft rejection (Yamataka *et al.*, 1993). Monoclonal anti-VLA-4 plus anti-LFA-1 prevents rat islet allograft rejection (from 6 day baseline to 60 days) (Yang *et al.*, 1995).

Combined anti-LFA-1 and anti-ICAM-1 treatment prevents graft rejection and diabetes in rat pancreas transplants (Uchikoshi *et al.*, 1995). Anti-LFA-1 mAb prevented acute graft rejection in rats (Lazarovits *et al.*, 1996).

Anti-endotoxin: Rejection of intestinal allografts may be complicated by gut translocation of bacteria resulting in elevation of serum IL-6 and TNF levels and anti-endotoxin therapy might be beneficial in this setting (Li *et al.*, 1994).

GRAFT VERSUS HOST DISEASE (GVHD)

Severe graft versus host disease (GVHD) is a major complication of bone marrow transplantation (BMT). Unrelated donor transplantation leads to more frequent, severe and resistant disease, however, significant GVHD is also seen using genoidentical donors. T cells, NK cells and TNF have been identified in the pathogenesis of GVHD (Imamura *et al.*, 1994). Elevated levels of IL-1, IL-6 and

TNF are increased in the peripheral circulation prior to onset of both acute and chronic GVHD and there is a positive correlation in severity and number of clinical complications with high levels of these proinflammatory cytokines (Barak *et al.*, 1995). Although IL-6 levels are increased in hyperacute GVHD, exogenous addition of IL-6 does not induce severe GVHD.

Clinical trials of mAbs have been performed as GVHD prophylaxis and in treatment of acute GVHD. Short term improvements have been seen with anti-CD5 treatment of acute GVHD. Promising results have been obtained with anti-TNF, and anti-IL-2 receptor mAbs.

Anti-CD5

An immunoconjugate mAb anti-CD5+ ricin A chain has been evaluated in several small clinical trials and in a large scale trial in prophylaxis of GVHD. Short term improvements (4 weeks) have been seen but no improvement has been found in 1 year survival or in prevention of the development of chronic GVHD. Anti-CD5 immunoconjugate has also been associated with significant toxicity and with serious viral infections.

An initial pilot study of anti-CD5 immunoconjugate (XomaZyme, H65, CD-5-Ricin) was performed in 8 patients as GVHD prophylaxis (Koehler *et al.*, 1994). Although engraftment occured in all patients, significant complications of virus infections occurred associated with low CD4 and CD8 T cells including 4 deaths (CMV=2, EBV=1, adenovirus=1) and recurrent infections in the other patients. A randomized trial of anti-CD5 immunoconjugate was performed in 22 patients who received unrelated donor BMT (Weisdorf *et al.*, 1993). Patients received immunoconjugate (0.1 mg/kg/day) plus either methotrexate and steroids (16 patients) or methotrexate and cyclosporine (6 patients) for 3 weeks following transplantation. Of 15 evaluable patients, 9 developed acute GVHD and of 8 evaluable patients, 6 developed chronic GVHD. There were 4 long term survivors (1.5 years). The combination of methotrexate, cyclosporine, immunoconjugate resulted in significant toxicity (renal, weight gain, edema). A second similar study was performed in 11 patients where all patients developed acute GVHD and the anti-CD5 did not inhibit circulating CD5+ and CD3+ lymphocytes (Przepiorka *et al.*, 1994).

A randomized, double-blind placebo controlled study was performed using immunoconjugate anti-CD5 (CD5Plus) in patients with acute GVHD after allogeneic bone marrow transplantation (Martin *et al.*, 1996). Two hundred forty three (243) patients were enrolled and randomized to receive either immunoconjugate anti-CD5 (0.1 mg/kg daily for 14 days) plus steroids or steroids alone. Early improvements in the first 4 weeks were noted with less severity of symptoms noted at 3, 4 and 5 weeks. Complete responses were seen at 4 weeks in 40% of the mAb group compared with 25% in the control group (p=0.019). However, there was no difference in 1 year survival (49% in the mAb group versus 45% in the controls), similarly clinically extensive chronic GVHD developed in patients in the two groups (65% in the mAb group versus 72% in the controls).

Anti-IL-2 receptor

Anti-IL-2 receptor therapy has been evaluated both in the prophylaxis and in the treatment of acute GVHD. Anti-IL-2 receptor mAb therapy did not change onset or severity of GVHD or survival (Belanger *et al.*, 1993; Ferrant *et al.*, 1995). More promising results have been obtained in a small trial of GVHD treatment with 2/3 patients becoming long term survivors (Herbelin *et al.*, 1994).

Sixty-four (64) patients were given mAb anti-IL-2 receptor mAb (33B31) 20 mg on days 1,2 and 10 mg/day on days 3–28 along with cyclosporine and methotrexate as part of a GVHD prophylaxis regimen (Belanger *et al.*, 1993). In comparison with 89 historical controls there was no difference in incidence and time of onset of GVHD, engraftment, relapse or survival. Twenty seven (27) patients with HLA-matched sibling marrow transplants were enrolled in an open label study of prophylaxis with anti IL-2 receptor mAb (LO-Tact-1, rat IgG2b) (Ferrant *et al.*, 1995). Patients received either cyclosporine + anti-IL-2 receptor (0.2 mg/kg/day) on days +7 to +28 or methotrexate + cyclosporine + anti IL-2 receptor (0.4 mg/kg/day) days -1 to day +28 but neither regimen altered survival rates.

Promising results were obtained in a small study of anti-IL-2 receptor mAb in treatment of GVHD (Herbelin *et al.*, 1994). Fifteen (15) children with steroid resistant acute GVHD were enrolled in an open label study of anti-IL-2 receptor mAb (BT 563, murine). Patients had received T cell-depleted marrow from partially matched related donor. Patients received 0.2 mg/kg anti-IL-2 receptor daily until GVHD was improved and steroids were tapered. Ten (10) of the 15 patients have been subsequently identified as long term survivors. Eleven (11) of the 15 patients achieved a complete remission, 2 had partial remissions and 2 did not improve. Relapses occurred in 6 of the 13 responders and were ameliorated by a second course of mAb. Six (6) patients developed chronic GVHD. One death was associated with GVHD infection.

Anti-TNF

Anti-TNF has been evaluated in a clinical trial of GVHD prophylaxis and in treatment of acute GVHD and results are promising. An open label study of anti-TNF fragment (MAK 195F) was performed in 21 high risk patients as part of pretransplant conditioning for BMT prophylaxis (Holler *et al.*, 1995). Anti-TNF therapy was well tolerated. In comparison with historic controls, anti-TNF treated patients had later onset of acute GVHD from day 15 to day 25 ($p < 0.05$) or from day 33 to day 53, depending upon the concomitant immunosuppressive regimen.

Combination therapy

Anti-CD2, anti-TNF and anti-IL-2 receptor mAbs were combined in treatment of steroid resistant acute GVHD (Racadot *et al.*, 1995). Fifteen (15) patients who received transplants of marrow from genotypically-identical siblings (7 patients), HLA-matched unrelated donors (5 patients) and partially-matched donors (3 patients) with steroid-resistant acute GVHD were enrolled. Initial treatment consisted of anti-TNF (B-C7, 10 mg/day/4days) and anti-CD2 (B-E2, 10 mg/day/10 days). On day 5 then every other day to day 50 patients were treated with anti-IL-2 receptor (B-B10). Three patients were free of GVHD and were considered long term survivors. On day 15, a complete remission was seen in 5 patients, partial response was seen in 6 patients and 4 patients failed to respond.

VASCULAR ISCHEMIA/ISCHEMIA REPERFUSION

Vascular ischemia is due to a variety of causes including platelet aggregation. Anti-platelet therapy is effective in preventing acute thromboembolic artery occlusions by inhibiting activation of platelets to release vasoactive mediators (Schror, 1995). Significant improvement in morbidity of patients undergoing coronary angioplasty has been seen using an anti-platelet Fab mouse/human chimeric mAb antibody (GPIIb/IIIa; EPIC Investigation, 1994: Topol, 1995; Coller *et al.*, 1995).

The consequences of ischemia result in activation of a variety of potentially harmful mediators including TNF, IL-1, IL-8 and increased immune cell adherance. Increased levels of TNF and other proinflammatory mediators such as soluble IL-2 receptors and IL-6 have been measured following ischemia-reperfusion injury and correlate with severity and outcome of the ischemia (Deng *et al.*, 1996; Liu *et al.*, 1994). Adherence of neutrophils to endothelium is a major initiating event leading to organ failure (Hartman *et al.*, 1995; Rabb *et al.*, 1995). Anti-TNF and anti-adhesion molecules (anti-CD11/CD18, ICAM-1, L-selectin and endothelial P-selectin) are being evaluated in animal models.

Anti-GPIIb/IIIa

A major issue in patients who undergo percutaneous transluminal coronary angioplasty (PTCA) is restenosis which occurs in 20–55% of patients and often requires additional revascularization procedures within 6 months (Califf, 1995). A major cause of ischemic complications and restenosis after PTCA is platelet aggregation. Inhibition of the activation of the platelet glycoprotein IIb/IIIa receptor can result in blocking of platelet aggregation. Inhibition of platelet aggregation using a Fab mouse/human chimeric monoclonal antibody (c7E3Fab, abciximab, ReoPro) has shown promising results in clinical situations predisposing to a high risk of vascular ischemia such as high risk angioplasty (Topol, 1995; Lefkovits and Topol, 1995; Coller *et al.*, 1995), unstable angina (Simoons *et al.*, 1994) and acute myocardial infarction (Lefkovitz and Topol, 1995).

Anti-GPIIb/IIIa (c7E3Fab) has been demonstrated to significantly reduce ischemic complications and clinical restenosis after high-risk angioplasty (Epic

Investigation, 1994; Topol, 1995). A multi-center randomized placebo controlled trial of the Fab fragment was performed in 2099 patients undergoing coronary angioplasty (Epic Investigation, 1994; Topol, 1995; Coller *et al.*, 1995). Patients were randomized to receive either the Fab fragment given as a bolus followed by 12 h infusion of placebo; Fab bolus plus 12 h Fab infusion or placebo bolus plus placebo infusion. Patients were scheduled for coronary angioplasty or atherectomy in high risk situations such as severe unstable angina, evolving acute myocardial infarction or high-risk coronary morphologic features. Primary endpoints included death, nonfatal myocardial infarction, unplanned surgical revascularization, unplanned repeat PTCA, unplanned implantation of coronary stent or insertion of an intraaortic balloon pump for refractory ischemia. Significant improvement was seen with the combination of Fab bolus/infusion (compared with placebo bolus/infusion), where a 35% reduction was seen at 30 days ($p=0.008$). However, Fab treatment resulted in a 2 fold increase in risk of major bleeding. A trend towards improvement was also seen for Fab bolus/placebo infusion compared with placebo bolus/infusion, where a 10% reduction was seen ($p=0.43$). In a follow-up study, the cohort of patients treated with Fab bolus/infusion was found to have a 23% reduction in acute ischemic complications at 6 months (Lefkovits *et al.*, 1996).

The Fab fragment was also studied in patients with refractory unstable angina (Simoons *et al.*, 1994). Sixty (60) patients were randomized to receive the Fab at 0.25 mg/kg bolus injection followed by Fab infusion at 10 μg/min for 18–24 h until 1 h after completion of second angiography and PTCA or placebo. Incidence of major complications was significantly decreased in the Fab group (3% versus 23%).

Anti-TNF

TNF has been implicated in the pathogenesis of vascular injury. TNF levels are elevated in myocardial infarction (Maury and Teppo, 1989), ischemia reperfusion (Yao *et al.*, 1986) and cardiac surgery (Deng *et al.*, 1996). Levels of TNF-alpha mRNA and protein are increased in ischemic neurons in a rat focal ischaemia middle cerebral artery occlusion model (Liu *et al.*, 1994). In a rat model, anti-TNF mAb (TN3, 20 mg/kg) pretreatment significantly reduced reperfusion injury in superior mesenteric artery occlusion model (Yao *et al.*, 1996).

Anti-adhesion mAbs

Mediators of cellular adherance have been implicated in the pathogenesis of vascular injury. mAbs that inhibit “adherance injury” protect animals from ischemia-reperfusion injury (Winn *et al.*, 1993; Ma *et al.*, 1993; Hartman *et al.*, 1995; Briedahl *et al.*, 1996) possibly by reducing neutrophil activity in vulnerable myocardium or vascular endothelium.

A number of animal studies demonstrate that anti-adhesion mAbs reduce ischemia reperfusion-induced injury. Rats pretreated with mAb to ICAM-1 had significantly decreased creatinine after renal artery occlusion (Rabb *et al.*, 1995). mAb against ICAM-1 (CL18/6) was tested in a dog model of ischemia reperfusion where treatment with the mAb resulted in cardioprotection (Hartman *et al.*, 1995). Anti-ICAM-1 was evaluated in a rabbit cerebral embolism (stroke) model (Bowes *et al.*, 1995). The antibody was given 5 minutes after embolization. There was no improvement in neurologic outcome at 15 or 30 minutes, however, combination of the antibody with tPA resulted in a significant improvement in outcome. Anti-ICAM mAbs reduce neutrophil infiltrates in a rabbit ischemic muscle model (Briedahl *et al.*, 1996). Rat hearts pretreated with anti-LFA-1 or anti-ICAM-1 prior to ischemia/reperfusion resulted in significant improvement in left ventricular developed pressure, coronary vascular resistance and cardiac energy status (Tamiya *et al.*, 1995). In a rabbit model of lung injury, animals treated with a mAb to CD-18 (R15.7) prevented fall in white blood cell count, however, the mAb did not protect against hypotension (Hill *et al.*, 1993). In a feline model of myocardial ischemia/reperfusion, mAb against L-selectin pretreatment resulted in significantly less myocardial necrosis, coronary endothelial damage and lower myeloperoxidase activity in the ischemic myocardium (Ma *et al.*, 1993).

Anti-IL-6

Levels of IL-6 correlate with duration of extracorporeal circulation, degree of hemodynamic support and surgical complications in cardiac surgery (Deng *et al.*, 1996). Levels of IL-6 are increased in peripheral blood and cerebrospinal fluid of patients following stroke and there is a significant correlation between early intrathecal production of IL-6 and size of the brain lesion (Tarkowski, 1995). However, IL-6 has pleiotropic effects and thus it is unclear if inhibition of IL-6 would be beneficial in ischemia-reperfusion.

Anti-C5

A reduction in cardiac reperfusion injury occurred with mAb to C5a in a pig model of ischemia induced by occlusion of the left anterior descending coronary artery (Amsterdam *et al.*, 1995). The mAb reduced C5a-stimulated neutrophil aggregation, chemotaxis, degranulation and superoxide generation and resulted in a reduction from 58 to 38% in infarct area.

OTHER ACUTE AND CHRONIC INFLAMMATORY DISORDERS

Multiple Sclerosis (MS)

Multiple sclerosis (MS) is the most common of the chronic demyelinating diseases and primarily afflicts young females ages 20–40. The illness is characterized by an unpredictable course of periods of disability (relapse) alternating with periods of recovery (remission), usually ultimately leading to severe disability. MS is thought to be due to an autoimmune disorder of activated T cells which initiate the inflammatory demyelination (rev in Brod *et al.*, 1996). Much of the MS research has been performed in animals using an experimental model of MS, experimental allergic encephalomyelitis (EAE), characterized by blood–brain barrier breakdown, lymphocyte infiltration and mediator release with demyelination, TNF present in the CNS lesions, and paralysis (Kent *et al.*, 1995). Newer strategies include mAbs to T cell epitopes and to proinflammatory cytokines such as TNF or ICAM-1 (Brod *et al.*, 1996; Baker *et al.*, 1994) and combination therapy with anti-LFA-1 and anti-ICAM-1 (Kobayashi *et al.*, 1995).

Anti-CD4

Two small, open label, uncontrolled trials with anti-CD4 or chimeric anti-CD4 have been performed with MS patients (Racadot *et al.*, 1993; Lindsey *et al.*, 1994, respectively). In a study of mAb anti-CD4 (B-F5, murine IgG1), 22 patients were treated and clinical improvements were noted in 4 of 17 evaluable patients (Racadot *et al.*, 1993). Side effects occurred during the first infusion possibly related to increased serum levels of IL-6 and TNF. An open label study of chimeric mAb anti-CD4 (49cM-T412) was performed in 29 patients with MS (Lindsey *et al.*, 1994). Treatment resulted in MRI scans showing less contrast enhancement one week after treatment. A long-lasting selective depletion of CD4 cells occurred similar to side effects seen in RA patients. Twenty-one (21) patients were subsequently treated with 2–4 doses of the mAb, resulting in long-lasting decreased CD4 counts but surprisingly few serious sequelae (Lindsey *et al.*, 1994a).

Anti-TNF

TNF has been implicated in the pathology of MS (Selmaj and Raine, 1995; Bake *et al.*, 1994). TNF levels are elevated in serum and cerebrospinal fluid in MS patients (Merrill *et al.*, 1989) and TNF levels are elevated in the brain (astrocytes) (Hofman, 1989). mAb anti-TNF (TN3.19.12) and bivalent human p55 and p75 TNF receptor-immunoglobulin (TNFR-Ig) fusion protein have been shown to be effective in EAE (Baker *et al.*, 1994). In this study repeated doses of anti-TNF or the fusion protein inhibited development of chronic EAE and the antibody or fusion protein was effective even after clinical symptoms appeared. The fusion proteins were effective at doses 10–100 fold less than the antibody.

Anti-adhesion molecules

Adhesion molecules have been implicated in the pathogensis of MS. Adhesion of leukocytes at the blood–brain barrier is thought to be an important step in the process of leukocyte migration to the CNS in experimental models of MS. Monoclonal anti-alpha 4-integrin (AN100226) treatment of guinea pigs protect animals from EAE (Kent *et al.*, 1995; 1995a). Guinea pigs were treated with anti-alpha-4-integrin resulting in prevention of leukocyte infiltration and decrease in acute EAE as measured by decreased cerebral edema, inflammation and demyelination. However, anti-ICAM-1 failed to be effective in a rat model of active and passive EAE (Willenbourg *et al.*, 1993). Combination therapy with anti-LFA-1 and anti-ICAM-1 has been evaluated in a rat EAE model (Kobayashi *et al.*, 1995). In this model, anti-LFA-1 and ICAM-1 treatment alone did not improve rat EAE clinical presentation, however, combined administration resulted in prevention of the cellular infiltrates and marked suppression of the disease progression.

Anti-CD3: An open label uncontrolled study was performed using a murine anti-CD3 mAb (OKT3) in 16 patients with multiple sclerosis (Weinshenker *et al.*, 1991). Patients were treated with OKT3, 5 mg/day for 10 days and no clinical improvement was noted. Severe side effects included hypotension and severe flu-like symptoms. Deterioration was seen in 4 patients and in two patients a transient but significant rise in interferon gamma and TNF occurred. Two patients had clinical improvement at 1 year. All patients developed antiimmunoglobulin antibodies.

Anti-CD2: An open label uncontrolled study of mAb anti-CD2 was performed in 12 MS patients given 0.2 mg/kg/day anti-CD2 for 5 days (Hafler and Weiner, 1988). There was no clinical efficacy reported and all patients developed anti-immunoglobulin antibodies.

Anti-nervous tissue: Other strategies include evaluation of anti-nervous tissue mAbs. For example, a mAb against spinal cord homogenates (SCH94.03) promoted central nervous system remyelination in a mouse chronic viral demyelination model (Asakura *et al.*, 1996).

Systemic Lupus Erythematosus

Systemic lupus erythematosus is an autoimmune disease of young adults (especially in the 20–40 year old range) characterized by malar rash, nephropathy, arthritis and other symptomatology (Ward *et al.*, 1996). More severe forms include life threatening nephritis, seizures and thrombocytopenia. CD4+ T cells and proinflammatory cytokines have been implicated in the pathogenesis of SLE. There is some correlation between serum levels of proinflammatory cytokines and complications of SLE such as the nephrotic syndrome and thrombocytopenia (al-Janadi *et al.*, 1993). Several mAb strategies have been attempted in animal and pilot clinical studies including treatment with mAb inhibitors of CD4, CD5, C5 or IFN-gamma.

Anti-CD4

Anti-CD4 has been evaluated in a mouse model of experimentally-induced (anti-DNA induction) SLE (Tomer *et al.*, 1994; Ruiz *et al.*, 1996) and in a mouse model of SLE-prone mice (Merino *et al.*, 1995) and these studies demonstrated that anti-CD4 protected mice from development of SLE. A small clinical trial with anti-CD4 (6H5) was performed in patients with SLE. Patients were treated with 0.3 mg/kg/day and improvement at 3 weeks was seen (Hiepe *et al.*, 1991). In a related disease, polychondritis, patients were treated with anti-CD4 (MT-412: 25 mg/day for 7 days; van der Lubbe *et al.*, 1991) or chimeric anti CD4(cMT-12; 50 mg monthly for 6 weeks; Choy *et al.*, 1991) and some clinical improvements were seen.

Anti-CD5

Anti-CD5 immunoconjugate has been evaluated in a small uncontrolled clinical trial of SLE patients and anti-CD5 showed partial activity (Stafford *et al.*, 1994). In this study, 6 patients were treated with 0.1 mg/kg of anti-CD5 immunoconjugate on 5 consecutive days and improvement was seen in 2 patients with nephritis but no effect on thrombocytopenia was seen. Nonresponders received a second course of anti-CD5 two months later. Similar to RA patients, SLE patients experienced significant reductions in numbers of circulating CD3+ cells where 34% reduction was seen at 6 months post therapy and a transient decrease in CD5+ B cells was seen.

Anti-gamma interferon: In a mouse model of SLE, mice were treated from 16 weeks until 35 weeks with soluble interferon-gamma receptor or anti-IFN-gamma mAb the mice and remained alive 4 weeks post therapy in contrast to control mice where 50% died over the same time interval (Ozmen *et al.*, 1995).

Anti-adhesion molecules: In a rat model of SLE, rats were treated with 500 μg anti-LFA-1 twice a week or 40 μg three times a week from age 5 months to age 10 months (Connolly *et al.*, 1994). Results were disappointing with little effect on longevity, however the high dose treatment inhibited production of autoantibodies to dsDNA.

Anti-C5: In an animal study using SLE prone mice, mAb anti-C5 treatment continuously for 6 months resulted in significant blockade of the expected course of glomerulonephritis and markedly enhanced survival (Wang *et al.*, 1996).

Psoriasis

Psoriasis is a heterogeneous skin disease characterized by often reversible epidermal proliferation and inflammation. Cytokines such as TNF and IL-6 are increasingly recognized as important in the pathogenesis of psoriasis (rev in Kemeny *et al.*, 1994). IL-6 and TGF-alpha may propagate the inflammatory response (Kapp, 1993). Small trials have been performed with anti-CD4 with some improvement noted.

Patients with psoriasis were treated with murine mAb anti-CD4 (B-B14) 0.2 to 0.8 mg/kg/day for 7 days and 1 month improvement was seen (Nicolas *et al.*, 1991; Morel *et al.*, 1992). In another small clinical study, patients received anti-CD4 plus cyclosporine A (Rizova *et al.*, 1994). In 3 psoriatic patients who experienced improvement subsequent to anti-CD4 therapy parakeratosis, papillomatosis and acanthosis improved. In the epidermis there was a decrease in both CD4 and CD8 positive cells, decreased ICAM-1 and HLA-DR positive keratinocytes.

Inflammatory Bowel Disease

T cell activation is pivotal in the pathogenesis of Crohn's disease and other inflammatory bowel diseases. Immunosuppressive regimens, such as steroids and cyclosporine are widely used and more specific therapies such as mAb anti-CD4 has been tried in small pilot clinical trials. TNF and adhesion molecules have been implicated in the pathogenesis of Crohn's disease. High levels of TNF are found in the mucosa and stools of patients with IBD (Braegger *et al.*, 1992; MacDonald *et al.*, 1990). Anti-TNF has been evaluated in several clinical trials of Crohn's disease patients and results are encouraging. Animal studies suggest anti-adhesion mAbs may also be useful in this disease.

Anti-TNF

TNF is considered to be pivotal in the pathogenesis of IBD, specifically Crohn's disease. Increased concentrations of TNF are found in the mucosa of patients with active disease. Clinical trials have been performed with humanized (CDP571, Stack *et al.*, 1996) and chimeric anti-TNF (cA2, van Dulleman *et al.*, 1995; McCabe *et al.*, 1996; Targon *et al.*, 1996) and results are encouraging.

Thirty (30) patients with Crohn's disease were randomized to receive either anti-TNF (CDP571, 5 mg/kg) or placebo and followed for 8 weeks (Stack *et al.*, 1996). At week 2 there were 6 patients in remission in the anti-TNF treated group and none in the placebo group. Similar improvements were also noted in other measures of disease activity. Although somewhat diminished by week 4, there was still improvement noted by week 4 in the treated group.

Ten (10) patients with refractory Crohn's disease were enrolled in an open label study of a single infusion of anti-TNF (cA2) (van Dullemen *et al.*, 1995). Eight (8) of the 10 patients exhibited improvement in disease (Crohn's Disease Activity Index score; colonoscopic evaluation of ulcerations) with the average response duration of 4 months. There were no adverse events related to anti-TNF therapy, however one patient had a "poor response" and one patient had a perforation related to the colonoscopy.

Two larger studies with cA2 demonstrated significant improvement in Crohn's disease (McCabe *et al.*, 1996; Targan *et al.*, 1996). A multicenter, open-label, dose escalation study of cA2 was performed in patients with active Crohn's disease (McCabe *et al.*, 1996). Twenty (20) patients were dosed with cA2 (1 to 20 mg/kg single injection) and followed for 3 months. Eighteen (18) of the 20 patients had a clinical response at 1 month and a marked reduction in endoscopic lesions was seen with the three highest doses. A multicenter, randomized, placebo-controlled trial of cA2 was performed in Crohn's disease patients (Targan *et al.*, 1996). One hundred and eight (108) patients were randomized to receive

cA2 (5, 10 or 20 mg/kg) or placebo as a single infusion and patients were followed for 12 weeks. Blinded results indicate that 51% of patients had a clinical response.
Additional large randomized trials with anti-TNF therapy are ongoing in Crohn's disease patients.

Anti-adhesion mabs

Antibodies to CD11b/CD18 (ED7 and OX42) have been evaluated in a rat model of TNBS (2,3,6-trinitrobenzene sulfonic acid) induced acute colitis (Palmen *et al.*, 1995). Rats were given the mAbs 2 hours before and 3 days after acute colitis. Four days later immunohistochemical evaluation of the colon demonstrated a reduction in colonic tissue damage, decrease in amount of submucosal infiltration with monocytes and leukocytes and decreased myeloperoxidase activity.

Diabetes

Due to the autoimmune nature of the development of type 1 diabetes, mAbs directed at T cell epitopes, cytokines or adhesion molecules are being developed for clinical use in the treatment of diabetes. Animal studies of anti-CD4 and anti-CD3 have been performed with partial improvement of diabetes seen. A small clinical trial has been peformed with immunoconjugate anti-CD5. ICAM-1 is thought to play a major role in the pathogenesis of diabetes by facilitating the interaction between T cells and their antigens or targets in the destruction of beta cells in the islets of Langerhans. Results of treatment of animal diabetic models with anti-ICAM-1 or anti-LFA-1 are promising.

Anti-CD5

An open label uncontrolled clinical study was performed with immunoconjugate anti-CD5 (CD5 Plus: anti-CD5+ ricin A chain) in patients with diabetes of 5 months duration (Skyler *et al.*, 1993). Fifteen (15) patients were treated with doses of anti-CD5 immunoconjugate 0.1 to 0.33 mg/kg/day for 5 days.

Anti-CD4

Anti-CD4 therapy has been evaluated in a non-obese diabetic (NOD) mouse model (Hayward *et al.*, 1993). NOD mice were treated from birth every two weeks to 6 months of age with mAb anti-CD4 and diabetes was prevented in all animals, with insulitis developing only later in some animals. However, injection of cyclophosphamide at 16 months resulted in 60% of animals becoming diabetic.

Anti-CD3

In a mouse model of NOD, treatment with anti-CD3 prevented the occurrence of cyclophosphamide acceleration of spontaneous diabetes (Chatenoud *et al.*, 1994).

In this study, anti-CD3 treatment of mice with diabetes resulted in complete remission of disease in 64–80% of mice for over 4 months.

Anti-Adhesion mAbs

Encouraging results have been obtained in animal models with anti-ICAM-1 and anti-LFA-1 treatment of diabetes. In animal models of diabetes using multidose streptozotocin induction, animals treated with anti-ICAM-1 and/or anti-LFA-1 do not develop diabetes (Herold *et al.*, 1994). In an adoptive transfer experiment, mAb anti-LFA-1 protected mice from adoptive transfer of type 1 diabetes (Fabien *et al.*, 1996): By day 60 post transfer only 3 of 11 mice developed diabetes versus diabetes present in 26 of 32 mice by 20 days posttransfer.

Infections and Other Inflammatory Disorders

mAbs are being developed for a variety of infectious disorders including, among others, malaria (Kwiatkowski *et al.*, 1993), HIV (Le Naour *et al.*, 1994; Looney *et al.*, 1994) CMV (Paar and Pollard, 1996) and for a variety of other inflammatory disorders such as uveitis (Thurau *et al.*, 1994). For example, an open label dose ranging study of murine mAb anti-TNF (CB0006) in children with cerebral malaria was performed (Kwiatkowski *et al.*, 1993). Forty one (41) patients received a single dose of 0.1, 1, 5 mg/kg anti-TNF or placebo. There was a dose dependent improvement in symptoms (fever), but no effect on mortality or on parasite clearance rates.

CONCLUSIONS

Evaluation of mAbs in acute and chronic inflammatory disorders have led to effective therapeutic approaches and to a greater understanding of the pathogenesis of the underlying inflammatory disease processes. mAb studies have aided in the identification of similarities between different inflammatory disorders, the uniqueness of different stages of given inflammatory disorders, and the specific cellular interactions in inflammatory diseases. mAb studies have resulted in identification of coexisting inflammatory mechanisms including immune adherance mechanisms, proinflammatory mediators, complement activation and vasoactive peptides.

mAb research has led to greater understanding of similarities between inflammatory disorders such as Crohn's disease and RA. mAb transplant studies have aided in defining differences between acute versus chronic transplant rejection. Anti-endotoxin and anti-TNF therapy has resulted in the identification of proinflammatory mediators in sepsis/SIRS versus later anti-inflammatory mediators such as IL-10. Therapies involving pan mAb depletion of overactive peripheral T cells has led to a greater understanding of the unique interactions of T cells, macrophages and synovial cells in the RA synovium. Combination mAb therapies are evolving subsequent to our understanding of the multifactorial nature of the inflammatory insult and the pleiotropic effects of destructive immune mediators.

Clinical mAb studies have been instrumental in defining realistic goals for clinical therapies with mAbs. mAb clinical studies have resulted in an understanding of the limitations of animal experimental models in predicting clinical effectiveness of mAbs. Timing of mAb administration relative to the inflammatory insult, antigenicity of specific mAbs and diversity of clinical populations are all challenges in effective mAb therapy. Additional challenges include the effects of concomitant medications, complications of compliance/side effect profiles and issues related to cost containment.

The future of mAb therapies lie in the successful integration of our knowledge of the inciting inflammatory events, our recognition of additional host inflammatory and homeostatic mechanisms, our ability to accurately predict the relevant patient subgroups who will benefit from a given therapy and finally the practical considerations of administering mAbs in clinical settings.

REFERENCES

Abraham, E.A., Wunderink, R., Silverman, H., Perl, T.M., Nasraway, S., Levy, H., Bone, R. *et al.* (1993) Efficacy and safety of monoclonal antibody to human tumor necrosis factor alpha in patients with sepsis syndrome: a randomized controlled, double-blind, multicenter clinical trial. *JAMA* **273**:934–941.

Abraham, E., Jesmok, G., Tuder, R., Allbee, J. and Chang, Y.H. (1995) Contribution of tumor necrosis factor-alpha to pulmonary cytkine expression and lung injury after hemorrhage and resuscitation. *Crit. Care Med.* **23**:1319–1326.

al-Janadi, M., al-Balla, S., al-Dalaan, A. and Raziuddin, S. (1993) Cytokine profile in systemic lupus erythematosus, rheumatoid arthritis, and other rheumatic diseases. *J. Clin. Immunol.* 193; **13**:58–67.

Amsterdam, E.A., Stahl, G.L., Pan, H.L., Rendig, S.V., Fletcher, M.P. and Longhurst, J.C. (1995) Limiation of reperfusion injury by a monoclonal antibody to C5a during myocardial infarction in pigs. *Am. J. Physiol.* **268**:H448–457.

American College of Chest Physicians/Society of Critical Care Medicine Consensus Committee. (1992) Definitions for sepsis and organ failures and guidelines for the use of innovative therapies in sepsis. *Chest* **101**:1658–1662.

Arend, W.P. and Dayer, J.M. (1995) Inhibition of the production and effects of interleukin-1 and tumor necrosis factor alpha in rheumatoid arthritis. *Arthritis Rheumatol* **38**:151–160.

Arvidson, N.G., Gudbjornsson, B., Elfman, L. *et al.* (1994) Circadian rhythm of serum interleukin-6 in rheumatoid arthritis. *Annals of the Rheumatic Diseases* **53**:521–524.

Asakura, K., Miller, D.J., Murray, K., Bansal, R., Pfeiffer, S.E. and Rodriguez, M. (1996) Monoclonal autoantibody SCH 94.03 which promotes central nervous system remyelination, recognizes an antigen on the surface of oligodendrocytes. *J. Neurosci. Res.* **43**:273–281.

Baker, D., Butler, D., Scallon, B.J., Oneill, J.K., Turk, J.L. and Feldmann, M. (1994) Control of established experimental allergic encephalomyelitis by inhibition of tumor necrosis factor (TNF) activity within the central nervous system using monoclonal antibodies and TNF receptor-immunoglobulin fusion proteins. *Eur. J. Immunol.* **24**:2040–2048.

Barak, V., Levi-Schaffer, F., Nisman, B. and Nagler, A. (1995) Cytokine dysregulation in chronic graft versus host disease. *Leuk Lymphoma* **17**:169–173.

Baumgartner, J.D. (1990) Monoclonal anti-endotoxin antibodies for the treatment of gram-negative bacteremia and septic shock. *Eur. J. Clin. Microbiol. Infect. Dis.* **9**:711–716.

Baumgartner, S., Moreland, L.W., Schiff, M.H., Tindall, E., Fleischmann, R.M., Weaver, A., Ettinger, R.E., Gruber, B.L., Katz, R.S., Skosey, J.L., Lies, R.B., Robison, A. and Blosch, C.M. (1996) Double-blind, placebo-controlled trial of tumor necrosis factor receptor (p80) fusion protein (TNFR:Fc) in active rheumatoid arthritis. *Arthritis and Rheumatol* **39**:S74 abstract #283.

Beilman, G.J., Shield, C.F. 3d, Hughes, J.D., Kelley, H.K., Ward, L.G. and Beck, D. (1993) The effects of intraoperative administration of OKT3 during renal transplantation. *Transplantation* **55**:490–493.

Belanger, C., Esperou-Bourdeau, H., Bordigoni, P., Jouet, J.P., Souillet, G., Milpied, N., Troussard, X., Kuentz, M., Herve, P., Reiffers, J. *et al.* (1993) Use of an anti-interleukin-2 receptor monoclonal antibody for GVHD prophylaxis in unrelated donor BMT. *Bone Marrow Transplant* **11**:293–297.

Beutler, B. (1993) Endotoxin, tumor necrosis factor and related mediators. New approaches to shock. *New Horizons* **1**:3–12.

Binder, J., Lehmann, M., Graser, E., Hancock, W.W., Watschinger, B., Onodera, K., Sayegh, M.H., Volk, H.D. and Kupiec-Weglinski, J.W. (1996) The effects of nondepleting CD4 targeted therapy in presensitized rat recipients of cardiac allografts. *Transplantation* **61**:804–811.

Blackwell, T.S. and Christman, J.W. (1996) Sepsis and cytokines: current status. *Br. J. Anaesthesia* **77**:110–117.

Boekstegers, P., Weidenhofer, S., Zell, R., Pilz, G., Holler, E., Ertel, W., Kapsner, T., Redl, H., Schlag, G., Kaul, M. *et al.* (1994) Repeated administration of a F(ab')2 fragment of an anti-tumor necrosis factor alpha monoclonal antibody in patients with severe sepsis: effects on the cardiovascular system and cytokine levels. *Shock* **1**:237–45.

Bolling, S.F., Kunkel, S.L. and Lin, H. (1992) Prolongation of cardiac allograft survival in rats by anti-TNF and cyclosporine combination therapy. *Transplantation* **53**:283–286.

Bone, R.C. (1991) The pathogeneiss of sepsis. *Ann. Intern. Med.* **115**:457–469.

Bone, R.C. (1994) Gram-positive Organisms and Sepsis. *Arch. Intern. Med.* **154**:26–34.

Bone, R.C., Balk, R.A., Fein, A.M., Perl, T.M., Wenzel, R.P., Renes, H.D., Quenzer, R.W., Iberti, T.J., Maclntyre, N. and Schein, R.M. (1995) A second large controlled clinical study of E5, a monoclonal antibody to endotoxin; results of a prospective, multicenter, randomized, controlled trial. The E5 Sepsis Study Group. *Crit. Care Med.* **23**:994–1006.

Bonnefoy-Berard, N. and Revillard, J.P. (1996) Mechanisms of immunosuppression induced by antithymocyte globulins and OKT3. *J. Heart Lung Transplant.* **15**:435–442.

Bonner, G.F. (1996) Current medical therapy for inflammatory bowel disease. *South Med. J.* **89**: 556–566.

Borden, E.C. and Chin, P. (1994) Interleukin-6: a cytokine with potential diagnostic and therapeutic roles. *J. Laboratory and Clincal Medicine* **123**:824–829.

Bowes, M.P., Rothlein, R., Fagan, S.C. and Zivin, J.A. (1995) Monoclonal antibodies preventing leukocyte activation reduce experimental neurologic inujury and enhance efficacy of thrombolytic therapy. *Neurology* **45**:815–819.

Braegger, C.P., Nicholls, S., Murch, S.H., Stephens, S. and MacDonald, T.T. (1992) Tumour necrosis factor alpha in stool as a marker of intestinal inflammation. *Lancet* **339**:89–91.

Breedveld, F.C. and van der Lubbe, P.A. (1995) Monoclonal antibody therapy of inflammatory rheumatic diseases. *Br. Med. Bull.* **51**:493–502.

Breidahl, A.F., Hickey, M.J., Stewart, A.G., Hayward, P.G. and Morrison, W.A. (1996) Effects of low dose intra-arterial monoclonal antibodies to ICAM-1 and CD11/CD18 on local and systemic consequences of ischaemia-reperfusion inujury in skeletal muscle. *Br. J. Plast. Surg.* **49**:202–209.

Brod, S.A., Lindsey, J.W. and Wolinsky, J.S. (1996) Multiple sclerosis: clinical presentation, diagnosis and treatment. *Am. Fam. Physician.* **54**:1301–1306,1309–1311.

Califf, R.M. (1995) Restenosis: the cost to society. *Am. Heart. J.* **130**:680–684.

Cannon, G.W., Marble, D.A., Griffiths, M.M., Cole, B.C., McCall, S., Schulman, S.F. and Strand, V. (1995) Immunologic assessment during treatment of rheumatoid arthritis with anti-CD5 immunoconjugate. *J. Rheumatol.* **22**:207–213.

Carey, G., Lisi, P.J. and Schroeder, T.J. (1995) The incidence of antibody formation to OKT3 consequent to its use in organ transplantation. *Transplantation* **60**:151–158.

Chandler, C. and Passaro, E. Jr. (1993) Transplant rejection. Mechanisms and treatment. *Arch. Surg.* **128**:279–283.

Chang, D.M. (1993) The role of cytokines in heat stroke. *Immunol. Invest.* **22**:553–561.

Chang, D.M., Ding, Y.A., Kuo, S.Y., Chang, M.L. and Wei, J. (1996) Cytokines and cell surface markers in prediction of cardiac allograft rejection. *Immunol. Invest.* **25**:13–21.

Chatenoud, L., Baudrihaye, M.F., Kreis, H., Goldstein, G., Schindler, J. and Bach, J.F. (1982) Human *in vivo* antigenic modulation induced by the anti-T cell OKT3 monoclonal antibody. *Eur. J. Immunol.* **12**:979–982.

Chatenoud, L. (1993) OKT3-induced cytokine-release syndrome: prevention effect of anti-tumor necrosis factor monoclonal antibody. *Transpl. Proc.* **25**(suppl.):47–51.

Chatenoud, L., Thervet, E., Primo, J. and Bach, J.F. (1994) Anti-CD3 antibody induces long-term remission of overt autoimmunity in nonobese diabetic mice. *PNAS* **91**:123–127.

Choy, E.H.S., Chikanxa, I.C., Kingsley, G.H. *et al.* (1991) Chimaeric anti-CD4 monoclonal antibody for relapsing polychondritis. *Lancet* **338**:450.

Chu, C.Q., Field, M., Feldmann, M. and Maini, R.N. (1991) Localization of tumor necrosis factor alpha in synovial tissues and at the cartilage-pannus junction in patients with rheumatoid arthritis. *Arthritis Rheum.* **34**:1125–1132.

Cohick, C.B., Furst, D.E., Quagliata, S., Corcoran, K.A., Steere, K.J., Yager, J.G. and Lindsley, H.B. (1994) Analysis of elevated serum interleukin-6 levels in rheumatoid arthritis: correlation with erythrocyte sedimentation rate or C-reactive protein. *J. Lab. Clin. Med.* **123**:721–727.

Coller, B.S., Anderson, K. and Weisman, H.F. (1995) New anti-platelet agents: platelet GPIIb/IIIa antagonists. *Thromb. Haemost.* **74**:302–308.

Connolly, M.K., Kitchens, E.A., Chan, B., Jardieu, P. and Wofsy, D. (1994) Treatment of murine lupus with monoclonal antibodies to lymphocyte function-associated antigen-1: dose-dependent inhibition of autoantibody production and blockade of the immune response to therapy. *Clin. Immunol. Immunopath.* **72**:198–203.

Cooper, W.O., Fava, R.A., Gates, C.A., Cremer, M.A. and Townes, A.S. (1992) Acceleration of onset of collagen-induced arthritis by intra-articuar injection of tumour necrosis factor or transforming growth factor-beta. *Clin. Exp. Immunol.* **89**:244–250.

Cope, A.P., Aderka, D., Doherty, M., Englemann, H., Gibbons, D., Jones, A.C., Brennan, F.M., Maini, R.N., Wallach, D. and Feldmann, M. (1992) Increased levels of soluble tumor necrosis factor receptors in the sera and synovial fluid of patients with rheumatoid diseases. *Arthritis Rheum.* **35**:1160–1169.

Crilly, A., McInness, I.B., McDonald, A.G., Watson, J., Capell, H.A. and Madhok, R. (1995) Interleukin 6 (IL-6) and soluble IL-2 receptor levels in patients with rheumatoid arthritis treated with low dose oral methotrexate. *J. Rheumatol.* **22**:224–226.

Cutolo, M., Sulli, A., Barone, A., Seriolo, B. and Accardo, S. (1993) Macrophages, synovial tissue and rheumatoid arthritis. *Clin. Exp. Rheumatol.* **11**:331–339.

Darby, C.R., Bushell, A., Morris, P.J. and Wood, K.J. (1994) Nondepleting anti-CD4 antibodies in transplantation. Evidence that modulation is far less effective than prolonged CD4 blockade. *Transplantation* **57**:1419–1426.

Davis, L.S., Kavanaugh, A.F., Nichols, L.A. and Lipsky, P.E. (1995) Induction of persistent T cell hyporesponsiveness *in vivo* by monoclonal antibody to ICAM-1 in patients with rheumatoid arthritis. *J. Immunol.* **154**:3525–3537.

Deng, M.C., Erren, M., Kammerling, L., Gunther, F., Kerber, S., Fahrenkamp, A., Assmann, G., Breithardt, G. and Scheld, H.H. (1995) The relationship of interleukin-6, tumor necrosis factor-alpha, IL-2 and IL-2 receptor levels to cellular rejection, allograft dysfunction, and clinical events early after cardiac transplantation. *Transplantation* **60**:1118–1124.

Deng, M.C., Dasch, B., Erren, M., Mollhoff, T. and Scheld, H.H. (1996) Impact of left ventricular dysfunction on cytokines, hemodynamics, and outcome in bypass grafting. *Ann. Thorac. Surg.* **62**:184–190.

Dhainaut, J.F., Vincent, J.L., Richard, C., Lejeune, P., Martin, C., Fierobe, L., Stephens, S., Ney, U.M. and Sopwith, M. (1995) CDP571, a humanized antibody to human tumor necrosis factor-alpha: safety, pharmacokinetics, immune response, and influence of the antibody on cytokine concentrations in patients with septic shock. CDP571 Sepsis Study Group. *Crit. Care Med.* **23**: 1461–1469.

Dinarello, C.A. (1989) The endogenous pyrogens in host defense interactions. *Hosp. Practice.* **24**: 111–128.

Duff, G.W. (1994) Cytokines and acute phase proteins in rheumatoid arthritis. *Scand. J. Rheumatol. Suppl.* **100**:9–19.

Duke, O., Panayi, G.S., Janossy, G. and Poulter, L.W. (1982) An immunohistological analysis of lymphocyte subpopulations and their microenvironment in the synovial membranes of patients with rheumatoid arthritis using monoclonal antibodies. *Clin. Exp. Immunol.* **49**:22–30.

Elliott, M.J., Maini, R.N., Feldmann, M., Long-Fox, A., Charles, P., Katsikis, P., Brennan, F.M., Walker, J., Bijl, H., Ghrayeb, J. *et al.* (1993) Treatment of rheumatoid arthritis with chimeric monoclonal antibodies to tumor necrosis factor alpha. *Arthritis Rheum.* **36**:1681–1690.

Elliott, M.J. and Maini, R.N. (1994) New directions for biological therapy in rheumatoid arthritis. *Int. Arch. Allergy. Immunol.* **104**:112–125.

Elliott, M.J., Maini, R.N., Feldmann, M., Long-Fox, A., Charles, P., Bijl, H. Woody, J.N. (1994a) Repeated therapy with monoclonal antibody to tumor necrosis factor alpha (cA2) in patients with rheumatoid arthritis. *Lancet* **344**:1125–1127.

Elliott, M.J., Feldmann, M. and Maini, R.N. (1995) TNF blockade in rheumatoid arthritis: rationale, clinical outcomes and mechanisms of action. *Int. J. Immunopharmacol.* **17**:141–145.

Elliott, M.J. and Maini, R.N. (1995) Anti-cytokine therapy in rheumatoid arthritis. *Bailliere's Clinical Rheumatology* **9**:633–649.

Elliott, M.J., Maini, R.N., Feldmann, M., Kalden, J.R., Antoni, C., Smolen, J.S., Leeb, B., Breedveld, F.C., Macfarlane, J.D., Bijl, H. *et al.* (1994) Randomized double-blind comparison of chimeric monoclonal antibody to tumour necrosis factor alpha (cA2) versus placebo in rheumatoid arthritis. *Lancet* **344**:1105–1110.

Elliott, T.B., Ledney, G.D. *et al.* (1995a) Mixed-filed neutrons and gamma photons induce different changes in ileal bacteria and correlated sepsis in mice. *Int. J. Radiat. Biol.* **68**:311–320.

Emerson, T.E., Lindsey, D.C., Jesmok, G.J., Duerr, M.L. and Fournel, M.A. (1992) Efficacy of monoclonal antibody against tumour necrosis factor alpha in an endotoxemic baboon model. *Circulation Shock* **38**:75–84.

Endo, S., Inada, K., Knoue, Y., Kuwata, Y., Suzuki, M., Yamashita, H., Hoshi, S. and Yoshida, M. (1992) Two types of septic shock classified by the plasma level of cytokines and endotoxin. *Circulatory Shock* **38**:264–274.

The EPIC Investigation. (1994) Use of a monoclonal antibody directed against the platelet glycoprotein IIb/IIIa receptor in high-risk coronary angioplasty. *N. Engl. J. Med.* **330**:956–961.

Fabien, N., Bergerot, I., Orgiazzi, J. and Thiovolet, C. (1996) Lymphocyte function associated antigen-1, integrin alpha 4, and L-selectin mediate T-cell homing to the pancreas in the model of adoptive transfer of diabetes in NOD mice. *Diabetes* **45**:1181–1186.

Farahat, M.N., Yanni, G., Poston, R. and Panayi, G.S. (1993) Cytokine expression in synovial membranes of patients with rheumatoid arthritis and osteoarthritis. *Ann. Rheum. Dis.* **52**:870–875.

Farges, O., Ericzon, B.G., Bresson-Hadni, S., Lynch, S.V., Hockerstedt, K., Houssin, D., Galmarini, D., Faure, J.L., Baldauf, C. and Bismuth, H. (1994) A randomized trial of OKT3-based versus cyclosporine-based immunoprophylaxis after liver transplantation. Long-term results of a European and Australian multicenter study. *Transplantation* **58**: 891–898.

Fekade, D., Knox, K., Hussein, K., Melka, A., Lalloo, D.G., Coxon, R.E. and Warrell, D.A. (1996) Prevention of Jarisch-Herxheimer reactions by treatment with antibodies against tumor necrosis factor alpha. *N. Engl. J. Med.* **335**:311–315.

Ferrant, A., Latinne, D., Bazin, H., Straetmans, N., Cornet, A., de la Parra, B. and Michaux, J.L. (1995) Prophylaxis of graft-versus-host disease in identical sibling donor bone marrow transplant by anti-IL-2 receptor monoclonal antibody LO-Tact-1. *Bone Marrow Transplant.* **16**:577–581.

Fiedler, V.B., Loof, I., Sander, E., Voehringer, V., Galanos, C. and Fournel, M.A. (1992) Monoclonal antibody to tumor necrosis factor-alpha prevents lethal endotoxin sepsis in adult rhesus monkeys. *J. Lab. Clin. Med.* **120**:574–588.

Field, M., Chu, C., Feldmann, M. and Maini, M. (1991) Interleukin-6 localization in the synovial membrane in rheumatoid arthritis. *Rheumatology International* **11**:45–50.

Fisher, C.J. Jr, Opal, S.M., Dhinaut, J.F., Stephens, S., Zimmerman, J.L., Nightingale, P., Harris, S.J., Schein, R.M., Panacek, E.A., Vincent, J.L. *et al.* (1993) Influence of an anti-tumor necrosis factor monoclonal antibody on cytokine levels in patients with sepsis. The CB0006 Sepsis Syndrome Study Group. *Crit. Care Med.* **21**:318–327.

Fisher, C.J. Jr. and Zheng, Y. (1996) Potential Strategies for Inflammatory Mediator Manipulaton: Retrospect and Prospect. *World J. Surg.* **20**:447–453.

Fishwild, D.M. and Strand, V. (1994) Administration of an anti-CD5 immunoconjugate to patients with rheumatoid arthritis: effect on peripheral blood mononuclear cells and *in vitro* immune function. *J. Rheumatol.* **21**:596–604.

The French National Registry of HA-1A (Centoxin) in septic shock. (1994) A cohort study of 600 patients. The National Committee for the Evaluation of Centoxin. *Arch. Intern. Med.* **154**:2484–2491.

Gianotti, L., Alexander, J.W. *et al.* (1996) Steroid therapy can modulate gut barrier function, host defense and survival in thermally injured mice. *J. Surg. Res.* **62**:53–58.

Gianotti, L., Braga, M., Vaiani, R., Almondo, F. and Di Carlo, V. (1996a) Experimental gut-derived endotoxaemia and bacteraemia are reduced by systemic administration of monoclonal anti-LPS antibodies. *Burns* **22**:120–124.

Greenberg, R.N., Wilson, K.M., Kunz, A.Y., Wedel, N.I. and Gorelick, K.J. (1991) Randomized, double-blind phase II study of anti-endotoxin antibody (E5) as adjuvant therapy in humans with serious gram-negative infections. *Prog. Clin. Bio. Res.* **367**:179–186.

Greenman, R.L., Schein, R.M.H., Martin, M.A., Wenzel, R.P., MacIntyre, N.R., Emmanuel, G., Chmel, H., Kohler, R.B., McCarthy, M., Plouffe, J. *et al.* (1991) A controlled clinical trial of E5 murine monoclonal IgM antibody to endotoxin in the treatment of gram-negative sepsis. The XOMA Sepsis Study Group. *JAMA.* **266**:1097–1102.

Hack, C.E., Ce Groot, E.R., Felt-Bersma, R.J., Nuijens, J.H., Strack Van Schijndel, R.J., Eerenberg-Belmer, A.J., Thijs, L.G. and Aarden, L.A. (1989) Increased plasma level of interleukin-6 in sepsis. *Blood* **74**:1704–1710.

Hafler, D.A. and Weiner, H.C. (1988) Anti-CD4 and anti-CD2 monoclonal antibody infusions in subjects with multiple sclerosis. Immunosuppressive effects and human antimouse responses. *Ann. NY. Acad. Sci.* **540**:557–559.

Hartman, J.C., Anderson, D.C., Wiltse, A.L., Lane, C.L., Rosenbloom, C.L., Manning, A.M., Humphrey, W.R., Wall, T.M. and Shebuski, R.J. (1995) Protection of ischemic/reperfused canine myocardium by CL 18/6, a monoclonal antibody to adhesion molecule ICAM-1. *Cardiovasc. Res.* **30**:47–54.

Hasler, F., van de Putte, L., Dumont, E., Kneer, J., Bock, J., Dickinson, S., Lesslauer, W. and Van der Auwera, P. (1996) Safety and efficacy of TNF neutralization by Lenercept (TNFp55-IgG1,Ro45-2081) in patients with rheumatoid arthritis exposed to a single dose. *Arthritis and Rheumatol.* **39**:S243 abstract #1291.

Hasler, F., van de Putte, L., Baudin, M., Ludin, E., Durrwell, L., McAuliffe, T. and Van der Auwera, P. (1996a) Chronic TNF neutralization (up to one year) by Lenercept (TNFp55-IgG1,Ro45-2081) in patients with rheumatoid arthritis: results of an open-label extension of a double-blind single dose phase I study. *Arthritis and Rheumatol.* **39**:S243 abstract #1292.

Haverty, T.P., Sanders, M. and Sheahan, M. (1993) OKT3 treatment of cardiac allograft rejection. *J. Heart. Lung. Transplant.* **12**:591–598.

Hayward, A.R., Shriber, M., Cooke, A. and Waldmann, H. (1993) Prevention of diabetes but not insulitis in NOD mice injected with antibody to CD4. *J. Autoimmun.* **6**:301–310.

Herbelin, A., Chatenoud, L., Roux-Lombard, P., De Groote, D., Legendre, C., Dayer, J.M., Descamps-Latscha, B., Kreis, H. and Bach, J.F. (1995) *In vivo* soluble tumor necrosis factor receptor release in OKT3-treated patients. Differential regulation of TNF-sR55 and TNF-sR75. *Transplantation* **59**:1470–1475.

Hervelin, C., Stephan, J.L., Donadieu, J., Le Deist, F., Racadot, E., Wijdenes, J. and Fischer, A. (1994) Treatment of steroid-resistant acute graft-versus-host disease with an anti-IL-2-receptor monoclonal antibody (BT 563) in children who received T cell-depleted, partially matched, related bone marrow transplants. *Bone Marrow Transplant.* **13**:563–569.

Herold, K.C., Vezys, V., Gage, A. and Montag, A.G. (1994) Prevention of autoimmune diabetes by treatment with anti-LFA-1 and anti-ICAM-1 monoclonal antibodies. *Cell Immunol.* **157**:489–500.

Hesse, C.J., van Gelder, T., Vaessen, L.M., Knoop, C.J., Balk, A.H., Yzermans, J.N., Jutte, N.H. and Weimar, W. (1995) Pharmacodynamics of prophylactic antirejection therapy with an anti interleukin-2 receptor monoclonal antibody (BT563) after heart and kidney transplantation. *Immunopharmacology* **30**:237–246.

Hiepe, F., Volk, H.D., Apostodoff, E., von Baehr, R. and Emmrich, F. (1991) Treatment of severe systemic lupus erythematosus with anti-CD4 monoclonal antibody [letter]. *Lancet* **338**:1529–30.

Hill, J., Lindsay, T., Valeri, C.R., Shepro, D. and Hechtman, H.B. (1993) A CD18 antibody prevents lung injury but not hypotension after intestional ischemia-reperfusion. *J. Appl. Physiol.* **74**:659–664.

Hinshaw, L.B., Emerson, T.E., Tyalor, F.B., Chang, A.C.K., Duerr, M., Peer, G.T., Flournoy, D.J., White, G.L., Kosanke, S.D., Murray, C.K., Xu, R., Passey, R.B. and Fournel, M.A. (1992) Lethal Staphylococcus aureus-induced shock in primates: prevention of death with anti-TNF antibody. *Journal of Trauma* **33**:568–573.

Hofman, F.M., Hinton, D.R., Johnson, K. and Merrill, J.E. (1989) TNF identified in multiple sclerosis brain. *J. Exp. Med.* **11**:1364–1365.

Holler, E., Kolb, J.H., Moller, A. *et al.* (1990) Increased serum levels of TNF precede major complications of bone marrow transplantation. *Blood* **75**:1011–16.

Holler, E., Kolb, H.J., Mittermuller, J., Kaul, M., Ledderose, G., Duell, T., Seeber, B., Schleuning, M., Hintermeier-Knabe, R., Ertl, B. *et al.* (1995) Modulation of acute graft-versus-host-disease after allogeneic bone marrow transplantation by tumor necrosis factor alpha (TNF) release in the course of pretransplant conditioning: role of conditioning regimens and prophylactic application of a monoclonal antibody neutralizing human TNF (MAK 195F). *Blood* **86**:890–899.

Horneff, G., Emmrich, F. Burmester, G.R. (1993) Advances in immunotherapy of rheumatoid arthritis: clinical and immunological findings following treatment with anti-CD4 antibodies. *Br. J. Rheumatol.* **32S**:39–47.

Horneff, G., Sack, U., Kalden, J.R., Emmrich, F. and Burmester, G.R. (1993a) Reduction of monocyte-macrophage activation markers upon anti-CD4 treatment. Decreased levels of IL-1, IL-6, neopterin and soluble CD14 in patients with rheumatoid arthritis. *Clin. Exp. Immunol.* **91**:207–213.

Houssiau, F.A., Devogelaer, J.-P., Fan Damme, J. *et al.* Interleukin-6 in synovial fluid and serum of patients with rheumatoid arthritis and other inflammatory arthritides. *Arthritis and Rheum.* **31**: 784–788.

Hughes, C.B., Gaber, L.W., Mohey el-Din, A.B., Grewal, H.P., Kotb, M., Mann, L. and Gaber, A.O. (1996) Inhibition of TNF improves survival in an experimental model of acute pancreatitis. *Am. Surg.* **62**:8–13.

Husby, G. and Williams, R.C. Jr. (1988) Synovial localization of tumor necrosis factor in patients with rheumatoid arthritis. *J. Autoimmun.* **1**:363–371.

Imagawa, D.K., Millis, J.M., Olthoff, K.M., Derus, L.J., Chia, D., Sugich, L.R., Ozawa, M., Dempsey, R.A., Iwaki, Y., Levy, P.J. *et al.* (1990) The role of tumor necrosis factor in allograft rejection I. Evidence that elevated levels of tumor necrosis factor-alpha predict rejection following orthotopic liver transplantation. *Transplantation* **50**:219–25.

Imagawa, D.K., Millis, J.M., Seu, P., Olthoff, K.M., Hart, J., Wasef, E., Dempsey, R.A., Stephens, S. and Busuttil, R.W. (1991) The role of TNF in allograft rejection III: Evidence that anti-TNF antibody therapy prolongs allograft survival in rats with acute rejection. *Transplantation* **51**:57–62.

Imamura, M., Tanaka, J., Hashino, S., Kobayashi, H., Hirano, S., Minagawa, T., Imai, K., Kobayashi, S., Fujii, Y., Kasai, M. *et al.* (1994) Cytokines involved in graft-versus-host disease. *Hokkaido Igaku Zasshi* **69**:1348–1353.

Isaacs, J.D., Manna, V.K., Rapson, N., Bulpitt, K.J., Hazleman, B.L., Matteson, E.L., St. Clair, E.W., Schnitzer, T.J. and Johnston, J.M. (1996) CAMPATH-1H in rheumatoid arthritis – an intravenous dose-ranging study. *Br. J. Rheumatol.* **35**:231–240.

Ishikawa, H., Hirata, S., Nishibayaski, Y., Imura, S., Kubo, H. and Ohno, O. (1994) The role of adhesion molecules in synovial pannus formation in rheumatoid arthritis. *Clin. Orthop.* **300**: 297–303.

Jasin, H.E., Lightfoot, E., Davis, L.S., Rothlein, R., Faanes, R.B. and Lipsky, P.E. (1992) Amelioraton of antigen-induced arthritis in rabbits treated with monoclonal antibodies to leukocyte adhesion molecules. *Arthritis Rheum* **35**:541–549.

Kalden, J.R. (1994) Biologic agents in the therapy of inflammatory rheumatic diseases, including therapeutic antibodies, cytokines, and cytokine antagonists. *Curr. Opin. Rheumatol.* **6**:281–286.

Kalden, J.R. and Manger, B. (1995) Biologic agents in the treatment of inflammatory rheumatic diseases. *Curr. Opin. Rheumatol.* **7**:191–197.

Kapp, A. (1993) The role of cytokines in the psoriatic inflammation. *J. Dermatol. Sci.* **5**:133–142.

Karsh, J., Klippel, J.H., Plotz, P.H., Decker, J.L., Wright, D.G. and Flye, M. (1981) Lymphapheresis in rheumatoid arthritis. *Arthritis Rheum.* **24**:867–873.

Kavanaugh, A.F., Davis, L.S., Nichols, L.A., Norris, S.H., Rothlein, R., Scharschmidt, L.A. and Lipsky, P.E. (1994) Treatment of refractory rheumatoid arthritis with a monoclonal antibody to intercellular adhesion molecule 1. *Arthritis Rheum.* **37**:992–999.

Kavanaugh, A.F., Cush, J.J., St. Clair, E.W., McCune, W.J., Braakman, T.A.J., Nichols, L.A. and Lipsky, P.E. (1996) Anti-TNF monoclonal antibody (mAb) treatment of rheumatoid arthritis (RA) patients with active disease on methotrexate (MTX); results of a double-blind placebo controlled multicenter trial. *Arthritis and Rheumatol.* **39**:S123 abstract #575.

Keffer, J., Probert, L., Cazlaris, H., Georgopoulous, S., Kaslaris, E., Kioussis, D. and Kollias, G. (1991) Transgenic mice expressing human tumor necrosis factor: A predictive genetic model of arthritis. *EMBOJ.* **10**:4025–4031.

Kemeny, L., Michel, G., Dobozy, A. and Ruzicka, T. (1994) Cytokine system as potential target for antipsoriatic therapy. *Exp. Dermatol.* **3**:1–8.

Kent, S.J., Karlik, S.J., Cannon, C., Hines, D.K., Yednock, T.A., Fritz, L.C. and Horner, H.C. (1995) A monoclonal antibody to alpha 4-integrin suppresses and reverses active experimental allergic encephalomyelitis. *J. Neuroimmunol.* **58**:1–10.

Kent, S.J., Karlik, S.J., Rice, G.P., Horner, H.C. (1995a) A monoclonal antibody to alpha 4-integrin reverses the detectable signs of experimental allergic encephalomyelitis in the guinea pig. *J. Magn. Reson. Imaging* **5**:535–540.

Kinne, R.W., Becker, W., Simon, G., Paganelli, G., Palombo-Kinne, E., Wolski, A., Bloch, S., Schwartz, A., Wolf, F. and Emmrich, F. (1993) Joint uptake and distribution of a technetium-99m-labeled anti-rat CD4 monoclonal antibody in rat adjuvant arthritis. *J. Nucl. Med.* **34**:92–98.

Kirkham, B., Chikanza, I., Pitzalis, C. *et al.* (1988) Response to monoclonal CD7 antibody in rheumatoid arthritis. *Lancet* **1**:589.

Kirkham, B.W., Pitzalis, C., Kingsley, G.H., Chikanza, I.C., Sabharwal, S., Barbatis, C., Grahame, R., Gibson, T., Amlot, P.L. and Panayi, G.S. (1991) Monoclonal antibody treatment in rheumatoid arthritis. The clinical and immunological effects of a CD7 monoclonal antibody. *Br. J. Rheumatol.* **30**:459–463.

Kirkham, B.W., Thien, F., Pelton, B.K., Pitzalis, C., Amlot, P., Denman, A.N. and Panayi, G.S. (1992) Chimeric CD7 monoclonal antibody therapy in rheumatoid arthritis. *J. Rheumatol.* **19**:1348–1352.

Kobayashi, H., Kawai, S., Sakayori, S., Kaneko, M., Ito, Y., Ujike, Y., Kobayashi, K., Imaizumi, H., Hoshi, S., Endo, S. *et al.* (1994) Phase II study of edobacomab (E5) in the treatment of gram-negative sepsis. *Kansenshogaku Zasshi* **68**:81–115.

Kobayashi, Y., Kawai, K., Honda, H., Tomida, S., Niimi, N., Tamatani, T., Miyasaka, M. and Yoshikai, Y. (1995) Antibodies against leukocyte function-associated antigen-1 and against intercellular adhesion molecule-1 together suppress the progression of experimental allergic encephalomyelitis. *Cell. Immunol.* **164**:295–305.

Kotzin, B.L., Strober, S., Engleman, E.G., Calin, A., Hoppe, R.T., Kansas, G.S., Terrell, C.P. and Kaplan, H.A. (1981) Treatment of intractable rheumatoid arthritis with total lymphoid irradiation. *N. Engl. J. Med.* **305**:969–976.

Krams, S.M., Asher, N.L. and Martinez, O.M. (1993) New insights into mechanisms of allograft rejection. *Gastroenterol. Clin. N. America* **22**:381–400.

Kuhns, D.B., Alvord, W.G. and Gallin, J.I. (1995) Increased circulating cytokines, cytokine antagonists, and E-selectin after intravenous administration of endotoxin in humans. *J. Inf. Diseases* **171**: 145–152.

Kumano, K., Irie, A., Mashimo, S., Endo, T. Koshiba, K. (1996) Long-term efficacy of OKT3 for steroid-resistant acute rejection in renal transplant patients. *Transplant. Proc.* **28**:1354–1355.

Kwiatkowski, D., Molyneux, M.E., Stephens, S., Curtis, N., Klein, N., Pointaire, P., Smit, M., Allan, R., Brewster, D.R., Grau, G.R. *et al.* (1993) Anti-TNF therapy inhibits fever in cerebral malaria. *Q. J. Med.* **86**:91–98.

Lazarovits, A.I. and Karsh, J. (1988) Decreased expression of CD7 occurs in rheumatoid arthritis. *Clin. Exp. Immunol.* **72**:470–475.

Lazarovits, A.I., White, M.J. and Karsh, J. (1992) CD7-T cells in rheumatoid arthritis. *Arthritis Rheum.* **35**:616–624.

Lazarovits, A.I., Rochon, J., Banks, L., Hollomby, D.J., Muirhead, N., Jevnikar, A.M., White, M.J., Amlot, P.L., Beauregard-Zollinger, L. and Stiller, C.R. (1993) Human mouse chimeric CD7 monoclonal antibody (SDZCHH380) for the prophylaxis of kidney transplant rejection. *J. Immunol.* **150**:5163–5174.

Lazarovits, A.I., Poppema, S., Zhang, Z., Khandaker, M., Le Feuvre, C.E., Singhal, S.K., Garcia, B.M., Ogasa, N., Jevnikar, A.M., White, M.H., Singh, G., Stiller, C.R. and Zhong, R.Z. (1996) Prevention and reversal of renal allograft rejection by antibody against CD45RB. *Nature* **380**:717–720.

Le Naour, R., Raoul, H., Mabondzo, A., Henin, Y., Bousseau, A. and Dormont, D. (1994) Treatment of human monocyte-derived macrophages with a TNF synthesis inhibitor prior to HIV1 infection: consequences of cytokine production and viral replication. *Res. Virol.* **145**:199–207.

Lefkovits, J. and Topol, E.J. (1995) Platelet glycoprotein IIb/IIIa receptor inhibitors in ischemic heart disease. *Curr. Opin. Cardiol.* **10**:420–426.

Lefkovits, J., Ivanhoe, R.J., Califf, R.M., Bergelson, B.A., Anderson, K.M., Stoner, G.L., Weisman, H.F. and Topol, E.J. (1996) Effects of platelet glycoprotein IIb/IIIa receptor blockade by a chimeric monoclonal antibody (abciximab) on acute and six-month outcomes after percutaneous transluminal coronary angioplasty for acute myocardial infarction. EPCI Investigators. *Am. J. Cardiol.* **77**: 1045–1051.

Levy, R., Weisman, M., Wiesenhutter, C., Yocum, D., Schnitzer, T., Goldman, A., Schiff, M., Breedveld, F., Solinger, A., MacDonald, B. and Lipani, J. (1996) Results of a placebo-controlled, multicenter trial using a primatized non-depleting, anti-CD4 monoclonal antibody in the treatment of rheumatoid arthritis. *Arthritis and Rheumatol.* **39**:S122 abstract #574.

Li, X.C., Zhong, R., Quan, D., Almawi, W., Jevnikar, A. and Grant, D. (1994) Endotoxin in the peripheral blood during acute intestinal allograft rejection. *Transpl. Int'l.* **7**:223–226.

Libert, C., Vink, A., Coulie, P., Brouchaert, P., Evaraerdt, B., van Snick, J. and Fiers, W. (1992) Limited involvement of interleukin-6 in the pathogenesis of lethal septic shock as revealed by the effect of monoclonal antibodies against interleukin-6 or its receptor in various murine models. *Eur. J. of Immunol.* **22**:2625–2630.

Lin, H., Chensue, S.W., Strieter, R.M., Remick, D.G., Gallagher, K.P., Bolling, S.F. and Kunkel, S.L. (1992) Antibodies against tumor necrosis factor prolong cardiac allograft survival in the rat. *J. Heart Lung Transplant.* **11**:330–335.

Lindsey, J.W., Hodgkinson, S., Mehta, R., Siegel, R.C., Mitchell, D.J., Lim, M., Piercy, C., Tram, T., Dorfman, L., Enzmann, D. *et al.* (1994) Phase I clinical trial of chimeric monoclonal anti-CD4 antibody in multiple sclerosis. *Neurology* **44**:413–419.

Lindsey, J.W., Hodgkinson, S., Mehta, R., Mitchell, D., Enzmann, D. and Steinman, L. (1994a) Repeated treatment with chimeric anti-CD4 antibody in mutiple sclerosis. *Ann. Neurol.* **36**:183–189.

Liu, T., Clark, R.K., McDonnell, P.C., Young, P.R., White, R.F., Barone, F.C. and Feuerstein, G.Z. (1994) Tumor necrosis factor-alpha expression in ischemic neurons. *Stroke* **25**:1481–1488.

Looney, J.E., Willinger, A., Lin, G., Pieber, E.P., Riethmuller, G. and Ghrayeb, J. (1994) Expression and characterization of cM-T413, a chimeric anti-CD4 antibody with *in vitro* immunosuppressive activity. *J. Immunother. Emphasis Tumor. Immunol.* **16**:36–46.

Luce, J.M. (1993) Introduction of new technology in critical care practice: A history of HA-1A human monoclonal antibody against endotoxin. *Crit. Care Med.* **21**:1233–1241.

Ma, X.L., Weyrich, A.S., Lefer, D.J., Buerke, M., Albertine, K.H., Kishimoto, T.K. and Lefer, A.M. (1993) Monoclonal antibody to L-seletin attenuates neutrophil accumulation and protects ischemic reperfused cat myocardium. *Circulation.* **88**:649–658.

Ma, S., Thompson, J., Hu, H., Neville, D.M. Jr. *et al.* (1996) Expression and characterization of a divalent chimeric anti-human CD3 single chain antibody. *Scand. J. Immunol.* **43**:134–139.

MacDonald, T.T., Hutchings, P., Choy, M., Murch, S. and Cooke, A. (1990) Tumour necrosis factor-alpha and interferon-gamma production measured at the single cell level in normal and inflamed human intestine. *Clin. Exp. Immunol.* **81**:301–305.

Maini, R.N., Elliott, M.J., Brennan, F.M., Williams, R.O., Chu, C.Q., Paleolog, E., Charles, P.J., Taylor, P.C. and Feldmann, M. (1995) Monoclonal anti-TNF antibody as a probe of pathogenesis and therapy of rheumatoid disease. *Imunol. Rev.* **144**:195–223.

Martin, C., Sauzx, P., Mege, J.L., Perrin, G., Papazian, L. and Gouin, F. (1994) Prognostic value of serum cytokines in septic shock. *Intensive Care Medicine* **20**:272–277.

Martin, P.J., Nelson, B.J., Appelbaum, F.R., Anasetti, C., Deeg, H.J., Hansen, J.A., McDonald, G.B., Nash, R.A., Sullivan, K.M., Witherspoon, R.P., Scannon, P.J., Friedmann, N. and Storb R. (1996) Evaluation of a CD5-specific immunotoxin for treatment of actue graft-versus-host disease after allogeneic marrow transplantation. *Blood* **88**:824–830.

Matteson, E.L., Yocum, D.E., St. Clair, E.W., Achkar, A.A., Thakor, M.S., Jacobns, M.R., Hays, A.E., Heitman, C.K. and Johnston, J.M. (1995) Treatment of active refractory rheumatoid arthritis with humanized monoclonal antibody CAMPATH-1H administered by daily subcutaneous injection. *Arthritis Rheum.* **38**:1187–1193.

Maury, C.P. and Teppo, A.M. (1987) Raised serum levels of cachecitn/TNF in renal allograft rejection. *J. Exp. Med.* **166**:1132–1137.

Maury, C.P. and Teppo, A.M. (1989) Circulating TNF (cachectin) in myocardial infarction. *J. Intern. Med.* **255**:333–336.

McCabe, R.P., Woody, J., van Deventer, S., Targan, S.R., Mayer, L., van Hogezand, R., Rutgeerts, P., Hanauer, S.B., Podolsky, D. and Elson, C.O. (1996) A multicenter trial of cA2 anti-TNF chimeric monoclonal antibody in patients with active Crohn's disease. *Gastroenterology* **110**:#4.

Merino, R., Fossati, L., Iwamoto, M., Takahashi, S., Lemoine, R., Ibnou-Zekri, N., Pugliatti, L., Merino, J. and Izui, S. (1995) Effect of long-term anti-CD4 or anti-CD8 treatment on the development of Ipr CD4–CD8-double negative T cells and of the autoimmune syndrome in MRL-lpr/lpr mice. *J. Autoimmun.* **8**:33–45.

Merrill, J.E., Strom, S.R., Ellison, G.W. and Meyhers, L.W. (1989) *In vitro* study of mediators of inflammation in multiple sclerosis. *J. Clin. Immunol.* **9**:84–96.

Michie, H.R., Manogue, K.R., Spriggs, D.R., Revhaug, A., O'Dwyer, S., Siarello, C.A., Cerami, A., Wolpe, S.A. and Wilmore, D.W. (1988) Detection of circulating tumour necrosis factor after endotoxin adminstration. *NEJM.* **318**:1481–1485.

Mihara, M., Ikuta, M., Koishihara, Y. and Ohsugi, Y. (1991) Interleukin-6 inhibits delayed-type hypersensitivity and the developent of adjuvant arthritis. *Eur. J. Immunol.* **21**:2327–2331.

Michie, H.R., Spriggs, D.R., Manogue, K.R., Sherman, M.L., Revhaug, A., O'Dwyer, S.T., Arthur, K., Dinarello, C.A., Cerami, A., Wolff, S.M. *et al.* (1988) Tumor necrosis factor and endotoxin induce similar metabolic responses in human beings. *Surgery* **104**:280–286.

Mochon, M., Kaiser, B., Palmer, J.A., Polinsky, M., Flynn, J.T., Caputo, G.C. and Baluarte, H.J. (1993) Evaluation of OKT3 monoclonal antibody and ATG in the treatment of steroid-resistant acute allograft rejection in pediatric renal transplants. *Pediatr. Nephrol.* **7**:259–262.

Morel, P., Revillard, J.P., Nicolas, J.F. *et al.* (1992) Anti-CD4 monoclonal antibody therapy in severe psoriasis. *J. Autoimmunity* **5**:465–77.

Moreland, L.W., Pratt, P.W., Sanders, M.E. and Koopman, W.J. (1993) Experience with a chimeric monoclonal anti-CD4 antibody in the treatment of refractory rheumatoid arthritis. *Clin. Exp. Rheumatol.* **11**(Suppl):S153–159.

Moreland, L.W., Bucy, R.P., Tilden, A., Pratt, P.W., LoBuglio, A.F., Khazaeli, M., Everson, M.P., Daddona, P., Ghrayeb, J., Kilgarriff, C. *et al.* (1993a) Use of a chimeric monoclonal anti-CD4 antibody in patients with refractory rheumatoid arthritis. *Arthritis. Rheum.* **36**: 307–318.

Moreland, L.W., Pratt, P.W., Bucy, R.P., Jackson, B.S., Feldman, J.W. and Koopman, W.J. (1994) Treatment of refractory rheumatoid arthritis with a chimeric anti-CD4 monoclonal antibody: long-term followup of CD4+ cell counts. *Arthritis. Rheum.* **37**:S834–838.

Moreland, L.W., Margolies, G.R., Heck, L.W. *et al.* (1994a) Soluble tumor necrosis factor receptor (STNFR): results of a phase I dose-escalation study in patients with rheumatoid arthritis. *Arthritis. and Rheumatism* **37**:S295.

Moreland, L.W., Pratt, P.W., Mayes, M.D., Postlethwaite, A., Weisman, M.H., Schnitzer, T., Lightfoot, R., Calabrese, L., Zelinger, D.J., Woody, J.N. *et al.* (1995) Double-blind, placebo-controlled multicenter trial using chimeric monoclonal anti-CD4 antibody, cM-T412 in rheumatoid arthritis patients receiving concomitant methotrexate. *Arthritis Rheum.* **38**:1581–1588.

Moreland, L.W., Bucy, R.P., Jackson, B., James, T. and Koopman, W.J. (1996) Long-term (5 years) follow-up of rheumatoid arthritis patients treated with a depleting anti-CD4 monoclonal antibody cM-T412. *Arthritis and Rheum.* **39**:S244 abstract #1299.

Moscovitz, H., Shofer, F., Mignott, H., Behrman, A. and Kilpatrick, L. (1994) Plasma cytokine determinations in emergency department patients as a predictor of bacteremia and infectious disease severity. *Crit. Care Med.* **22**:1102–1107.

Nashan, B., Schwinzer, R., Schlitt, H.J., Wonigeit, K. and Pichlmayr, R. (1995) Immunological effects of the anti-IL-2 receptor monoclonal antibody BT 563 in liver allografted patients. *Transpl. Immunol.* **3**:203–211.

Nashan, B., Schlitt, H.J., Schwinzer, R., Ringe, B., Kuse, E., Tusch, G., Wonigeit, K. and Pichlmayr, R. (1996) Immunoprophylaxis with a monoclonal anti-IL-2 receptor antibody in liver transplant patients. *Transplantation* **61**:546–554.

Nathan, C. and Sporn, M. (1991) Cytokines in context. *J. Cell. Biol.* **113**:981–986.

Nicolas, J.F., Chamechick, N., Thiovolet, J. *et al.* (1991) CD4 antibody treatment of severe psoriasis. *Lancet* **338**:321.

Norman, D.J., Kahana, L., Stuart, F.P. Jr, Thistlethwaite, J.R., Shield, C.F. 3rd, Monaco, A., Dehlinger, J., Wu, S.C., Van Horn, A. and Haverty, T.P. (1993) A randomized clinical trial of induction therapy with OKT3 in kidney transplantation. *Transplantation* **55**:44–50.

Norman, D.J. (1995) Mechanisms of action and overview of OKT3. *Ther. Drug. Monit.* **17**:625–620.

Olson, A.D., DelBuono, E.A., Bitar, K.N. and Remick, D.G. (1995) Antiserum to tumor necrosis factor and failure to prevent murine colitis. *J. Pediatr. Gastroenterol. Nutr.* **21**:410–418.

Olsen, N.J., Brooks, R.H., Cush, J.J., Lipsky, P.E., St. Clair, E.W., Matteson, E.L., Gold, K.N., Cannon, G.W., Jackson, C.G., MuCune, W.J., Fox, D.A., Nelson, B., Lorenz, T. and Strand V. (1996) A double-blind placebo-controlled study of anti-C5 immunoconjugate in patients with rheumatoid arthritis. The Xoma RA Investigator Group. *Arthritis Rheum.* **39**:1102–1108.

Ozmen, L., Roman, D., Fountoulakis, M., Schmid, G., Ryffel, B. and Garotta, G. (1995) Experimental therapy of systemic lupus erytematosus: the treatment of NZB/W mice with mouse soluble interferon-gamma receptor inhibits the onset of glomerulonephritis. *Eur. J. Immunol.* **25**:6–12.

Paar, D.P. and Pollard, R.B. (1996) Immunotherapy of CMV infections. *Adv. Exp. Med. Biol.* **394**:145–151.

Paleolog, E.M., Hunt, M., Elliott, M.J., Feldmann, M., Maini, R.N. and Woody, J.N. (1996) Deactivation of vascular endothelium by monoclonal anti-tumor necrosis factor alpha antibody in rheumatoid arthritis. *Arthritis Rheum.* **39**:1082–1091.

Palmen, M.J., Dijkstra, C.D., van der Ende, M.B., Pena, A.S. and van Rees, E.P. (1995) Anti-CD11b/CD18 antibodies reduce inflammation in acute colitis in rats. *Cln. Exp. Immunol.* **101**:351–356.

Parillo, J.E. (1991) Management of septic shock: present and future. *Ann. Intern. Med.* **115**:491–493.

Parlevliet, K.J. and Schellekens, P.T. (1992) Monoclonal antibodies in renal transplantation: a review. *Transpl. Int.* **5**:234–246 .

Parlevliet, K.J., Bemelman, F.J., Yong, S.L., Hack, C.E., Surachno, J., Wilmink, J.M. ten Berge, I.J. and Schellekens, P.T. (1995) Toxicity of OKT3 increases with dosage: a controlled study in renal transplant recipients. *Transpl. Int.* **8**:141–146.

Panayi, G.S., Choy, E.H.S., Connolly, D.J.A., Regan, T., Manna, V.K., Rapson, N., GH Kingsley, G.H. and Johnson, J.M. (1996) T cell hypothesis in rheumatoid arthritis (RA) tested by humanized non-depleting anti-CD4 monoclonal antibody (mAb) treatment I:suppression of disease activity and acute phase response. *Arthritis. Rheum.* **39**:S244 abstract #1300.

Paulus, H.E., Machleder, H.I., Levine, S., Yu, D.T. and MacDonald, N. Lymphocyte involvement in rheumatoid arthritis. (1977) Studies during thoracic duct drainage. *Arthritis. Rheum.* **20**:1249–1262.

Petrie, J.J., Rigby, R.J., Hawley, C.M., Suranyi, M.G., Whitby, M., Wall, D. and Hardie, I.R. (1995) Effect of OKT3 in steroid-resistant renal transplant rejection. *Transplantation* **59**:347–352.

Pinsky, M.R., Vincent, J.L., Deviere, J., Alegre, M., Kahn, R.J. Dupont, E. (1993) Serum cytokine levels in human septic shock. Relation to multiple-system organ failure and mortality. *Chest* **103**:565–575.

Pitzalis, C., Choy, E. and Kingsley, G. (1994) Monoclonal antibody therapy in rheumatic disease. *Presse. Med.* **23**:532–539.

Przepiorka, D., LeMaistre, C.F., Huh, Y.O., Luna, M., Saria, E.A., Brown, C.T. and Champlin, R.E. (1994) Evaluation of anti-CD5 ricin A chain immunoconjugate for prevention of acute graft-vs.-host disease after HLA-identical marrow transplantation. *Ther. Immunol.* **1**:77–82.

Plater-Zyberk, C., Taylor, P.C., Blaylock, M.G. and Maini, R.N. (1994) Anti-CD5 therapy decreases severity of established disease in collagen type I induced arthritis in DBA/1 mice. *Clin. Exp. Immunol.* **98**:442–447.

Plater-Zyberk, C. and Bonnefoy, J.Y. (1995) Marked amelioration of established collagen-induced arthritis by treatment with antibodies to CD23 *in vivo*. *Nat. Med.* **1**:781–785.

Powelson, J.A., Knowles, R.W., Delmonico, F.L., Kawai, T., Mourad, G., Preffer, F.K., Colvin, R.B. and Cosimi, A.B. (1994) CDR-grafted OKT4A monoclonal antibody in cynomolgus renal allograft recipients. *Transplantation* **57**:788–793.

Preiser, J.C., Schmartz, D., van der Linden, P., Content, J., vanden Bussche, P., Buurman, W., Sebald, W., Dupont, E., Pinsky, M.R., Vincent, J.L. (1991) Interleukin-6 administration has no acute hemodynamic or hematologic effects in the dog. *Cytokine* **3**:1.

Pulito, V.L., Roberts, V.A., Adair, J.R., Rothermel, A.C., Collins, A.M., Varga, S.S., Martocello, C., Bodmer, M., Joliffe, L.K. and Zivin, R.A. (1996) Humanization and molecular modeling of the anti-CD4 monoclonal antibody OKT4A. *J. Immunol.* **156**:2840–2850.

Quezado, Z.M., Natanson, C., Alling, D.W., Banks, S.M., Koev, C.A., Elin, R.J., Hosseini, J.M., Bacher, J.D., Danner, R.L. and Hoffman, W.D. (1993) A controlled trial of HA-1A in a canine model of gram-negative septic shock. *JAMA*. **269**:2221–2227.

Rabb, H., Mendiola, C.C., Saba, S.R., Dietz, J.R., Smith, C.W., Bonventre, J.V. and Ramirez, G. (1995) Antibodies to ICAM-1 protect kidneys in severe ischemic reperfusion injury. *Bochem. Biophys. Res. Commun.* **211**:67–73.

Racadot, E., Wijdenes, J. and Wendling, D. (1992) Immunological follow-up of 17 patients with rheumatoid arthritis treated *in vivo* with an anti-T CD4+ monoclonal antibody (B-F5). *Clin. Exp. Rheumatol.* **10**:365–374.

Racadot, E., Rumbach, L., Bataillard, M., Galmiche, J., Henlin, J.L., Truttmann, M., Herve, P. and Wijdenes, J. (1993) Treatment of multiple sclerosis with anti-CD4 monoclonal antibody. A preliminary report on B-F5 in 21 patients. *J. Autoimmun.* **6**:771–786.

Racadot, E., Milpied, N., Bordigoni, P., Cahn, J.Y., Plouvier, E., Lioure, B., Lutz, P., Wijdenes, J. and Herve, P. (1995) Sequential use of three monoclonal antibodies in corticosteroid-resistant acute GVHD: a multicentric pilot study including 15 patients. *Bone Marrow Transplant.* **15**: 669–677.

Ranges, G.E., Sriram, S. and Cooper, S.M. (1985) Prevention of type II collagen-induced arthritis by *In vivo* treatment with anti-L3T4. *J. Exp. Med.* **162**:1105–1110.

Rankin, E.C., Choy, E.H., Kassimos, D., Kingsley, G.H., Sopwith, A.M., Isenberg, D.A. and Panayi, G.S. (1995) The therapeutic effects of an engineered human anti-tumour necrosis factor alpha antibody (CDP571) in rheumatoid arthritis. *Br. J. Rheumatol.* **34**:334–342.

Rau, R., Sander, O., Schattenkirchner, M., Baudin, M., Siehr, U., Bisschops, C. and van der Ausera, P. (1996) Monthly vs bimonthly dosing of Lenercept (TNFp55-IgG1, Ro 45-2081) in patients with rheumatoid arthritis treated for 3 months: results of a double-blind controlled phase II trial. *Arthritis and Rheumatol.* **39**:S243 abstract #1293.

Reding, R., Veuz, H., de Ville de Goyet, J., Sokal, E., de Hemptinne, B., Latinne, D., Rahier, J., Jamart, J., Vincenzotto, C., Cormont, F. *et al.* (1993) Monoclonal antibodies in prophylactic immunosuppression after liver transplantation. A randomized controlled trial comparing OKT3 and anti-IL-2 receptor monoclonal antibody LO-Tact-1. *Transplantation* **55**:534–541.

Reece, R.J., Veale, D.J., Orgles, C.S., O'Connor, P.J., Ridgeway, J., Gibbon, W. and Emery, P. (1996) Intra-articular injection of the knee with human anti-CD4 monoclonal antibody in rheumatoid arthritis: an arthroscopic and magnetic resonance study of efficacy and safety. *Arthritis and Rheumatol.* **39**:S245 abstract #1304.

Remick, D., Manohar, P., Bolgos, G., Rodriguez, J., Moldawer, L. and Wollenberg, G. (1995) Blockade of tumor necrosis factor reduces lipopolysaccharide lethality, but not the lethality of cecal ligation and puncture. *Shock* **4**:89–95.

Reynolds, J.V., Murchan, P., Leonard, N., Gough, D.B., Clarke, P., Keane, F.B. and Tanner, W.A. (1995) High-dose interleukin-2 promotes bacterial translocation from the gut. *Br. J. Cancer.* **72**:634–636.

Riechmann, L., Clark, M., Waldmann, H. and Winter, G. (1988) Reshaping human antibodies for therapy. *Nature* **332**:323–327.

Rizova, H., Nicolas, J.F., Morel, P., Kanitakis, J., Demidem, A., Revillard, J.P., Wijdenes, J., Thiovolet, J. and Schmitt, D. (1994) The effect of anti-CD4 monoclonal antibody treatment on immunopathological changes in psoriatic skin. *J. Dermatol. Sci.* **7**:1–13.

Romulo, R.L., Palardy, J.E. and Opal, S.M. (1993) Efficacy of anti-endotoxin monoclonal antibody E5 alone or in combination with cirpofloxacin in neutropenic rats with Pseudomonas sepsis. *J. Infect. Dis.* **167**:126–130.

Roumen, R.M., Hendriks, T., van der Ven-Jongekrijg, J., Nieuwenhuijzen, G.A., Sauerwein, R.W., van der Meer, J.W. and Goris, R.J. (1993) Cytokine patterns in patients after major vascular surgery, hemorrhagic shock, and severe blunt trauma. Relation with subsequent adult respiratory distress syndrome and multiple organ failure. *Ann. Surg.* **218**:769–776.

Ruderman, E.M., Weinblatt, M.E., Thrumond, L.M., Pinkus, G.S. and Gravallese, E.M. (1995) Synovial tisssue response to treatment with CAMPATH-1H. *Arthritis Rheum.* **38**:254–258.

Ruiz, P.J., Zinger, H. and Mozes, E. (1996) Effect of injection of anti-CD4 and anti-CD8 monoclonal antibodies on the development of experimental systemic lupus erythematosus in mice. *Cell. Immunol.* **167**:30–37.

Saxne, T., Palladino, M.A. Jr, Heinegard, D., Talal, N. and Wollheim, F.A. (1988) Detection of tumor necrosis factor alpha but not tumor necrosis factor beta in rheumatoid arthritis synovial fluid and serum. *Arthritis Rheum.* **31**:1041–1045.

Schindler, R., Mancilla, J., Endres, S., Ghorbani, R., Clark, S.C. and Dinarello, C.A. (1990) Correlations and interactions in the production of interleukin-6 (IL-6), IL-1 and tumour necrosis factor (TNF) in human blood mononuclear cells: IL-6 suppresses IL-1 and TNF. *Blood* **75**:40–47.

Schror, K. (1995) Anti-platelet drugs. A comparitive review. *Drugs* **50**:7–28.

Selmaj, K.W. and Rane, C.S. (1995) Experimental autoimmune encephalomyelitis: immunotherapy with anti-tumor necrosis factor antibodies and soluble tumor necrosis factor receptors. *Neurology* **45** (Suppl. 6):S44–49.

Seu, P., Imagawa, D.K., Wasef, E., Olthoff, K.M., Hare, J., Stephen, S.S., Dempsey, R.A. and Busuttil, R.W. (1991) Monoclonal anti-TNF treatment of rat cardiac allografts: Synergism with low dose cyclosporine and immuno-histological studies. *J. Surg. Res.* **50**:520–528.

Sgro, C. (1995) Side-effects of a monoclonal antibody, muromonab CD3/orthoclone OKT3: bibliographic review. *Toxicology* **105**:23–29.

Shaddy, R.E., Bullock, E.A., Morwessel, N.J., Hannon, D.W., Renlund, D.G., Karwande, S.V., McGough, E.C. and Hawkins, J.A. (1993) Murine monoclonal CD3 antibody (OKT3)-based early rejection prophylaxis in pediatric heart transplantation. *J. Heart Lung Transplant.* **12**:434–439.

Shennib, H., Massard, G., Reynaud, M. and Noirclerc, M. (1994) Efficacy of OKT3 therapy for acute rejection in isolated lung transplantation. *J. Heart Lung Transplant.* **13**:514–519.

Sherman, M.L., Spriggs, D.R., Arthur, K.A., Imamura, K., Frei, E. 3rd and Kufe, D.W. (1988) Recombinant human tumor necrosis factor administered as a five-day continuous infusion in cancer patients: Phase I toxicity and effects on lipid metabolism. *J. Clin. Oncol.* **6**:344–350.

Simoons, M.L., de Boer, M.J., van der Brand, M.J., van Miltenburg, A.J., Hoorntje, J.C., Heyndricks, G.R., van der Wieken, L.R., de Bono, D., Rutsch, W., Schaible, T.F. *et al.* (1994) Randomized trial of a GPIIb/IIIa platelet receptor blocker in refractory unstable angina. *European Cooperative Study Group. Circulation* **89**:596–603.

Skyler, J.S., Lorenz, T.Z., Schwartz, S., Elsenbarth, G.S., Einhorn, D., Palmer, J.P., Marks, J.B., Greenbaum, C., Saria, E.A. and Byers, V. (1993) Effects of an anti-CD5 immunoconjugate (CD5-plus) in recent onset type I diabetes mellitus: a preliminary investigation. The CD5 Diabetes Project Team. *J. Diabetes Complications* **7**:224–32.

Sprung, C.L., Finch, R.G., Thigs, L.G. and Glauser, M.P. (1996) International sepsis trial (INTERSEPT): role and impact of a clinical evaluation committee. *Crit. Care Med.* **24**:1441–1447.

Stack, A.M., Saladino, R.A., Thompson, C., Sattler, F., Weiner, D.L., Parsonnet, J., Nariuchi, H., Siber, G.R. and Fleisher, G.R. (1995) Failure of prophylactic and therapeutic use of a murine anti-tumor necrosis factor monoclonal antibody in Escherichia coli sepsis in the rabbit. *Crit. Care Med.* **23**:1512–1518.

Stack, W., Mann, S., Roy, A., Heath, P., Sopwith, M., Freeman, J., Holmes, G., Long, R., Forbes, A., Kamm, M. and Hawkey, C. (1996) The effects of CDP571, an engineered human IgG4 anti-TNF antibody in Crohn's disease. Gastroenterology 110 #4.

Stafford, F.J., Fleisher, T.A., Lee, G., Brown, M., Strand, V., Austin, H.A. 3rd, Balow, J.E. and Klippel, J.A. (1994) A pilot study of anti-CD5 ricin A chain immunoconjugate in systemic lupus erythematosus. *J. Rheumatol.* **21**:2068–2070.

Stevens, D.L., Bryant, A.E., Hackett, S.P., Chang, A., Peer, G., Kosanke, S., Emerson, T. and Hinshaw, L. (1996) Group A streptococcal bacteremia: the role of tumor necrosis factor in shock and organ failure. *J. Infect. Dis.* **173**:619–626.

Stock, P.G., Lantz, M., Light, S. and Vincenti, F. (1996) *In vivo* (phase I) trial and *in vitro* efficacy of humanized anti-Tac for the prevention of rejection in renal transplant recipients. *Transplant. Proc.* **28**:915–916.

Strand, V. and Lee, M.L. (1993) Differential patterns of response in patients with rheumatoid arthritis following administration of an anti-CD5 immunoconjugate. *Clin. Exp. Rheumatol.* **11**:S161–163.

Strieter, R.M., Kunkel, S.L. and Bone, R.C. (1993) Role of tumour necrosis factor-alpha in disease states and inflammation. *Crit. Care Med.* **21**:S447–463.

Suffredini, A.F., Fromm, R.E., Parker, M.M., Brenner, N., Kovacs, H.A., Wesley, R.A. and Parrillo, J.E. (1989) The cardiovascular response of normal humans to the administration of endotoxin. *New Engl. J. Med.* **321**:280–287.

Suffredini, A.F. (1994) Current prospects for the treatment of clinical sepsis. *Crit. Care Med.* **22**:S12–S18.

Suffredini, A.F., Reda, D., Banks, S.M., Tropea, M., Agosti, J.M. and Miller R. (1995) Effects of recombinant dimeric TNF receptor on human inflammatory responses following intravenous endotoxin administration. *J. Immunol.* **155**:5038–5045.

Swant, G.M. and Deitch, E.A. (1996) Role of the gut in multiple organ failure: bacterial translocation and permeability changes. *World J. Surg.* **20**:411–417.

Tak, P.P., van der Lubbe, P.A., Cauli, A., Daha, M.R., Smeets, T.J., Kluin, P.M., Meinders, A.E., Yanni, G., Panayi, G.S. and Breedveld, F.C. (1995) Reduction of synovial inflammation after anti-CD4 monoclonal antibody treatment in early rheumatoid arthritis. *Arthritis Rheum.* 1995; **38**:1457–1465.

Tak, P.P., Taylor, P.C., Breedveld, F.C., Smeets, T.J., Daha, M.R., Kluin, P.M., Meinders, A.E. and Maini, R.N. (1996) Decrease in cellularity and expression of adhesion molecules by anti-tumor necrosis factor alpha monoclonal antibody treatment in patients with rheumatoid arthritis. *Arthritis Rheum* **39**:1077–1081.

Talan, D.A. (1993) Recent developments in our understanding of sepsis: evaluation of anti-endotoxin antibodies and biological response modifiers. *Ann. Emerg. Med.* **22**:1871–1890.

Talento, A., Nguyen, M., Blake, T., Sirotina, A., Fioravnati, C., Burkholder, D., Gibson, R., Sigal, N.H., Springer, M.S. and Koo, G.C. (1993) A single administration of LFA-1 antibody confers prolonged allograft survival. *Transplantation* **55**:418–422.

Tamiya, Y., Yamamota, N. and Uede, T. (1995) Protective effects of monoclonal antibodies against LFA-1 and ICAM-1 on myocardial reperfusion injury following global ischemia in rat hearts. *Immunopharmacology* **29**:53–63.

Tanabe, K., Takahashi, K., Sonda, K., Tokumoto, T., Koga, S., Nakazawa, H., Goya, N., Yagisawa, T., Fuchinoue, S., Kawai, T., Toma, H. and Ota, K. (1996) Long-term results of OKT3 treated renal transplant recipients. *Transplant. Proc.* **28**:1350–1351.

Targon, A.R., Rutgeerts, P., Hanauer, S.B., van Deventer, S.J.H., Mayer, L., Present, D.H., Braakman T.A.J., Woody, J.N. and the Crohn's Disease cA2 study group. (1996) A multicenter trial of anti-tumor necrosis factor (TNF) antibody (cA2) for treatment of patients with active Crohn's disease. *Gastroenterology* **110#4**.

Tarkowski, E., Rosengren, L., Blomstrand, C., Wikkelso, C., Jensen, C., Ekholm, S. and Tarkowski, A. (1995) Early intrathecal production of interleukin-6 predicts the size of brain lesion in "stroke". *Stroke* **26**:1393–1398.

Thurau, S.R., Wildner, G., Reiter, C., Riethmuller, G. and Lund, O.E. (1994) Treatment of endogenous uveitis with anti-CD4 monoclonal antibody: first report. *Ger. J. Opthalmol.* **3**:409–413.

Tomer, Y., Blank, M. and Shoenfeld, Y. (1994) Suppression of experimental antiphospholipid syndrome and systemic lupus erythematosus in mice by anti-CD4 monoclonal antibodies. *Arthritis Rheum* **37**:1236–1244.

Topol, E.J. (1995) Prevention of cardiovascular ischemic complications with new platelet glycoprotein IIb/IIIa inhibitors. *Am. Heart J.* **130**:666–672.

Tracey, K.L. and Cerami, A. (1993) Tumor necrosis factor: An updated review of its biology. *Crit. Care Med.* **21**:S415–422.

Trentham, D.E., Belli, J.A., Anderson, R.J., Buckley, J.A., Goetzl, E.J., David, J.R. and Austen, K.F. (1988) Clinical and immunological effects of fractionated total lymphoid irradiation in refractory rheumatoid arthritis. *N. Engl. J. Med.* **305**:976–982.

Uchida, K., Namii, Y., Tominaga, Y., Haba, T., Tanaka, H., Ichimori, T., Uemura, O., Morozumi, K., Hayashi, S., Yokoyama, J. and Takagi, H. (1996) OKT3 rescue therapy for 63 refractory rejections in 405 renal allografts. *Transplant. Proc.* **28**:1358–1359.

Uchikoshi, F., Ito, T., Kamiike, W., Moriguchi, A., Nozaki, S., Ito, A., Kuhara, A., Miyata, M., Matsuda, H., Miyasaka, M. *et al.* (1995) Anti-ICAM-1/LFA-1 monoclonal antibody therapy prevents graft rejection and IDDM recurrence in BB rat pancreas transplantation. *Transplant. Proc.* **27**:1527–1528.

van der Lubbe, P.A., Miltenburg, A.M. and Breedveld, F.C. (1991) Anti-CD4 monoclonal antibody for relapsing polychondritis. *Lancet* **337**:1349.

van der Lubbe, P.A., Reiter, C., Breedveld, F.C., Kruger, K., Schattenkirchner, M., Sanders, M.E. and Riethmuller, G. (1993) Chimeric CD4 monoclonal antibody cM-T412 as a therapeutic approach to rheumatoid arthritis. *Arthritis Rheum* **36**:1375–1379.

van der Lubbe, P.A., Reiter, C., Miltenburg, A.M., Kruger, K., de Ruyter, A.N., Rieber, E.P., Bijl, J.A., Riethmuller, G. and Breedveld, F.C. (1994) Treatment of rheumatoid arthritis with a chimeric CD4 monoclonal antibody (cM-T412): immunopharmacological aspects and mechanisms of action. *Scand. J. Immunol.* **39**:286–294.

van der Lubbe, P.A., Dijkmans, B.A., Markusse, H.M., Nassander, U. and Breedveld, F.C. (1995) A randomized, double-blind, placebo-controlled study of CD4 monoclonal antibody therapy in early rheumatoid arthritis. *Arthritis Rheum.* **38**:1097–1106.

van der Poll, T., Buller, H.R. and ten Cate, H.T. (1990) Activation of coagulation after administration of TNF to normal subjects. *N. Engl. J. Med.* **322**:1622–1627.

van der Poll, T., Levi, M., van Deventer, S.J., ten Cate, H., Haagmans, B.L., Biemond, B.J., Buller, H.R., Hack, C.E. and ten Cate, J.W. (1994) Differential effects of anti-tumor necrosis factor monoclonal antibodies on systemic inflammatory responses in experimental endotoxemia in chimpanzees. *Blood* **83**:446–451.

van der Poll, T., Levi, M., Hack, C.E., ten Vate, H., van Deventer, S.J., Eerenberg, A.J., de Groot, E.R., Jansen, J., Gallati, H., Buller, H.R., ten Cate, J.W. and Aarden, L.A. (1994a) Elimination of interleukin-6 attenuate coagulation activation in experimental endotoxemia in chimpanzees. *J. Exp. Med.* **179**:1253–1259.

van Dullemen, H.M., van Deventer, S.J., Hommes, D.W., Bijl, H.A., Jansen, J., Tytgat, G.N. and Woody, J. (1995) Treatment of Crohn's disease with anti-tumor necrosis factor chimeric monoclonal antibody (cA2). *Gastroenterology* **109**:129–135.

van Gelder, T., Zietse, R., Mulder, A.H., Yzermans, J.N., Hesse, C.J., Vaessen, L.M. and Weimar, W. (1995) A double-blind placebo-controlled study of monoclonal anti-interleukin-2 receptor antibody (BT563) administration to prevent acute rejection after kidney transplantation. *Transplantation* **60**:248–252.

van Gelder, T., Balk, A.H. and Jonkman, F.A., Zietse, R., Zondervan, P., Hesse, C.J., Vaessen, L.M., Mochtar, B. and Weimar, W. (1996) A randomized trial comparing safety and efficacy of OKT3 and a monoclonal anti-interleukin-2 receptor antibody (BT563) in the prevention of acute rejection after heart transplantation. *Transplantation* **62**:51–55.

Van Leeuwen, P.A.M., Boermeester, M.A. *et al.* (1994) Clinical significance of translocation. *Gut.* S1:S28–S34.

Vreugdenhil, G., Lowenberg, B., van Eijk, H.G. and Swaak, A.J.G. (1992) Tumor necrosis factor alpha is associated with disease activity and the degree of anemia in patients with rheumatoid arthritis. *Eur. J. Clin. Invest.* **22**:488–493.

Waage, A., Halstensen, A. and Espevik, T. (1987) Association between tumour necrosis factor in serum and fatal outcome in patients with meningogococcal disease. *Lancet* **1**:355–359.

Wall, W.J., Ghent, C.N., Roy, A. and McAlister, V.C., Grant, D.R. and Adams, P.C. (1995) Use of OKT3 monoclonal antibody as induction therapy for control of rejection in liver transplantation. *Dig. Dis. Sci.* **40**:52–57.

Wang, Y.M., Rollins, S.A., Madri, J.A. and Matis, L.A. (1995) Anti-C5 monoclonal antibody therapy prevents collagen-induced arthritis and ameliorates established disease. *Proc. Natl. Acad. Sci.* **92**:8955–8959.

Wang, Y., Hu, Q., Madri, J.A., Rollins, S.A., Chodera, A. and Matis, L.A. (1996) Amelioration of lupus-like autoimmune disease in NZB/WF1 mice after treatment with a blocking monoclonal antibody specific for complement component C5. *Proc. Natl. Acad. Sci.* **93**:8563–8568.

Ward, M.M., Pyun, E. and Studenski. (1996) Mortality risks associated with specific clinical manifestations of systemic lupus erythematosus. *Arch. Intern. Med.* **156**:1337–1344.

Watts, R.A., Issacs, J.D., Hale, G., Hazleman, B.L. and Waldmann, H. (1993) High dose (200 mg) CAMPATH-1H therapy in rheumatoid arthritis. *Br. J. Rheumatol.* **32** (suppl 1):54.

Wedel, N. (1992) Update on the use of E5, antiendotoxin monoclonal antibody, in the treatment of Gram negative sepsis. Presented at New Advances in the Treatment of Endotoxemia and Sepsis: Second annual meeting of international business communications, June 22–23, Philadelphia, PA.

Wei, R.Q., Lin, H., Chen, G.H., Beer, D.G., Kunkel, S.L. and Bolling, S.F. (1994) Inhibition of tumor necrosis factor production by lymphocytes from anti-TNF antibody-treated, cardiac-allografted rats. *J. Surg. Res.* **56**:601–5.

Weinblatt, M.E., Maddison, P.J., Bulpitt, K.J., Hazleman, B.L., Urowitz, M.B., Sturrock, R.D., Coblyn, J.S., Maier, A.L., Spreen, W.R., Manna, V.K. *et al.* (1995) CAMPATH-1H, a humanized monoclonal antibody, in refractory rheumatoid arthritis. An intravenous dose-escalation study. *Arthritis Rheum* **38**:1589–1594.

Weinblatt, M.E., Coblyn, J.S., Maier, A.L., Anderson, R., Helfgott, S., Thurmond, L.M. and Johnston, J.M. (1996) Sustained lymphoctye suppression after single dose CAMPATH-1H (C-1H) infusion: a long term follow-up. *Arthritis and Rheum.* **39**:S245 abstract #1306.

Weinshenker, B.G., Bass, B., Karlik, S., Ebers, G.C. and Rice, G.P. (1991) An open trial of OKT3 in patients with multiple sclerosis. *Neurology* **41**:1047–1052.

Weisdorf, D., Filipovich, A., McGlave, P., Ramsay, N., Kersey, J., Miller, W. and Blazar, B. (1993) Combination graft-versus-host disease prophylaxis using immunotoxin (anti-CD5–RTA [Xomazyme-CD5]) plus methotrexate and cyclosporine or prednisone after unrelated donor marrow transplantation. *Bone Marrow Transplant.* **12**:531–536.

Wendling, D., Racodot, E., Morel-Fourrier, B. and Wijdenes, J. (1992) Treatment of rheumatoid arthritis with anti-CD4 monoclonal antibody. Open study of 25 patients with the B-F5 clone. *Clin. Rheumatol.* **11**:542–547.

Wendling, D., Racadot, E. and Wijdenes J. (1993) Treatment of severe rheumatoid arthritis by anti-interleukin-6 monoclonal antibody. *J. Rheumatol.* **20**:259–262.

Wendling, D., Racadot, E. and Wijdenes, J. (1994) Serum levels of IL6, CRP, IL6 receptor and cortisol under anti-IL6 monoclonal antibody therapy in rheumatoid arthritis. *Arthritis and Rheum.* **37**:S382.

Wendling, D., Racadot, E., Widenes, J. and the French Investigators Group. (1996) Randomized, double-blind, placebo-controlled multicenter trial of murine anti-CD4 monoclonal antibody therapy in rheumatoid arthritis. *Arthritis and Rheumatol.* **39**:S245 abstract #1303.

Wherry, J.C., Pennington, J.E. and Wenzel, R.P. (1993) Tumor necrosis factor and the therapeutic potential of anti-tumor necrosis factor antibodies. *Crit. Care. Med.* **21**:S436–440.

Whitcup, S.M., Hikita, N., Shirao, M., Miyasaka, M., Tamatani, T., Mochizuki, M., Nussenblatt, R.B. and Chan, C.C. (1995) Monoclonal antibodies against CD54 (ICAM-1) and CD11a (LFA-1) prevent and inhibit endotoxin-induced uveitis. *Exp. Eye Res.* **60**:597–601.

Willenbourg, D.O., Simmons, R.D., Tamatani, T. and Miyasaka, M. (1993) ICAM-1-dependent pathway is not critically involved in the inflammatory process of autoimmune encephalomyelitis or in cytokine-induced inflammation of the central nervous system. *J. Neuroimmunol.* **45**:147–154.

Williams, R.O., Mason, L.J., Feldmann, M. and Maini, R.N. (1994) Synergy between anti-CD4 and anti-tumor necrosis factor in the amelioration of established collegen-induced arthritis. *Proc. Natl. Acad. Sci.* **91**:2762–2766.

Williams, R.O., Ghrayeb, J., Feldmann, M. and Maini, R.N. (1995) Successful therapy of collagen-induced arthritis with TNF receptor-IgG fusion protein and combination with anti-CD4. *Immunology* **84**:433–439.

Windsor, A.C., Mullen, P.G., Walsh, C.J., Fisher, B.J., Blocker, C.R., Jesmok, G., Fowler, A.A. 3rd and Sugerman, H.J. (1994) Delayed tumor necrosis factor alpha blockade attenuates pulmonary dysfunction and metabolic acidosis associated with experimental gram-negative sepsis. *Arch. Surg.* **129**:80–89.

Winn, R.K., Mihelicic, D., Vedder, N.B., Sharar, S.R. and Harlan, J.M. (1993) Monoclonal antibodies to leukocyte and endothelial adhesion molecules attenuate ischemia-reperfusion injury. *Behring. Inst. Mitt.* **92**:229–237.

Wooley, P.H., Dutcher, U., Widmer, M.B. and Gillis, S. (1993) Influence of a recombinant human soluble tumor necrosis factor receptor FC fusion protein on type II collagen-induced arthritis in mice. *J. Immunol.* **151**: a6602–6607.

Yamataka, T., Kobayashi, H., Yagita, H., Okumura, K., Tamatani, T. and Miyasaka, M. (1993) The effect of anti-ICAM-1 monoclonal antibody treatment on the transplantation of the small bowel in rats. *J. Pediatr. Surg.* **28**:1451–1457.

Yang, H., Issekutz, T.B. and Wright, J.R. Jr (1995) Prolongation of rat islet allograft survival by treatment with monoclonal antibodies against VLA-4 and LFA-1. *Transplantation* **60**:71–76.

Yao, Y.M., Yu, Y., Sheng, Z.Y., Tian, H.M., Wang, Y.P., Lu, L.R. and Yu, Y. (1995) Role of gut-derived endotoxaemia and bacterial translocation in rats after thermal injury: effects of selective decontamination of the digestive tract. *Burns* **21**:580–585.

Yao, Y.M., Bahrami, S., Redl, H. and Schlag, G. (1996) Monoclonal antibody to tumor necrosis factor-alpha attenuates hemodynamic dysfunction secondary to intestinal ischemia/reperfusion in rats. *Crit. Care Med.* **24**:1547–1553.

Ziegler, E.J., Fisher, C.J. Jr, Sprung, C.L. *et al.* (1991) Treatment of gram-negative bacteremia and septic shock with HA-1A human monoclonal antibody against endotoxin: a randomized, double-blind placebo-controlled trial. The HA-1A Sepsis Study Group. *N. Engl. J. Med.* **324**:429–436.

8. MARROW PURGING AND STEM CELL PREPARATION

DENIS CLAUDE ROY, NADINE BEAUGER and MARTIN GYGER

Division of Hematology-Immunology, Maisonneuve-Rosemont Hospital; Department of Medicine, Universite de Montreal, Quebec, Canada

INTRODUCTION

Aggressive therapeutic modalities using high doses of chemotherapy and/or radiation therapy are now commonplace treatments, and are no longer restricted to hematologic malignancies. Indeed, these intensive approaches can induce major responses and remissions in patients with relapsed and even refractory diseases, ultimately contributing to improve the disease-free-survival of patients with high risk hematologic malignancies, and, hopefully, solid tumors. The widespread use of bone marrow transplantation (BMT), probably better defined as progenitor cell transplantation (PCT), also reflects the major improvements in patient management that have resulted in significant lowering of the treatment-related morbidity and mortality. It is difficult to envision this rapid expansion in PCT without the recent biotechnological advances, in particular the development of (i) growth factors, which not only facilitate stem cell collection, but also decrease the duration of aplasia, and (ii) monoclonal antibodies (mAbs), which permit accurate identification and selection and/or elimination of specific cell populations responsible for engraftment, graft-versus-host disease (GVHD) and relapse.

Monoclonal antibodies have enabled an accurate definition of each of the different steps involved in the ontogeny of normal and malignant cells. Thus, it is not surprising that these agents are amongst the most popular to purge blood or marrow grafts of (i) their malignant cells prior to autologous PCT, or (ii) their T cells prior to allogeneic PCT. In this chapter, we will review the principal purging methodologies using mAbs either for the elimination of malignant cells or defined normal cell subsets, and discuss clinical applications and outcomes. We will focus primarily on the purging of hematologic malignancies, but will also discuss important recent findings associated with the purging of solid tumors. This chapter will be divided in two sections corresponding to the two main strategies of treatment of the progenitor cell graft: (a) negative selection procedures to eliminate unwanted cells and (b) positive selection procedures to isolate normal progenitor cells.

Relapse after Autologous Transplantation: What is the Contribution of Progenitor Cell Graft Contamination to this Event?

Relapse is the primary limitation of autologous PCT (Ritz *et al.*, 1994). Relapse can originate either from malignant cells in the host escaping elimination by the preparative regimen or from residual tumor cells in the bone marrow graft reinfused at the time of PCT (Gribben *et al.*, 1991b). However, the source of relapsing cells after autologous PCT has never been clearly defined. Nevertheless, bone marrow is the primary site of disease either at diagnosis or relapse in almost all patients with acute leukemia, a large proportion of patients with lymphoma and numerous solid tumors like neuroblastomas and small cell lung cancer (SCLC). Furthermore, even when these patients reach complete remission, residual malignant cells are detectable using either mAbs (Stahel *et al.*, 1985; Uckun *et al.*, 1992), culture assays (Estrov *et al.*, 1986; Sharp *et al.*, 1992), or sensitive molecular techniques such as the polymerase chain reaction (PCR) (Yamada *et al.*, 1990; Jonsson *et al.*, 1990; Ngan *et al.*, 1989; Gribben *et al.*, 1991a; Hetu *et al.*, 1994). In order to clearly determine whether or not malignant cells from the bone marrow graft contribute to relapse, Brenner *et al.* have transferred the neomycin-resistance gene into bone marrow graft cells from patients with acute myeloid leukemia and neuroblastoma (Brenner *et al.*, 1993). Malignant cells from patients relapsing following autologous PCT were shown to possess the retrovirally-introduced gene. These results indicate that occult malignant cells are not only present in the marrow graft but most likely contribute to disease recurrence. Consistent with this finding, Gorin *et al.* reported that patients with acute myeloid leukemia transplanted with a purged marrow have a better disease-free-survival than patients receiving unpurged marrow grafts (Gorin *et al.*, 1990). Furthermore, in patients with acute leukemia (Forman *et al.*, 1987; Rowley *et al.*, 1989; Uckun *et al.*, 1992) and non-Hodgkin's lymphoma (Gribben *et al.*, 1991b), a better prognosis is associated with lower numbers of detectable residual tumor cells in the marrow graft following purging. Although these observations are quite convincing, the genuine necessity for a "clean" stem cell graft remains to be confirmed in controlled randomized clinical trials. One of the reasons hampering the implementation of such trials has been the need to develop selection procedures that (i) are highly effective in eliminating malignant cells, (ii) preserve normal hematopoietic progenitor cells, and (iii) involve widely applicable methodologies. During the last fifteen years, most efforts have focused on the first two aspects, resulting in the development of increasingly effective and less harmful purging approaches. We may now be very close to answering the third requirement, which is needed to finally make purging accessible and, most importantly, evaluable in large randomized trials.

• NEGATIVE SELECTION •

PURGING OR NEGATIVE SELECTION PROCEDURES

A variety of agents and techniques have been developed to purge selective cell populations, mainly from remission marrow; these include primarily physical, pharmacologic, photodynamic and immunologic methods. Physical methods are most diverse, and include centrifugation-based methodologies (albumin density gradients (Dicke *et al.*, 1979) and counterflow elutriation (Wagner *et al.*, 1990; Preijers *et al.*, 1994)), lectins (soybean agglutination for T-cell depletion (Young *et al.*, 1992), peanut agglutinin to eliminate normal and malignant plasma cells (Lazzaro *et al.*, 1995)), and hyperthermia either alone or in combination with lipids (Min *et al.*, 1994). Pharmacologic methods have been particularly favored, using 4-hydroperoxycyclophosphamide or mafosfamide, either alone, or in combination with other chemotherapeutic agents or immunotoxins (Douay *et al.*, 1989; Uckun *et al.*, 1987; Rowley *et al.*, 1991). Also new agents like cortivazol, a glucocorticoid with an unusual structure (Juneja *et al.*, 1995), molecules for photodynamic therapy (Traul *et al.*, 1995; Jamieson *et al.*, 1993; Pal *et al.*, 1996), and cellular effector cells like NK cells or T cells (Scheffold *et al.*, 1995; Cesano *et al.*, 1996), activated with IL-2 and/or other cytokines (Beaujean *et al.*, 1995; Verma *et al.*, 1994; Arbour *et al.*, 1996) have expanded our horizon and represent other appealing purging alternatives. Molecular approaches, primarily using anti-sense oligonucleotides also show potential for the eradication of leukemia cells (Bishop *et al.*, 1994).

Nevertheless, mAbs, which were among the first purging agents developed, in association with complement (Stepan *et al.*, 1984; Zola *et al.*, 1987) or immunomagnetic beads (Kvalheim *et al.*, 1988; Reading *et al.*, 1987), and also in the form of immunotoxin conjugates (Casellas *et al.*, 1985; Stong *et al.*, 1985) still remain among the most practical and widely used. Preclinical *in vitro* studies have demonstrated that virtually all of these techniques are capable of eliminating at least 3 to 4 logarithms (logs) of malignant cells from normal human marrow. While non-immunologic methods usually show advantages in terms of reagent availability, standardization and spectrum of activity against numerous different tumour cells, they are often burdened by non-specific toxicity against normal hematopoietic progenitors. In addition, although some agents could eliminate specific cell populations, like T cells, for the prevention of graft-versus-host disease (GVHD), their activity was neither selective nor effective (Rosenfeld *et al.*, 1995). Monoclonal antibodies have different limitations, such as the need to ascertain the presence of the targeted antigen on the surface of malignant cells, and potential heterogeneity in antigen expression of neoplastic cells. However, the specificity of mAb-mediated targeting, which enables extensive elimination of distinct cell types and spares the majority of normal progenitor cells, represents a major advantage explaining their widespread use and potential for future applications.

Antibodies Directed against Myeloid Antigens

Acute myelogenous leukemia

Autologous bone marrow transplantation is playing an increasing role in the treatment of acute myelogenous leukemia (AML) (Zittoun *et al.*, 1995; Imrie *et al.*, 1996). Most impressive results were first achieved when high risk patients with AML, in second and subsequent remission, were treated by autologous PCT using marrows purged with 4-hydroperoxycyclophosphamide or mafosfamide

(Yeager *et al.*, 1986; Korbling *et al.*, 1989), but no randomized trial has been completed. For first remission AML, a non-randomized study indicates that patients transplanted within the first 6 months after remission induction have an improved disease-free-survival (DFS) when they receive purged rather than unpurged marrow grafts (Gorin *et al.*, 1990). Purging with cyclophosphamide derivatives is yet a delicate procedure that must take into account the hematocrit, the nucleated cell concentration, the nature of cells and even the concentration of plasma (Giarratana *et al.*, 1995). It is also associated with non-specific elimination of normal hematopoietic progenitors, and some problems with platelet reconstitution were observed (Yeager *et al.*, 1986; Korbling *et al.*, 1989). Fortunately, some of these obstacles could be alleviated with agents like amifostine, which was recently shown to protect normal progenitor cells from mafosfamide toxicity (Douay *et al.*, 1995). There are concerns over the role of purging in inducing cytogenetic abnormalities and secondary malignancies after autologous PCT with mafosfamide or 4-HC purged marrows (Perot *et al.*, 1993; Shah *et al.*, 1993; Bassan *et al.*, 1995). However, as secondary myelodysplastic syndromes and leukemias were documented following unpurged PCT (Rohatiner, 1994), such a proposition is not substantiated at the present time.

Elimination of malignant cells with mAbs should have the advantage of specificity and decreased toxicity. However, in acute myelogenous leukemia (AML), selection of the optimal cell marker for purging is particularly difficult for numerous reasons. First, antigens present on AML cells are extremely diverse, with their patterns of expression being usually associated with their FAB subtype, and few antigens being present on a majority of AML cases (Griffin *et al.*, 1986; Wang *et al.*, 1995). In addition, within individual patients, leukemia cells can exhibit considerable heterogeneity of antigenic expression (Griffin *et al.*, 1981, 1983b; Neame *et al.*, 1986; Sabbath *et al.*, 1985). Moreover, antigens present on detectable AML cells may not correspond to those expressed by AML progenitor cells (Griffin *et al.*, 1983b; Sabbath *et al.*, 1985). Indeed, the maturation process, which prevails in AML cells, generates a majority of AML cells which are differentiated and non-dividing. In order to address this problem, several indirect approaches, using mainly *in vitro* clonogenic assays, were developed (Sabbath *et al.*, 1985; Lange *et al.*, 1984; Griffin *et al.*, 1983a). Finally, selection of the appropriate target is complicated by the sharing of antigens on malignant and normal myeloid progenitor cells (Sabbath *et al.*, 1985; Griffin and Lowenberg, 1986). Thus, elimination of AML cells through a poorly selected target antigen could impair and even prevent progenitor cell engraftment.

The mAbs PM-81 and AML-2-23, which react with differentiation antigens CD15 and CD14, respectively, were used by Ball *et al.* in association with rabbit complement (C) for purging of AML cells (Ball *et al.*, 1986). PM-81 and AML-2-23 were chosen for mAb purging because they react with the majority of cells from 91% and 77% of cases with AML, respectively; they can also bind to a majority of clonogenic AML progenitors (L-CFC) in 65% and 30% of AML patients, respectively (Griffin *et al.*, 1986; Griffin and Lowenberg, 1986). PM-81 binds to a highly immunogenic pentasaccharide, lacto-N-fucopentose-III (LNF-III), and AML-2-23, primarily to FAB-M4 and -M5 blasts. Interestingly, CD15 expression on AML cells has been shown to increase following treatment with neuraminidase (Ball *et al.*, 1991), and to result in higher depletion of leukemia cells (Ball *et al.*, 1990b). CD15 is also present on approximately 50% of normal CFU-GM progenitors, but neuraminidase has no effect on LNF-III expression in normal progenitor cells (Ball and Howell, 1988), and does not impair engraftment (Ball *et al.*, 1990b).

This combination of mAbs and C was used to treat marrow grafts from 30 patients in first (6 patients), second (18 patients) or third (6 patients) complete remission (Ball *et al.*, 1990a). Median time to neutrophil engraftment*[1] was 30 days and time to platelet engraftment*[2] was 45 days. These patients experienced delayed T-cell immune reconstitution, that did not translate into significant clinically-related toxicity (Ericson *et al.*, 1992). The relapse-free-survival (RFS) at 3 years was of 67% for patients in CR1, 25% in CR2 and 18% in CR3 patients. These patients were followed for up to 9 years and additional patients entered in a multicenter trial, to evaluate the curative potential of this treatment strategy. A total of 63 patients with AML were thus transplanted while in CR1 (7 pts), CR2 (44 pts), CR3 (1 pt), and first relapse (R1: 11 pts) (Selvaggi *et al.*, 1994). The preparative regimen of the first 36 patients consisted of cyclophosphamide (Cy: 120 mg/kg) and fractionated total body irradiation (TBI: 1200 cGy), and in the subsequent patients of busulfan (Bu: 16 mg/kg) and Cy. Engraftment occurred at a median of 37 days for neutrophils and 44 days for platelets ($>20\times10^9$/L). The 4- to 5-year disease-free-survival (DFS) was estimated at 53% for the 7 CR1 patients, 45% for R1 patients and 30% for CR2/3 patients. Interestingly, patients receiving Bu/Cy as preparative regimen had an improved DFS over patients receiving Cy/TBI. This could be attributable to a lower toxicity of Bu/Cy, particularly in the older patients, but other confounding prognostic factors such as CD15 expression on AML blasts prevent definitive conclusions.

Fabritiis *et al.* used another mouse mAb (S4-7) reacting with CD15 on human myelomonocytic cells and selected acute myelogenous leukemia (AML) cells. Seven AML patients, 6 in first complete remission (CR) and one in second CR underwent autologous PCT with S4-7+C purged marrow after BAVC conditioning regimen (De Fabritiis *et al.*, 1989). Granulopoiesis recovered rapidly at a median of 20 days post-PCT. The patient transplanted in CR2 relapsed 3 months after autologous PCT, and of the 6 patients transplanted in 1st CR, 3 remained in continuous CR at 35, 47 and 57 months. It is noteworthy that AML cells from 2 patients, studied at the time of relapse, had a significant percentage of S4-7 negative cells that, in both instances, were not detectable either at diagnosis or previous relapse, an observation that confirms the importance of eliminating the earliest clonogenic AML cells.

CD33 is a surface glycoprotein of 67 kd, member of the immunoglobulin superfamily, and homologous to sialoadhesin, myelin-associated glycoprotein and CD22 antigen, and a particularly appealing antigen to target AML cells

---

[1] In this section, neutrophil engraftment will be defined as the time to greater than 0.5×10^9 neutrophils/L.

[2] In this section, platelet engraftment will be defined as the time to greater than 20×10^9 platelets/L.

(Simmons and Seed, 1988; Freeman *et al.*, 1995; Griffin *et al.*, 1984; Takahashi *et al.*, 1992). Indeed, greater than 80% of AML cells express this sialic acid-dependent cell adhesion molecule (Freeman *et al.*, 1995; Griffin *et al.*, 1984; 1986). Moreover, it is found on the majority of leukemia progenitor cells (L-CFC) in almost all AML patients evaluated (95%) (Griffin and Lowenberg, 1986). MY9, a mAb with reactivity against CD33, in combination with C is capable of eliminating 3.6 logarithms (logs) of clonogenic HL-60 cells (Griffin *et al.*, 1984; Roy *et al.*, 1991). When its effect was measured against normal hematopoietic progenitors, MY9+C inhibited the growth of 99% of day 7 and 14 CFU-GM and 53% of CFU-GEMM (Roy *et al.*, 1991; Griffin *et al.*, 1984; Griffin and Lowenberg, 1986). Following purging, hematologic reconstitution would thus probably originate from CFU-GEMM progenitors, which were only partially eliminated by such treatment, and from the hematopoietic progenitor cells that do not express CD33 (Griffin, 1987). With these results, Robertson *et al.* used CD33+C to treat marrow grafts from 12 patients with AML in CR2 (10 pts), CR1 (1 pt) and CR3 (1 pt). Neutrophil engraftment* was achieved at a median interval of 43 days, and platelet engraftment* at 92 days following autologous PCT. The DFS at 3 to 4 years was estimated at 33% with 4 patients in continuous CR at 3 to 5 years post PCT. These results clearly demonstrate that it is possible to target AML clonogenic cells for purging, and that radical eradication of CFU-GM and even partial elimination of CFU-GEMM do not irreversibly impair engraftment.

Combinations of different purging techniques could increase treatment efficacy. Indeed, the addition of chemotherapeutic agents 4-hydroperoxycyclophosphamide (4-HC) and VP-16 to mAbs M195 (anti-CD33) and F23 (anti-CD13) with C, increased leukemia cell elimination by approximately 3 logs, while sparing hematoietic progenitors earlier than CFU-GM (Lemoli *et al.*, 1991). Similarly, ara-C and VP-16 synergize with MY9+C to increase purging efficacy (Stiff *et al.*, 1991). It is noteworthy that M195+C induced much less elimination of AML cells than MY9+C, underlining the importance of mAb selection in C mediated purging (Gee and Boyle, 1988). As mAb and C purging implies sequential incubations and washes, methods were developed to increase not only treatment efficacy, but also its efficiency. Continuous infusion of complement by an automated cell processor, to replace early complement components which are rapidly exhausted (Martin, 1987), was shown to increase cytotoxicity, and reduce treatment duration and reagents costs (Howell *et al.*, 1989). Decay-accelerating factor (DAF)(CD55) is a membrane protein protecting host cells from damage by autologous C. In an attempt to increase C-mediated cytotoxicity, an anti-CD55 mAb was used to block DAF activity; it was found to markedly increase target cell lysis by human complement (Zhong *et al.*, 1995). All the methods described above take advantage of rabbit complement to bind the murine mAbs and cause cell lysis. Complement is, however, associated with numerous drawbacks including limited availability, variable efficacy and non-specific toxicity, need for appropriate standardization, not to mention the difficult compliance with regulatory issues (See chapter on Regulatory issues) (Roy *et al.*, 1990a; Gee and Boyle, 1988).

In an attempt to facilitate the use of mAbs by rendering them directly cytotoxic, several groups have developed immunoconjugates. Thus, M195, an anti-CD33 mAb, conjugated with iodine 131 was able to target marrow cells and demonstrated significant leukemic cytoreduction after *in vivo* administration (Schwartz *et al.*, 1993). However, with a range of β-emission of approximately 50 cell diameters, *ex vivo* purging with such radioactive conjugates would most likely cause irreversible damage to bystander-stem cells, preventing their *in vitro* use. Immunotoxins (ITs) represent another alternative, and Myers *et al.* were the first to produce an IT directed against AML cells by conjugating AML-2-23 (anti-CD14) and MCS-2 (anti-CD13) mAbs to the ribosome-inactivating phytotoxin, ricin (Myers *et al.*, 1988). Ricin consists of two subunits, the A and B chains, which are linked by a single disulfide bond as well as noncovalent interactions (Vitetta *et al.*, 1987; Blakey and Thorpe, 1988). The A chain is an enzyme which inactivates the 60S subunit of eukaryotic ribosomes, and the B chain binds to galactose-terminated oligosaccharides that are ubiquitous on eukaryotic cell surfaces. Thus, lactose was added *in vitro* to block non-specific binding of the whole ricin fraction of the ITs to normal cells (Myers *et al.*, 1988). Both ITs selectively bound to target cells, inhibited protein synthesis, and prevented the clonogenic growth of fresh marrow blasts from AML patients as well as KG-1 (AML) cells. Tecce *et al.* rather used saporin emitoxin (SAP) conjugates, which demonstrated low cytotoxicity in unconjugated form, and highly specific cytotoxicity and favorable pharmacokinetic properties once conjugated to LAM3 and LAM7 mAbs (Tecce *et al.*, 1991). These ITs yet suffered from a narrow spectrum of activity against monocytes and M5b AML. Nevertheless, the fact that immunotoxins do not necessitate the complicated standardization procedures associated with complement use, nor the sophisticated apparatus required for magnetic purging methods, warrants further developments in this field (Arbour *et al.*, 1996).

Lambert *et al.* had developed a technique to chemically block the two galactose binding sites of the ricin B chain, thereby preventing non-specific linkage to eukaryotic cells (Lambert *et al.*, 1991a;b). An anti-MY9-blocked-ricin (anti-MY9–bR) IT, comprised of anti-MY9 mAb conjugated to a modified whole ricin, eliminated several logs of AML cell lines and patient cells (Roy *et al.*, 1991). As expected, MY9–bR was toxic to normal CFU-GM, which express CD33 and also to BFU-E and CFU-GEMM, although to a lesser extent. When compared to anti-MY9+C, anti-MY9–bR could be used in conditions which provided more effective depletion of AML cells with substantially less depletion of normal CFU-GM. Therefore, MY9–bR was used at high IT concentrations for short incubation periods and evaluated for *in vitro* purging of AML cells from autologous marrow.

A phase I/II clinical trial using anti-MY9–bR for purging of the marrow graft was initiated in 2 centers, Maisonneuve-Rosemont Hospital in Montreal and the Dana-Farber Cancer Institute in Boston. Twenty-six adult patients without an HLA-matched donor underwent autologous PCT for AML. Median age at PCT was 40 years (range: 18–57). At the time of harvest, 11 patients were in CR1, 13 patients were in CR2 and 1 patient in CR3. Patient bone marrow was harvested just prior to autologous PCT and purged with anti-MY9–bR. All patients received Bu/Cy for preparative regimen. This IT eliminated 40% of erythroid progenitors and 60% of myeloid progenitors. Median time to neutrophil engraftment* was 40 days (range: 17–64) and to platelet engraftment* was 63 days (range: 16–307). Patients transplanted with anti-MY9−bR experienced significantly more rapid neutrophil engraftment compared to patients receiving anti-MY9+C purged marrows ($p=0.0005$ for time to achieve a neutrophil count greater than 0.1×10^9/L) (Robertson *et al.*, 1994). Interestingly, 9 of the 11 patients transplanted in first remission are alive and in continuous unmaintained complete remission, with a median follow-up of 28 months post-PCT. Five of 13 patients transplanted in second or subsequent CR are alive and leukemia-free +18, +20, +21, +44 and +45 months post-PCT.

Hematopoietic engraftment is notoriously slow in patients with AML, even when unpurged marrows are used (Lowenberg *et al.*, 1990). The above mentioned studies demonstrate that the selective elimination of early myeloid progenitor cells, even those that are CD33 positive, may delay, but not prevent stem cell engraftment. Thus, methodologies such as *in vitro* expansion may be useful to induce maturation of progenitor cells prior to their administration and to shorten time to engraftment. Nevertheless, slow engraftment was not associated with a higher incidence of procedure-related toxicity. Moreover, more than 60% of patients with AML in first complete remission and treated by autologous mAb purged marrow transplants have a prolonged DFS. Even more compelling is the 30% long-term DFS that we and others reported for patients transplanted in second or subsequent remission, patients with dismal prognosis when treated by standard approaches. These results warrant randomized clinical trials to clearly define the contribution of mAb purging of progenitor cell grafts to long-term survival.

Chronic myelogenous leukemia

Results from 8 major centers compiled by McGlave *et al.* show that autologous transplants can be performed successfully in patients with chronic myelogenous leukemia (CML). Indeed, autologous PCT results in low procedure-related toxicity, adequate engraftment, and most impressively a survival curve with a plateau (McGlave *et al.*, 1994). Not surprisingly, contamination of the progenitor cell graft by CML cells is of significant concern, and several centers use marrow purging techniques. Options for purging of CML cells were previously described, ranging from long-term culture systems (Barnett *et al.*, 1994), to interferon administered *ex vivo* (McGlave *et al.*, 1990) or *in vivo* (Carella *et al.*, 1996), and incubation with anti-sense oligonucleotides (Tari *et al.*, 1994), 5-fluorouracil (Jazwiec *et al.*, 1995), eilatin (Einat *et al.*, 1995), etc. However, few studies have used mAbs to purge CML cells, mostly because of the difficulty in specifically targeting the very early CML progenitor cells. When antigens such as erythropoietin receptor and HLA-DR were identified on the surface of CD34+ CML cells by Wognum *et al.* and Verfaillie *et al.*, respectively, a major step forward has been made (Wognum *et al.*, 1992; Verfaillie *et al.*, 1992). It may now be possible to identify benign primitive hematopoietic progenitor cells that can be distinguished from their malignant CML counterparts. However, limitations such as the number of CD34+DR- progenitor cells, and the need for *in vitro* culture and expansion may preclude the immediate application of these methods, but maybe not for long! (Verfaillie, 1994)

Antibodies Directed against B-lineage Antigens

Neoplastic cells from 80–85% of acute lymphoblastic leukemia (ALL) and non-Hodgkin's lymphoma (NHL) patients originate from B-lineage cells, the rest deriving primarily from T cells. The majority of B-lineage ALL cells express one or more of the following antigens: Ia, CD19, CD10, CD9 and CD24 (Abramson *et al.*, 1981; Kersey *et al.*, 1981; Ritz *et al.*, 1980; Freedman *et al.*, 1988), which are surface markers found on the surface of normal pre-B cells. Several lines of evidence suggest that these leukemia cells not only share surface markers with normal B-lineage lymphoblasts but that these ALL cells are in fact derived from stages of pre-B cell differentiation (Foa *et al.*, 1989).

Acute lymphoblastic leukemia

B-lineage leukemia cells consistently express the same combinations of surface antigens and can be classified according to their maturation stage (Nadler *et al.*, 1984). Early ALL cells, expressing HLA-DR alone or HLA-DR+CD19+CD10−, tend to be found in infants and young children and carry a poor prognosis. HLA-DR+CD19+CD10+ ALL represent approximately one third of ALL and are most frequently encountered in children. ALL with an HLA-DR+CD19+CD20+ phenotype are detected predominantly overall (48% of ALL). Patients with the more mature B-cell ALL phenotype (HLA-DR+CD19+CD20+sIg+) are usually adults and present a particularly aggressive disease (Copelan and McGuire, 1995).

Culture conditions were developed to study target antigens most frequently found on clonogenic ALL cells. Interestingly, the antigens detected on the majority of ALL blasts, namely Ia, CD19, CD10 and CD9 were also found to be expressed on the proliferating fraction of cells in this disease (Hudson *et al.*, 1989; Uckun and Ledbetter, 1988; Freedman, 1996). This finding suggests that surface markers present on the whole population of ALL cells, whether they are dividing or not, also characterize the clonogenic precursor cells (Hudson *et al.*, 1989).

Several groups reported on the use of autologous PCT to treat patients with ALL who do not have a histocompatible donor. As shown in Table 1, autologous PCT was performed at all stages of the disease, sometimes in high risk first remission, but usually in second or subsequent remission and rarely at relapse (Kersey *et al.*, 1987; Janossy *et al.*, 1988; Simonsson *et al.*, 1989; Rizzoli *et al.*, 1989; Gorin *et al.*, 1989; Sallan *et al.*, 1989; Schroeder *et al.*, 1991; Carey *et al.*, 1991; Cahn *et al.*, 1991; Gilmore *et al.*, 1991; Schmid *et al.*, 1993; Uckun *et al.*, 1993; Soiffer *et al.*, 1993; Doney *et al.*, 1993; Morishima *et al.*, 1993; Fiere *et al.*, 1993). Most studies used purging procedures, mAbs and C being the preferred option.

Table 1 Autologous bone marrow transplantation for patients with ALL

Authors	*Remission Status*				*N*	*Age Range*	*Preparative Regimen*	*Purging*	*Relapse*	*DFS (3–4 Y)*
	1	*2*	*3+*	*Rel*						
Kersey *et al.*, 87	3	17	25	0	45	0–50	CY+TBI±6MP±MTX	B: Anti-CD9+CD10+CD24+C	79%	20%
								T: Anti-CD3±CD5±CD18+Ricin		
Janossy *et al.*, 88	10	23	3	0	36	6–42	multiple CT+TBI	Anti-CD10±CD19±CD7+C		90%CR1
										57% CR2+
Simonsson *et al.*, 89	21	29	3	0	54	3–55	TBI+multiple CT	B: Anti-CD10±CD19+C		65% CR1
								T: Anti-CD7+C		31% CR2+
Rizzoli *et al.*, 89	37	45	0	0	82	2–57	CY+TBI	Maf or no purging		38% CR1
										27% CR2
Gorin *et al.*, 89	233	205	0	0	438	1–55	Multiple CT±TBI	mAb+C or Maf or no purging		41% CR1
										29% CR2
Sallan *et al.*, 89	1	11	5	1	44	1–14	CY+Ara C+VM26+TBI	Anti-CD9+CD10+C		29%
Schroeder *et al.*, 91	0	17	7	0	24	0–16	Mel+TBI	Campath-1 or no purging		45%
Carey *et al.*, 91	15	0	0	0	15	18–51	Mel+TBI	No purging	52%	48%
Cahn *et al.*, 91	6	16	3	1	26	2–38	TBI+Ara C+ Mel	mAbs+C±Maf	62%	28%

Gilmore *et al.*, 91	27	0	0	0	27	11–45	CY+TBI+Ara C	B: Anti-CD10±CD19+C	65%	32%
								T: Anti-CD7+C		
Schmid *et al.*, 93	0	13	8	1	22	1–16	VP16+TBI	B: Anti-CD10+CD19+CD24+IB	80%	18%
								T: Anti-CD2+CD3+CD5+CD7+IB		
Uckun *et al.*, 93	11	47	25	0	83	1–48	TBI±AraC±	B: Anti-CD9+CD10+CD24+C+4HC	85%	15%
							VP-16±CTX	or Anti-CD19-PAP+4HC		
								T: Anti-CD5 +CD7-Ricin+4HC		
Soiffer *et al.*, 93	1	11	7	3	22	18–54	CY±AraC±VM26+TBI	Anti-CD9+CD10+C	75%	20%
Doney *et al.*, 93	10	27	25	27	89	2–47	CY±VP16+TBI	mAb+C or 4HC	27%	50%-CR1
									69%	50%-CR2+
									90%	0%-Rel
Morishima *et al.*, 93	8	5	2	2	17	4–51	multiple CT+TBI	Anti-CD10+C		75%-CR1
										20%-CR2
										0%-CR3
Fière *et al.*, 93	63	0	0	0	63	15–50	CY+TBI	B: Anti-CD10+CD19+C±Maf		51%
								T: Anti-CD2+CD5+CD7±Maf		

Abbreviations: N: number of patients, Rel: relapse, 6MP: 6-mercaptopurine, MTX: methotrexate, TBI: total body irradiation, CY: cyclophosphamide, Bu: busulfan, B: B-cells, T: T-cells, C: complement, IB: immunomagnetic beads, Maf: Mafosfamide, Mel: Melphalan, CT: chemotherapy, 4HC: 4-hydroperoxycyclophosphamide.

Monoclonal antibodies varied between centers, yet the targeted antigens were mainly CD9, CD10 and CD19 for B-lineage ALL cells, and CD5 and CD7 for T-ALL cells. When autologous PCT was performed in CR1, the DFS was in the range of 50% at 3–4 years post-PCT. As observed in AML, the field of bone marrow transplantation suffers from the lack of randomized controlled studies. Fiere *et al.* conducted one of the few large scale randomized studies comparing autologous PCT and chemotherapy. They were not able to find a difference in DFS between autologous PCT and chemotherapy arms, but late relapses occurred mainly in the chemotherapy arm, and reanalysis after further follow-up will be interesting (Fiere *et al.*, 1993). For patients in second or subsequent remission, the prolonged DFS was in the range of 20–30%, results highly similar to those reported for allogeneic PCT in such high risk patients (Kersey *et al.*, 1987; Fiere *et al.*, 1993). In fact, relapse was the main cause for treatment failure after autologous PCT, and also a major determinant after allogeneic PCT (Kersey *et al.*, 1987; Fiere *et al.*, 1993). When added to the poor results of lymphocyte infusions after allogeneic PCT for relapsed ALL, these results suggest that the GVL effect may not be so active against ALL cells (Kolb *et al.*, 1995).

The high relapse rates after allogeneic PCT and the identification of leukemic burden as an important predictor of outcome show that present preparative regimens are insufficient to eradicate residual host ALL cells (Kersey *et al.*, 1987; Uckun *et al.*, 1992; Fiere *et al.*, 1993). Also, the observation that leukemic progenitor cells can be detected in the purged graft in numerous cases, coupled with the highly variable efficacy of different purging methods, suggest that the most potent purging procedures will have to be identified and optimized (Uckun *et al.*, 1992; 1993; Roy *et al.*, 1990a; Soiffer *et al.*, 1993). However, methods for the detection of patient ALL cells remain either complex and time consuming, or difficult to reproduce. Ph+ ALL offers a unique opportunity to easily monitor residual leukemia cells. Martin *et al.* found that a cocktail of anti-CD9, -CD10 and AB4 mAbs coupled to immunomagnetic beads was consistently able to eradicate PCR detectable cells from all peripheral blood progenitor cell grafts, confirming that mAb purging approaches can be most effective (Martin *et al.*, 1995). In terms of clinical studies to evaluate purging, the European Bone Marrow Transplantation Group has reported the results of autologous PCT in 560 patients with ALL (Gorin *et al.*, 1989). Patients with ALL were treated with heterogeneous conditioning regimen and a majority of these patients had marrow purging according to various protocols. In this retrospective analysis, a comparison of disease-free probability for patients receiving purged versus unpurged marrows showed a trend in favor of marrow purging.

In summary, results of most studies of autologous PCT have demonstrated that this type of intensive treatment is a reasonnable approach to the treatment of high risk first remission patients and can provide effective salvage therapy for patients who have failed standard treatment. In most studies results of autologous PCT are similar to those obtained following allogeneic PCT in patients who have HLA-matched donors. Finally, the great majority of studies evaluating autologous PCT in adult and pediatric patients and reporting interesting survival have used purging methods to eliminate residual neoplastic cells in the marrow grafts.

Non-Hodgkin's lymphoma

Autologous PCT is usually performed in patients with NHL who have relapsed following optimal chemotherapy and/or radiation therapy regimens. Indeed, salvage chemotherapy in these relapsed patients can often induce CR, but these remissions are usually of short duration and only a few of these patients can be cured (Philip *et al.*, 1987). In contrast, autologous PCT in this high risk patient population can result in prolonged disease-free-survival rates ranging from 30 to 50% (Freedman *et al.*, 1990; Colombat *et al.*, 1990; Petersen *et al.*, 1990; Philip *et al.*, 1995; Stahel *et al.*, 1995). These impressive results were obtained with different conditioning regimens, using either purged or unpurged marrows. Unmanipulated marrow grafts are almost uniformly used in patients without marrow involvement at the time of harvest, and preferentially at any time since diagnosis (Philip *et al.*, 1995; Stahel *et al.*, 1995). Purging of the marrow graft is based on the histologic documentation of bone marrow infiltration at the time of diagnosis or relapse in a large proportion of patients with NHL. In addition, clonogenic methodologies have shown that lymphoma cells can be cultured from histologically negative bone marrows (Philip and Favrot, 1988; Sharp *et al.*, 1992). These observations, coupled with the finding that molecular rearrangements found in lymphoma cells were also detected by polymerase chain reaction (PCR) in all relapsed patients, even after achieving a CR, suggest that bone marrow purging should probably be performed even in patients with NHL who show no evidence of marrow infiltration by conventional techniques (Gribben *et al.*, 1991a). Similar observations were also made in peripheral blood stem cell grafts (McCann *et al.*, 1996). Nevertheless, the clinical implications of the persistence of NHL cells in progenitor cell grafts is still debated.

Among purging agents, some groups use 4-hydroperoxycyclophosphamide or mafosfamide (Gulati *et al.*, 1992; Colombat *et al.*, 1990), but most centers rely on mAb-mediated purging methodologies. The nature of the clonogenic cell in NHL is poorly defined, but the mature phenotype of NHL cells suggest that its clonogenic precursor is probably more mature than its ALL clonogenic counterpart, and therefore amenable to immunologic purging (Freedman, 1996). At the Dana-Farber Cancer Institute, Boston, 114 patients with relapsed NHL and marrows presenting a 14;18 translocation detectable by PCR had their bone marrows purged with either anti-CD20 mAb+C, or anti-CD10, -CD20 and -B5+C, which could eliminate more than 3–4 logs of NHL cells (Gribben *et al.*, 1991b). PCR detectable cells were eliminated from half of the patients. These patients had a significantly better DFS than the other half of patients who received a persistently PCR positive marrow graft. This was true for patients with up to 5% infiltration of the intratrabecular space, but all patients with higher lymphoma cell content had a poor outcome. This suggests that 5% infiltration by lymphoma cells was the upper limit for efficacy of this purging procedure, or corresponded to the highest lymphoma burden that could be eliminated by the preparative regimen. Studies of the Nebraska group, that did not use stem cell purging, are also supporting a negative impact of progenitor cell graft contamination by lymphoma cells (Sharp *et al.*, 1996). Indeed, when lymphoma cells could be grown from marrow grafts, these patients had a lower DFS than those receiving grafts without detectable clonogenic lymphoma cells.

In follicular lymphomas, where marrow infiltration is prominent, Freedman *et al.* reported a 4-year DFS of approximately 45% after autologous purged PCT (Freedman *et al.*, 1991). Rohatiner *et al.* deployed similar efforts to purge grafts from 75 patients with follicular NHL, and obtained an identical 4–5 year overall DFS (Johnson *et al.*, 1994). In the latter study, only 14% of marrow samples tested negative by PCR for the t(14;18) after purging with a single CD20 mAb+C, while marrow purging by the Boston group resulted in 50% of samples becoming negative. This higher purging efficacy could be attributed to the multiple mAb approach of the latter group, yet other unexplored factors, such as the pre-purge lymphoma burden, could explain this difference.

Other mAb purging methods include primarily immunomagnetic beads, which can eliminate 3 to 5 logs of lymphoma cells after 2 cycles of treatment with multiple mAbs (Kvalheim *et al.*, 1988; 1989). Monoclonal antibodies are comprised of IgG and/or IgM, and they can target even weakly expressed cell surface antigens (Kvalheim *et al.*, 1988, 1989). Importantly, when utilized in small numbers of patients for purging of peripheral blood progenitor cell grafts, such immunomagnetic beads did not affect engraftment (Dreger *et al.*, 1995b; Straka *et al.*, 1995). Immunotoxins were also tested for the elimination of lymphoma cells and shown to deplete similar numbers of clonogenic lymphoma cells (Roy *et al.*, 1995b; Lambert *et al.*, 1991a; Uckun and Reaman, 1995). Although not exclusively, several ITs target CD19, an antigen expressed on all pre-B and B cells, and nearly all B-cell leukemias (both acute and chronic) and lymphomas (Freedman, 1996). Anti-CD19 mAbs were mostly linked to either whole blocked ricin (anti-B4 – bR), deglycosylated ricin-A chain or other toxin moieties such as PAP (B43-PAP) (Uckun and Reaman, 1995; Ghetie *et al.*, 1995; Roy *et al.*, 1995b), and primarily administered *in vivo* to patients with B-cell NHL and also ALL in clinical phase I–II trials (Uckun and Reaman, 1995; Sausville *et al.*, 1995; Grossbard *et al.*, 1993a;b). Anti-B4 – bR is particularly appealing for purging because only a few hours of incubation are necessary to induce very high levels of cytotoxic activity, a factor most important in progenitor cell purging as it decreases manipulation time, lowers chances of contamination by infectious agents and facilitates laboratory work. A phase I trial of autologous PCT using anti-B4 – bR as a purging agent was initiated in patients with high risk NHL (Roy *et al.*, 1995a). Preliminary results on 41 patients show that all patients evaluable demonstrated engraftment with purged marrow, without signs of delayed neutrophil or platelet engraftment. In addition, the 3- to 4-year DFS for these high risk relapsed patients was 62%. This result, as well as those of other studies reported above, justifies efforts to further evaluate the role of purging in PCT for patients with NHL. Although it may be best evaluated in controlled randomized trials, this patient population is highly heterogeneous and such studies may benefit from stratification according to the level of contamination of the progenitor cell graft. Furthermore, PCR positivity may not be sufficient because it can detect an extremely large range of lymphoma cells. Fortunately, semi-quantitative and quantitative PCR measurements of the numbers of malignant cells are now available and should permit accurate estimations of the lymphoma cell burden present before and after purging (Hetu *et al.*, 1994; Meijerink *et al.*, 1993).

Other hematologic malignancies

Chronic lymphocytic leukemia (CLL) and multiple myeloma (MM) are other B cell malignancies where autologous PCT is gaining ground. CLL cells are usually CD19+, CD20+, CD23+, CD5+, CD22± and FMC7± (Matutes *et al.*, 1994). Rabinowe *et al.* purged marrow grafts of CLL patients with a combination of mAbs directed against CD10, CD20, and B5 in association with rabbit complement (Rabinowe *et al.*, 1993). Anti-CD10 mAb is consistently absent from CLL cells, and was used with the intent of eliminating theoretical early clonogenic CLL precursors. Khouri *et al.* rather used a single anti-CD19 mAb with sheep antimouse IgG1-conjugated magnetic beads which resulted in a median depletion of leukemic cells of 1.3 log (Khouri *et al.*, 1994). In both studies, CLL cells were easily detectable by flow-cytometry or pathological examination in a majority of patients before purging of the marrow graft. After transplantation, CLL cells were not identified in numerous patients by Southern blot analysis and/or immunophenotyping, for periods extending up to 2 years. Although interesting, survival curves are impossible to interpret because of the short follow-up period.

Similar approaches were used for purging of multiple myeloma, using either mAbs directed at CD10, CD20 and PCA-1 with complement (Anderson *et al.*, 1993), or CD19 and sheep antimouse IgG1 conjugated magnetic beads (Dimopoulos *et al.*, 1993). In both instances, malignant cells could not be detected by flow-cytometry after purging, yet PCR for CDR3 showed tumor cell involvement in 17 of 28 patients before purging, and 10 patients after purging. Another alternative is positive selection of CD34+ cells, as multiple myeloma cells may not express CD34 (Vescio *et al.*, 1994). The role of purging in these diseases is especially difficult to define as the role of autologous PCT itself remains uncertain. It is probable that with such traditionally "incurable diseases", transplantation, and most likely purging, may offer a better chance of cure in patients with favorable prognostic features such as stage I disease, minimal exposure to chemotherapy, and complete remission (Anderson, 1995).

Elimination of malignant T cells, either leukemia or lymphoma, from progenitor cell grafts can be performed effectively using immunotoxin conjugates, immunomagnetic microspheres, or combinations of ITs and pharmacologic methods (predominantly using 4-hydroperoxycyclophosphamide or mafosfamide) (Casellas *et al.*, 1985; Stong *et al.*, 1985; Uckun *et al.*, 1987; Bertolini *et al.*, 1997b). However, T-cell purging of autologous grafts was associated with an increased incidence of lymphoproliferative disorders, a complication that significantly hampered its widespread implementation (Anderson *et al.*, 1990).

Antibodies Directed against Antigens Found on Solid Tumors

Neuroblastoma is one of the first diseases for which marrow purging techniques were developed. Although the necessity for marrow purging was never conclusively demonstrated, the rationale of marrow treatment is based on a number of indirect findings. In neuroblastoma, up to 50% of patients have marrow infiltration detectable by light microscopy examination, and this number increases when more sensitive immunofluorescence techniques are used (Johnson and Goldman, 1993; Moss *et al.*, 1991). In addition, the infusion of marrow grafts contaminated by neuroblastoma cells has been associated with the development of pulmonary metastases, an infrequent finding in neuroblastoma patients (Glorieux *et al.*, 1986). Moreover, large numbers of tumor cells in the marrow graft have been shown to correlate with marrow relapse (Matthay *et al.*, 1993). Furthermore, ongoing studies by Rill *et al.* show that when patients with neuroblastoma relapse after autologous bone marrow transplantation, the malignant cells harbor the genetic marker originally introduced in the marrow graft (Rill *et al.*, 1994). Thus, in a majority of clinical trials using autologous bone marrow transplantation for the treatment of patients with neuroblastoma, purging is performed, usually with mAbs and immunomagnetic microspheres (Graham-Pole *et al.*, 1991; Matthay *et al.*, 1993). Several mAbs are reactive against antigens found on the surface of neuroblastoma cells, and were used in combination to try to bypass escape mechanisms, such as low antigen density, antigenic modulation, and inter- and intra-patient antigenic heterogeneity (Graham-Pole *et al.*, 1991; Matthay *et al.*, 1993; Ladenstein *et al.*, 1994). Recently, the antigen CD56 was found to display high levels of expression on neuroblastoma cells and its targeting with a blocked ricin conjugate (N901-bR) resulted in specific and effective cytotoxicity, suggesting that it may be one of the best target antigens to deplete neuroblastoma cells (Roy *et al.*, 1996).

After PCT with purged marrow grafts, hematologic reconstitution seemed delayed, with neutrophil engraftment extending up to 100 days. While marrow purging procedures were only marginally implicated, pre-existing factors such as cumulative chemotherapy and interval from last chemotherapy to harvest were shown to be associated with slow engraftment (Graham-Pole *et al.*, 1991). There were also concerns that the high relapse rates observed after autologous purged PCT could be attributed to the persistence of malignant cells in the graft at levels below detectable thresholds. Although possible, the above purging procedures could remove more than 3–4 logs of tumor cells, resulting in contamination of marrow grafts by less than 1 tumor cell in 10^5 marrow cells (Graham-Pole *et al.*, 1991; Ladenstein *et al.*, 1994; Matthay *et al.*, 1993; Roy *et al.*, 1996). In addition, similar relapse rates were observed following allogeneic PCT, rather suggesting that neuroblastoma cells in the host escape elimination by the preparative regimen (Matthay *et al.*, 1994; Ladenstein *et al.*, 1994).

Several mAbs were also developed against solid tumors, including breast cancer and small cell lung cancer (SCLC) (Stahel *et al.*, 1994). Conjugation of these mAbs with toxins or their association with magnetic microspheres produces agents capable of eliminating more than 3–5 logs of malignant cells from progenitor cell grafts (Myklebust *et al.*, 1994; Zangemeister-Wittke *et al.*, 1994; Roy *et al.*, 1996). The fact that progenitor cell treatment methodologies are now effective, accessible and close to standardization, along with the extremely rapid expansion of

PCT for the treatment of patients with solid tumors, will most likely favor the implantation of purging methodologies in this field.

T-cell depletion in allogeneic progenitor cell transplantation

Graft-versus-host disease (GVHD) is a major obstacle to successful allogeneic PCT, and results from an alloreaction of T-lymphocytes against normal host cells, primarily skin, liver and gut. Thus, T-cell depletion of the marrow graft is theoretically the most logical approach to decrease the incidence and severity of GVHD. Most groups have used mAbs for pan-T cell depletion, and shown that the incidence of both acute and chronic GVHD was significantly decreased (Mitsuyasu *et al.*, 1986; Henslee *et al.*, 1987; Pollard *et al.*, 1986; Goldman *et al.*, 1988; Gratwohl *et al.*, 1993). However, the troublesome finding of a higher incidence of leukemic relapse indicates that T-cell depletion also eliminates donor cells with graft-versus-leukemia (GVL) activity, i.e. capable of eradicating host leukemia cells (Mitsuyasu *et al.*, 1986; Henslee *et al.*, 1987; Pollard *et al.*, 1986; Goldman *et al.*, 1988; Gratwohl *et al.*, 1993). In fact, relapse rates increased to such an extent that the survival advantage afforded by the decreased severity of GVHD was absent or even decreased in comparison to patients receiving unmanipulated marrow grafts. Thus, the most pertinent question became: is it possible to dissociate GVHD and GVL effectors?

Bortin *et al.* have demonstrated that, in AKR mice, the GVL effect can be increased by alloimmunization, without increasing GVHD (Bortin *et al.*, 1979). In addition, Truitt *et al.* were able to generate clones of cytotoxic T lymphocytes (CTL) directed specifically against either leukemia cells only, normal cells only, or against these 2 cell types (Truitt *et al.*, 1983). These results suggest that GVL effect and GVHD can occur as two separate phenomenons and that in these instances the effector cells are probably distinct (Bortin *et al.*, 1973). Therefore, the ideal option would be to selectively deplete grafts of T cell subpopulations only responsible for graft-versus-host disease (GVHD), and preserve T cells with GVL activity.

Depletion of CD6+ (anti-T12) T-cell was attempted by Ritz *et al.* in order to eliminate "mature" T lymphocytes of the marrow graft prior to allogeneic PCT (Roy *et al.*, 1990b; Soiffer *et al.*, 1992). Such an approach spares NK cells and a small proportion of T-cells, CD4+ or CD8+ (Rohatiner *et al.*, 1986; Rasmussen *et al.*, 1994). Champlin *et al.* rather opted to eliminate CD8+ T-cells from the marrow graft, sparing CD4+ T cells and a majority of NK cells (Champlin *et al.*, 1990). In both instances, such efforts at selective T-cell depletion resulted in a low incidence of GVHD (Champlin *et al.*, 1990; Soiffer *et al.*, 1997). In addition, graft rejection, which was previously found to be a major problem in non-selective T-cell depleted transplants, occurred with an incidence comparable to that observed in transplants with unmanipulated marrow grafts (Soiffer *et al.*, 1992; Champlin *et al.*, 1990; Voltarelli *et al.*, 1990; Soiffer *et al.*, 1997). Mixed chimerism was documented in 51% of patients receiving CD6-depleted marrow grafts, and it correlated with lower numbers of CD4+ cells early post-PCT, but not with relapse, survival and disease-free-survival (Roy *et al.*, 1990b). These results suggest that the subset of cells responsible for the GVL effect is distinct from those that may be responsible for preventing recurrence of normal recipient hematopoiesis.

Pichert *et al.* explored the respective roles of T-cell depletion, ablative regimen, and GVHD, in the persistence of minimal residual disease after allogeneic PCT (Pichert *et al.*, 1995). Ninety-two (92) patients with CML either received T-depleted marrow grafts using CD6 mAb+C as sole method of GVHD prophylaxis, or unmanipulated grafts with cyclosporine and methotrexate (von Bueltzingsloewen *et al.*, 1993). Patients received either cyclophosphamide and total body irradiation (Cy/TBI), or busulfan and cyclophosphamide (Bu/Cy) as preparative regimen. During the first six months post-PCT, the majority of patients (80–83%) in both patient populations had PCR-detectable CML cells (PCR+), suggesting that both preparative regimens were incapable of totally eradicating malignant cells (Pichert *et al.*, 1995). However, between 6 and 24 months post-PCT, the majority of patients receiving unmodified marrows converted to PCR negativity, while the proportion of PCR+ patients remained unchanged in those receiving T-cell depleted marrow. Most interestingly, this decrease in PCR detectable CML cells post-PCT followed the development of chronic GVHD, suggesting an association between GVHD and GVL activity. This was further strengthened by noting that persistent PCR- and intermittent PCR- results were clearly associated with both acute and chronic GVHD. Nevertheless, a significant proportion of patients (44%) without GVHD also had intermittent or persistent PCR- assays, a result providing evidence that although GVL is frequently linked to GVHD, it can happen in its absence (Pichert *et al.*, 1995).

The difficulty in identifying specific antigens that would discriminate GVHD from GVL effector cells remains an obstacle to selective T-cell depletion. An alternative to more selective T-cell purging would be to add post-transplant strategies to enhance the GVL activity, e.g. with cytokines such as interleukin-2 (IL-2) (Soiffer *et al.*, 1994; Massumoto *et al.*, 1996). These maneuvers, along with the rapid expansion of PCT using mismatched and especially unrelated marrow donors, which are associated with a very high incidence of severe GVHD, have rejuvenated interest in T-cell depletion of the graft. Humanized Campath mAbs, which are attractive because of their capacity to activate human complement directly, can be used in conditions that fulfill these requirements (Naparstek *et al.*, 1995; Hale and Waldmann, 1994). Immunomagnetic separation procedures and immunotoxins also offer appealing alternatives to rabbit complement for mAb-mediated T-cell depletions (Lamb *et al.*, 1994; Filipovich *et al.*, 1990). Also, in patients with non-malignant diseases, T-cell purging of progenitor cells decreases the incidence and severity of GVHD, and although there remains an increased propension for graft rejection, it may be controlled by additional host immunosuppression (Hale and Waldmann, 1994). Finally, in the context of transplantation with unrelated or mismatched donors, recent studies demonstrated that T-cell depletions were able to lower the incidence of severe GVHD, and preserve sufficient GVL activity to prevent relapse (Spencer *et al.*, 1995; Hessner *et al.*, 1995).

Another application for T-cell depletion has appeared with the revolutionary infusion of donor lymphocytes (Kolb *et al.*, 1990). Indeed, patients with relapsed

CML after allogeneic PCT could be induced back into complete remission and even molecular negativity by simply administering peripheral blood lymphocytes of the donor. More recently, several groups have used this approach with some success for the treatment of patients with other relapsing malignancies, such as acute leukemias and even myelodysplastic syndromes (Bar *et al.*, 1993; Porter *et al.*, 1994; Kolb *et al.*, 1995). These results clearly demonstrate that unsensitized peripheral blood donor lymphocytes have the potential to recognize and eventually eliminate recipient malignant cells *in vivo*, but significant GVHD can be encountered. The infusion of CD8-depleted lymphocytes or reduced numbers of lymphocytes were able to induce GVL activity, without as much GVHD (Mackinnon *et al.*, 1995; Giralt *et al.*, 1994; Alyea *et al.*, 1995). It thus seems that lymphocyte preparations selectively T-cell depleted may be useful to treat relapses following allogeneic PCT. This "adoptive" form of GVL immunotherapy could also be implemented to prevent disease recurrence (Naparstek *et al.*, 1995).

• POSITIVE SELECTION •

POSITIVE SELECTION OF NORMAL HEMATOPOIETIC PROGENITOR CELLS

Selective elimination of contaminating malignant cells or normal T cells may not be the only approach to purge autologous or allogeneic progenitor cell grafts. A potentially simpler alternative, circumventing the difficulty in eliminating all unwanted malignant or T cells, would be to identify and select normal hematopoietic stem cells (Spangrude, 1991). For a number of years, the difficulty of this approach has resided in defining the nature of the hematopoietic stem cell. However, the pursuit of the phenotypic and functional characteristics of this elusive cell is now yielding dividends.

Despite the fact that normal hematopoietic progenitors only constitute a fraction (less than 3%) of the total BM cells (Civin *et al.*, 1984; Baum *et al.*, 1992), they can be identified and selected by taking advantage of their unique surface markers and physical properties, such as size and density (Lasky and Zanjani, 1985; Humblet *et al.*, 1988). However, density separation methods are not very specific nor practical for daily application (Lasky and Zanjani, 1985; Humblet *et al.*, 1988). Cell fractionation by counterflow centrifugal elutriation can result in collection of progenitor cells, but T cells are recovered in the same fraction, since they have the same sedimentation rate (Gao *et al.*, 1987; Noga *et al.*, 1986). Therefore, physical separation methods give rise to a low degree of purification of stem cells.

CD34 Expression According to Hematopoietic Maturation

Hematopoietic cells at different stages of the differentiation process express different cell surface antigens (Krause *et al.*, 1996; Beverley *et al.*, 1980;

Fitchen *et al.*, 1981; Civin and Loken, 1987). The quest to identify these antigens has led to the development of mAbs recognizing CD34, a cell surface marker almost exclusively expressed on immature hematopoietic cells (Fina *et al.*, 1990; Civin *et al.*, 1984; Andrews *et al.*, 1986; Uchansak-Ziegler *et al.*, 1989; Watt *et al.*, 1987; Young *et al.*, 1995). The CD34 molecule is a heavily glycosylated type I transmembrane protein which is a member of the sialomucin family of surface antigens (Civin *et al.*, 1984; Andrews *et al.*, 1986; Watt *et al.*, 1987). This 115 kDa molecule is expressed on 1–3% of human normal bone marrow cells, including almost all committed progenitor cells along with the more primitive progenitors, such as long-term culture-initiating cells (LTC-IC) (Civin *et al.*, 1984; Loken *et al.*, 1987; Berenson *et al.*, 1988; Andrews *et al.*, 1986). CD34 is also expressed on some types of nonhematopoietic cells such as small-vessel endothelial cells (Fina *et al.*, 1990; Young *et al.*, 1995) and embryonic fibroblasts (Brown *et al.*, 1991). The combination of CD34 purification, colony-forming assays and flow cytometric analysis has provided insights into the expression of CD34 along the hematopoietic maturational pathway. The earliest hematopoietic progenitors, such as LTC-IC, CFU-blast and CFU-GEMM, were primarily identified within the highly fluorescent CD34 fraction ($CD34^{bright}$) (Strauss *et al.*, 1986; Krause *et al.*, 1994; Bernstein *et al.*, 1991; Leary and Ogawa, 1987; Krause *et al.*, 1996). The more mature CFU-G, CFU-GM, BFU-E and CFU-megakaryocytes could also be recovered in the $CD34^{bright}$ fraction, but most mature lineage-committed progenitors were $CD34^{dim}$ (Andrews *et al.*, 1986; Krause *et al.*, 1994; Civin and Loken, 1987; Andrews *et al.*, 1989).

Further studies showed that a single CD34+ progenitor cell should have the potential to give rise to all the lymphoid and myeloid progeny, and to sustain long-term hematopoiesis (Berenson *et al.*, 1988; Baum *et al.*, 1992; Huang and Terstappen, 1992). Single-cell liquid culture assays demonstrated a high level of immunophenotypic heterogeneity among CD34+ cells, which translated into distinct clonogenic properties (Huang and Terstappen, 1994). CD34+ cells could be divided into four populations based on co-expression of HLA-DR and/or CD38. The majority (90%) of CD34+ cells also express the antigens CD38 and HLA-DR (CD34+CD38+DR+) and present the lowest self-renewal potential. These cells only generate myeloid colonies and do not have the potential to induce second generation progeny. They are lineage-specific (lin+), thus bearing various differentiation markers such as CD33, CD14 and CD19. However, they could be involved in early hematopoietic reconstitution occurring during the first weeks post-transplant (Lapointe *et al.*, 1996). CD34+CD38+DR− and CD34+CD38−DR− cells each represent approximately 4% of the whole CD34+ population: both populations only generate cells of the myeloid lineage in short-term culture assays. Finally, few CD34+ cells (3%) are CD38− while being HLA-DR+ (CD34+CD38−DR+). This population includes the only cells able to reconstitute long-term hematopoiesis, as evidenced by their potential to induce long-term cultures (5 weeks and more). These CD34+CD38−DR+ cells can generate lymphoid and myeloid colonies, express the Thy-1 antigen (present on early undifferentiated progenitors) and do not bear differentiation markers (lin−) (CD5−, CD10−, CD33−, CD71−) (Huang and Terstappen, 1994).

CD34 expression has also been detected on lymphoid cells rearranging either immunoglobulin genes (B lymphocytes) (Loken *et al.*, 1987) or T-cell antigen receptor genes (T lymphocytes) (Gore *et al.*, 1991), as consistent with previous studies showing expression of CD34 on primitive B- and T-lymphocyte precursors (Ryan *et al.*, 1986). Furthermore, CD34 expression can be detected early in the B cell differentiation arm on CD19+CD10+ cells, but it disappears as CD20 is expressed (Strauss *et al.*, 1986).

CD34 Expression on Malignant Cells

In addition to its expression on normal hematopoietic cells, the CD34 antigen was found to be expressed on their malignant counterparts (Krause *et al.*, 1996). For instance, approximately 40% of cases of acute myeloid leukemia (AML) express CD34 (Civin *et al.*, 1984; Borowitz *et al.*, 1989; Soligo *et al.*, 1991; Vaughan *et al.*, 1988; Geller *et al.*, 1990). CD34 expression can also be detected *in vitro* on colony-forming progenitor cells of chronic myelogenous leukemia (CML) (Silvestri *et al.*, 1992; Katz *et al.*, 1986). In addition, in patients with myelodysplasia, increased numbers of CD34+ cells could be used as a prognostic indicator of blast crisis (Guyotat *et al.*, 1990). Moreover, CD34 was found to be expressed in approximately 70% of cases of childhood B-lineage ALL and in a minority of cases of T-lineage ALL (Gore *et al.*, 1991). In patients with follicular lymphoma, bcl-2-IgH rearranged lymphoma cells were detectable in the CD34+CD19+ fraction, but rarely in the CD34+CD19− fraction (Macintyre *et al.*, 1995).

Unlike leukemic cell types previously mentioned, no CD34 expression has been detected to date on chronic lymphocytic leukemia cells or on multiple myeloma cells (Krause *et al.*, 1996; Vescio *et al.*, 1994). Also, most solid tumors do not express CD34. Indeed, CD34 expression has not been detected on breast cancer, ovarian cancer and neuroblastoma cells (Berenson *et al.*, 1991). However, CD34 expression was detected on tumors of vascular endothelium, like angiosarcoma and Kaposi's sarcoma (Sankey *et al.*, 1990; Krause *et al.*, 1996).

Monoclonal Antibodies Directed against CD34

Numerous anti-CD34 mAbs have been described (Civin, 1989), including My10 (Civin *et al.*, 1984), BI.3C5 (Katz *et al.*, 1985), 12.8 and 115.2 (Andrews *et al.*, 1986), ICH3 (Watt *et al.*, 1987) and TUK3 (Uchansak-Ziegler *et al.*, 1989), which were generated following immunization with KG1/KG1a cell lines. Immunization with human placental endothelial cells produced the anti-CD34 mAb QBEND10 (Fina *et al.*, 1990). All these mAbs, which bind to epitopes located on the extracellular domain of CD34, were classified into 3 different classes according to the differential sensitivity of their epitopes to various enzymes (Sutherland *et al.*, 1992). Class I mAbs (My10, BI.3C5, 12.8, ICH3) are directed against epitopes which are sensitive to neuraminidase, chymopapain and Pasteurella hemolytica-derived glycoprotease (PHDG)(specific cleavage of sialylated O-linked glycans).

Class II mAbs (QBEND10) recognize an epitope sensitive to chymopapain and PHDG but resistant to neuraminidase. Class III mAbs (TUK3, 115.2) bind to epitopes which are insensitive to the enzymes previously mentioned.

Sources of CD34+ Cells and Mobilization Procedures

CD34+ cells are particularly abundant in BM, yet only 1–3% of these CD34+ cells are capable of long-term marrow reconstitution (Huang and Terstappen, 1994; Krause *et al.*, 1996; Hardwick *et al.*, 1992). However, the presence of both committed and stem cell populations seems to be important for rapid and sustained engraftment. Indeed, committed and non-committed progenitor cells were isolated by counterflow elutriation of BM from lethally irradiated mice and transplanted separately. The fraction containing the less primitive progenitors gave transient, early engraftment while the stem cell fraction produced delayed, but durable engraftment (Jones *et al.*, 1989; 1990).

Hematopoietic progenitor cells are present in the peripheral blood, and they are capable of both early and late hematopoietic reconstitution (Goldman *et al.*, 1978; Korbling *et al.*, 1981). However, when compared to BM, the number of progenitor cells normally present in the peripheral circulation is small (Kessinger *et al.*, 1986; Bender *et al.*, 1991). Fortunately, human peripheral blood progenitor cell concentration has been shown to vary with exercise, diurnal cycle, prior cell harvests, administration of steroids and other hormones, endotoxins, dextran and most importantly with chemotherapy and/or hematopoietic growth factors (Lasky, 1991).

High levels of committed and non-committed progenitor cells were mobilized into the peripheral blood (PB) after the administration of a single high dose of cyclophosphamide (Kotasek *et al.*, 1992; To *et al.*, 1990b). Alternatively, growth factors such as G-CSF and GM-CSF were successful at increasing the CD34 cell content of PB (Gianni *et al.*, 1989; Sheridan *et al.*, 1992). Currently, several mobilization regimens with IL-3, IL-6, PIXY321, and other cytokine combinations are evaluated in clinical trials (Guillaume *et al.*, 1993).

Mobilization with chemotherapy alone or single growth factors can increase the CD34+ cell concentration by 10-fold or greater and limit the number of leukophereses to 3 or 4 (Siena *et al.*, 1989; Dreger *et al.*, 1993). However, by combining intensive chemotherapy and cytokine administration during hematopoietic recovery, the steady state PB progenitor cell concentration could be multiplied by up to 1000, and a single collection was shown to provide sufficient numbers of CD34+ cells for durable engraftment (Siena *et al.*, 1991; Gianni *et al.*, 1989; Pettengell *et al.*, 1993; Negrin *et al.*, 1995). Nevertheless, there are legitimate concerns that the mobilization procedure itself may recruit not only the desired hematopoietic progenitor cells, but also malignant cells into the peripheral blood. Indeed, patients with multiple myeloma, breast cancer, small cell lung cancer and also recently NHL were found to have increased numbers of malignant cells in their blood following mobilization, and breast tumors recovered were found to be clonogenic *in vitro* (Lemoli *et al.*, 1996; Ross *et al.*, 1992; Brugger *et al.*, 1994a; Leonard *et al.*, 1997).

CLINICAL USES OF CD34

Quantification of Progenitor Cells in the Graft

Neither leukocyte counts nor morphologic assessment provide a reliable estimate of the numbers of progenitor cells necessary for a safe engraftment. The number, morphologic and functional characteristics of hematopoietic progenitor cells can be evaluated in short- and long-term colony-forming assays, long-term culture-initiating cells (LTC-IC) probably providing the best qualitative assessment of the hematopoietic stem cell potential of a graft (Pettengell *et al.*, 1994; Hirao *et al.*, 1994). However, because of the delay in obtaining results, the main utility of such assays is in retrospective analysis.

CD34+ cell counts provide an objective and precise measurement of hematopoietic progenitor cells, particularly when higher numbers are attained following mobilization (Siena *et al.*, 1993; 1989; Sutherland *et al.*, 1994; Pettengell *et al.*, 1993). In addition, CD34 content can be determined within a few hours, allowing immediate adjustments in the harvesting and transplant schedules, and it can even be predicted by preceding day PB sampling (Elliott *et al.*, 1996). Nevertheless, the low numbers of CD34+ cells mandate careful control of flow cytometric procedures. Thus, major efforts are presently developed to standardize procedures, and to implant large scale quality controls.

There is a general agreement that grafts with a threshold number of more than 2 million CD34+ mobilized blood cells/kg patient weight, usually result in adequate and sustained engraftment of neutrophils and platelets (Bensinger *et al.*, 1995; Haas *et al.*, 1995; Haynes *et al.*, 1995; Weaver *et al.*, 1995). CD34+ cell counts in the peripheral blood before apheresis also predict the yield of progenitor cells that can be collected by leukopheresis (Sutherland *et al.*, 1994; Haas *et al.*, 1994). Nevertheless, because of the varying cloning efficiencies and heterogeneity of CD34+ cells, these numbers must be individually established for each combination of cytokines used. In addition, caution must be exercised as the composition of CD34+ cell subsets may vary between individuals and between harvests. For example, CD34+CD19+ B lymphoid progenitors, which do not contribute to generation of neutrophils and platelets, could represent a high proportion of CD34+ cells (Gorin *et al.*, 1995). These findings justify efforts to identify cell surface molecules present on normal cells and responsible for hematopoietic engraftment, such as CD34+Thy1+ or CD34+CD38− cells (Baum *et al.*, 1992; Huang and Terstappen, 1994).

Purification of Stem and Progenitor Cells

In addition to their role in the quality control of progenitor cell grafts, anti-CD34 mAbs are being used to specifically purify normal stem and progenitor cells administered to rescue hematopoiesis in myeloablated patients (Shpall *et al.*, 1994). This process decreases the amount of cells to be stored and allows removal of unwanted cells that can be occult cancer cells, or cells of a specific hematopoietic compartment, like T cells. This may be especially important in PCT, to eliminate contaminating tumor cells released in PB after growth factor or chemotherapy mobilization (Lemoli *et al.*, 1996; Ross *et al.*, 1992; Brugger *et al.*, 1994a).

Selection methods include flow cytometry (Korbling *et al.*, 1994) and solid phase techniques such as avidin-biotin column chromatography (Korbling *et al.*, 1994; De Bruyn *et al.*, 1995; Dreger *et al.*, 1995a; David *et al.*, 1995; Dewynter *et al.*, 1995; Mahe *et al.*, 1995; Link *et al.*, 1995; Brugger *et al.*, 1995; 1994b; Bohbot *et al.*, 1996; Shpall *et al.*, 1994; Gorin *et al.*, 1995; Lemoli *et al.*, 1996), immunomagnetic separation (Straka *et al.*, 1995; Dreger *et al.*, 1995a; David *et al.*, 1995; Lane *et al.*, 1995; Dewynter *et al.*, 1995; Bensinger *et al.*, 1995a; Cornetta *et al.*, 1995; Williams *et al.*, 1996) and panning (Dewynter *et al.*, 1995; Cardoso *et al.*, 1995; 1993; Holyoake *et al.*, 1994). Although high-speed flow sorting allows selection of single cells harboring specific antigen characteristics, its clinical use is limited primarily by considerations of availability of the apparatus and sterility (Verfaillie *et al.*, 1996). Table 2 provides results on the efficiency of the different methods in terms of quality of the graft: median yield of CD34+ cells, purity, and level of tumor cell and T-cell depletion; and time to hematopoietic reconstitution.

Avidin–biotin chromatography

Avidin–biotin chromatography takes advantage of the capacity of avidin to bind to biotinylated CD34 mAbs to recuperate CD34+ cells (Berenson *et al.*, 1986; Lemoli *et al.*, 1988, 1989). Cells are first exposed to biotinylated anti-CD34 mAbs, and then, to avidin-coated polyacrylamide beads. CD34− cells flow through the column and are removed by continuous column washing. Bound CD34+ cells are eluted from the column by mechanical agitation theoretically *via* breakage of the weak link between antigen and antibody. Thus, high affinity of binding between avidin and biotin prevents release of the mAb, which is not administered to the patient among the antigen-positive cells (Auditore-Hargreaves *et al.*, 1994).

This method was used to select CD34+ cells from the peripheral blood of patients with various hematologic malignancies and solid tumors (Korbling *et al.*, 1994; De Bruyn *et al.*, 1995; Dreger *et al.*, 1995a; Lill *et al.*, 1994; David *et al.*, 1995; Dewynter *et al.*, 1995; Mahe *et al.*, 1995; Link *et al.*, 1995; Brugger *et al.*, 1994b; Bohbot *et al.*, 1996; Shpall *et al.*, 1994; Lemoli *et al.*, 1996). As shown in Table 2, the purity of CD34+ cells selected by immunoadsorption columns was approximately 70%, and between 30 and 60% of CD34+ cells were recovered. T-cell depletion was in the range of 1.5 to 3 logs, which may be sufficient to decrease the risk of GVHD (Dreger *et al.*, 1995a; Bensinger *et al.*, 1995a; Cornetta *et al.*, 1995). Neutrophil engraftment* occurred between days 10 and 22, while the time to more than 50×10^9 platelets/L ranged from 12 to 24 days.

Immunomagnetic selection

Immunomagnetic separation relies on the binding of CD34+ cells first to anti-CD34 mAb and then to ferric oxide-treated polystyrene beads coated with sheep anti-mouse IgG (Hardwick *et al.*, 1992; Krause *et al.*, 1996; Auditore-Hargreaves *et al.*, 1994). Magnets applied to the side of the column pull out bead-rosetted CD34+ cells and free CD34− cells are washed out by gravity flow. Chymopapain was first used to cleave a peripheral domain of the CD34 molecule, and separate CD34+ cells from beads and mAbs. As chymopapain is potentially toxic, Pasteurella haemolytica glycoprotease, a non-cytotoxic enzyme, has been used to cleave CD34 epitopes (Marsh *et al.*, 1992). This strategy permitted recovery of up to 78% of CD34+ cells, with a purity of up to 95%. More recently, enzymes were replaced by a releasing agent that competes for antibody binding (PR-34) and eliminates mAb and beads from the surface of CD34+ cells (Hansen *et al.*, 1995). The physiologic role of CD34 epitopes and the impact of such deletions being unknown, this last procedure has the significant advantage of releasing CD34+ cells without altering the CD34 molecule.

Recovery of CD34+ cells by immunomagnetic selection ranged from 5 to 65%; purity, from 26 to 94%; and T-cell elimination, from 3 to 4 logs (Table 2). These results could yet be influenced by the nature of previous mobilization regimens and experience with the procedure. In terms of hematopoietic reconstitution, initial clinical studies showed no significant differences in engraftment between immunomagnetically separated and unseparated marrow cells (Dreger *et al.*, 1995a). Finally, concerns about reinfusion of residual beads, which could lead to the elimination of cells attached by the reticuloendothelial system, or other side effects did not yet materialize (Kemshead, 1991).

Selection by panning

An alternative to immunoadsorption and immunomagnetic separation is positive selection by panning. In this methodology, anti-CD34 mAbs are attached to a solid matrix (polystyrene flasks), and cells released by mechanical agitation. The isolation is performed in two steps: panning with soybean agglutinin to first decrease the number of unwanted T cells, followed by panning with anti-CD34 antibody (ICH3) (Okarma, 1992; Lebkowski *et al.*, 1992).

Panning has been used *in vitro* for the selection of CD34+ cells from either human bone marrow, peripheral blood and umbilical cord blood (Cardoso *et al.*, 1995, 1993; Holyoake *et al.*, 1994; Dewynter *et al.*, 1995). In general, this technique has yielded CD34+ recoveries ranging from 17 to 74%, and with a purity estimated between 32 and 94%. Clinical results on transplantation and engraftment of patients with cells selected by panning have not been published yet. This may be related to the technical problems associated with the sequential enrichment steps, and large surface areas needed to accomodate the high numbers of CD34+ cells necessary for engraftment.

Clinical Application of CD34 Positive Selection

Before considering widespread application of this procedure, there were definite obstacles to consider. Foremost were concerns that the selection process would result in the loss of cells, either CD34+ or CD34−, directly or indirectly

Table 2 Results of CD34 positive selection procedures

METHODS	*Source of CD34+ cells*	*Origin*	*N*	*Mobilization*	*Yield*	*Purity*	*T cell depletion*	*Neutrophils* ($>0.5\times10^9$/L)	*Platelets* ($20–50\times10^9$/L)	*Authors*
Avidin–Biotin	BM	P	5	None		83%				Korbling *et al.*, 1994
Chromatography										
	UCB	D	15	None	28%	77%				DeBruyn *et al.*, 1995
	BM	D	6	None	27%	92%				
	PB	D+P	6	G-CSF	27%	65%	3 logs			Dreger *et al.*,1995
	PB	P	6	G-CSF		71%				Mahé *et al.*, 1995
	UCB	D	1	None		62%				David *et al.*, 1995
	PB	D	5	Chemo+G-CSF		85%				
	UCB	D	3	None	54%	56%				DeWynter *et al.*, 1995
	BM	D	3	None	40%	75%				
	PB	P	7	chemo+G-CSF	30%	84%				
	BM±PB±GF	P	44	Chemo+G-CSF	PB:52%	PB:42%		10–34 days	9–156 days	Shpall *et al.*, 1994
					BM:42%	BM:72%				
	PB	P	15	Chemo+G-CSF	73%	61%	3 logs	12 days	15 days	Brugger *et al.*, 1994
	PB[†]	D	5	G-CSF	PB:30%	PB:70%	PB:3 logs	15 days†	24 days[†]	Link *et al.*, 1995
	PB+BM	D	5	G-CSF	BM:48%	BM:66%	BM:2 logs	15 days	41 days	
	PB	P	10	Chemo+G-CSF	64%	72%	11-15 days*	14 days*		Brugger *et al.*, 1995
	BM+GF	P	25	None		49%		15 days	23 days	Gorin *et al.*, 1995
	PB	P	10	Chemo+G-CSF	58%	90%	10 days	12 days		Lemoli *et al.*, 1996
	PB[0]	P	5	Chemo+G-CSF	92%	87%		11 days[00]	22 days[00]	Bohbot *et al.*, 1996

Immunomagnetic separation	PB	D	8	G-CSF	36%	94%	4 logs			Dreger *et al*., 1995
	UCB	D	3	None		35%				David *et al*., 1995
	PB	P	3	Chemo+G-CSF		38%				
	PB	D	13	G-CSF ± GM-CSF	48–79%	22–81%	3 logs			Lane *et al*., 1995
	UCB	D	2	None	8%	13%				DeWynter *et al*., 1995
	BM	D	6	None	3%	30%				
	PB	P	3	Chemo+G-CSF	5%	35%				
	PB	D	7	G-CSF			4 logs	15 days	10 days	Bensinger *et al*., 1995
	BM	D	9		49%	59%	3 logs	10 days	20 days	Cornetta *et al*., 1995
	PB	P	9	Chemo+G-CSF		91%		8 days*	10 days*	Williams *et al*., 1996
Panning	UCB	D		None	49%	93%				Cardoso *et al*., 1993
	BM	D	13		61%	79%				Holyoake *et al*., 1994
	BM	D	7	None	66%	94%				Cardoso *et al*., 1995
	UCB	D	4	None	4%	31%				DeWynter *et al*., 1995
	BM	D	7	None	9%	34%				
	PB	P	4	Chemo+G-CSF	39%	31%				

Abbreviations: PB: peripheral blood, UCB: umbilical cord blood, BM: bone marrow, D: normal donor, P: patient, N: number of samples, Neutro: neutrophils, *: infusion of expanded ± unexpanded CD34+ selected cells, †: transplantation of PB CD34+ cells along with unselected BM cells, [0]: selection of CD34+ cells on thawed cells, [00]: data on only one patient, GF: growth factor administration after infusion of this graft.

responsible for short- and long-term engraftment. However, a first study by Berenson *et al.* showed that autologous CD34+ cells from the marrow of patients with breast cancer and neuroblastoma infused after marrow ablative therapy resulted in successful engraftment in all evaluable patients (Berenson *et al.*, 1991). Subsequently, with the development of standardized methodologies for the selection of CD34+ cells, engraftment was documented with CD34+ cells from bone marrow and/or peripheral blood (Table 2). Time to granulocyte and platelet reconstitution was 10–14 days and 14–15 days, respectively, in most patients. Also, Brugger *et al.* showed that in patients with breast cancer, transfusion of the same numbers of positively selected CD34+ cells or unseparated PB progenitor cells after high-dose chemotherapy and mobilization with G-CSF, led to identical times to neutrophil and platelet recovery (Brugger *et al.*, 1994b). However, patients transplanted with CD34+ cells from peripheral blood seemed to engraft more rapidly than CD34+ selected cells from bone marrow (Shpall *et al.*, 1994). Thus, fears of impaired early engraftment did not materialize, and although it is too early to conclude about the quality of long-term reconstitution, a few years of follow-up do not indicate any reason for worry.

Other problems associated with the use of CD34 selection include requirements in terms of time, equipment and expense for this procedure. In addition, the efficacy of purging must be sufficient to eradicate malignant cells from the graft. To date, this aspect has not been thoroughly addressed, but purity levels attained imply that residual malignant cells will persist in CD34 selected grafts. Nevertheless, CD34 selection was shown capable of eliminating several logs of various malignant cells and of depleting breast cancer cells from progenitor cell grafts below detectable levels (Shpall *et al.*, 1994).

Among the other advantages of CD34 selection is the reduction in number of cells to be stored and administered, resulting in a decrease in the size of freezing containers and in requirements for storage space. The decreased amount of potentially toxic cryopreservatives such as dimethylsulfoxide also translates in decreased toxicity to the patient at the time of administration of the graft. Interestingly, one study shows that it would even be possible to do the CD34 selection after thawing of unselected peripheral blood cells (Bohbot *et al.*, 1996). In the five patients evaluated, the selection yielded 92% of CD34+ cells with a purity of 87%; and in the one patient who received a post-thawing CD34 selected graft, neutrophil and platelet engraftment were documented at 11 and 22 days, respectively. Furthermore, in the field of gene therapy, enrichment for cells with long-term clonogenic potential could facilitate transfection procedures and enhance chances for stable expression of transduced genes (see chapter on Gene Therapy) (Korbling, 1995).

Allogeneic PCT

In allogeneic PCT transplantation, the severity of GVHD stems in part from the T cells present in the graft and mainly from the degree of mismatch between donor and recipient (Martin, 1990; 1992). Allogeneic PBPCs can be mobilized with hematopoietic growth factors in healthy BM donors (Tjonnfjord *et al.*, 1994;

Arseniev *et al.*, 1994). However, the allogeneic recipients of PBPCs are exposed to a high risk of acute graft-versus-host disease (aGVHD) because of the high content of immunocompetent T cells in the apheresis product. In order to optimize allogeneic PBPCT, positive selection of CD34+ cells could reduce T-cell content by 2- to 3-logs (Dreger *et al.*, 1995a; Brugger *et al.*, 1994b; Link *et al.*, 1995; Bensinger *et al.*, 1995a; Cornetta *et al.*, 1995). However, in contrast to what was feared, unselected peripheral blood progenitor cell grafts with their high levels of T cells did not generate a higher incidence of GVHD than bone marrow grafts (Bensinger *et al.*, 1995b; Korbling *et al.*, 1995; Schmitz *et al.*, 1995). The induction of a predominant Th2 phenotype by G-CSF would be a plausible explanation for this result (Pan *et al.*, 1995). Thus, these findings suggest that CD34 selection of PB progenitor cells may represent a particularly attractive option to prevent GVHD in high risk patients and enable transplants of patients across the major histocompatibility barrier. Indeed, in the few patients where PB progenitor cells were CD34 selected, acute GVHD seldom occurred (Link *et al.*, 1995).

Combining Positive and Negative Selection

Taken together, negative and positive selection procedures present their limitations. In both positive and negative selection, tumor cells can remain after the procedure. Even if the purity of CD34 selections could be increased by several passages on the columns, the final fraction could still contain CD34− clonogenic precursors. Therefore, in order to generate grafts of the highest quality, it is likely that future approaches will lie in the combination of positive and negative selection procedures. This may be true for both autologous and allogeneic progenitor cell grafts (Bertolini *et al.*, 1997b).

Ex vivo expansion of progenitor cells

Positive, negative or combinations of these selection procedures to eliminate malignant cells may yet cause significant depletion of normal hematopoietic progenitor cells, particularly when antigens on early stem cells are targeted as is the case for CML cells, and result in delayed engraftment. As the number of CFU-GM and/or CD34+ cells in progenitor cell grafts correlate with the rate of recovery for both neutrophils and platelets (To *et al.*, 1992; 1990a), it was postulated that the rate of hematopoietic reconstitution could be enhanced by increasing the number of progenitor cells infused at the time of transplant (Haylock *et al.*, 1992). Indeed, previous studies in mice have shown that *ex vivo* expanded bone marrow cells led to a more rapid recovery than unexpanded BM cells (Muench and Moore, 1992).

Interestingly, CD34 selection has favored progenitor cell growth *in vitro*, most likely because some CD34− marrow or blood cells limited their expansion (Auditore-Hargreaves *et al.*, 1994). Several growth factor combinations are being tested to increase the number of CD34+ cells and also to induce differentiation (Brugger *et al.*, 1993). CD34+ cells cultured in a mixture of IL-3, GM-CSF, and

G-CSF resulted in an 18-fold increase in cell number after only 10 to 12 days in culture (Smith *et al.*, 1993). Various combinations of Stem Cell Factor (SCF), IL-3, -4, -6, -11, -12, G-CSF, GM-CSF, and other cytokines, such as FLT3L and PIXY321, also gave rise to major expansion of stem and progenitor cells (Ogawa, 1993; Elwood *et al.*, 1996; Shah *et al.*, 1996; Gabbianelli *et al.*, 1995; Shapiro *et al.*, 1996; Smith *et al.*, 1993; Williams *et al.*, 1996). Such cytokines may also alter "homing" characteristics (Tavassoli *et al.*, 1991; Zanjani *et al.*, 1992). Optimum growth conditions can even be defined to promote the expansion of various cell lineages such as megakaryocytes or lymphocytes (Bertolini *et al.*, 1997a; Rosenzweig *et al.*, 1996). However, studies on murine hematopoietic cells suggested that IL-1, IL-3 or *ex vivo* expansion itself may impair the long term reconstituting ability of stem cells (Yonemura *et al.*, 1996; Peters *et al.*, 1996). Results from clinical studies should thus yield crucial information on the long term reconstituting potential of such treated cells. To date, their use in humans for *ex vivo* expansion of PB grafts harboring less than 10 percent of the usual number of CD34+ cells has resulted in rapid engraftment (Brugger *et al.*, 1995). In addition, Williams *et al.* recently demonstrated that after expansion with PIXY321 alone, the final product was comprised of about 70% granulocytic precursors. Reinfusion of these expanded myeloid progenitors did not lead to acute toxicity and hematopoietic recovery was prompt although not shortened when compared to historical cases (Williams *et al.*, 1996). Furthermore, in a phase I clinical study conducted on patients with nonmyeloid malignancies, Alcorn *et al.* demonstrated that even with cryopreserved PBPCs, CD34 selection, *ex vivo* expansion and reinfusion can be performed successfully (Alcorn *et al.*, 1996). Nevertheless, the capacity of progenitor cell expansion to enhance engraftment and to eliminate the toxic effects of positive and/or negative purging on normal hematopoietic progenitor cells remains to be determined. Moreover, the absolute effect on both transient and sustained engraftment remains to be confirmed in clinical trials using intensive preparative regimens.

Residual tumor cells can be detected in the CD34-selected fraction of a number of autologous harvests (Brugger *et al.*, 1994b; Shpall *et al.*, 1994; Willems *et al.*, 1996). If *ex vivo* culture of stem cells could support the growth of normal progenitors, while inhibiting the growth of malignant cells, as demonstrated with long term culture of CML and AML cells, it could have an additional purging effect (Barnett *et al.*, 1994). Widmer *et al.* have recently addressed this question in patients with non-Hodgkin's lymphoma (Widmer *et al.*, 1996). Using a competitive PCR titration assay, this study determined the number of residual lymphoma cells before and after selection and *ex vivo* expansion in the presence of SCF, IL-1β, IL-3 and IL-6. Their results show that in a majority of cases, residual lymphoma cells do not proliferate under conditions which allow expansion of CD34+ cells.

Ex vivo expansion of CD34+ selected cells has the potential to limit the size of progenitor cell collections, decrease the engraftment period and facilitate graft manipulation, including gene therapy approaches. It could also enable optimization of the T-cell dose, to prevent the rejection of T-cell depleted allografts, and even enable "cell farming" for transfusion purposes.

CONCLUSIONS

Monoclonal antibodies, with their capacity to target specific cell populations, are now implanted in the field of progenitor cell transplantation. The need for bone marrow and peripheral blood progenitor cell purging either by positive or negative selection, and most likely both approaches in sequence, is becoming increasingly evident. In addition, the numerous other applications of mAbs, which include progenitor cell enumeration and expansion, have set deep roots for their usage. Moreover, their prominent advantages for genetic therapies promise that the coming years will be most exciting.

BIBLIOGRAPHY

Abramson, C.S., Kersey, J.H. and LeBien, T.W. (1981) A monoclonal antibody (BA-1) reactive with cells of human B lymphocyte lineage. *J. Immunol.*, **126**, 83–88.

Alcorn, M.J., Holyoake, T.L., Richmond, L., Pearson, C., Farrell, E., Kyle, B., Dunlop, D.J., Fitzsimons, E., Steward, W.P., Pragnell, I.B. and Franklin, I.M. (1996) CD34-positive cells isolated from cryopreserved peripheral-blood progenitor cells can be expanded *ex vivo* and used for transplantation with little or no toxicity. *J. Clin. Oncol.*, **14**, 1839–1847.

Alyea, E., Soiffer, R., Murray, C., Bartett-Pandite, L., Collins, H., Pickett, C., Wang, Y., Chartier, S., Anderson, K. and Ritz, J. (1995) Adoptive immunotherapy following allogeneic bone marrow transplantation with donor lymphocytes depleted of CD8+ T cells. *Blood*, **85 Suppl. 1**, 1158–1150.

Anderson, K.C., Soiffer, R., DeLage, R., Takvorian, T., Freedman, A.S., Rabinowe, S.L., Nadler, L.M., Dear, K., Heflin, L. and Mauch, P. (1990) T-cell-depleted autologous bone marrow transplantation therapy: analysis of immune deficiency and late complications. *Blood*, **76**, 235–244.

Anderson, K.C., Andersen, J., Soiffer, R., Freedman, A.S., Rabinowe, S.N., Robertson, M.J., Spector, N., Blake, K., Murray, C. and Freeman, A. (1993) Monoclonal antibody-purged bone marrow transplantation therapy for multiple myeloma. *Blood*, **82**, 2568–2576.

Anderson, K.C. (1995) Who benefits from high-dose therapy for multiple myeloma? *J. Clin. Oncol.*, **13**, 1291–1296.

Andrews, R.G., Singer, J.W. and Bernstein, I.D. (1986) Monoclonal antibody 12-8 recognizes a 115-kd molecule present on both unipotent and multipotent hematopoietic colony-forming cells and their precursors. *Blood*, **67**, 842–845.

Andrews, R.G., Singer, J.W. and Bernstein, I.D. (1989) Precursors of colony-forming cells in humans can be distinguished from colony-forming cells by expression of the CD33 and CD34 antigens and light scatter properties. *J. Exp. Med.*, **169**, 1721–1731.

Arbour, S., Toupin, S., Belanger, R., Gyger, M., Halle, J.P., Perreault, C. and Roy, D.C. (1996) Phenotypic and functional characterization of peripheral blood and bone marrow natural killer cells prior to autologous transplantation. *Bone Marrow Transplant.*, **17**, 315–322.

Arseniev, L., Tischler, H.J., Battmer, K., Sudmeier, I., Casper, J. and Link, H. (1994) Treatment of poor marrow graft function with allogeneic CD34+ cells immunoselected from G-CSF-mobilized peripheral blood progenitor cells of the marrow donor. *Bone Marrow Transplant.*, **14**, 791–797.

Auditore-Hargreaves, K., Heimfeld, S. and Berenson, R.J. (1994) Selection and transplantation of hematopoietic stem and progenitor cells. *Bioconjugate Chemistry*, **5**, 287–300.

Ball, E.D., Mills, L.E., Coughlin, C.T., Beck, J.R. and Cornwell, G.G. (1986) Autologous bone marrow transplantation in acute myelogenous leukemia: *in vitro* treatment with myeloid cell-specific monoclonal antibodies. *Blood*, **68**, 1311–1315.

Ball, E.D., Mills, L.E., Cornwell, G.G., Davis, B.H., Coughlin, C.T., Howell, A.L., Stukel, T.A., Dain, B.J., McMillan, R., Spruce, W., Miller, W.E. and Thompson, L. (1990a) Autologous bone marrow transplantation for acute myeloid leukemia using monoclonal antibody-purged bone marrow. *Blood*, **75**, 1199–1206.

Ball, E.D., Vredenburgh, J.J., Mills, L.E., Cornwell, G.G., Schwarz, L., Howell, A.L. and Troy, K. (1990b) Autologous bone marrow transplantation for acute myeloid leukemia following *in vitro* treatment with neuraminidase and monoclonal antibodies. *Bone Marrow Transplant.*, **6**, 277–280.

Ball, E.D., Schwarz, L.M. and Bloomfield, C.D. (1991) Expression of the CD15 antigen on normal and leukemic myeloid cells: effects of neuraminidase and variable detection with a panel of monoclonal antibodies. *Mol. Immunol.*, **28**, 951–958.

Ball, E.D. and Howell, A.L. (1988) Monoclonal antibodies to carbohydrate antigens in autologous bone marrow transplantation. *J. Cell. Biochem.*, **36**, 445–452.

Bar, B.M., Schattenberg, A., Mensink, E.J., Geurts Van Kessel, A., Smetsers, T.F., Knops, G.H., Linders, E.H. and de Witte, T. (1993) Donor leukocyte infusions for chronic myeloid leukemia relapsed after allogeneic bone marrow transplantation. *J. Clin. Oncol.*, **11**, 513–519.

Barnett, M.J., Eaves, C.J., Phillips, G.L., Gascoyne, R.D., Hogge, D.E., Horsman, D.E., Humphries, R.K., Klingemann, H.G., Lansdorp, P.M. and Nantel, S.H. (1994) Autografting with cultured marrow in chronic myeloid leukemia: results of a pilot study. *Blood*, **84**, 724–732.

Bassan, R., Cortelazzo, S., Rambaldi, A., Cornelli, P., Borleri, G., Bellavita, P., Biondi, A. and Barbui, T. (1995) Autologous PBSC transplant for late onset AML after mafosfamide-purged and TBI-containing autologous BMT. *Bone Marrow Transplant*, **15**, 791–793.

Baum, C.M., Weissman, I.L., Tsukamoto, A.S., Buckle, A.M. and Peault, B. (1992) Isolation of a candidate human hematopoietic stem-cell population. *Proc. Natl. Acad. Sci. USA*, **89**, 2804–2808.

Beaujean, F., Bernaudin, F., Kuentz, M., Lemerle, S., Cordonnier, C., Le Forestier, C., Reinert, P., Duedari, N., Brandely, M. and Vernant, J.P. (1995) Successful engraftment after autologous transplantation of 10-day cultured bone marrow activated by interleukin 2 in patients with acute lymphoblastic leukemia. *Bone Marrow Transplant*, **15**, 691–696.

Bender, J.G., Unverzagt, K.L., Walker, D.E., Lee, W., Van Epps, D.E., Smith, D.H., Stewart, C.C. and To, L.B. (1991) Identification and comparison of CD34-positive cells and their subpopulations from normal peripheral blood and bone marrow using multicolor flow cytometry. *Blood*, **77**, 2591–2596.

Bensinger, W., Appelbaum, F., Rowley, S., Storb, R., Sanders, J., Lilleby, K., Gooley, T., Demirer, T., Schiffman, K. and Weaver, C. (1995) Factors that influence collection and engraftment of autologous peripheral-blood stem cells. *J. Clin. Oncol.*, **13**, 2547–2555.

Bensinger, W.I., Rowley, S., Appelbaum, F.R., Mils, B., Oldham, F., Chauncey, T. and Buckner, C.D. (1995a) CD34 selected allogeneic peripheral blood stem cell (PBSC) transplantation in older patients with advanced hematologic malignancies. *Blood*, **86** (Suppl. 1), 97a.

Bensinger, W.I., Weaver, C.H., Appelbaum, F.R., Rowley, S., Demirer, T., Sanders, J., Storb, R. and Buckner, C.D. (1995b) Transplantation of allogeneic peripheral blood stem cells mobilized by recombinant human granulocyte colony-stimulating factor. *Blood*, **85**, 1655–1658.

Berenson, R.J., Bensinger, W.I. and Kalamasz, D. (1986) Positive selection of viable cell populations using avidin-biotin immunoadsorption. *J. Imm. Meth.*, **91**, 11–19.

Berenson, R.J., Andrews, R.G., Bensinger, W.I., Kalamasz, D., Knitter, G., Buckner, C.D. and Bernstein, I.D. (1988) Antigen CD34+ marrow cells engraft lethally irradiated baboons. *J. Clin. Invest.*, **81**, 951–955.

Berenson, R.J., Bensinger, W.I., Hill, R.S., Andrews, R.G., Garcia-Lopez, J., Kalamasz, D.F., Still, B.J., Spitzer, G., Buckner, C.D., Bernstein, I.D. and Thomas, E.D. (1991) Engraftment after infusion of CD34+ marrow cells in patients with breast cancer or neuroblastoma. *Blood*, **77**, 1717–1722.

Bernstein, I.D., Leary, A.G., Andrews, R.G. and Ogawa, M. (1991) Blast colony-forming cells and precursors of colony-forming cells detectable in long-term marrow culture express the same phenotype (CD33– CD34+). *Exp. Hematol.*, **19**, 680–682.

Bertolini, F., Battaglia, M., Pedrazzoli, P., Daprada, G.A., Lanza, A., Soligo, D., Caneva, L., Sarina, B., Murphy, S., Thomas, T. and Dellacuna, G.R. (1997a) Megakaryocytic progenitors can be generated *ex vivo* and safely administered to autologous peripheral blood progenitor cell transplant recipients. *Blood*, **89**, 2679–2688.

Bertolini, F., Thomas, T., Battaglia, M., Gibelli, N., Pedrazzoli, P. and Dellacuna, G.R. (1997b) A new two step procedure for 4.5 log depletion of T and B cells in allogeneic transplantation and of neoplastic cells in autologous transplantation. *Bone Marrow Transplant*, **19**, 615–619.

Beverley, P.C., Linch, D. and Delia, D. (1980) Isolation of human haematopoietic progenitor cells using monoclonal antibodies. *Nature*, **287**, 332–333.

Bishop, M.R., Warkentin, P.I., Jackson, J.D., Bayever, E., Iversen, P.L., Whalen, V.I., Lastovica, J., Haines, K. and Kessinger, A. (1994) Antisense oligonucleotide OL (1) p53 for *in vitro* purging of autologous bone marrow in acute myelogenous leukemia. *Prog. Clin. Biol. Res.*, **389**, 183–187.

Blakey, D.C. and Thorpe, P.E. (1988) An overview of therapy with immunotoxins containing ricin or its A chain. *Antibody Immunoconjugates and Radiopharmaceuticals*, **1**, 1–16.

Bohbot, A., Lioure, B., Faradji, A., Schmitt, M., Cuillerot, J.M., Laplace, A. and Oberling, F. (1996) Positive selection of CD34(+) cells from cryopreserved peripheral blood stem cells after thawing – technical aspects and clinical use. *Bone Marrow Transplant*, **17**, 259–264.

Borowitz, M.J., Gockerman, J.P., Moore, J.O., Civin, C.I., Page, S.O., Robertson, J. and Bigner, S.H. (1989) Clinicopathologic and cytogenic features of CD34 (My 10)-positive acute nonlymphocytic leukemia. *American Journal of Clinical Pathology*, **91**, 265–270.

Bortin, M.M., Rimm, A.A., Saltzstein, E.C. and Rodey, G.E. (1973) Graft versus leukemia. 3. Apparent independent antihost and antileukemia activity of transplanted immunocompetent cells. *Transplantation*, **16**, 182–188.

Bortin, M.M., Truitt, R.L., Rimm, A.A. and Bach, F.H. (1979) Graft-versus-leukaemia reactivity induced by alloimmunisation without augmentation of graft-versus-host reactivity. *Nature*, **281**, 490–491.

Brenner, M.K., Rill, D.R., Holladay, M.S., Heslop, H.E., Moen, R.C., Buschle, M., Krance, R.A., Santana, V.M., Anderson, W.F. and Ihle, J.N. (1993) Gene marking to determine whether autologous marrow infusion restores long-term haemopoiesis in cancer patients. *Lancet*, **342**, 1134–1137.

Brown, J., Greaves, M.F. and Molgaard, H.V. (1991) The gene encoding the stem cell antigen, CD34, is conserved in mouse and expressed in haemopoietic progenitor cell lines, brain, and embryonic fibroblasts. *International Immunology*, **3**, 175–184.

Brugger, W., Mocklin, W., Heimfeld, S., Berenson, R.J., Mertelsmann, R. and Kanz, L. (1993) *Ex vivo* expansion of enriched peripheral blood CD34+ progenitor cells by stem cell factor, interleukin-1 beta (IL-1 beta), IL-6, IL-3, interferon-gamma, and erythropoietin. *Blood*, **81**, 2579–2584.

Brugger, W., Bross, K.J., Glatt, M., Weber, F., Mertelsmann, R. and Kanz, L. (1994a) Mobilization of tumor cells and hematopoietic progenitor cells into peripheral blood of patients with solid tumors. *Blood*, **83**, 636.

Brugger, W., Henschler, R., Heimfeld, S., Berenson, R.J., Mertelsmann, R. and Kanz, L. (1994b) Positively selected autologous blood CD34+ cells and unseparated peripheral blood progenitor cells mediate identical hematopoietic engraftment after high-dose VP16, ifosfamide, carboplatin, and epirubicin. *Blood*, **84**, 1421–1426.

Brugger, W., Heimfeld, S., Berenson, R.J., Mertelsmann, R. and Kanz, L. (1995) Reconstitution of hematopoiesis after high-dose chemotherapy by autologous progenitor cells generated *ex vivo*. *N. Engl. J. Med.*, **333**, 283–287.

Cahn, J.Y., Bordigoni, P., Souillet, G., Pico, J.L., Plouvier, E., Reiffers, J., Benz-Lemoine, E., Bergerat, J.P., Lutz, P. and Colombat, P. (1991) The TAM regimen prior to allogeneic and autologous bone marrow transplantation for high-risk acute lymphoblastic leukemias: a cooperative study of 62 patients. *Bone Marrow Transplant*, **7**, 1–4.

Cardoso, A.A., Li, M.L., Batard, P., Sansilvestri, P., Hatzfeld, A., Levesque, J.P., Lebkowski, J.S. and Hatzfeld, J. (1993) Human umbilical cord blood CD34+ cell purification with high yield of early progenitors. *Journal of Hematotherapy*, **2**, 275–279.

Cardoso, A.A., Watt, S.M., Batard, P., Li, M.L., Hatzfeld, A., Genevier, H. and Hatzfeld, J. (1995) An improved panning technique for the selection of CD34+ human bone marrow hematopoietic cells with high recovery of early progenitors. *Exp. Hematol.*, **23**, 407–412.

Carella, A.M., Chimirri, F., Podesta, M., Pitto, A., Piaggio, G., Dejana, A., Lerma, E., Pollicardo, N., Vassallo, F., Soracco, M., Benvenuto, F., Valbonesi, M., Carlier, P., Vimercati, R., Prencipe, E., Gatti, A.M., Ferrara, R.A., Incagliato, M., Florio, G. and Frassoni, F. (1996) High-dose chemoradiotherapy followed by autologous Philadelphia chromosome-negative blood progenitor cell transplantation in patients with chronic myelogenous leukemia. *Bone Marrow Transplant*, **17**, 201–205.

Carey, P.J., Proctor, S.J., Taylor, P. and Hamilton, P.J. (1991) Autologous bone marrow transplantation for high-grade lymphoid malignancy using melphalan/irradiation conditioning without marrow purging or cryopreservation. The Northern Regional Bone Marrow Transplant Group. *Blood*, **77**, 1593–1598.

Casellas, P., Canat, X., Fauser, A.A., Gros, O., Laurent, G., Poncelet, P. and Jansen, F.K. (1985) Optimal elimination of leukemic T cells from human bone marrow with T101-ricin A-chain immunotoxin. *Blood*, **65**, 289–297.

Cesano, A., Pierson, G., Visonneau, S., Migliaccio, A.R. and Santoli, D. (1996) Use of a lethally irradiated major histocompatibility complex nonrestricted cytotoxic T-cell line for effective purging of marrows containing lysis-sensitive or -resistant leukemic targets. *Blood*, **87**, 393–403.

Champlin, R., Ho, W., Gajewski, J., Feig, S., Burnison, M., Holley, G., Greenberg, P., Lee, K., Schmid, I. and Giorgi, J. (1990) Selective depletion of CD8+ T lymphocytes for prevention of graft-versus-host disease after allogeneic bone marrow transplantation. *Blood*, **76**, 418–423.

Civin, C.I., Strauss, L.C., Brovall, C., Fackler, M.J., Schwartz, J.F. and Shaper, J.H. (1984) Antigenic analysis of hematopoiesis. III. A hematopoietic progenitor cell surface antigen defined by a monoclonal antibody raised against KG-1a cells. *J. Immunol.*, **133**, 157–165.

Civin, C.I. (1989) Reducing the cost of the cure in childhood leukemia. *N. Engl. J. Med.*, **321**, 185–187.

Civin, C.I. and Loken, M.R. (1987) Cell surface antigens on human marrow cells: dissection of hematopoietic development using monoclonal antibodies and multiparameter flow cytometry. *Int. J. Cell Cloning*, **5**, 267–288.

Colombat, P., Gorin, N.C., Lemonnier, M.P., Binet, C., Laporte, J.P., Douay, L., Desbois, I., Lopez, M., Lamagnere, J.P. and Najman, A. (1990) The role of autologous bone marrow transplantation in 46 adult patients with non-Hodgkin's lymphomas. *J. Clin. Oncol.*, **8**, 630–637.

Copelan, E.A. and McGuire, E.A. (1995) The biology and treatment of acute lymphoblastic leukemia in adults. *Blood*, **85**, 1151–1168.

Cornetta, K., Gharpure, V., Hromas, R., Abonour, R., Broun, E.R., Cox, E., Wyman, N., Menke, C., Baute, J., Mills, B., Oldham, F. and Srour, E.F. (1995) Sibling-matched allogeneic bone marrow transplantation using CD34 cells obtained by immunomagnetic bead separation. *Blood*, **86** (Suppl. 1), 389a.

David, S., Rice, A., Vianes, I., Duperray, V., Dupouy, M. and Reiffers, J. (1995) Expansion of blood CD34 positive cells - committed precursors expansion does not affect immature hematopoietic progenitors. *Nouvelle Revue Francaise d'Hematologie*, **37**, 343–349.

De Bruyn, C., Delforge, A., Bron, D., Bernier, M., Massy, M., Ley, P., de Hemptinne, D. and Stryckmans, P. (1995) Comparison of the coexpression of CD38, CD33 and HLA-DR antigens on CD34+ purified cells from human cord blood and bone marrow. *Stem Cells*, **13**, 281–288.

De Fabritiis, P., Ferrero, D., Sandrelli, A., Tarella, C., Meloni, G., Pulsoni, A., Pregno, P., Badoni, R., De Felice, L., Gallo, E., Amadori, S., Mandelli, F. and Pileri, A. (1989) Monoclonal antibody purging and autologous bone marrow transplantation in acute myelogenous leukemia in complete remission. *Bone Marrow Transplant*, **4**, 669–674.

Dewynter, E.A., Coutinho, L.H., Pei, X., Marsh, J.C.W., Hows, J., Luft, T. and Testa, N.G. (1995) Comparison of purity and enrichment of CD34(+) cells from bone marrow, umbilical cord and peripheral blood (primed for apheresis) using five separation systems. *Stem Cells*, **13**, 524–532.

Dicke, K.A., Zander, A., Spitzer, G., Verma, D.S., Peters, L., Vellekoop, L., McCredie, K.B. and Hester, J. (1979) Autologous bone-marrow transplantation in relapsed adult acute leukaemia. *Lancet*, **1**, 514–517.

Dimopoulos, M.A., Alexanian, R., Przepiorka, D., Hester, J., Andersson, B., Giralt, S., Mehra, R., van Besien, K., Delasalle, K.B., Reading, C., Deisseroth, A.B. and Champlin, R.E. (1993) Thiotepa, busulfan, and cyclophosphamide: a new preparative regimen for autologous marrow or blood stem cell transplantation in high-risk multiple myeloma. *Blood*, **82**, 2324–2328.

Doney, K., Buckner, C.D., Fisher, L., Petersen, F.B., Sanders, J., Appelbaum, F.R., Anasetti, C., Badger, C., Bensinger, W. and Deeg, H.J. (1993) Autologous bone marrow transplantation for acute lymphoblastic leukemia. *Bone Marrow Transplant*, **12**, 315–321.

Douay, L., Mary, J.Y., Giarratana, M.C., Najman, A. and Gorin, N.C. (1989) Establishment of a reliable experimental procedure for bone marrow purging with mafosfamide (ASTA Z 7557). *Exp. Hematol.*, **17**, 429–432.

Douay, L., Hu, C., Giarratana, M.C., Bouchet, S., Conlon, J., Capizzi, R.L. and Gorin, N.C. (1995) Amifostine improves the antileukemic therapeutic index of mafosfamide: implications for bone marrow purging. *Blood*, **86**, 2849–2855.

Dreger, P., Marquardt, P., Haferlach, T., Jacobs, S., Mulverstedt, T., Eckstein, V., Suttorp, M., Loffler, H., Muller-Ruchholtz, W. and Schmitz, N. (1993) Effective mobilisation of peripheral blood progenitor cells with 'Dexa-BEAM' and G-CSF: timing of harvesting and composition of the leukapheresis product. *British Journal of Cancer*, **68**, 950–957.

Dreger, P., Viehmann, K., Steinmann, J., Eckstein, V., Muller-Ruchholtz, W., Loffler, H. and Schmitz, N. (1995a) G-CSF-mobilized peripheral blood progenitor cells for allogeneic transplantation: comparison of T cell depletion strategies using different CD34+ selection systems or CAMPATH-1. *Exp. Hematol.*, **23**, 147–154.

Dreger, P., Vonneuhoff, N., Suttorp, M., Loffler, H. and Schmitz, N. (1995b) Rapid engraftment of peripheral blood progenitor cell grafts purged with B cell-specific monoclonal antibodies and immunomagnetic beads. *Bone Marrow Transplant*, **16**, 627–629.

Einat, M., Lishner, M., Amiel, A., Nagler, A., Yarkorli, S., Rudi, A., Kashman, Y., Markel, D. and Fabian, I. (1995) Eilatin – a novel marine alkaloid inhibits *in vitro* proliferation of progenitor cells in chronic myeloid leukemia patients. *Exp. Hematol.*, **23**, 1439–1444.

Elliott, C., Samson, D.M., Armitage, S., Lyttelton, M.P., McGuigan, D., Hargreaves, R., Giles, C., Abrahamson, G., Abboudi, Z., Brennan, M. and Kanfer, E.J. (1996) When to harvest peripheral-blood stem cells after mobilization therapy: prediction of CD34-positive cell yield by preceding day CD34-positive concentration in peripheral blood. *J. Clin. Oncol.*, **14**, 970–973.

Elwood, N.J., Zogos, H., Willson, T. and Begley, C.G. (1996) Retroviral transduction of human progenitor cells: use of granulocyte colony-stimulating factor plus stem cell factor to mobilize progenitor cells *in vivo* and stimulation by flt3/flk-2 ligand *in vitro*. *Blood*, **88**, 4452–4462.

Ericson, S.G., Colby, E., Welch, L. and Ball, E.D. (1992) Engraftment of leukocyte subsets following autologous bone marrow transplantation in acute myeloid leukemia using anti-myeloid (CD14 and CD15) monoclonal antibody-purged bone marrow. *Bone Marrow Transplant*, **9**, 129–137.

Estrov, Z., Grunberger, T., Dube, I.D., Wang, Y.P. and Freedman, M.H. (1986) Detection of residual acute lymphoblastic leukemia cells in cultures of bone marrow obtained during remission. *N. Engl. J. Med.*, **315**, 538–542.

Fiere, D., Lepage, E., Sebban, C., Boucheix, C., Gisselbrecht, C., Vernant, J.P., Varet, B., Broustet, A., Cahn, J.Y., Rigal-Huguet, F., Witz, F., Michaux, J.-L., Michallet, M. and Reiffers, J. (1993) Adult acute lymphoblastic leukemia: a multicentric randomized trial testing bone marrow transplantation as postremission therapy. *J. Clin. Oncol.*, **11**, 1990–2001.

Filipovich, A.H., Vallera, D., McGlave, P., Polich, D., Gajl-Peczalska, K., Haake, R., Lasky, L., Blazar, B., Ramsay, N.K. and Kersey, J. (1990) T cell depletion with anti-CD5 immunotoxin in histocompatible bone marrow transplantation. The correlation between residual CD5 negative T cells and subsequent graft-versus-host disease. *Transplantation*, **50**, 410–415.

Fina, L., Molgaard, H.V., Robertson, D., Bradley, N.J., Monaghan, P., Delia, D., Sutherland, D.R., Baker, M.A. and Greaves, M.F. (1990) Expression of the CD34 gene in vascular endothelial cells. *Blood*, **75**, 2417–2426.

Fitchen, J.H., Foon, K.A. and Cline, M.J. (1981) The antigenic characteristics of hematopoietic stem cells. *N. Engl. J. Med.*, **305**, 17–25.

Foa, R., Migone, N., Francia di Celle, P., Fierro, M.T., Tassinari, A., Lo Coco, F., Casorati, G. and Gavosto, F. (1989) Ontogeny, gene rearrangements and immunophenotype of acute leukaemias. *Bone Marrow Transplant*, **4** (Suppl. 1), 66–69.

Forman, S.J., O'Donnell, M.R., Nademanee, A.P., Snyder, D.S., Bierman, P.J., Schmidt, G.M., Fahey, J.L., Stein, A.S., Parker, P.M. and Blume, K.G. (1987) Bone marrow transplantation for patients with Philadelphia chromosome-positive acute lymphoblastic leukemia. *Blood*, **70**, 587–588.

Freedman, A.S., Griffin, J.D. and Nadler, L.M. (1988) Leukemias and the Ontogeny of Leukocytes. *Diagnostic Immunopathology* (ed. by R.B.Colvin, A.K.Bhan and R.T.McCluskey), p.245. Raven Press, New York.

Freedman, A.S., Takvorian, T., Anderson, K.C., Mauch, P., Rabinowe, S.N., Blake, K., Yeap, B., Soiffer, R., Coral, F., Heflin, L., Ritz, J. and Nadler, L.M. (1990) Autologous bone marrow

transplantation in B-cell non-Hodgkin's lymphoma: very low treatment-related mortality in 100 patients in sensitive relapse. *J. Clin. Oncol.*, **8**, 784–791.
Freedman, A.S., Ritz, J., Neuberg, D., Anderson, K.C., Rabinowe, S.N., Mauch, P., Takvorian, T., Soiffer, R., Blake, K. and Yeap, B. (1991) Autologous bone marrow transplantation in 69 patients with a history of low-grade B-cell non-Hodgkin's lymphoma. *Blood*, **77**, 2524–2529.
Freedman, A.S. (1996) Cell surface antigens in leukemias and lymphomas. *Cancer Invest.*, **14**, 252–276.
Freeman, S.D., Kelm, S., Barber, E.K. and Crocker, P.R. (1995) Characterization of CD33 as a new member of the sialoadhesin family of cellular interaction molecules. *Blood*, **85**, 2005–2012.
Gabbianelli, M., Pelosi, E., Montesoro, E., Valtieri, M., Luchetti, L., Samoggia, P., Vitelli, L., Barberi, T., Testa, U., Lyman, S. *et al.* (1995) Multi-level effects of flt3 ligand on human hematopoiesis: expansion of putative stem cells and proliferation of granulomonocytic progenitors/monocytic precursors. *Blood*, **86**, 1661–1670.
Gao, I.K., Noga, S.J., Wagner, J.E., Cremo, C.A., Davis, J. and Donnenberg, A.D. (1987) Implementation of a semiclosed large scale counterflow centrifugal elutriation system. *Journal of Clinical Apheresis*, **3**, 154–160.
Gee, A.P. and Boyle, M.D. (1988) Purging tumor cells from bone marrow by use of antibody and complement: a critical appraisal. *J. Natl. Cancer Inst.*, **80**, 154–159.
Geller, R.B., Zahurak, M., Hurwitz, C.A., Burke, P.J., Karp, J.E., Piantadosi, S. and Civin, C.I. (1990) Prognostic importance of immunophenotyping in adults with acute myelocytic leukaemia: the significance of the stem-cell glycoprotein CD34 (My10). *Br. J. Haematol.*, **76**, 340–347.
Ghetie, V., Engert, A., Schnell, R. and Vitetta, E.S. (1995) The *in vivo* anti-tumor activity of immunotoxins containing two versus one deglycosylated ricin A chains. *Cancer Lett.*, **98**, 97–101.
Gianni, A.M., Siena, S., Bregni, M., Tarella, C., Stern, A.C., Pileri, A. and Bonadonna, G. (1989) Granulocyte-macrophage colony-stimulating factor to harvest circulating haemopoietic stem cells for autotransplantation. *Lancet*, **2**, 580–585.
Giarratana, M.C., Gorin, N.C. and Douay, L. (1995) Plasma interacts with mafosfamide toxicity to normal haematopoietic progenitor cells: impact on *in vitro* marrow purging. *Nouvelle Revue Francaise d'Hematologie*, **37**, 125–130.
Gilmore, M.J., Hamon, M.D., Prentice, H.G., Katz, F., Slaper-Cortenbach, I.C., Hunter, A.E., Gandhi, L., Brenner, M.K., Hoffbrand, A.V. and Mehta, A.B. (1991) Failure of purged autologous bone marrow transplantation in high risk acute lymphoblastic leukaemia in first complete remission. *Bone Marrow Transplant*, **8**, 19–26.
Giralt, S., Hester, J., Huh, Y., Hirsch-Ginsberg, C., Rondon, G., Guo, J., Lee, M., Gajewski, J., Talpaz, M., Kantarjian, H., Fischer, H., Deisseroth, A. and Champlin, R. (1994) CD8+ depleted donor lymphocyte infusion as treatment for relapsed chronic myelogenous leukemia after allogeneic bone marrow transplantation: graft vs leukemia without graft vs host disease. *Blood*, **84** (Suppl. 1), 538a.
Glorieux, P., Bouffet, E., Philip, I., Biron, P., Holzapfel, L., Floret, D., Bouvier, R., Vitrey, D., Pinkerton, R. and Brunat-Mentigny, M. (1986) Metastatic interstitial pneumonitis after autologous bone marrow transplantation. A consequence of reinjection of malignant cells? *Cancer*, **58**, 2136–2139.
Goldman, J.M., Th'ng, K.H. and Park, D.S. (1978) Collection, cryopreservation and subsequent viability of haemopoietic stem cells intended for treatment of chronic granulocytic leukaemia in transformation. *Br. J. Haematol.*, **40**, 185–195.
Goldman, J.M., Gale, R.P., Horowitz, M.M., Biggs, J.C., Champlin, R.E., Gluckman, E., Hoffmann, R.G., Jacobsen, S.J., Marmont, A.M. and McGlave, P.B. (1988) Bone marrow transplantation for chronic myelogenous leukemia in chronic phase. Increased risk for relapse associated with T-cell depletion. *Ann. Intern. Med.*, **108**, 806–814.
Gore, S.D., Kastan, M.B. and Civin, C.I. (1991) Normal human bone marrow precursors that express terminal deoxynucleotidyl transferase include T-cell precursors and possible lymphoid stem cells. *Blood*, **77**, 1681–1690.
Gorin, N.C., Aegerter, P. and Auvert, B. (1989) Autologous bone marrow transplantation for acute leukemia in remission: an analysis of 1322 cases. *Bone Marrow Transplant*, **4** (Suppl. 2), 3–5.
Gorin, N.C., Aegerter, P., Auvert, B., Meloni, G., Goldstone, A.H., Burnett, A., Carella, A., Korbling, M., Herve, P., Maraninchi, D., Lowenberg, R., Verdonck, L., dePlanque, M., Hermans, J., Helbig, W.,

Porcellini, A., Rizzoli, V., Alesandrino, E.P., Franklin, I.M., Reiffers, J., Colleselli, P. and Goldman, J.M. (1990) Autologous bone marrow transplantation for acute myelocytic leukemia in first remission: a European survey of the role of marrow purging. *Blood*, **75**, 1606–1614.

Gorin, N.C., Lopez, M., Laporte, J.P., Quittet, P., Lesage, S., Lemoine, F., Berenson, R.J., Isnard, F., Grande, M. and Stachowiak, J. (1995) Preparation and successful engraftment of purified CD34+ bone marrow progenitor cells in patients with non-Hodgkin's lymphoma. *Blood*, **85**, 1647–1654.

Graham-Pole, J., Gee, A., Emerson, S., Gallo, J., Lee, C., Luzins, J., Janssen, W.E., Pick, T., Worthington-White, D. and Elfenbein, G. (1991) Myeloablative chemoradiotherapy and autologous bone marrow infusions for treatment of neuroblastoma: factors influencing engraftment. *Blood*, **78**, 1607–1614.

Gratwohl, A., Hermans, J., Niederwieser, D., Frassoni, F., Arcese, W., Gahrton, G., Bandini, G., Carreras, E., Vernant, J.P. and Bosi, A. (1993) Bone marrow transplantation for chronic myeloid leukemia: long-term results. Chronic Leukemia Working Party of the European Group for Bone Marrow Transplantation. *Bone Marrow Transplant*, **12**, 509–516.

Gribben, J.G., Freedman A.S., Woo, S.D., Blake, K., Shu, R.S., Freeman, G., Longtine, J.A., Pinkus, G.S. and Nadler, L.M. (1991a) All advanced stage non-Hodgkin's lymphomas with a polymerase chain reaction amplifiable breakpoint of bcl-2 have residual cells containing the bcl-2 rearrangement at evaluation and after treatment. *Blood*, **78**, 3275–3280.

Gribben, J.G., Freedman, A.S., Neuberg, D., Roy, D.C., Blake, K.W., Woo, S.D., Grossbard, M.L., Rabinowe, S.N., Coral, F., Freeman, G.J., Ritz, J. and Nadler, L.M. (1991b) Immunologic purging of marrow assessed by PCR before autologous bone marrow transplantation for B-cell lymphoma. *N. Engl. J. Med.*, **325**, 1525–1533.

Griffin, J.D., Ritz, J., Nadler, L.M. and Schlossman, S.F. (1981) Expression of myeloid differentiation antigens on normal and malignant myeloid cells. *J. Clin. Invest.*, **68**, 932–941.

Griffin, J.D., Larcom, P. and Schlossman, S.F. (1983a) Use of surface markers to identify a subset of acute myelomonocytic leukemia cells with progenitor cell properties. *Blood*, **62**, 1300–1303.

Griffin, J.D., Mayer, R.J., Weinstein, H.J., Rosenthal, D.S., Coral, F.S., Beveridge, R.P. and Schlossman, S.F. (1983b) Surface marker analysis of acute myeloblastic leukemia: identification of differentiation-associated phenotypes. *Blood*, **62**, 557–563.

Griffin, J.D., Linch, D., Sabbath, K., Larcom, P. and Schlossman, S.F. (1984) A monoclonal antibody reactive with normal and leukemic human myeloid progenitor cells. *Leukemia Research*, **8**, 521–534.

Griffin, J.D., Davis, R., Nelson, D.A., Davey, F.R., Mayer, R.J., Schiffer, C., McIntyre, O.R. and Bloomfield, C.D. (1986) Use of surface marker analysis to predict outcome of adult acute myeloblastic leukemia. *Blood*, **68**, 1232–1241.

Griffin, J.D. (1987) The use of monoclonal antibodies in the characterization of myeloid leukemias. *Hematologic Pathology*, **1**, 81–91.

Griffin, J.D. and Lowenberg, B. (1986) Clonogenic cells in acute myeloblastic leukemia. *Blood*, **68**, 1185–1195.

Grossbard, M.L., Gribben, J.G., Freedman, A.S., Lambert, J.M., Kinsella, J., Rabinowe, S.N., Eliseo, L., Taylor, J.A., Blattler, W.A. and Epstein, C.L. (1993a) Adjuvant immunotoxin therapy with anti-B4-blocked ricin after autologous bone marrow transplantation for patients with B-cell non-Hodgkin's lymphoma. *Blood*, **81**, 2263–2271.

Grossbard, M.L., Lambert, J.M., Goldmacher, V.S., Spector, N.L., Kinsella, J., Eliseo, L., Coral, F., Taylor, J.A., Blattler, W.A., Epstein, C.L. and Nadler, L.M. (1993b) Anti-B4-blocked ricin: a phase I trial of 7-day continuous infusion in patients with B-cell neoplasms. *J. Clin. Oncol.*, **11**, 726–737.

Guillaume, T., D'Hondt, V. and Symann, M. (1993) IL-3 and peripheral blood stem cell harvesting. *Stem Cells*, **11**, 173–181.

Gulati, S., Yahalom, J., Acaba, L., Reich, L., Motzer, R., Crown, J., Toia, M., Igarashi, T., Lemoli, R. and Hanninen, E. (1992) Treatment of patients with relapsed and resistant non-Hodgkin's lymphoma using total body irradiation, etoposide, and cyclophosphamide and autologous bone marrow transplantation. *J. Clin. Oncol.*, **10**, 936–941.

Guyotat, D., Campos, L., Thomas, X., Vila, L., Shi, Z.H., Charrin, C., Gentilhomme, O. and Fiere, D. (1990) Myelodysplastic syndromes: a study of surface markers and *in vitro* growth patterns. *Am. J. Hematol.*, **34**, 26–31.

Haas, R., Mohle, R., Fruhauf, S., Goldschmidt, H., Witt, B., Flentje, M., Wannenmacher, M. and Hunstein, W. (1994) Patient characteristics associated with successful mobilizing and autografting of peripheral blood progenitor cells in malignant lymphoma. *Blood*, **83**, 3787–3794.

Haas, R., Witt, B., Mohle, R., Goldschmidt, H., Hohaus, S., Freuhauf, S., Wannenmacher, M. and Hunstein, W. (1995) Sustained long-term hematopoiesis after myeloablative therapy with peripheral blood progenitor cell support. *Blood*, **85**, 3754–3761.

Hale, G. and Waldmann, H. (1994) Control of graft-versus-host disease and graft rejection by T cell depletion of donor and recipient with Campath-1 antibodies. Results of matched sibling transplants for malignant diseases. *Bone Marrow Transplant*, **13**, 597–611.

Hansen, M., Yacob, D., Schaeffer, A., Jain, R., Guha, M. and Deans, R. (1995) A comparison of expansion potential and phenotype of cultured CD34+ cells released from immunomagnetic beads by chymopapain or a peptide epitope release reagent. *Blood*, **86** (Suppl. 1), 666a.

Hardwick, R.A., Kulcinski, D., Mansour, V., Ishizawa, L., Law, P. and Gee, A.P. (1992) Design of large-scale separation systems for positive and negative immunomagnetic selection of cells using superparamagnetic microspheres. *Journal of Hematotherapy*, **1**, 379–386.

Haylock, D.N., To, L.B., Dowse, T.L., Juttner, C.A. and Simmons, P.J. (1992) *Ex vivo* expansion and maturation of peripheral blood CD34+ cells into the myeloid lineage. *Blood*, **80**, 1405–1412.

Haynes, A., Hunter, A., McQuaker, G., Anderson, S., Bienz, N. and Russell, N.H. (1995) Engraftment characteristics of peripheral blood stem cells mobilised with cyclophosphamide and the delayed addition of G-CSF. *Bone Marrow Transplant*, **16**, 359–363.

Henslee, P.J., Thompson, J.S., Romond, E.H., Doukas, M.A., Metcalfe, M., Marshall, M.E. and MacDonald, J.S. (1987) T cell depletion of HLA and haploidentical marrow reduces graft-versus-host disease but it may impair a graft-versus-leukemia effect. *Transplantation Proceedings*, **19**, 2701–2706.

Hessner, M.J., Endean, D.J., Casper, J.T., Horowitz, M.M., Keever-Taylor, C.A., Roth, M., Flomenberg, N. and Drobyski, W.R. (1995) Use of unrelated marrow grafts compensates for reduced graft-versus-leukemia reactivity after T-cell-depleted allogeneic marrow transplantation for chronic myelogenous leukemia. *Blood*, **86**, 3987–3996.

Hetu, F., Coutlee, F. and Roy, D.C. (1994) A non-isotopic nested polymerase chain reaction method to quantitate minimal residual disease in patients with non-Hodgkin's lymphoma. *Mol. Cell Probes*, **8**, 449–457.

Hirao, A., Kawano, Y., Takaue, Y., Suzue, T., Abe, T., Sato, J., Saito, S., Okamoto, Y., Makimoto, A., Kawahito, M. and Kuroda, Y. (1994) Engraftment potential of peripheral and cord blood stem cells evaluated by a long-term culture system. *Exp. Hematol.*, **22**, 521–526.

Holyoake, T.L., Alcorn, M.J., Richmond, L.J., Freshney, M.G., Pearson, C., Fitzsimons, E., Steward, W.P., Dunlop, D.J. and Pragnell, I.B. (1994) Efficient isolation of human CD34 positive hemopoietic progenitor cells by immune panning. *Stem Cells*, **12**, 114–124.

Howell, A.L., Fogg-Leach, M., Davis, B.H. and Ball, E.D. (1989) Continuous infusion of complement by an automated cell processor enhances cytotoxicity of monoclonal antibody sensitized leukemia cells. *Bone Marrow Transplant*, **4**, 317–322.

Huang, S. and Terstappen, L.W. (1992) Formation of haematopoietic microenvironment and haematopoietic stem cells from single human bone marrow stem cells. *Nature*, **360**, 745–749.

Huang, S. and Terstappen, L.W. (1994) Lymphoid and myeloid differentiation of single human CD34+, HLA-DR+, CD38- hematopoietic stem cells. *Blood*, **83**, 1515–1526.

Hudson, A.M., Makrynikola, V., Kabral, A. and Bradstock, K.F. (1989) Immunophenotypic analysis of clonogenic cells in acute lymphoblastic leukemia using an *in vitro* colony assay. *Blood*, **74**, 2112–2120.

Humblet, Y., Lefebvre, P., Jacques, J.L., Bosly, A., Feyens, A.M., Sekhavat, M., Agaliotis, D. and Symann, M. (1988) Concentration of bone marrow progenitor cells by separation on a Percoll gradient using the Haemonetics model 30. *Bone Marrow Transplant*, **3**, 63–67.

Imrie, K., Dicke, K.A. and Keating, A. (1996) Autologous bone marrow transplantation for acute myeloid leukemia. *Stem Cells*, **14**, 69–78.

Jamieson, C., Richter, A. and Levy, J.G. (1993) Efficacy of benzoporphyrin derivative, a photosensitizer, in selective destruction of leukemia cells using a murine tumor model. *Exp. Hematol.*, **21**, 629–634.

Janossy, G., Campana, D., Burnett, A., Coustan-Smith, E., Timms, A., Bekassy, A.N., Hann, I., Alcorn, M.J., Totterman, T. and Simonsson, B. (1988) Autologous bone marrow transplantation in acute lymphoblastic leukemia-preclinical immunologic studies. *Leukemia*, **2**, 485–495.

Jazwiec, B., Mahon, F.X., Pigneux, A., Pigeonnier, V. and Reiffers, J. (1995) 5-Fluorouracil-resistant CD34(+) cell population from peripheral blood of CML patients contains bcr-abl-negative progenitor cells. *Exp. Hematol.*, **23**, 1509–1514.

Johnson, F.L. and Goldman, S. (1993) Role of autotransplantation in neuroblastoma. *Hematol. Oncol. Clinics of N. America*, **7**, 647–662.

Johnson, P.W., Price, C.G., Smith, T., Cotter, F.E., Meerabux, J., Rohatiner, A.Z., Young, B.D. and Lister, T.A. (1994) Detection of cells bearing the t(14;18) translocation following myeloablative treatment and autologous bone marrow transplantation for follicular lymphoma. *J. Clin. Oncol.*, **12**, 798–805.

Jones, R.J., Celano, P., Sharkis, S.J. and Sensenbrenner, L.L. (1989) Two phases of engraftment established by serial bone marrow transplantation in mice. *Blood*, **73**, 397–401.

Jones, R.J., Wagner, J.E., Celano, P., Zicha, M.S. and Sharkis, S.J. (1990) Separation of pluripotent haematopoietic stem cells from spleen colony-forming cells. *Nature*, **347**, 188–189.

Jonsson, O.G., Kitchens, R.L., Scott, F.C. and Smith, R.G. (1990) Detection of minimal residual disease in acute lymphoblastic leukemia using immunoglobulin hypervariable region specific oligonucleotide probes. *Blood*, **76**, 2072–2079.

Juneja, H.S., Harvey, W.H., Brasher, W.K. and Thompson, E.B. (1995) Successful *in vitro* purging of leukemic blasts from marrow by cortivazol, a pyrazolosteroid: a preclinical study for autologous transplantation in acute lymphoblastic leukemia and non-Hodgkin's lymphoma. *Leukemia*, **9**, 1771–1778.

Katz, F.E., Tindle, R., Sutherland, D.R. and Greaves, M.F. (1985) Identification of a membrane glycoprotein associated with haemopoietic progenitor cells. *Leuk. Res.*, **9**, 191–198.

Katz, F.E., Watt, S.M., Martin, H., Lam, G., Capellaro, D., Goldman, J.M. and Greaves, M.F. (1986) Co-ordinate expression of BI.3C5 and HLA-DR antigens on haemopoietic progenitors from chronic myeloid leukaemia. *Leuk. Res.*, **10**, 961–971.

Kemshead, J.T. (1991) The immunomagnetic manipulation of bone marrow. *Bone marrow processing and purging* (ed. by A.P. Gee), p. 293. CRC Press, Boca Raton, Florida.

Kersey, J.H., LeBien, T.W., Abramson, C.S., Newman, R., Sutherland, R. and Greaves, M. (1981) P-24: a human leukemia-associated and lymphohemopoietic progenitor cell surface structure identified with monoclonal antibody. *J. Exp. Med.*, **153**, 726–731.

Kersey, J.H., Weisdorf, D., Nesbit, M.E., LeBien, T.W., Woods, W.G., McGlave, P.B., Kim, T., Vallera, D.A., Goldman, A.I., Bostrom, B., Hurd, D. and Ramsay, N.K.C. (1987) Comparison of autologous and allogeneic bone marrow transplantation for treatment of high-risk refractory acute lymphoblastic leukemia. *N. Engl. J. Med.*, **317**, 461–467.

Kessinger, A., Armitage, J.O., Landmark, J.D. and Weisenburger, D.D. (1986) Reconstitution of human hematopoietic function with autologous cryopreserved circulating stem cells. *Exp. Hematol.*, **14**, 192–196.

Khouri, I.F., Keating, M.J., Vriesendorp, H.M., Reading, C.L., Przepiorka, D., Huh, Y.O., Andersson, B.S., van Besien, K.W., Mehra, R.C., Giralt, S.A., Ippoliti, C., Marshall, M., Thomas, M.W., O'brien, S., Robertson, L.E., Deisseroth, A.B. and Champlin, R.E. (1994) Autologous and allogeneic bone marrow transplantation for chronic lymphocytic leukemia: preliminary results. *J. Clin. Oncol.*, **12**, 748–758.

Kolb, H.J., Mittermuller, J., Clemm, C., Holler, E., Ledderose, G., Brehm, G., Heim, M. and Wilmanns, W. (1990) Donor leukocyte transfusions for treatment of recurrent chronic myelogenous leukemia in marrow transplant patients. *Blood*, **76**, 2462–2465.

Kolb, H.J., Schattenberg, A., Goldman, J.M., Hertenstein, B., Jacobsen, N., Arcese, W., Ljungman, P., Ferrant, A., Verdonck, L., Niederwieser, D. *et al.* (1995) Graft-versus-leukemia effect of donor lymphocyte transfusions in marrow grafted patients. *Blood*, **86**, 2041–2050.

Korbling, M., Burke, P., Braine, H., Elfenbein, G., Santos, B.W. and Kaizer, H. (1981) Successful engraftment of blood derived normal hemopoietic stem cells in chronic myelogenous leukemia. *Exp. Hematol.*, **9**, 684–690.

Korbling, M., Hunstein, W., Fliedner, T.M., Cayeux, S., Dorken, B., Fehrentz, D., Haas, R., Ho, A.D., Keilholz, U., Knauf, W., Konig, A., Mende, U., Pezzutto, A., von Reumont, J., Wolf, G.K., Wannenmacher, M., zum Winkel, K. and Rother, K. (1989) Disease-free survival after autologous bone marrow transplantation in patients with acute myelogenous leukemia. *Blood*, **74**, 1898–1904.

Korbling, M., Drach, J., Champlin, R.E., Engel, H., Huynh, L., Kleine, H.D., Berenson, R., Deisseroth, A.B. and Andreeff, M. (1994) Large-scale preparation of highly purified, frozen/thawed CD34+, HLA-DR- hematopoietic progenitor cells by sequential immunoadsorption (CEPRATE SC) and fluorescence-activated cell sorting: implications for gene transduction and/or transplantation. *Bone Marrow Transplant*, **13**, 649–654.

Korbling, M. (1995) Blood stem cell transplantation and gene therapy of cancer. *Stem Cells*, **13**, 106–113.

Korbling, M., Huh, Y.O., Durett, A., Mirza, N., Miller, P., Engel, H., Anderlini, P., van Besien, K., Andreeff, M. and Przepiorka, D. (1995) Allogeneic blood stem cell transplantation: peripheralization and yield of donor-derived primitive hematopoietic progenitor cells (CD34+ Thy-1dim) and lymphoid subsets, and possible predictors of engraftment and graft-versus-host disease. *Blood*, **86**, 2842–2848.

Kotasek, D., Shepherd, K.M., Sage, R.E., Dale, B.M., Norman, J.E., Charles, P., Gregg, A., Pillow, A. and Bolton, A. (1992) Factors affecting blood stem cell collections following high-dose cyclophosphamide mobilization in lymphoma, myeloma and solid tumors. *Bone Marrow Transplant*, **9**, 11–17.

Krause, D.S., Ito, T., Fackler, M.J., Smith, O.M., Collector, M.I., Sharkis, S.J. and May, W.S. (1994) Characterization of murine CD34, a marker for hematopoietic progenitor and stem cells. *Blood*, **84**, 691–701.

Krause, D.S., Fackler, M.J., Civin, C.I. and May, W.S. (1996) CD34 – Structure, biology, and clinical utility. *Blood*, **87**, 1–13.

Kvalheim, G., Sorensen, O., Fodstad, O., Funderud, S., Kiesel, S., Dorken, B., Nustad, K., Jakobsen, E., Ugelstad, J. and Pihl, A. (1988) Immunomagnetic removal of B-lymphoma cells from human bone marrow: a procedure for clinical use. *Bone Marrow Transplant*, **3**, 31–41.

Kvalheim, G., Fjeld, J.G., Pihl, A., Funderud, S., Ugelstad, J., Fodstad, O. and Nustad, K. (1989) Immunomagnetic removal of B-lymphoma cells using a novel mono-sized magnetizable polymer bead, M-280, in conjunction with primary IgM and IgG antibodies. *Bone Marrow Transplant*, **4**, 567–574.

Ladenstein, R., Lasset, C., Hartmann, O., Klingebiel, T., Bouffet, E., Gadner, H., Paolucci, P., Burdach, S., Chauvin, F. and Pinkerton, R. (1994) Comparison of auto versus allografting as consolidation of primary treatments in advanced neuroblastoma over one year of age at diagnosis: report from the European Group for Bone Marrow Transplantation. *Bone Marrow Transplant*, **14**, 37–46.

Lamb, L., Metherell, J., Lee, C., Brouillette, M., Thompson, J., Brown, S., Henslee-Downey, P.J. and Gee, A. (1994) Development of an immunomagnetic separation procedure to deplete V alpha beta-positive T cells from marrow for use in partially mismatched related donor (PMRD) transplantation. *Prog. Clin. Biol. Res.*, **389**, 541–550.

Lambert, J.M., Goldmacher, V.S., Collinson, A.R., Nadler, L.M. and Blattler, W.A. (1991a) An immunotoxin prepared with blocked ricin: a natural plant toxin adapted for therapeutic use. *Cancer Research*, **51**, 6236–6242.

Lambert, J.M., McIntyre, G., Gauthier, M.N., Zullo, D., Rao, V., Steeves, R.M., Goldmacher, V.S. and Blattler, W.A. (1991b) The galactose-binding sites of the cytotoxic lectin ricin can be chemically blocked in high yield with reactive ligands prepared by chemical modification of glycopeptides containing triantennary N-linked oligosaccharides. *Biochemistry*, **30**, 3234–3247.

Lane, T.A., Law, P., Maruyama, M., Young, D., Burgess, J., Mullen, M., Mealiffe, M., Terstappen, L.W., Hardwick, A. and Moubayed, M. (1995) Harvesting and enrichment of hematopoietic progenitor cells mobilized into the peripheral blood of normal donors by granulocyte-macrophage colony-stimulating factor (GM-CSF) or G-CSF: potential role in allogeneic marrow transplantation. *Blood*, **85**, 275–282.

Lange, B., Ferrero, D., Pessano, S., Palumbo, A., Faust, J., Meo, P. and Rovera, G. (1984) Surface phenotype of clonogenic cells in acute myeloid leukemia defined by monoclonal antibodies. *Blood*, **64**, 693–700.

Lapointe, C., Forest, L., Lussier, P., Busque, L., Lagace, F., Perreault, C., Roy, D.C. and Gyger, M. (1996) Sequential analysis of early hematopoietic reconstitution following allogeneic bone marrow transplantation with fluorescence in situ hybridization (FISH). *Bone Marrow Transplant*, **17**, 1143–1148.

Lasky, L.C. (1991) Peripheral blood stem cell collection and use. *Cellular and humoral immunotherapy and apheresis* (ed. by R.A. Sacher, D.B. Brubaker, D.O. Kasprisin and L.J. McCarthy), p.73. American Association of Blood Banks, Arlington.

Lasky, L.C. and Zanjani, E.D. (1985) Size and density characterization of human committed and multipotent hematopoietic progenitors. *Exp. Hematol.*, **13**, 680–684.

Lazzaro, G.E., Meyer, B.F., Willis, J.I., Erber, W.N., Herrmann, R.P. and Davies, J.M. (1995) The synthesis of a peanut agglutinin-ricin A chain conjugate-Potential as an *in vitro* purging agent for autologous bone marrow in multiple myeloma. *Exp. Hematol.*, **23**, 1347–1352.

Leary, A.G. and Ogawa, M. (1987) Blast cell colony assay for umbilical cord blood and adult bone marrow progenitors. *Blood*, **69**, 953–956.

Lebkowski, J.S., Schain, L.R., Okrongly, D., Levinsky, R., Harvey, M.J. and Okarma, T.B. (1992) Rapid isolation of human CD34 hematopoietic stem cells-purging of human tumor cells. *Transplantation*, **53**, 1011–1019.

Léonard, B.M., Hétu, F., Busque, L., Gyger, M., Bélanger, R., Perreault, C. and Roy, D.C. (1998) Lymphoma Cell Burden in Progenitor Cell Grafts Measured by Competitive Polymerase Chain Reaction: Less Than One Log Difference Between Bone Marrow and Peripheral Blood Sources. *Blood*, **91**, 331–339.

Lemoli, R.M., Gobbi, M., Tazzari, P.L., Grassi, G., Dinota, A., Gherlinzoni, F., Mazza, P. and Tura, S. (1988) Avidin-biotin immunoadsorption for *ex vivo* bone marrow purging: effect of percentage of target cells and flow rate. *Haematologica*, **73**, 183–185.

Lemoli, R.M., Gobbi, M., Tazzari, P.L., Tassi, T.C., Dinota, A., Visani, G., Grassi, G., Mazza, P., Cavo, M. and Tura, S. (1989) Bone marrow purging for multiple myeloma by avidin-biotin immunoadsorption. *Transplantation*, **47**, 385–387.

Lemoli, R.M., Gasparetto, C., Scheinberg, D.A., Moore, M.A., Clarkson, B.D. and Gulati, S.C. (1991) Autologous bone marrow transplantation in acute myelogenous leukemia: *in vitro* treatment with myeloid-specific monoclonal antibodies and drugs in combination. *Blood*, **77**, 1829–1836.

Lemoli, R.M., Fortuna, A., Motta, M.R., Rizzi, S., Giudice, V., Nannetti, A., Martinelli, G., Cavo, M., Amabile, M., Mangianti, S., Fogli, M., Conte, R. and Tura, S. (1996) Concomitant mobilization of plasma cells and hematopoietic progenitors into peripheral blood of multiple myeloma patients-positive selection and transplantation of enriched CD34(+) cells to remove circulating tumor cells. *Blood*, **87**, 1625–1634.

Lill, M.C., Lynch, M., Fraser, J.K., Chung, G.Y., Schiller, G., Glaspy, J.A., Souza, L., Baldwin, G.C. and Gasson, J.C. (1994) Production of functional myeloid cells from CD34-selected hematopoietic progenitor cells using a clinically relevant *ex vivo* expansion system. *Stem Cells*, **12**, 626–637.

Link, H., Arseniev, L., Bahre, O., Berenson, R.J., Battmer, K., Kadar, J.G., Jacobs, R., Casper, J., Kuhl, J. and Schubert, J. (1995) Combined transplantation of allogeneic bone marrow and CD34+ blood cells. *Blood*, **86**, 2500–2508.

Loken, M.R., Shah, V.O., Dattilio, K.L. and Civin, C.I. (1987) Flow cytometric analysis of human bone marrow: I. Normal erythroid development. *Blood*, **69**, 255–263.

Lowenberg, B., Verdonck, L.J., Dekker, A.W., Willemze, R., Zwaan, F.E., de Planque, M., Abels, J., Sonneveld, P., van der Lelie, J., Goudsmit, R., van Putten, W.L.J., Sizoo, W., Hagenbeek, A. and de Gast, G.C. (1990) Autologous bone marrow transplantation in acute myeloid leukemia in first remission: results of a Dutch prospective study. *J. Clin. Oncol.*, **8**, 287–294.

Macintyre, E.A., Belanger, C., Debert, C., Canioni, D., Turhan, A.G., Azagury, M., Hermine, O., Varet, B., Flandrin, G. and Schmitt, C. (1995) Detection of clonal CD34(+)19(+) progenitors in bone marrow of bcl-2-IgH-positive follicular lymphoma patients. *Blood*, **86**, 4691–4698.

Mackinnon, S., Papadopoulos, P., Carabasi, M.H., Reich, L. *et al.* (1995) Adoptive immunotherapy evaluating escalating doses of donor leukocytes for relapse of chronic myeloid leukemia following bone marrow transplantation: Separation of graft-versus-leukemia responses from graft-versus-host disease. *Blood*, **86**, 1261–1268.

Mahe, B., Menard, A., Accard, F., Pineau, D., Robillard, N. and Hermouet, S. (1995) *In vitro* expansion of CD34+ cells from peripheral blood of myeloma and lymphoma patients. *Nouvelle Revue Francaise d'Hematologie*, **37**, 335–341.

Marsh, J.C., Sutherland, D.R., Davidson, J., Mellors, A. and Keating, A. (1992) Retention of progenitor cell function in CD34+ cells purified using a novel O-sialoglycoprotease. *Leukemia*, **6**, 926–934.

Martin, H., Atta, J., Zumpe, P., Eder, M., Elsner, S., Rode, C., Wassmann, B., Bruecher, J. and Hoelzer, D. (1995) Purging of peripheral blood stem cells yields bcr-abl-negative autografts in patients with bcr-abl-positive acute lymphoblastic leukemia. *Exp. Hematol.*, **23**, 1612–1618.

Martin, P.J. (1987) T cell purging with antibody – the Seattle experience. *Bone Marrow Transplant*, **2** (Suppl. 2), 53–57.

Martin, P.J. (1990) The role of donor lymphoid cells in allogeneic marrow engraftment. *Bone Marrow Transplant*, **6**, 283–289.

Martin, P.J. (1992) Animal experimentation relevant to human marrow transplantation. *Current Opinion in Oncology*, **4**, 239–246.

Massumoto, C., Benyunes, M.C., Sale, G., Beauchamp, M., York, A., Thompson, J.A., Buckner, C.D. and Fefer, A. (1996) Close simulation of graft-versus-host disease by interleukin-2 administered after autologous bone marrow transplantation for hematologic malignancy. *Bone Marrow Transplant*, **17**, 351–356.

Matthay, K.K., Atkinson, J.B., Stram, D.O., Selch, M., Reynolds, C.P. and Seeger, R.C. (1993) Patterns of relapse after autologous purged bone marrow transplantation for neuroblastoma: a Childrens Cancer Group pilot study. *J. Clin. Oncol.*, **11**, 2226–2233.

Matthay, K.K., Seeger, R.C., Reynolds, C.P., Stram, D.O., O'Leary, M.C., Harris, R.E., Selch, M., Atkinson, J.B., Haase, G.M. and Ramsay, N.K. (1994) Allogeneic versus autologous purged bone marrow transplantation for neuroblastoma: a report from the Childrens Cancer Group. *J. Clin. Oncol.*, **12**, 2382–2389.

Matutes, E., Owusu-Ankomah, K., Morilla, R., Garcia Marco, J., Houlihan, A., Que, T.H. and Catovsky, D. (1994) The immunological profile of B-cell disorders and proposal of a scoring system for the diagnosis of CLL. *Leukemia*, **8**, 1640–1645.

McCann, J.C., Kanteti, R., Shilepsky, B., Miller, K.B., Sweet, M. and Schenkein, D.P. (1996) High degree of occult tumor contamination in bone marrow and peripheral blood stem cells of patients undergoing autologous transplantation for non-Hodgkin's lymphoma. *Biol. Blood and Marrow Transplant*, **2**, 37–43.

McGlave, P.B., Arthur, D., Miller, W.J., Lasky, L. and Kersey, J. (1990) Autologous transplantation for CML using marrow treated *ex vivo* with recombinant human interferon gamma. *Bone Marrow Transplant*, **6**, 115–120.

McGlave, P.B., De Fabritiis, P., Deisseroth, A., Goldman, J., Barnett, M., Reiffers, J., Simonsson, B., Carella, A. and Aeppli, D. (1994) Autologous transplants for chronic myelogenous leukaemia: results from eight transplant groups. *Lancet*, **343**, 1486–1488.

Meijerink, J.P., Smetsers, T.F., Raemaekers, J.M., Bogman, M.J., de Witte, T. and Mensink, E.J. (1993) Quantitation of follicular non-Hodgkin's lymphoma cells carrying t(14;18) by competitive polymerase chain reaction. *Br. J. Haematol.*, **84**, 250–256.

Min, W., Kim, D., Lee, J., Park, J. and Kim, C. (1994) Autologous bone marrow rescue for patients with acute myelogenous leukemia: purging with hyperthermia and ether lipid *in vitro*. *Prog. Clin. Biol. Res.*, **389**, 197–203.

Mitsuyasu, R.T., Champlin, R.E., Gale, R.P., Ho, W.G., Lenarsky, C., Winston, D., Selch, M., Elashoff, R., Giorgi, J.V., Wells, J., Terasaki, P., Billing, R. and Feig, S. (1986) Treatment of donor bone marrow with monoclonal anti-T-cell antibody and complement for the prevention of graft-versus-host disease. A prospective, randomized, double-blind trial. *Ann. Intern. Med.*, **105**, 20–26.

Morishima, Y., Miyamura, K., Kojima, S., Ueda, R., Morishita, Y., Sao, H., Tanimoto, M., Ohno, R., Sobue, R. and Hirano, M. (1993) Autologous BMT in high risk patients with CALLA-positive ALL: possible efficacy of *ex vivo* marrow leukemia cell purging with monoclonal antibodies and complement. *Bone Marrow Transplant*, **11**, 255–259.

Moss, T.J., Reynolds, C.P., Sather, H.N., Romansky, S.G., Hammond, G.D. and Seeger, R.C. (1991) Prognostic value of immunocytologic detection of bone marrow metastases in neuroblastoma. *N. Engl. J. Med.*, **324**, 219–226.

Muench, M.O. and Moore, M.A. (1992) Accelerated recovery of peripheral blood cell counts in mice transplanted with *in vitro* cytokine-expanded hematopoietic progenitors. *Exp. Hematol.*, **20**, 611–618.

Myers, D.E., Uckun, F.M., Ball, E.D. and Vallera, D.A. (1988) Immunotoxins for *ex vivo* marrow purging in autologous bone marrow transplantation for acute nonlymphocytic leukemia. *Transplantation*, **46**, 240–245.

Myklebust, A.T., Godal, A., Juell, S., Pharo, A. and Fodstad, O. (1994) Comparison of two antibody-based methods for elimination of breast cancer cells from human bone marrow. *Cancer Research*, **54**, 209–214.

Nadler, L.M., Korsmeyer, S.J., Anderson, K.C., Boyd, A.W., Slaughenhoupt, B., Park, E., Jensen, J., Coral, F., Mayer, R.J. and Sallan, S.E. (1984) B cell origin of non-T cell acute lymphoblastic leukemia. A model for discrete stages of neoplastic and normal pre-B cell differentiation. *J. Clin. Invest.*, **74**, 332–340.

Naparstek, E., Or, R., Nagler, A., Cividali, G., Engelhard, D., Aker, M., Gimon, Z., Manny, N., Sacks, T., Tochner, Z., Weiss, L., Samuel, S., Brautbar, C., Hale, G., Waldmann, H., Steinberg, S.M. and Slavin, S. (1995) T-cell-depleted allogeneic bone marrow transplantation for acute leukaemia using Campath-1 antibodies and post-transplant administration of donor's peripheral blood lymphocytes for prevention of relapse. *Br. J. Haematol.*, **89**, 506–515.

Neame, P.B., Soamboonsrup, P., Browman, G.P., Meyer, R.M., Benger, A., Wilson, W.E., Walker, I.R., Saeed, N. and McBride, J.A. (1986) Classifying acute leukemia by immunophenotyping: a combined FAB-immunologic classification of AML. *Blood*, **68**, 1355–1362.

Negrin, R.S., Kusnierz-Glaz, C.R., Still, B.J., Schriber, J.R., Chao, N.J., Long, G.D., Hoyle, C., Hu, W.W., Horning, S.J. and Brown, B.W. (1995) Transplantation of enriched and purged peripheral blood progenitor cells from a single apheresis product in patients with non-Hodgkin's lymphoma. *Blood*, **85**, 3334–3341.

Ngan, B.Y., Nourse, J. and Cleary, M.L. (1989) Detection of chromosomal translocation t(14;18) within the minor cluster region of bcl-2 by polymerase chain reaction and direct genomic sequencing of the enzymatically amplified DNA in follicular lymphomas. *Blood*, **73**, 1759–1762.

Noga, S.J., Donnenberg, A.D., Schwartz, C.L., Strauss, L.C., Civin, C.I. and Santos, G.W. (1986) Development of a simplified counterflow centrifugation elutriation procedure for depletion of lymphocytes from human bone marrow. *Transplantation*, **41**, 220–229.

Ogawa, M. (1993) Differentiation and proliferation of hematopoietic stem cells. *Blood*, **81**, 2844–2853.

Okarma, T.B. (1992) Stem cell selection for autologous bone marrow transplantation. *Seminars in Hematology*, **29**, 9–20.

Pal, P., Zeng, H.L., Durocher, G., Girard, D., Li, T.C., Gupta, A.K., Giasson, R., Blanchard, L., Gaboury, L., Balassy, A., Turmel, C., Laperriere, A. and Villeneuve, L. (1996) Phototoxicity of some bromine-substituted rhodamine dyes-synthesis, photophysical properties and application as photosensitizers. *Photochemistry and Photobiology*, **63**, 161–168.

Pan, L., Delmonte, J., Jr., Jalonen, C.K. and Ferrara, J.L. (1995) Pretreatment of donor mice with granulocyte colony-stimulating factor polarizes donor T lymphocytes toward type-2 cytokine production and reduces severity of experimental graft-versus-host disease. *Blood*, **86**, 4422–4429.

Perot, C., van den Akker, J., Laporte, J.P., Douay, L., Lopez, M., Stachowiak, J., Isnard, F., Taillemite, J.L., Najman, A. and Gorin, N.C. (1993) Multiple chromosome abnormalities in patients with acute leukemia after autologous bone marrow transplantation using total body irradiation and marrow purged with mafosfamide. *Leukemia*, **7**, 509–515.

Peters, S.O., Kittler, E.L., Ramshaw, H.S. and Quesenberry, P.J. (1996) *Ex vivo* expansion of murine marrow cells with interleukin-3 (il-3), il-6, il-11, and stem cell factor leads to impaired engraftment in irradiated hosts. *Blood*, **87**, 30–37.

Petersen, F.B., Appelbaum, F.R., Hill, R., Fisher, L.D., Bigelow, C.L., Sanders, J.E., Sullivan, K.M., Bensinger, W.I., Witherspoon, R.P., Storb, R., Clift, R.A., Fefer, A., Press, O.W., Weiden, P.L., Singer, J., Thomas, E.D. and Buckner, C.D. (1990) Autologous marrow transplantation for malignant lymphoma: a report of 101 cases from Seattle. *J. Clin. Oncol.*, **8**, 638–647.

Pettengell, R., Morgenstern, G.R., Woll, P.J., Chang, J., Rowlands, M., Young, R., Radford, J.A., Scarffe, J.H., Testa, N.G. and Crowther, D. (1993) Peripheral blood progenitor cell transplantation in lymphoma and leukemia using a single apheresis. *Blood*, **82**, 3770–3777.

Pettengell, R., Luft, T., Henschler, R., Hows, J.M., Dexter, T.M., Ryder, D. and Testa, N.G. (1994) Direct comparison by limiting dilution analysis of long-term culture-initiating cells in human bone marrow, umbilical cord blood, and blood stem cells. *Blood*, **84**, 3653–3659.

Philip, I. and Favrot, M. (1988) [Cell cultures in Burkitt's lymphoma. Detection of residual disease and application to autologous graft purging]. [French]. *Pathologie Biologie*, **36**, 79–82.

Philip, T., Armitage, J.O., Spitzer, G., Chauvin, F., Jagannath, S., Cahn, J.Y., Colombat, P., Goldstone, A.H., Gorin, N.C., Flesh, M., Laporte, J.P., Maraninchi, D., Pico, J., Bosly, A., Anderson, C., Schots, R., Biron, P., Cabanillas, F. and Dicke, K. (1987) High-dose therapy and autologous bone marrow transplantation after failure of conventional chemotherapy in adults with intermediate-grade or high-grade non-Hodgkin's lymphoma. *N. Engl. J. Med.*, **316**, 1493–1498.

Philip, T., Guglielmi, C., Hagenbeek, A., Somers, R., Van der Lelie, H., Bron, D., Sonneveld, P., Gisselbrecht, C., Cahn, J.Y., Harousseau, J.L., Coiffier, B., Biron, P., Mandelli, F. and Chauvin, F. (1995) Autologous bone marrow transplantation as compared with salvage chemotherapy in relapses of chemotherapy-sensitive non-Hodgkin's lymphoma. *N. Engl. J. Med.*, **333**, 1540–1545.

Pichert, G., Roy, D.C., Gonin, R., Alyea, E.P., Belanger, R., Gyger, M., Perreault, C., Bonny, Y., Lerra, I., Murray, C., Soiffer, R.J. and Ritz, J. (1995) Distinct patterns of minimal residual disease associated with graft-versus-host disease after allogeneic bone marrow transplantation for chronic myelogenous leukemia. *J. Clin. Oncol.*, **13**, 1704–1713.

Pollard, C.M., Powles, R.L., Millar, J.L., Shepherd, V., Milan, S., Lakhani, A., Zuiable, A., Treleaven, J. and Helenglass, G. (1986) Leukaemic relapse after Campath 1 treated bone marrow transplantation for leukaemia. *Lancet*, **2**, 1404.

Porter, D.L., Roth, M.S., McGarigle, C., Ferrara, J.L. and Antin, J.H. (1994) Induction of graft-versus-host disease as immunotherapy for relapsed chronic myeloid leukemia. *N. Engl. J. Med.*, **330**, 100–106.

Preijers, F., Ruijs, P., Schattenberg, A. and de Witte, T. (1994) T-cell depletion from allogeneic bone marrow by counterflow centrifugation is not associated with a substantial loss of CD34-positive cells. *Prog. Clin. Biol. Res.*, **389**, 339–344.

Rabinowe, S.N., Soiffer, R.J., Gribben, J.G., Daley, H., Freedman, A.S., Daley, J., Pesek, K., Neuberg, D., Pinkus, G., Leavitt, P.R., Spector, N.A., Grossbard, M.L., Anderson, K., Robertson, M.J., Mauch, P., Chayt-Marcus, K., Ritz, J. and Nadler, L.M. (1993) Autologous and allogeneic bone marrow transplantation for poor prognosis patients with B-cell chronic lymphocytic leukemia. *Blood*, **82**, 1366–1376.

Rasmussen, R.A., Counts, S.L., Daley, J.F. and Schlossman, S.F. (1994) Isolation and characterization of CD6-T cells from peripheral blood. *J. Immunol.*, **152**, 527–536.

Reading, C.L., Thomas, M.W., Hickey, C.M., Chandran, M., Tindle, S., Ball, E.D., Poynton, C.H. and Dicke, K.A. (1987) Magnetic affinity colloid (MAC) cell separation of leukemia cells from autologous bone marrow aspirates. *Leukemia Research*, **11**, 1067–1077.

Rill, D.R., Santana, V.M., Roberts, W.M., Nilson, T., Bowman, L.C., Krance, R.A., Heslop, H.E., Moen, R.C., Ihle, J.N. and Brenner, M.K. (1994) Direct demonstration that autologous bone marrow transplantation for solid tumors can return a multiplicity of tumorigenic cells. *Blood*, **84**, 380–383.

Ritz, J., Pesando, J.M., Notis-McConarty, J., Lazarus, H. and Schlossman, S.F. (1980) A monoclonal antibody to human acute lymphoblastic leukaemia antigen. *Nature*, **283**, 583–585.

Ritz, J., Ramsay, N.K. and Kersey, J.H. (1994) Autologous bone marrow transplantation for acute lymphoblastic leukemia. *Bone marrow transplantation* (ed. by S.J. Forman, K.G. Blume and E.D. Thomas), p.731. Blackwell Scientific Publications, Boston.

Rizzoli, V., Mangoni, L., Carella, A.M., Aglietta, M., Porcellini, A., Coleselli, P., Angrilli, F., Alessandrino, E.P., Madon, E., Locatelli, F., Mozzana, A., DeLaurenzi, A., Greco, M.M., Bernabei, P.A. and Tarella, C. (1989) Drug-mediated marrow purging: mafosfamide in adult acute leukemia in remission. The experience of the Italian study group. *Bone Marrow Transplant*, **4** (Suppl. 1), 190–194.

Robertson, M.J., Roy, D.C., Stone, R.M. and Ritz, J. (1994) Use of CD33 monoclonal antibodies in bone marrow transplantation for acute myeloid leukemia. *Prog. Clin. Biol. Res.*, **389**, 47–63.

Rohatiner, A., Gelber, R., Schlossman, S.F. and Ritz, J. (1986) Depletion of T cells from human bone marrow using monoclonal antibodies and rabbit complement. A quantitative and functional analysis. *Transplantation*, **42**, 73–80.

Rohatiner, A. (1994) Myelodysplasia and acute myelogenous leukemia after myeloablative therapy with autologous stem-cell transplantation [editorial]. *J. Clin. Oncol.*, **12**, 2521–2523.

Rosenfeld, C.S., Thiele, D.L., Shadduck, R.K., Zeigler, Z.R. and Schindler, J. (1995) *Ex vivo* purging of allogeneic marrow with L-Leucyl-L-leucine methyl ester. A phase I study. *Transplantation*, **60**, 678–683.

Rosenzweig, M., Marks, D.F., Zhu, H., Hempel, D., Mansfield, K.G., Sehgal, P.K., Kalams, S., Scadden, D.T. and Johnson, R.P. (1996) *In vitro* T lymphopoiesis of human and rhesus CD34+ progenitor cells. *Blood*, **87**, 4040–4048.

Ross, A.A., Cooper, B.W., Lazarus, H.M., Mackay, W. and Moss, T.J. (1992) Detection and viability of tumor cells in peripheral blood stem cell collections from breast cancer patients using immunocytochemical and clonogenic assay techniques. *Blood*, **10**, 936–941.

Rowley, S.D., Jones, R.J., Piantadosi, S., Braine, H.G., Colvin, O.M., Davis, J., Saral, R., Sharkis, S., Wingard, J., Yeager, A.M. and Santos, G.W. (1989) Efficacy of *ex vivo* purging for autologous bone marrow transplantation in the treatment of acute nonlymphoblastic leukemia. *Blood*, **74**, 501–506.

Rowley, S.D., Miller, C.B., Piantadosi, S., Davis, J.M., Santos, G.W. and Jones, R.J. (1991) Phase I study of combination drug purging for autologous bone marrow transplantation. *J. Clin. Oncol.*, **9**, 2210–2218.

Roy, D.C., Felix, M., Cannady, W.G., Cannistra, S. and Ritz, J. (1990a) Comparative activities of rabbit complements of different ages using an *in vitro* marrow purging model. *Leukemia Research*, **14**, 407–416.

Roy, D.C., Tantravahi, R., Murray, C., Dear, K., Gorgone, B., Anderson, K.C., Freedman, A.S., Nadler, L.M. and Ritz, J. (1990b) Natural history of mixed chimerism after bone marrow transplantation with CD6-depleted allogeneic marrow: a stable equilibrium. *Blood*, **75**, 296–304.

Roy, D.C., Griffin, J.D., Belvin, M., Blattler, W.A., Lambert, J.M. and Ritz, J. (1991) Anti-MY9-blocked-ricin: an immunotoxin for selective targeting of acute myeloid leukemia cells. *Blood*, **77**, 2404–2412.

Roy, D.C., Bélanger, R., Perreault, C., Bonny, Y., Busque, L., Kassis, J., Guertin, M.J., Esseltine, D. and Gyger, M. (1995a) Autologous bone marrow transplantation for patients with non-Hodgkin's lymphoma using anti-B4-bR immunotoxin purging. *Proc. Fourth Internat. Symp. Immunotoxins.*, 158a.

Roy, D.C., Perreault, C., Bélanger, R., Gyger, M., Le Houillier, C., Blättler, W.A., Lambert, J.M. and Ritz, J. (1995b) Elimination of B-lineage leukemia and lymphoma cells from bone marrow grafts using anti-B4-blocked-ricin immunotoxin. *J. Clin. Immunol.*, **15**, 51–57.

Roy, D.C., Ouellet, S., Le Houillier, C., Ariniello, P.D., Perreault, C. and Lambert, J.M. (1996) Elimination of neuroblastoma and small cell lung cancer cells with an anti-neural cell adhesion molecule immunotoxin. *J. Natl. Cancer Inst.*, **88**, 1136–1145.

Ryan, D., Kossover, S., Mitchell, S., Frantz, C., Hennessy, L. and Cohen, H. (1986) Subpopulations of common acute lymphoblastic leukemia antigen-positive lymphoid cells in normal bone marrow identified by hematopoietic differentiation antigens. *Blood*, **68**, 417–425.

Sabbath, K.D., Ball, E.D., Larcom, P., Davis, R.B. and Griffin, J.D. (1985) Heterogeneity of clonogenic cells in acute myeloblastic leukemia. *J. Clin. Invest.*, **75**, 746–753.

Sallan, S.E., Niemeyer, C.M., Billett, A.L., Lipton, J.M., Tarbell, N.J., Gelber, R.D., Murray, C., Pittinger, T.P., Wolfe, L.C., Bast, R.C., Jr. and Ritz, J. (1989) Autologous bone marrow transplantation for acute lymphoblastic leukemia. *J. Clin. Oncol.*, **7**, 1594–1601.

Sankey, E.A., More, L. and Dhillon, A.P. (1990) QBEnd/10: a new immunostain for the routine diagnosis of Kaposi's sarcoma. *Journal of Pathology*, **161**, 267–271.

Sausville, E.A., Headlee, D., Stetler-Stevenson, M., Jaffe, E.S., Solomon, D., Figg, W.D., Herdt, J., Kopp, W.C., Rager, H. and Steinberg, S.M. (1995) Continuous infusion of the anti-CD22 immunotoxin IgG-RFB4-SMPT-dgA in patients with B-cell lymphoma: a phase I study. *Blood*, **85**, 3457–3465.

Scheffold, C., Brandt, K., Johnston, V., Lefterova, P., Degen, B., Schontube, M., Huhn, D., Neubauer, A. and Schmidt-Wolf, I.G. (1995) Potential of autologous immunologic effector cells for bone marrow purging in patients with chronic myeloid leukemia. *Bone Marrow Transplant*, **15**, 33–39.

Schmid, H., Henze, G., Schwerdtfeger, R., Baumgarten, E., Besserer, A., Scheffler, A., Serke, S., Zingsem, J. and Siegert, W. (1993) Fractionated total body irradiation and high-dose VP-16 with purged autologous bone marrow rescue for children with high risk relapsed acute lymphoblastic leukemia. *Bone Marrow Transplant*, **12**, 597–602.

Schmitz, N., Dreger, P., Suttorp, M., Rohwedder, E.B., Haferlach, T., Löffler, H., Hunter, A. and Russell, N.H. (1995) Primary transplantation of allogeneic peripheral blood progenitor cells mobilized by filgrastim (granulocyte colony-stimulating factor). *Blood*, **85**, 1666–1672.

Schroeder, H., Pinkerton, C.R., Powles, R.L., Meller, S.T., Tait, D., Milan, S. and McElwain, T.J. (1991) High dose melphalan and total body irradiation with autologous marrow rescue in childhood acute lymphoblastic leukaemia after relapse. *Bone Marrow Transplant*, **7**, 11–15.

Schwartz, M.A., Lovett, D.R., Redner, A., Finn, R.D., Graham, M.C., Divgi, C.R., Dantis, L., Gee, T.S., Andreeff, M. and Old, L.J. (1993) Dose-escalation trial of M195 labeled with iodine 131 for cytoreduction and marrow ablation in relapsed or refractory myeloid leukemias. *J. Clin. Oncol.*, **11**, 294–303.

Selvaggi, K.J., Wilson, J.W., Mills, L.E., Cornwell, G.G., Hurd, D., Dodge, W., Gingrich, R., Martin, S.E., McMillan, R., Miller, W. and Ball, E.D. (1994) Improved outcome for high-risk acute myeloid leukemia patients using autologous bone marrow transplantation and monoclonal antibody-purged bone marrow. *Blood*, **83**, 1698–1705.

Shah, A.J., Smogorzewska, E.M., Hannum, C. and Crooks, G.M. (1996) Flt3 ligand induces proliferation of quiescent human bone marrow cd34+cd38- cells and maintains progenitor cells *in vitro*. *Blood*, **87**, 3563–3570.

Shah, N.K., Wingard, J.R., Piantadosi, S., Rowley, S., Santos, G. and Griffin, C.A. (1993) Chromosome abnormalities in patients treated with 4-hydroperoxycyclophosphamide-purged autologous bone marrow transplantation. *Cancer Genetics and Cytogenetics*, **65**, 135–140.

Shapiro, F., Pytowski, B., Rafii, S., Witte, L., Hicklin, D.J., Yao, T.J. and Moore, M.A.S. (1996) The effects of Flk-2/flt3 ligand as compared with c-kit ligand on short-term and long-term proliferation of CD34+ hematopoietic progenitors elicited from human fetal liver, umbilical cord blood, bone marrow, and mobilized peripheral blood. *Journal of Hematotherapy*, **5**, 655–662.

Sharp, J.G., Joshi, S.S., Armitage, J.O., Bierman, P., Coccia, P.F., Harrington, D.S., Kessinger, A., Crouse, D.A., Mann, S.L. and Weisenburger, D.D. (1992) Significance of detection of occult non-Hodgkin's lymphoma in histologically uninvolved bone marrow by a culture technique. *Blood*, **79**, 1074–1080.

Sharp, J.G., Kessinger, A., Mann, S., Crouse, D.A., Armitage, J.O., Bierman, P. and Weisenburger, D.D. (1996) Outcome of high-dose therapy and autologous transplantation in non-Hodgkin's lymphoma based on the presence of tumor in the marrow or infused hematopoietic harvest. *J. Clin. Oncol.*, **14**, 214–219.

Sheridan, W.P., Begley, C.G., Juttner, C.A., Szer, J., To, L.B., Maher, D., McGrath, K.M., Morstyn, G. and Fox, R.M. (1992) Effect of peripheral-blood progenitor cells mobilised by filgrastim (G-CSF) on platelet recovery after high-dose chemotherapy. *Lancet*, **339**, 640–644.

Shpall, E.J., Jones, R.B., Bearman, S.I., Franklin, W.A., Archer, P.G., Curiel, T., Bitter, M., Claman, H.N., Stemmer, S.M., Purdy, M., Myers, S.E., Hami, L., Taffs, S., Heimfeld, S., Hallagan, J. and Berenson, R.J. (1994) Transplantation of enriched CD34-positive autologous marrow into breast cancer patients following high-dose chemotherapy: influence of CD34-positive peripheral-blood progenitors and growth factors on engraftment. *J. Clin. Oncol.*, **12**, 28–36.

Siena, S., Bregni, M., Brando, B., Ravagnani, F., Bonadonna, G. and Gianni, A.M. (1989) Circulation of CD34+ hematopoietic stem cells in the peripheral blood of high-dose cyclophosphamide-treated patients: enhancement by intravenous recombinant human granulocyte-macrophage colony-stimulating factor. *Blood*, **74**, 1905–1914.

Siena, S., Bregni, M., Brando, B., Belli, N., Ravagnani, F., Gandola, L., Stern, A.C., Lansdorp, P.M., Bonadonna, G. and Gianni, A.M. (1991) Flow cytometry for clinical estimation of circulating hematopoietic progenitors for autologous transplantation in cancer patients. *Blood*, **77**, 400–409.

Siena, S., Bregni, M., Bonsi, L., Sklenar, I., Bagnara, G.P., Bonadonna, G. and Gianni, A.M. (1993) Increase in peripheral blood megakaryocyte progenitors following cancer therapy with high-dose cyclophosphamide and hematopoietic growth factors. *Exp. Hematol.*, **21**, 1583–1590.

Silvestri, F., Banavali, S., Baccarani, M. and Preisler, H.D. (1992) The CD34 hemopoietic progenitor cell associated antigen: biology and clinical applications. *Haematologica*, **77**, 265–273.

Simmons, D. and Seed, B. (1988) Isolation of a cDNA encoding CD33, a differentiation antigen of myeloid progenitor cells. *J. Immunol.*, **141**, 2797–2800.

Simonsson, B., Burnett, A.K., Prentice, H.G., Hann, I.H., Brenner, M.K., Gibson, B., Grob, J.P., Lonnerholm, G., Morrison, A. and Smedmyr, B. (1989) Autologous bone marrow transplantation with monoclonal antibody purged marrow for high risk acute lymphoblastic leukemia. *Leukemia*, **3**, 631–636.

Smith, S.L., Bender, J.G., Maples, P.B., Unverzagt, K., Schilling, M., Lum, L., Williams, S. and Van Epps, D.E. (1993) Expansion of neutrophil precursors and progenitors in suspension cultures of CD34+ cells enriched from human bone marrow. *Exp. Hematol.*, **21**, 870–877.

Soiffer, R.J., Murray, C., Mauch, P., Anderson, K.C., Freedman, A.S., Rabinowe, S.N., Takvorian, T., Robertson, M.J., Spector, N. and Gonin, R. (1992) Prevention of graft-versus-host disease by selective depletion of CD6-positive T lymphocytes from donor bone marrow. *J. Clin. Oncol.*, **10**, 1191–1200.

Soiffer, R.J., Roy, D.C., Gonin, R., Murray, C., Anderson, K.C., Freedman, A.S., Rabinowe, S.N., Robertson, M.J., Spector, N., Pesek, K., Mauch, P., Nadler, L.M. and Ritz, J. (1993) Monoclonal antibody-purged autologous bone marrow transplantation in adults with acute lymphoblastic leukemia at high risk of relapse. *Bone Marrow Transplant*, **12**, 243–251.

Soiffer, R.J., Murray, C., Gonin, R. and Ritz, J. (1994) Effect of low-dose interleukin-2 on disease relapse after T-cell-depleted allogeneic bone marrow transplantation. *Blood*, **84**, 964–971.

Soiffer, R.J., Fairclough, D., Robertson, M., Alyea, E., Anderson, K., Freedman, A., Bartlettpandite, L., Fisher, D., Schlossman, R.L., Stone, R., Murray, C., Freeman, A., Marcus, K., Mauch, P., Nadler, L. and Ritz, J. (1997) CD6-depleted allogeneic bone marrow transplantation for acute leukemia in first complete remission. *Blood*, **89**, 3039–3047.

Soligo, D., Delia, D., Oriani, A., Cattoretti, G., Orazi, A., Bertolli, V., Quirici, N. and Deliliers, G.L. (1991) Identification of CD34+ cells in normal and pathological bone marrow biopsies by QBEND10 monoclonal antibody. *Leukemia*, **5**, 1026–1030.

Spangrude, G.J. (1991) Hematopoietic stem-cell differentiation. *Curr. Opin. Immunol.*, **3**, 171–178.

Spencer, A., Szydlo, R.M., Brookes, P.A., Kaminski, E., Rule, S., van Rhee, F., Ward, K.N., Hale, G., Waldmann, H. and Hows, J.M. (1995) Bone marrow transplantation for chronic myeloid leukemia with volunteer unrelated donors using *ex vivo* or *in vivo* T-cell depletion: major prognostic impact of HLA class I identity between donor and recipient. *Blood*, **86**, 3590–3597.

Stahel, R.A., Mabry, M., Skarin, A.T., Speak, J. and Bernal, S.D. (1985) Detection of bone marrow metastasis in small-cell lung cancer by monoclonal antibody. *J. Clin. Oncol.*, **3**, 455–461.

Stahel, R.A., Gilks, W.R. and Schenker, T. (1994) Antigens of lung cancer: results of the Third International Workshop on Lung Tumor and Differentiation Antigens. *J. Natl. Cancer Inst.*, **86**, 669–672.

Stahel, R.A., Jost, L.M., Pichert, G. and Widmer, L. (1995) High-dose chemotherapy and autologous bone marrow transplantation for malignant lymphomas. *Cancer Treat. Rev.*, **21**, 3–32.

Stepan, D.E., Bartholomew, R.M. and LeBien, T.W. (1984) *In vitro* cytodestruction of human leukemic cells using murine monoclonal antibodies and human complement. *Blood*, **63**, 1120–1124.

Stiff, P.J., Schulz, W.C., Bishop, M. and Marks, L. (1991) Anti-CD33 monoclonal antibody and etoposide/cytosine arabinoside combinations for the *ex vivo* purification of bone marrow in acute nonlymphocytic leukemia. *Blood*, **77**, 355–362.

Stong, R.C., Uckun, F., Youle, R.J., Kersey, J.H. and Vallera, D.A. (1985) Use of multiple T cell-directed intact ricin immunotoxins for autologous bone marrow transplantation. *Blood*, **66**, 627–635.

Straka, C., Drexler, E., Mitterer, M., Langenmayer, I., Pfefferkorn, L., Stade, B., Koll, R. and Emmerich, B. (1995) Autotransplantation of B-cell purged peripheral blood progenitor cells in B-cell lymphoma. *Lancet*, **345**, 797–798.

Strauss, L.C., Rowley, S.D., La Russa, V.F., Sharkis, S.J., Stuart, R.K. and Civin, C.I. (1986) Antigenic analysis of hematopoiesis. V. Characterization of My-10 antigen expression by normal lympho-hematopoietic progenitor cells. *Exp. Hematol.*, **14**, 878–886.

Sutherland, D., Marsh, J., Davidson, J., Baker, M., Keating, A. and Mellors, A. (1992) Differential sensitivity of CD34 epitopes to cleavage by *Pasteurella haemolytica* glycoprotease: Implications for purification of CD34-positive progenitor cells. *Exp. Hematol.*, **20**, 590.

Sutherland, D.R., Keating, A., Nayar, R., Anania, S. and Stewart, A.K. (1994) Sensitive detection and enumeration of CD34+ cells in peripheral and cord blood by flow cytometry. *Exp. Hematol.*, **22**, 1003–1010.

Takahashi, M., Maruyama, S., Moriyama, Y. and Shibata, A. (1992) Applicability of antimyeloid monoclonal antibodies (L4F3, 1G10) to autologous bone marrow transplantation for patients with acute myelogenous leukemia. *Transplantation Proceedings*, **24**, 416–418.

Tari, A.M., Tucker, S.D., Deisseroth, A. and Lopez-Berestein, G. (1994) Liposomal delivery of methylphosphonate antisense oligodeoxynucleotides in chronic myelogenous leukemias. *Blood*, **84**, 601–607.

Tavassoli, M., Konno, M., Shiota, Y., Omoto, E., Minguell, J.J. and Zanjani, E.D. (1991) Enhancement of the grafting efficiency of transplanted marrow cells by preincubation with interleukin-3 and granulocyte-macrophage colony-stimulating factor. *Blood*, **77**, 1599–1606.

Tecce, R., Fraioli, R., De Fabritiis, P., Sandrelli, A., Savarese, A., Santoro, L., Cuomo, M. and Natali, P.G. (1991) Production and characterization of two immunotoxins specific for M5b ANLL leukaemia. *Int. J. Cancer.*, **49**, 310–316.

Tjonnfjord, G.E., Steen, R., Evensen, S.A., Thorsby, E. and Egeland, T. (1994) Characterization of CD34+ peripheral blood cells from healthy adults mobilized by recombinant human granulocyte colony-stimulating factor. *Blood*, **84**, 2795–2801.

To, L.B., Haylock, D.N., Dyson, P.G., Thorp, D., Roberts, M.M. and Juttner, C.A. (1990a) An unusual pattern of hemopoietic reconstitution in patients with acute myeloid leukemia transplanted with autologous recovery phase peripheral blood. *Bone Marrow Transplant*, **6**, 109–114.

To, L.B., Shepperd, K.M., Haylock, D.N., Dyson, P.G., Charles, P., Thorp, D.L., Dale, B.M., Dart, G.W., Roberts, M.M., Sage, R.E. and Juttner, C.A. (1990b) Single high doses of cyclophosphamide enable the collection of high numbers of hemopoietic stem cells from the peripheral blood. *Exp. Hematol.*, **18**, 442–447.

To, L.B., Roberts, M.M., Haylock, D.N., Dyson, P.G., Branford, A.L., Thorp, D., Ho, J.Q., Dart, G.W., Horvath, N., Davy, M.L., Olweny, C.L.M., Abdi, E. and Juttner, C.A. (1992) Comparison of haematological recovery times and supportive care requirements of autologous recovery phase peripheral blood stem cell transplants, autologous bone marrow transplants and allogeneic bone marrow transplants. *Bone Marrow Transplant*, **9**, 277–284.

Traul, D.L., Anderson, G.S., Bilitz, J.M., Krieg, M. and Sieber, F. (1995) Potentiation of merocyanine 540-mediated photodynamic therapy by salicylate and related drugs. *Photochemistry and Photobiology*, **62**, 790–799.

Truitt, R.L., Shih, C.Y., Lefever, A.V., Tempelis, L.D., Andreani, M. and Bortin, M.M. (1983) Characterization of alloimmunization-induced T lymphocytes reactive against AKR leukemia *in vitro* and correlation with graft-vs-leukemia activity *in vivo*. *J. Immunol.*, **131**, 2050–2058.

Uchansak-Ziegler, B., Petrasch, S., Michel, J. and Ziegler, A. (1989) Characterization of the CD34-specific monoclonal antibody TUK3. *Tissue Antigens*, **33**, 230

Uckun, F.M., Gajl-Peczalska, K., Meyers, D.E., Ramsay, N.C., Kersey, J.H., Colvin, M. and Vallera, D.A. (1987) Marrow purging in autologous bone marrow transplantation for T-lineage acute lymphoblastic leukemia: efficacy of *ex vivo* treatment with immunotoxins and 4-hydroperoxycyclophosphamide against fresh leukemic marrow progenitor cells. *Blood*, **69**, 361–366.

Uckun, F.M., Kersey, J.H., Haake, R., Weisdorf, D. and Ramsay, N.K. (1992) Autologous bone marrow transplantation in high-risk remission B-lineage acute lymphoblastic leukemia using a cocktail of three monoclonal antibodies (BA-1/CD24, BA-2/CD9, and BA-3/CD10) plus complement and 4-hydroperoxycyclophosphamide for *ex vivo* bone marrow purging. *Blood*, **79**, 1094–1104.

Uckun, F.M., Kersey, J.H., Haake, R., Weisdorf, D., Nesbit, M.E. and Ramsay, N.K. (1993) Pretransplantation burden of leukemic progenitor cells as a predictor of relapse after bone marrow transplantation for acute lymphoblastic leukemia. *N. Engl. J. Med.*, **329**, 1296–1301.

Uckun, F.M. and Ledbetter, J.A. (1988) Immunobiologic differences between normal and leukemic human B-cell precursors. *Proc. Natl. Acad. Sci. USA*, **85**, 8603–8607.

Uckun, F.M. and Reaman, G.H. (1995) Immunotoxins for treatment of leukemia and lymphoma. *Leuk. Lymphoma*, **18**, 195–201.

Vaughan, W.P., Civin, C.I., Weisenburger, D.D., Karp, J.E., Graham, M.L., Sanger, W.G., Grierson, H.L., Joshi, S.S. and Burke, P.J. (1988) Acute leukemia expressing the normal human hematopoietic stem cell membrane glycoprotein CD34 (MY10). *Leukemia*, **2**, 661–666.

Verfaillie, C.M., Miller, W.J., Boylan, K. and McGlave, P.B. (1992) Selection of benign primitive hematopoietic progenitors in chronic myelogenous leukemia on the basis of HLA-DR antigen expression. *Blood*, **79**, 1003–1010.

Verfaillie, C.M. (1994) Can human hematopoietic stem cells be cultured ex vivo? *Stem Cells*, **12**, 466–476.

Verfaillie, C.M., Bhatia, R., Miller, W., Mortari, F., Roy, V., Burger, S., McCullough, J., Stieglbauer, K., Dewald, G., Heimfeld, S., Miller, J.S. and McGlave, P.B. (1996) BCR/ABL-negative primitive progenitors suitable for transplantation can be selected from the marrow of most early-chronic phase but not accelerated-phase chronic myelogenous leukemia patients. *Blood*, **87**, 4770–4779.

Verma, U.N., Bagg, A., Brown, E. and Mazumder, A. (1994) Interleukin-2 activation of human bone marrow in long-term cultures: an effective strategy for purging and generation of anti-tumor cytotoxic effectors. *Bone Marrow Transplant*, **13**, 115–123.

Vescio, R.A., Hong, C.H., Cao, J., Kim, A., Schiller, G.J., Lichtenstein, A.K., Berenson, R.J. and Berenson, J.R. (1994) The hematopoietic stem cell antigen, CD34, is not expressed on the malignant cells in multiple myeloma. *Blood*, **84**, 3283–3290.

Vitetta, E.S., Fulton, R.J., May, R.D., Till, M. and Uhr, J.W. (1987) Redesigning nature's poisons to create anti-tumor reagents. *Science*, **238**, 1098–1104.

Voltarelli, J.C., Corpuz, S. and Martin, P.J. (1990) *In vitro* comparison of two methods of T cell depletion associated with different rates of graft failure after allogeneic marrow transplantation. *Bone Marrow Transplant*, **6**, 419–423.

von Bueltzingsloewen, A., Belanger, R., Perreault, C., Bonny, Y., Roy, D.C., Lalonde, Y., Boileau, J., Kassis, J., Lavallee, R., Lacombe, M. and Gyger, M. (1993) Acute graft-versus-host disease prophylaxis with methotrexate and cyclosporine after busulfan and cyclophosphamide in patients with hematologic malignancies. *Blood*, **81**, 849–855.

Wagner, J.E., Santos, G.W., Noga, S.J., Rowley, S.D., Davis, J., Vogelsang, G.B., Farmer, E.R., Zehnbauer, B.A., Saral, R. and Donnenberg, A.D. (1990) Bone marrow graft engineering by counterflow centrifugal elutriation: results of a phase I-II clinical trial. *Blood*, **75**, 1370–1377.

Wang, J.C., Beauregard, P., Soamboonsrup, P. and Neame, P.B. (1995) Monoclonal antibodies in the management of acute leukemia. *Am. J. Hematol.*, **50**, 188–199.

Watt, S.M., Karhi, K., Gatter, K., Furley, A.J., Katz, F.E., Healy, L.E., Altass, L.J., Bradley, N.J., Sutherland, D.R. and Levinsky, R. (1987) Distribution and epitope analysis of the cell membrane glycoprotein (HPCA-1) associated with human hemopoietic progenitor cells. *Leukemia*, **1**, 417–426.

Weaver, C.H., Hazelton, B., Birch, R., Palmer, P., Allen, C., Schwartzberg, L. and West, W. (1995) An analysis of engraftment kinetics as a function of the CD34 content of peripheral blood progenitor cell collections in 692 patients after the administration of myeloablative chemotherapy. *Blood*, **86**, 3961–3969.

Widmer, L., Pichert, G., Jost, L.M. and Stahel, R.A. (1996) Fate of contaminating t(14; 18)+ lymphoma cells during *ex vivo* expansion of CD34-selected hematopoietic progenitor cells. *Blood*, **88**, 3166–3175.

Willems, P., Croockewit, A., Raymakers, R., Holdrinet, R., van Der Bosch, G., Huys, E. and Mensink, E. (1996) CD34 selections from myeloma peripheral blood cell autografts contain residual tumour cells due to impurity, not to CD34+ myeloma cells. *Br. J. Hematol.*, **93**, 613–622.

Williams, S.F., Lee, W.J., Bender, J.G., Zimmerman, T., Swinney, P., Blake, M., Carreon, J., Schilling, M., Smith, S., Williams, D.E., Oldham, F. and Van Epps, D. (1996) Selection and expansion of peripheral blood CD34(+) cells in autologous stem cell transplantation for breast cancer. *Blood*, **87**, 1687–1691.

Wognum, A.W., Krystal, G., Eaves, C.J., Eaves, A.C. and Lansdorp, P.M. (1992) Increased erythropoietin-receptor expression on CD34-positive bone marrow cells from patients with chronic myeloid leukemia. *Blood*, **79**, 642–649.

Yamada, M., Wasserman, R., Lange, B., Reichard, B.A., Womer, R.B. and Rovera, G. (1990) Minimal residual disease in childhood B-lineage lymphoblastic leukemia. Persistence of leukemic cells during the first 18 months of treatment. *N. Engl. J. Med.*, **323**, 448–455.

Yeager, A.M., Kaizer, H., Santos, G.W., Saral, R., Colvin, O.M., Stuart, R.K., Braine, H.G., Burke, P.J., Ambinder, R.F. and Burns, W.H. (1986) Autologous bone marrow transplantation in patients with acute nonlymphocytic leukemia, using *ex vivo* marrow treatment with 4-hydroperoxycyclophosphamide. *N. Engl. J. Med.*, **315**, 141–147.

Yonemura, Y., Ku, H., Hirayama, F., Souza, L.M. and Ogawa, M. (1996) Interleukin 3 or interleukin 1 abrogates the reconstituting ability of hematopoietic stem cells. *Proc. Natl. Acad. Sci. USA*, **93**, 4040–4044.

Young, J.W., Papadopoulos, E.B., Cunningham, I., Castro-Malaspina, H., Flomenberg, N., Carabasi, M.H., Gulati, S.C., Brochstein, J.A., Heller, G., Black, P. and Collins, N.H. (1992) T-cell-depleted allogeneic bone marrow transplantation in adults with acute nonlymphocytic leukemia in first remission. *Blood*, **79**, 3380–3387.

Young, P.E., Baumhueter, S. and Lasky, L.A. (1995) The sialomucin CD34 is expressed on hematopoietic cells and blood vessels during murine development. *Blood*, **85**, 96–105.

Zangemeister-Wittke, U., Collinson, A.R., Frosch, B., Waibel, R., Schenker, T. and Stahel, R.A. (1994) Immunotoxins recognising a new epitope on the neural cell adhesion molecule have potent cytotoxic effects against small cell lung cancer. *British Journal of Cancer*, **69**, 32–39.

Zanjani, E.D., Ascensao, J.L., Harrison, M.R. and Tavassoli, M. (1992) *Ex vivo* incubation with growth factors enhances the engraftment of fetal hematopoietic cells transplanted in sheep fetuses. *Blood*, **79**, 3045–3049.

Zhong, R.K., Kozii, R. and Ball, E.D. (1995) Homologous restriction of complement-mediated cell lysis can be markedly enhanced by blocking decay-accelerating factor. *Br. J. Haematol.*, **91**, 269–274.

Zittoun, R.A., Mandelli, F., Willemze, R., de Witte, T., Labar, B., Resegotti, L., Leoni, F., Damasio, E., Visani, G. and Papa, G. (1995) Autologous or allogeneic bone marrow transplantation compared with intensive chemotherapy in acute myelogenous leukemia. *N. Engl. J. Med.*, **332**, 217–223.

Zola, H., Potter, A., Neoh, S.H., Juttner, C.A., Haylock, D.N., Rice, A.M., Favaloro, E.J., Kabral, A. and Bradstock, K.F. (1987) Evaluation of a monoclonal IgM antibody for purging of bone marrow for autologous transplantation. *Bone Marrow Transplant*, **1**, 297–301.

INDEX